Edwin Hewitt

MATHEMATICS OF FINITE-DIMENSIONAL CONTROL SYSTEMS

PURE AND APPLIED MATHEMATICS

A Program of Monographs, Textbooks, and Lecture Notes

Contributions to *Lecture Notes in Pure and Applied Mathematics* are reproduced by direct photography of the author's typewritten manuscript. Potential authors are advised to submit preliminary manuscripts for review purposes. After acceptance, the author is responsible for preparing the final manuscript in camera-ready form, suitable for direct reproduction. Marcel Dekker, Inc. will furnish instructions to authors and special typing paper. Sample pages are reviewed and returned with our suggestions to assure quality control and the most attractive rendering of your manuscript. The publisher will also be happy to supervise and assist in all stages of the preparation of your camera-ready manuscript.

LECTURE NOTES
IN PURE AND APPLIED MATHEMATICS

1. *N. Jacobson,* Exceptional Lie Algebras
2. *L.-Å. Lindahl and F. Poulsen,* Thin Sets in Harmonic Analysis
3. *I. Satake,* Classification Theory of Semi-Simple Algebraic Groups
4. *F. Hirzebruch, W. D. Newmann, and S. S. Koh,* Differentiable Manifolds and Quadratic Forms
5. *I. Chavel,* Riemannian Symmetric Spaces of Rank One
6. *R. B. Burckel,* Characterization of C(X) Among Its Subalgebras
7. *B. R. McDonald, A. R. Magid, and K. C. Smith,* Ring Theory: Proceedings of the Oklahoma Conference
8. *Y.-T. Siu,* Techniques of Extension of Analytic Objects
9. *S. R. Caradus, W. E. Pfaffenberger, and B. Yood,* Calkin Algebras and Algebras of Operators on Banach Spaces
10. *E. O. Roxin, P.-T. Liu, and R. L. Sternberg,* Differential Games and Control Theory
11. *M. Orzech and C. Small,* The Brauer Group of Commutative Rings
12. *S. Thomeier,* Topology and Its Applications
13. *J. M. López and K. A. Ross,* Sidon Sets
14. *W. W. Comfort and S. Negrepontis,* Continuous Pseudometrics
15. *K. McKennon and J. M. Robertson,* Locally Convex Spaces
16. *M. Carmeli and S. Malin,* Representations of the Rotation and Lorentz Groups: An Introduction
17. *G. B. Seligman,* Rational Methods in Lie Algebras
18. *D. G. de Figueiredo,* Functional Analysis: Proceedings of the Brazilian Mathematical Society Symposium
19. *L. Cesari, R. Kannan, and J. D. Schuur,* Nonlinear Functional Analysis and Differential Equations: Proceedings of the Michigan State University Conference
20. *J. J. Schäffer,* Geometry of Spheres in Normed Spaces
21. *K. Yano and M. Kon,* Anti-Invariant Submanifolds
22. *W. V. Vasconcelos,* The Rings of Dimension Two
23. *R. E. Chandler,* Hausdorff Compactifications
24. *S. P. Franklin and B. V. S. Thomas,* Topology: Proceedings of the Memphis State University Conference
25. *S. K. Jain,* Ring Theory: Proceedings of the Ohio University Conference
26. *B. R. McDonald and R. A. Morris,* Ring Theory II: Proceedings of the Second Oklahoma Conference
27. *R. B. Mura and A. Rhemtulla,* Orderable Groups
28. *J. R. Graef,* Stability of Dynamical Systems: Theory and Applications
29. *H.-C. Wang,* Homogeneous Banach Algebras
30. *E. O. Roxin, P.-T. Liu, and R. L. Sternberg,* Differential Games and Control Theory II
31. *R. D. Porter,* Introduction to Fibre Bundles
32. *M. Altman,* Contractors and Contractor Directions Theory and Applications
33. *J. S. Golan,* Decomposition and Dimension in Module Categories
34. *G. Fairweather,* Finite Element Galerkin Methods for Differential Equations
35. *J. D. Sally,* Numbers of Generators of Ideals in Local Rings
36. *S. S. Miller,* Complex Analysis: Proceedings of the S.U.N.Y. Brockport Conference
37. *R. Gordon,* Representation Theory of Algebras: Proceedings of the Philadelphia Conference
38. *M. Goto and F. D. Grosshans,* Semisimple Lie Algebras
39. *A. I. Arruda, N. C. A. da Costa, and R. Chuaqui,* Mathematical Logic: Proceedings of the First Brazilian Conference
40. *F. Van Oystaeyen,* Ring Theory: Proceedings of the 1977 Antwerp Conference

41. *F. Van Oystaeyen and A. Verschoren,* Reflectors and Localization: Application to Sheaf Theory
42. *M. Satyanarayana,* Positively Ordered Semigroups
43. *D. L. Russell,* Mathematics of Finite-Dimensional Control Systems

Other Volumes in Preparation

MATHEMATICS OF FINITE-DIMENSIONAL CONTROL SYSTEMS

THEORY AND DESIGN

David L. Russell
University of Wisconsin-Madison
Madison, Wisconsin

MARCEL DEKKER INC. New York and Basel

Library of Congress Cataloging in Publication Data

Russell, David L.
Mathematics of finite-dimensional control systems.

(Lecture notes in pure and applied mathematics; v. 43)

Includes index.
1. Control theory. I. Title
QA402.3.R87 629.8'312 79-9847
ISBN 0-8247-6869-8

MARCEL DEKKER, INC.
270 Madison Avenue, New York, New York 10016

Current printing (last digit):
10 9 8 7 6 5 4 3 2 1

PRINTED IN THE UNITED STATES OF AMERICA

PREFACE

These lecture notes - and that, despite the imposing title, is what they are - have developed over a period of six years or so in a course on linear systems theory which I have taught at the University of Wisconsin, Madison. They are also the outgrowth of about fourteen years of consulting experience with Honeywell, Inc., Minneapolis. They may be regarded as a personal view of the subject; the natural product of a protracted endeavor to understand and appreciate the actual problems which control engineers meet and solve, with all of their uncertainty of definition and complexity of solution, and to express the essentials of the theory and methods which have been developed in standard mathematical language - the "lingua franca" of modern science. I have tried to avoid altogether methods of reasoning meaningful only from the viewpoint of one inhabiting a particular technological environment.

In this presentation we may very well fail to please, much less impress, the control theory specialist. With ample cause he may well ask, "Why yet another book primarily devoted to the linear-quadratic methodology?". That question is not an easy one to answer. Aside from the satisfaction of personal vanity in seeing one's own ideas in print, I have to offer only the experience of having long searched in vain for a text which explains those aspects of modern control theory which have actually been put to use along with adequate explanation of why they meet a particular need and how they are justified mathematically. What has been required, from my point of view, is an exposition accessible to a general audience - scientifically literate but without extensive preparation in engineering and innocent of most mathematics beyond elementary analysis and linear algebra.

I do not mean to imply that good books are not to be found, or even that they are particularly hard to find. There are many excellent presentations of control theory; among the many a few stand out in my mind because of the repeated use I have made of them over the years. I am particularly indebted to the pioneering mathematical presentation of the subject, "Foundations of Optimal Control Theory", by E. B. Lee and L.W. Markus. It is generally more broad in coverage and more mathematically advanced than what we present here. In fact it is fair to say that, had they given a more extensive treatment of the linear-quadratic methodology which is now beginning to dominate multivariable control design, this book might never have been written. In the engineering literature two expository treatments stand out prominently: "Linear Optimal Control" by B. D. O. Anderson and J. B. Moore and "Linear Optimal Control Systems" by H. Kwakernaak and R. Sivan. I have found both of them virtually indispensable. Between the mathematical and engineering traditions we have W. M. Wonham's monumental "Linear Multivariable Control: A Geometric Approach" - perhaps a little austere but enormously informative. On almost any given day it will be found, well worn, sitting on my desk. Finally, relatively new and not yet as thumb smudged as the others, I would mention "Deterministic and Stochastic Optimal Control" by W. Fleming and R. Rishel. It has filled many a gap in the literature, particularly as regards stochastic control. There remain many others which I really should cite here but space and the reader's patience are alike finite.

Because I have been teaching a mixed background, generally first year graduate, audience I have taken pains to provide proofs which are as elementary as the inherent content of the propositions permit. The various equations and formulae of controllability, observability and stabilizability and those peculiar to linear quadratic optimal control can usually be derived in a number of different ways. In the deterministic case certainly, and to some extent even in the stochastic case, the theorems of the linear quadratic theory can be obtained quite readily from either the Pontryagin Maximum Principle, Dynamic Programming, or, for that matter, from classical treatments of the Problem of Lagrange. But I feel that all of these are, properly, part of a course in the Calculus of Variations and, except for brief descriptive material in the case of Dynamic Programming as it relates to the Method of Liapounov, I have not wished either to develop such material here or to assume it as background. Optimality is, in the last analysis, one of the tools of control theory and not its raison d'etre. Consequently I have relied almost entirely on algebraic arguments of the completion of squares variety and on the algebraic theory of linear matrix equations $AX + XB = C$ to carry the weight of the argument. Perhaps the best instance of this is the almost purely algebraic proof of the separation theorem which appears in VI-6. I have also made an attempt to develop the properties of that strange entity, "white noise", in what I hope is a credible manner as the limit of piecewise constant disturbances with temporally uncorrelated elements.

If this book is used as a text for a one semester course, I would recommend that the course material per se should consist of the first six chapters, leaving Chapter VII largely for individual reading, perhaps in conjunction with some design problems assigned to the students as computational projects. It is certainly more difficult to give a compact and unified treatment of Chapter VII than to do so for the earlier chapters. Chapter VII has developed as the result of numerous design problems which I have encountered over the years as a consultant and which I have discussed at length with engineers and mathematicians at Honeywell and elsewhere.

There are many to whom personal acknowledgments of one kind or another are due. On the academic side I am happy to recognize a continuing debt to my "mentor" in control theory, Prof. L. W. Markus of the University of Minnesota. He first taught me control theory as a mathematical discipline and has, over the years, given encouragement and been available for counsel and advice (not always heeded, of course). In a similar manner I also wish to thank Prof. E. B. Lee, an associate of mine both at the University of Minnesota and at Honeywell, Inc. At Honeywell there are almost too many names to mention but I must acknowledge the years of very pleasant and enormously educational association with (in no particular order)

C. A. Harvey, C. R. Stone, G. Stein, K. D. Graham, E. E. Yore, F. Konar, J. Hauge, C. Mueller and R. Pope along with many others not recorded here but by no means forgotten. In addition let me add sincere appreciation of my very valued friend, Prof. D. L. Lukes of the University of Virginia, who first acquainted me with the peculiar merits of the linear quadratic theory - with the hope that he will accept my treatment of his magnum opus in IV - 5 with charitable understanding.

I want to thank Mrs. Grace Krewson for superb handling of a very challenging job of technical typing - including uncomplaining acceptance of many corrections which were my fault rather than hers. Those errors remaining are also mine. Marcel Dekker, Inc., deserves my gratitude for the finest cooperation and almost infinite patience in waiting for a manuscript due long before this.

Finally, I dedicate this book with greatest affection to my wife, Rebecca, without whom nothing would have been possible.

Madison, Wisconsin
July 15, 1978

TABLE OF CONTENTS

MATHEMATICS OF FINITE-DIMENSIONAL CONTROL SYSTEMS

CHAPTER I

CONTROL SYSTEMS: PRELIMINARIES

1. Control System Formulation

Throughout this book we will consider control systems which can be represented by a system of vector differential equations

$$\dot{x} = f(t, x, u, v) . \tag{1.1}$$

Here $\cdot$ denotes d/dt, t is a scalar time variable, x is a vector in n dimensional space and u, v are vectors in E^m, E^p. Frequently u and v will appear as vector functions of t, x; viz $= u(t, x)$, $v(t, x)$. In principle then (1.1) would be rewritten as $\dot{x} = \tilde{f}(t, x)$ but, since part of the problem, particularly with respect to u, involves the determination of this functional form, we retain the representation shown. Unless explicitly indicated otherwise, all functions appearing in (1.1) and other formulae will be assumed continuously differentiable (at least). Ordinarily all functions and vectors are taken to be real. Nevertheless most results, particularly for linear systems, continue to be valid for complex functions and vectors and, as every student of differential equations knows, complex analysis is often necessary for a complete understanding of systems originally posed in an entirely real setting.

The vector x appearing in (1.1) is identified as the "state" of our control system and a vector function $x(t)$ satisfying this equation is a "state trajectory" or "system trajectory". The vector u is the "control", used by the plant "operator" (human or otherwise) to influence the evolution of system trajectories or to modify the dynamical characteristics of the system generally. The vector v is used in a number of ways: as an external forcing term; as a "disturbance"; as a "noise" term, etc. We will be more precise later.

Associated with (1.1) we have several auxiliary notions. The first of these is that of an <u>observation</u>

$$\omega = h(t,x,u,v), \qquad \omega \in E^r,$$

which represents instantaneous information available to the plant operator through various measuring devices. Usually the dimension, r, of the observation vector ω is less than n; thus, even if t, u, v are known, only certain components of the system state x are available through direct measurement. A second notion is that of an <u>output</u> or <u>response</u>

$$y = g(t,x,u,v), \qquad y \in E^q .$$

The vector y usually represents certain aspects of the system configuration which are most directly of concern to the plant operator; often, but not always, the components of y are certain functions of the system state x, in which case we would write $y = g(x)$. It is frequently the case that the observation, h, and response, g, are the same functions.

Very succinctly (and this is, of course, an over-simplification) we can state the basic control problem as follows: given the system (1.1), to determine how the control u available to the plant operator should be employed, based on knowledge of the observation ω, so as to cause the response y to behave in a "desirable" manner.

In our treatment of control theory we are going to rely on a small number of fairly concrete (but still somewhat idealized) examples in order to present the main ideas more clearly. We will introduce one of these now so as to provide examples of what we mean by "system", "state", "control", "disturbance", "observation", and "response". In Figure I-1 we show, schematically, a crane used for lifting and placing heavy objects. For this

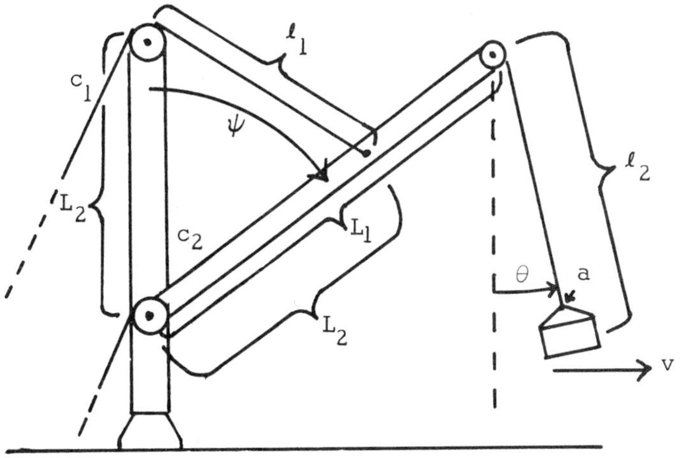

Figure I-1. A Crane

example we will assume that the vertical support and the crane arm are rigid and that the two cables c_1 and c_2 used for moving the arm and for raising and lowering the load are inextensible. We also assume the diameters of the cable pulleys to be negligible.

We recognize that, in terms of the mechanics of the system, we have specified the state completely if we give

(i) the boom angle ψ;

(ii) the boom angular velocity $\dot{\psi}$;

(iii) the load swing angle θ;

(iv) the load swing angular velocity $\dot{\theta}$;

(v) the length, ℓ_2, of the portion of the cable c_2 joining the boom to the load;

(vi) the rate, $\dot{\ell}_2$, at which the length ℓ_2 is being changed.

We are going to take $\ddot{\psi}$ and $\dot{\ell}_2$ as our initial choice for control functions. This choice is based upon the consideration that when $\psi(t)$ and $\ell_2(t)$ are given functions of time, the equation satisfied by $\theta(t)$ takes the form (cf. [1])

$$\ddot{\theta} = \frac{1}{\ell_2}[-g \sin\theta + L_1 \sin(\theta + \psi)(\dot{\theta} + \dot{\psi})\dot{\psi} - 3\dot{\ell}_2\dot{\theta}$$

$$+ L_1 \sin(\theta + \psi)\dot{\psi} - L_1 \cos(\theta + \psi)\ddot{\psi} + \frac{v}{m}\cos\theta]$$

$$\equiv f^2(\theta, \dot{\theta}, \psi, \dot{\psi}, \ell_2, \dot{\ell}_2, \ddot{\psi}, v) \tag{1.2}$$

so that $\dot{\ell}_2$ and $\ddot{\psi}$ appear as independent quantities. To obtain a first order system of the form (1.1) we let $x^1 = \theta$, $x^2 = \dot{\theta}$, $x^3 = \psi$, $x^4 = \dot{\psi}$, $x^5 = \ell_2$, $u^1 = \dot{\ell}_2$, $u^2 = \ddot{\psi}$ and we have

$$\frac{d}{dt}\begin{pmatrix} x^1 \\ x^2 \\ x^3 \\ x^4 \\ x^5 \end{pmatrix} = \begin{pmatrix} x^2 \\ f^2(x^1, x^2, x^3, x^4, x^5, u^1, u^2, v) \\ x^4 \\ u^2 \\ u^1 \end{pmatrix} \tag{1.3}$$

In the case of a crane the quantities of primary concern to the operator are the displacement and velocity of the load in the horizontal direction and the displacement and velocity of the load in the vertical direction. Denoting these by y^1, y^2, y^3, y^4, respectively, we have a four dimensional response vector with components

$$\begin{aligned} y^1 &= L_1 \sin\psi + \ell_2 \sin\theta \equiv g^1(x^1, x^3, x^5), \\ y^2 &= L_1 \cos\psi\dot{\psi} \quad \ell_2 \sin\theta + \ell_2 \cos\theta\dot{\theta} \equiv g^2(x^1, x^2, x^3, x^4, x^5, u^1), \\ y^3 &= L_1 \cos\psi - \ell_2 \cos\theta \equiv g^3(x^1, x^3, x^5), \\ y^4 &= -L_1 \sin\psi\dot{\psi} - \dot{\ell}_2 \cos\theta + \ell_2 \sin\theta\dot{\theta} \equiv g^4(x^1, x^2, x^3, x^4, x^5, u^1) \end{aligned} \tag{1.4}$$

It is slightly more difficult to decide what the observation vector ω should be for this example - since it is hard to specify measurements which would be realistic in practice. Since $\dot{\ell}_2$ and $\ddot{\psi}$ are (presumably known) controls, ℓ_2, ψ and ψ can, at least in principle, be obtained by integration.

But, as we will see later, they are of no use in obtaining information about the swing angle θ. Various possibilities suggest themselves but, to avoid a lengthy discussion, let us assume a rather sophisticated crane is at our disposal with an accelerometer mounted at point a in the diagram, set so as to measure the acceleration in the direction normal to the cable c_2 at that point. When the boom angular velocity $\dot{\psi}$ is nonzero this is not the same as $\ddot{\theta}$. A little calculation shows it to be given by the rather complicated expression

$$\begin{aligned}
&\cos\theta(L_1\cos\psi\ddot{\psi} - L_1\sin\psi(\dot{\psi})^2 + \ddot{\ell}_2\sin\theta + 2\dot{\ell}_2\cos\theta\dot{\theta} + \ell_2\cos\theta\ddot{\theta} \\
&\quad - \ell_2\sin\theta(\dot{\theta})^2) + \sin\theta(-L_1\sin\psi\ddot{\psi} - L_1\cos\psi(\dot{\psi})^2 - \ddot{\ell}_2\cos\theta \\
&\quad + 2\dot{\ell}_2\sin\theta\dot{\theta} + \ell_2\sin\theta\ddot{\theta} + \ell_2\cos\theta(\dot{\theta})^2) \\
&\quad = L_1\cos(\theta+\psi)\ddot{\psi} - L_1\sin(\theta+\psi)(\dot{\psi})^2 + 2\dot{\ell}_2\dot{\theta} + \ell_2\ddot{\theta}.
\end{aligned}$$

We use (1.2) to express $\ell_2\ddot{\theta}$ as

$$\ell_2\ddot{\theta} = \ell_2 f^2(\theta, \dot{\theta}, \psi, \dot{\psi}, \ell_2, \dot{\ell}_2, \ddot{\psi}, v)$$

and thus have, finally, the accelerometer measurement

$$\begin{aligned}
\omega^3 &\equiv \cos(\theta+\psi)\ddot{\psi} - L_1\sin(\theta+\psi)(\dot{\psi})^2 + 2\dot{\ell}_2\dot{\theta} \\
&\quad + \ell_2 f^2(\theta, \dot{\theta}, \psi, \dot{\psi}, \ell_2, \dot{\ell}_2, \ddot{\psi}, v) \equiv h^3(\theta, \dot{\theta}, \psi, \dot{\psi}, \ell_2, \dot{\ell}_2, \ddot{\psi}, v). \qquad (1.5)
\end{aligned}$$

We take as our observation vector, then,

$$\omega = \begin{pmatrix} \omega^1 \\ \omega^2 \\ \omega^3 \end{pmatrix} = \begin{pmatrix} \dot{\ell}_2 \\ \ddot{\psi} \\ h^3(\theta, \dot{\theta}, \psi, \dot{\psi}, \ell_2, \dot{\ell}_2, \ddot{\psi}, v) \end{pmatrix} = \begin{pmatrix} u^1 \\ u^2 \\ h^3(x^1, x^2, x^3, x^4, x^5, u^1, u^2, v) \end{pmatrix}$$

(1.6)

A control system may operate throughout the whole range of possible states or may at times be restricted to a smaller subset of these states by the application of constraints of one kind or another. Each such restriction defines a mode of operation. We illustrate this concept by describing the "level hold" mode of operation of the crane. The load is kept at a constant height or "level" y^3 (cf. (1.4)) except for the (normally small) variations induced by the swinging of the load. That is, y^3 is constant for $\theta(t) \equiv 0$. This is accomplished by simultaneously varying ℓ_2 and ψ so that the relationship

$$\ell_2(t) = L_1 \cos \psi(t) - y_0^3 \tag{1.7}$$

is maintained, y_0^3 being the desired load level. Mathematically this permits elimination of $\ell_2(t) \equiv x^5(t)$ and

$$u^1(t) \equiv \dot{\ell}_2(t) = - L_1 \sin \psi(t)\dot{\psi}(t) \tag{1.8}$$

from (1.3), leaving a four dimensional system with a single control $u^2(t) \equiv \ddot{\psi}(t)$:

$$\frac{d}{dt}\begin{pmatrix} x^1 \\ x^2 \\ x^3 \\ x^4 \end{pmatrix} = \begin{pmatrix} x^2 \\ \hat{f}^2(x^1, x^2, x^3, x^4, u^2, v) \\ x^4 \\ u^2 \end{pmatrix} \tag{1.9}$$

The function $\hat{f}^2$ is obtained by substituting (1.7) and (1.8) into (1.2), viz. :

$$\hat{f}^2(\theta, \dot{\theta}, \psi, \dot{\psi}, \ddot{\psi}, v)$$

$$= \frac{1}{L_1 \cos\psi - y_0^3} [-g \sin \theta + L_1 \sin(\theta + \psi)(\dot{\theta} + \dot{\psi})\dot{\psi}$$

$$+ 3L_1 \sin \psi\dot{\psi}\dot{\theta} + L_1 \sin(\theta + \psi)\dot{\psi} - L_1 \cos(\theta + \psi)\ddot{\psi} + \frac{v}{m} \cos \theta]. \tag{1.10}$$

The response and observation equations (1.4), (1.5) change similarly. We now have as the output equations

$$
\begin{aligned}
y^1 &= L_1 \sin\psi + (L_1 \cos\psi - y_0^3)\sin\theta \\
y^2 &= L_1 \cos\psi\dot{\psi} - L_1 \sin\psi\dot{\psi} \sin\theta + (L_1 \cos\psi - y_0^3)\cos\theta\dot{\theta} \\
y^3 &= L_1 \cos\psi - (L_1 \cos\psi - y_0^3)\cos\theta \\
y^4 &= - L_1 \sin\psi\dot{\psi} + L_1 \sin\psi\dot{\psi} \cos\theta + (L_1 \cos\psi - y_0^3)\sin\theta\dot{\theta}\,.
\end{aligned} \tag{1.11}
$$

The accelerometer reading becomes

$$
\begin{aligned}
\omega^3 &\equiv \cos(\theta + \psi)\ddot{\psi} - L_1 \sin(\theta + \psi)(\dot{\psi})^2 \\
&\quad - 2L_1 \sin\psi\dot{\psi}\dot{\theta} \\
&\quad + (L_1 \cos\psi - y_0^3)\hat{f}^2(\theta, \dot{\theta}, \psi, \dot{\psi}, \ddot{\psi}, v) \\
&\equiv \hat{h}^3(\theta, \dot{\theta}, \psi, \dot{\psi}, \ddot{\psi}, v),
\end{aligned} \tag{1.12}
$$

$\hat{f}^2$ being given by (1.10). Since $\dot{\ell}_2$ is now given in terms of $\psi, \dot{\psi}$, which can be obtained from integration of $\ddot{\psi}$, we now have a two dimensional observation vector

$$
\begin{pmatrix} \omega^2 \\ \omega^3 \end{pmatrix} = \begin{pmatrix} \ddot{\psi} \\ \hat{h}^3(\theta, \dot{\theta}, \psi, \dot{\psi}, \ddot{\psi}, v) \end{pmatrix} = \begin{pmatrix} u^2 \\ \hat{h}^3(x^1, x^2, x^3, x^4, u^2, v) \end{pmatrix}. \tag{1.13}
$$

The "level hold" mode of crane operation is thus quite different from the full mode. The dimension of the system state is reduced from five to four, the dimension of the control is reduced from two to one and the dimension of the observation vector may, without loss of information, be reduced from three to two.

2. Linear Systems and Linearization

The system (1.1) is linear if the vector function $f(t, x, u, v)$ has the form

$$
f(t, x, u, v) = A(t)x + B(t)u + C(t)v
$$

where $A(t)$, $B(t)$, $C(t)$ are $n \times n$, $n \times m$ and $n \times p$ matrices. Here, and unless specifically stated otherwise in the sequel, we will assume these matrices are at least continuous with respect to t, by which we mean that the various entries of these matrices are continuous scalar functions of t.

A word here about notation. We designate rows of matrices by lower case latin letters and superscripts. Thus

$$A = \begin{pmatrix} a^1 \\ a^2 \\ \vdots \\ a^n \end{pmatrix}.$$

We designate columns of matrices by lower case latin letters and subscripts. Thus also

$$A = (a_1, a_2, \ldots, a_n).$$

This notation extends to vectors as well. If x is a column vector

$$x = \begin{pmatrix} x^1 \\ x^2 \\ \vdots \\ x^n \end{pmatrix}$$

while if ξ is a row vector

$$\xi = (\xi_1, \xi_2, \ldots, \xi_n).$$

Extending this convention to the row vectors $a^1, a^2, \ldots, a^n$ which are the rows of A and to the column vectors $a_1, a_2, \ldots, a_n$ which are the columns of A we have

$$A = \begin{pmatrix} a_1^1 & a_2^1 & \cdots & a_n^1 \\ a_1^2 & a_2^2 & \cdots & a_n^2 \\ \vdots & \vdots & & \vdots \\ a_1^n & a_2^n & \cdots & a_n^n \end{pmatrix} \equiv (a_j^i).$$

This notation is consistent with the index conventions of tensor analysis and, at least in a linear context, rarely presents any difficulties. If we want to raise the i-th component of the column vector x to the k-th power we write $(x^i)^k$.

A linear control system then takes the form

$$\dot{x} = A(t)x + B(t)u + C(t)v . \qquad (2.1)$$

Associated with it are linear observations and outputs

$$\omega = H(t)x + J(t)u + L(t)v , \qquad (2.2)$$

$$y = F(t)x + G(t)u + M(t)v , \qquad (2.3)$$

where the continuous matrix functions $H(t)$, $J(t)$, $L(t)$, $F(t)$, $G(t)$ and $M(t)$ have dimensions $r \times n$, $r \times m$, $r \times p$, $q \times n$, $q \times m$, $q \times p$, respectively.

Linear control systems are usually obtained from nonlinear ones in the following way. Let x_0, u_0, v_0 denote "nominal" values of the indicated vector variables. Expanding the function $f(t, x, u, v)$ about these nominal values we have

$$\begin{aligned}\dot{x} = {} & f(t, x_0, u_0, v_0) + f_x(t, x_0, u_0, v_0)(x - x_0) \\ & + f_u(t, x_0, u_0, v_0)(u - u_0) + f_v(t, x_0, u_0, v_0)(v - v_0) \\ & + \mathcal{O}(\|(x - x_0, u - u_0, v - v_0)\|) . \end{aligned} \qquad (2.4)$$

Here f_x denotes the Jacobian matrix

$$f_x = \begin{pmatrix} \frac{\partial f^1}{\partial x^1} & \frac{\partial f^1}{\partial x^2} & \cdots & \frac{\partial f^1}{\partial x^n} \\ \frac{\partial f^2}{\partial x^1} & \frac{\partial f^2}{\partial x^2} & \cdots & \frac{\partial f^2}{\partial x^n} \\ \vdots & \vdots & & \vdots \\ \frac{\partial f^n}{\partial x^n} & \frac{\partial f^n}{\partial x^2} & \cdots & \frac{\partial f^n}{\partial x^n} \end{pmatrix}$$

and f_u, f_v are defined similarly. Letting

$$A(t) = f_x(t, x_0, u_0, v_0), \quad B(t) = f_u(t, x_0, u_0, v_0), \quad C(t) = f_v(t, x_0, u_0, v_0) \tag{2.5}$$

and noting that $\dot{x}_0 = 0$ since x_0 is constant we have

$$\begin{aligned}(\dot{x - x_0}) = {} & f(t, x_0, u_0, x_0) + A(t)(x - x_0) + B(t)(u - u_0) + C(t)(v - v_0) \\ & + \mathcal{O}(\|(x - x_0, u - u_0, v - v_0)\|).\end{aligned} \tag{2.6}$$

(In (2.4) and (2.6) $\mathcal{O}$ refers to a function which becomes arbitrarily small with respect to the indicated quantity $(\|(x - x_0, u - u_0, v - v_0)\|)$ as $(x, u, v) \to (x_0, u_0, v_0)$.) That is, $h(x, u, v) = \mathcal{O}(\|(x - x_0, u - u_0, v - v_0\|)$ if

$$\lim_{(x, u, v) \to (x_0, u_0, v_0)} h(x, u, v) / \|(x - x_0, u - u_0, v - v_0)\| = 0.$$

An equilibrium point for (1.7) is a point (x_0, u_0, v_0) where $f(t, x_0, u_0, v_0) \equiv 0$. It has the property that it is itself a (constant) solution of (1.1). If we suppose (x_0, u_0, v_0) to be an equilibrium point, (2.6) becomes

$$(\dot{x - x_0}) = A(t)(x - x_0) + B(t)(u - u_0) + C(t)(v - v_0) + \mathcal{O}(\|(x - x_0, u - u_0, v - v_0)).$$

The "linearized" system is obtained by dropping the $\mathcal{O}$ term, which is small relative to the linear terms near (x_0, u_0, v_0). Then renaming $x - x_0$, $u - u_0$ and $v - v_0$ as x, u and v we have (2.1). The observation ω and the output y are treated in much the same way. Letting

$$\omega_0(t) = h(t, x_0, u_0, v_0), \qquad y_0(t) = g(t, x_0, u_0, v_0)$$

we have

$$\begin{aligned}\omega(t) - \omega_0(t) = {} & h_x(t, x_0, u_0, v_0)(x - x_0) + h_u(t, x_0, u_0, v_0)(u - u_0) \\ & + h_v(t, x_0, u_0, v_0)(v - v_0) + \mathcal{O}(\|x - x_0, u - u_0, v - v_0)\|)\end{aligned}$$

and

$$y(t)-y_0(t) = g_x(t,x_0,u_0,v_0)(x-x_0)+g_u(t,x_0,u_0,v_0)(u-u_0)$$
$$+ g_v(t,x_0,u_0,v_0)(v-v_0) + \mathcal{O}(\|(x-x_0,u-u_0,v-v_0)\|) .$$

Renaming $\omega-\omega_0$ and $y-y_0$ as ω, y, respectively, and defining $H(t)$, $J(t)$, $L(t)$, $F(t)$, $G(t)$, $M(t)$ in the obvious way, we have (2. 2), (2. 3).

In passing from the original system to (2. 1), (2. 2), (2. 3) we are exercising a lot of "faith" in the proposition that the behavior of $x-x_0$, $\omega-\omega_0$ and $y-y_0$ near x_0, ω_0, y_0 is "adequately described" by the last three equations. For the time being we ask the reader to share this "faith" with us. In Chapter II we will provide a certain amount of mathematical justification upon which such confidence may be based. (See also [2], [3].)

The primary motivation for replacing the original system by the linearized system is, of course, the relative simplicity of behavior of linear systems of differential equations. We will review some of the relevant theorems in Section 3 of this chapter (which may be skipped by those already proficient in linear differential equations).

We will next undertake the linearization of the example system (1. 3) and its output and observation expressions, (1. 4) and (1. 6).

It is physically evident that our crane system has a large number of equilibria. They are described by

$$\theta = \dot{\theta} = 0$$
$$\psi = \psi_0 \text{ (constant)}, \ \dot{\psi} = \ddot{\psi} = 0$$
$$\ell_2 = \ell_{2,0} \text{ (constant)}, \ \dot{\ell}_2 = 0$$
$$v = 0 . \tag{2. 7}$$

As only one nonlinear component occurs on the right side of (1. 3) the linearization is quite direct. We have as the linearization of $f^2(\theta,\dot{\theta},\psi,\dot{\psi},\ell_2,\dot{\ell}_2,\ddot{\psi},v)$ about $(0,0,\psi_0,0,\ell_{2,0},0,0,0)$ the function

$$-\frac{g}{\ell_{2,0}}\theta - \frac{L_1}{\ell_{2,0}}\cos(\psi_0)\ddot{\psi} + \frac{v}{m\ell_{2,0}} .$$

Letting $\theta, \dot{\theta}, \psi-\psi_0, \dot{\psi}, \ell_2-\ell_{2,0}, \ell_2, \ddot{\psi}$, v be re-designated as $x^1, x^2, x^3, x^4, x^5, u^1, u^2$, v, the linearization of (1. 3) about the equilibrium point (2. 7) is

$$\begin{pmatrix} \dot{x}^1 \\ x^2 \\ x^3 \\ x^4 \\ x^5 \end{pmatrix} = \begin{pmatrix} 0 & 1 & 0 & 0 & 0 \\ -g/\ell_{2,0} & 0 & 0 & 0 & 0 \\ 0 & 0 & 0 & 1 & 0 \\ 0 & 0 & 0 & 0 & 0 \\ 0 & 0 & 0 & 0 & 0 \end{pmatrix} \begin{pmatrix} x^1 \\ x^2 \\ x^3 \\ x^4 \\ x^5 \end{pmatrix}$$

$$+ \begin{pmatrix} 0 & 0 \\ 0 & -\dfrac{L_1}{\ell_{2,0}}\cos(\psi_0) \\ 0 & 0 \\ 0 & 1 \\ 1 & 0 \end{pmatrix} \begin{pmatrix} u^1 \\ u^2 \end{pmatrix} + \begin{pmatrix} 0 \\ \dfrac{1}{m\ell_{2,0}} \\ 0 \\ 0 \\ 0 \end{pmatrix} v . \qquad (2.8)$$

The linearization of the output expression (1. 4) is

$$\begin{pmatrix} y^1 \\ y^2 \\ y^3 \\ y^4 \end{pmatrix} = \begin{pmatrix} \ell_{2,0} & 0 & L_1 \cos\psi_0 & 0 & 0 \\ 0 & \ell_{2,0} & 0 & L_1 \cos\psi_0 & 0 \\ 0 & 0 & -L_1 \sin\psi_0 & 0 & -1 \\ 0 & 0 & 0 & -L_1 \sin\psi_0 & 0 \end{pmatrix} \begin{pmatrix} x^1 \\ x^2 \\ x^3 \\ x^4 \\ x^5 \end{pmatrix} + \begin{pmatrix} 0 & 0 \\ 0 & 0 \\ 0 & 0 \\ -1 & 0 \end{pmatrix} \begin{pmatrix} u^1 \\ u^2 \end{pmatrix}$$

(2. 9)

and the linearization of the observation expression (1. 6) is

$$\begin{pmatrix} \omega^1 \\ \omega^2 \\ \omega^3 \end{pmatrix} = \begin{pmatrix} 0 & 0 & 0 & 0 & 0 \\ 0 & 0 & 0 & 0 & 0 \\ -g & 0 & 0 & 0 & 0 \end{pmatrix} \begin{pmatrix} x^1 \\ x^2 \\ x^3 \\ x^4 \\ x^5 \end{pmatrix} + \begin{pmatrix} 1 & 0 \\ 0 & 1 \\ 0 & -L_1 \cos\psi \end{pmatrix} \begin{pmatrix} u^1 \\ u^2 \end{pmatrix} + \begin{pmatrix} 0 \\ 0 \\ 1 \end{pmatrix} v. \tag{2.10}$$

This system, along with other example systems, will be studied extensively in the sequel.

The linearization just discussed is not the only type of linearization of importance. One may define a linearization of the original nonlinear system about any solution. Let us suppose that $x_0(t)$, $u_0(t)$, $v_0(t)$ together satisfy $\dot{x} = f(t,x,u,v)$ (equation (1.1)). Writing

$$\begin{aligned} f(t,x(t),\ u(t),\ v(t) &= f(t,x_0(t),u_0(t),v_0(t)) + f_x(t,x_0(t),u_0(t),v_0(t))(x(t)-x_0(t)) \\ &+ f_u(t,x_0(t),u_0(t),v_0(t))(u(t)-u_0(t)) \\ &+ f_v(t,x_0(t),u_0(t),v_0(t))(v(t)-v_0(t)) \\ &+ \mathcal{O}(\|(x(t)-x_0(t),\ u(t)-u_0(t),\ v(t)-v_0(t)|\|) \end{aligned}$$

we find, using $\dot{x}_0(t) = f(t, x_0(t), u_0(t), v_0(t))$, that

$$\begin{aligned} (x(t)-x_0(t))^{\boldsymbol{\cdot}} &= \tilde{A}(t)(x(t)-x_0(t)) + \tilde{B}(t)(u(t)-u_0(t)) \\ &+ \tilde{C}(t)(v(t)-v_0(t)) + \mathcal{O}(\|(x(t)-x_0(t),\ u(t)-u_0(t),\ v(t)-v_0(t))\|) \end{aligned} \tag{2.11}$$

where now

$$\tilde{A}(t) = f_x(t,\ x_0(t),\ u_0(t),\ v_0(t))$$

and $\tilde{B}(t)$, $\tilde{C}(t)$ are defined analogously. In the same way we have

$$(\omega(t) - \omega_0(t)) = \tilde{H}(t)(x(f)-x_0(t)) + \tilde{J}(t)(u(t)-u_0(t)) + \tilde{L}(t)(v(t) - v_0(t)) + \mathcal{O}(\cdots) \tag{2.12}$$

$$(y(t) - y_0(t)) = \tilde{F}(t)(x(t)-x_0(t)) + \tilde{G}(t)(u(t)-u_0(t)) + M(t)(v(t)-v_0(t)) + \mathcal{O}(\cdots) . \tag{2.13}$$

The difference lies in the fact that the Jacobians are now evaluated at $(t, x_0(t), u_0(t), v_0(t))$ rather than at (t, x_0, u_0, v_0). To see how this can lead to a rather different linearized system let us return to (1.3), (1.4), (1.6). We consider the solution which corresponds to simply raising or lowering the load from one level to another at a steady rate without moving the beam. This solution can be expressed as

$$\theta = \dot{\theta} = 0$$

$$\psi = \psi_0, \ \dot{\psi} = \ddot{\psi} = 0$$

$$\ell_2 = \ell_{2,0} + \dot{\ell}_{2,0} t \equiv \ell_{2,0}(t)$$

$$\dot{\ell}_2 = \dot{\ell}_{2,0}, \quad \text{constant}$$

$$v = 0 .$$

This changes (2.8), (2.9), (2.10) to the extent that the linearization now takes place about $(0, 0, \psi_0, 0, \ell_{2,0}(t), \dot{\ell}_{2,0}, 0, 0)$ rather than about $(0, 0, \psi_0, 0, \ell_{2,0}, 0, 0, 0)$. The resulting linearized system is now (2.8), (2.9), (2.10) with the following changes. In (2.8) the second equation becomes

$$\dot{x}^2 = \frac{1}{\ell_2(t)}[-gx^1 - 3\dot{\ell}_{2,0}x^2 + L_1 \sin\psi_0 x^4 - L_1 \cos\psi_0 u^2 + \frac{v}{m}] \tag{2.14}$$

In the linearized output equations (2.9) the second and fourth equations are replaced by

$$y^2 = \dot{\ell}_{2,0}x^1 + \ell_2(t)x^2 + L_1 \cos(\psi_0)x^4 \tag{2.15a}$$

$$y^4 = L_1 \sin(\psi_0) x^4 - u_1 . \tag{2.15b}$$

In the linearized observation equations, (2.10), the third equation is replaced by

$$\omega^3 = -g x^1 + L_1 \sin(\psi_0) x^4 . \tag{2.16}$$

Linearization about this "moving" solution rather than about the earlier equilibrium solution thus leads to a linearized system with additional coefficients which have a very definite impact upon the dynamics of the system. Also, the linearized system is now time varying (since $\ell_2 = \ell_2(t)$ is now a function of t) whereas the earlier linearized system (2.8), (2.9), (2.10) is a constant coefficient system.

There is an important point which we wish to make now. Given an original system such as (1.3), (1.4), (1.5) it has associated with it, in general, an infinite number of linearized systems - indeed, as we have seen, there is one for each solution of the original system. The result is a collection of linear control systems, each one describing small variations about a basic solution, or "operating condition" in aircraft control systems the term "flight condition" is current and quite descriptive). Linear systems, such as (2.8), (2.9), (2.10) represent the behavior of small variations about a static operating condition while other systems, such as (2.8), (2.9), (2.10) modified via (2.14), (2.15), (2.16) have reference to a moving or "dynamic" operating condition.

Control problems are customarily treated in the context of a linearization about some fixed "nominal" operating condition. But once the control strategy has been worked out for the nominal system many questions remain to be asked. For example: Will this control strategy give adequate performance for other operating conditions? Will it be adequate in terms of the original nonlinear plant (another word often used in place of "system")? Will the plant operator be "comfortable" with this control configuration? The list can be extended almost indefinitely.

What are the goals of a control strategy (by which we mean a set of rules for operating the controls so as to achieve performance which is "adequate" or at least "improved" in terms of some criterion)? Perhaps an example explains as well as anything else. In the case of the crane, various disturbances, such as the wind, movement of the boom, etc., cause the load to oscillate, i.e. the swinging motion described by $\theta(t)$ is excited. The evolution of the swing amplitude, whether it gets worse or is attenuated, etc., all depend on the operating condition in which the crane is being used. The goal of a control strategy might be (a) to damp or attenuate the swinging motion; (b) to make the swinging motion uniform and reduce its variability from one operating condition to another; (c) to bring the swinging to rest in a fixed, finite time period. Numerous alternate goals might be envisioned. Our goal in this text is to develop, in a methodical way, certain control design techniques which have proved useful in devising control strategies to meet goals such as these and to provide the relevant mathematical background and tools which are needed to implement these design techniques in practice.

We complete this chapter with a review of the theory of linear systems of differential equations. Many readers will wish to proceed directly to Chapter II.

3. Basic Properties of Linear Systems

In this section we will review the basic theory pertaining to a linear differential equation in E^n:

$$\dot{x} = A(t)x + f(t) . \tag{3.1}$$

We will assume that the $n \times n$ matrix $A(t)$ is continuous on some non-empty open interval (α, β) of the real line. Concerning the inhomogeneous, or "forcing", term, $f(t)$, we will suppose that for each closed bounded subinterval $[a, b] \subseteq (\alpha, \beta)$

$$f \in L_n^2([a, b]) .$$

This requires some explanation, of course. The notation $L_n^2([a, b])$ refers to the infinite dimensional vector space of all measurable (see [4],

[5] for definition and properties of measurable functions) functions $w(t)$ defined for $t \in [a,b]$ with values $w(t) \in E^n$, $t \in [a,b]$, such that

$$\int_a^b \|w(t)\|^2_{E^n} \, dt < \infty .$$

In $L^2_n([a,b])$ the inner product

$$(w_1, w_2)_{L^2_n[a,b]} = \int_a^b (w_1(t), w_2(t))_{E^n} \, dt \tag{3.2}$$

is well defined, $(w_1(t), w_2(t))_{E^n}$ referring to the usual scalar or dot product in the finite dimensional space E^n:

$$(w_1(t), w_2(t))_{E^n} = w_2(t)^* w_1(t)$$

$$\equiv (\overline{w_2^1(t)}, \overline{w_2^2(t)}, \ldots, \overline{w_2^n(t)}) \begin{pmatrix} w_1^1(t) \\ w_1^2(t) \\ \vdots \\ w_1^n(t) \end{pmatrix} = \sum_{k=1}^n \overline{w_2^k(t)} \, w_1^k(t),$$

$\overline{}$ denoting the conjugate if the complex space E^n is considered. With the inner product (3.2), $L^2_n([a,b])$ becomes what is known as a Hilbert space ([4], [5]). Each function $w \in L^2_n[a,b]$ has associated with it a <u>norm</u>, or magnitude,

$$\|w\|_{L^2_n[a,b])} = [(w,w)_{L^2_n([a,b])}]^{\frac{1}{2}} = [\int_a^b (w(t), w(t))_{E^n} dt]^{\frac{1}{2}} = [\int_a^b \|w(t)\|^2_{E^n} dt]^{\frac{1}{2}} .$$

Here $\|w(t)\|_{E^n} = [(w(t), w(t))_{E^n}]^{\frac{1}{2}}$ is the usual Euclidean norm in E^n. The <u>distance</u> between any two functions w_1, w_2 in $L^2_n([a,b])$ is defined to be $\|w_1 - w_2\|_{L^2_n([a,b])}$.

The reader is likely aware that, strictly speaking, the elements of

$L_n^2([a,b])$ are not functions at all but, rather, equivalence classes $\tilde{w}$ of functions, w_1 and w_2 belonging to the same equivalence class $\tilde{w}$ just in case w_1 and w_2 differ only on a "set of measure zero" or, equivalently,

$$\int_a^b \|w_1(t) - w_2(t)\|^2_{E^n} \, dt = 0 .$$

In most applications the distinction between a function and its equivalence class plays no role so we will not stress the difference here. In particular, we will say that $w = 0$ in $L_n^2([a,b])$ if

$$\int_a^b \|w(t)\|^2_{E^n} \, dt = 0 .$$

Practically all of the functions which we will deal with in this book will not exhibit the wide variety of discontinuities and singular behavior on small sets which is permitted for general functions in $L_n^2([a,b])$. Indeed, we will for the most part be concerned with continuous functions only. It is the inner product structure of $L_n^2([a,b])$ which is of greatest value to us and not the generality of functions permitted.

With reference to this inner product, perhaps the most important properties are those expressed by

<u>Proposition 3.1.</u> <u>Let</u> $w_1, w_2 \in L_n^2([a,b])$. <u>Then</u> $w_1 + w_2 \in L_n^2([a,b])$ <u>and</u>

$$\|w_1 + w_2\|_{L_n^2[a,b]} \le \|w_1\|_{L_n^2[a,b])} + \|w_2\|_{L_n^2([a,b])} \quad \left(\begin{matrix}\text{Minkowski}\\ \text{inequality}\end{matrix}\right). \tag{3.3}$$

<u>Further</u>:

$$|(w_1, w_2)_{L_n^2([a,b])}| \le \|w_1\|_{L^2[a,b]} \|w_2\|_{L_n^2[a,b]} \quad \left(\begin{matrix}\text{Schwarz}\\ \text{inequality}\end{matrix}\right) \tag{3.4}$$

<u>with equality holding in</u> (3.4) <u>if and only if there are scalars</u> λ_1, λ_2, <u>not both zero, such that</u>

$$\lambda_1 w_1 + \lambda_2 w_2 = 0 \quad \text{in} \quad L_n^2([a,b]) .$$

Since this is a basic result in function space theory and proved in almost any text on the subject, a proof here would be superfluous (see [4], e.g.).

The following property of integrals of vector functions is frequently useful.

$$\left\| \int_a^b f(t)dt \right\| \le \int_a^b \|f(t)\| dt = \int_a^b (1 \times \|f(t)\|)dt \le \text{(Schwarz inequality)}$$

$$\le \left[\int_a^b 1\, dt\right]^{\frac{1}{2}} \left[\int_a^b \|f(t)\|^2 dt\right]^{\frac{1}{2}} = \sqrt{b-a}\, \|f\|_{L_n^2[a,b]} \tag{3.5}$$

Here and in the sequel $\| \ \|$ and $(\ , \ )$ will mean the norm and inner product in E^n; the subscript $L_n^2([a,b])$ will be appended to these symbols when they refer to the norm and inner product in that space. The standard properties of $\| \ \|$ and $(\ , \ )$ will be used without further discussion. However, one notion which we probably should mention is that of the norm of a matrix. If M is an $n \times m$ matrix, $\|M\|$ is the smallest nonnegative number with the property

$$\|My\|_{E^n} \le \|M\| \, \|y\|_{E^m}, \qquad y \in E^m . \tag{3.6}$$

It can be shown that, so defined,

$$\|M\| \le \left[\sum_{j=1}^{n} \sum_{k=1}^{m} |m_k^j|^2\right]^{\frac{1}{2}} .$$

The basic existence and uniqueness theorem for solutions of (3.1) is the following.

<u>Theorem 3.2.</u> <u>Let the</u> $n \times n$ <u>matrix</u> $A(t)$ <u>be continuous on the interval</u> $I = (\alpha, \beta)$ <u>and let</u> $f \in L_n^2(J)$ <u>for every closed bounded subinterval</u> $J \subseteq I$. <u>Given</u> $t_0 \in I$ <u>and</u> $x_0 \in E^n$ <u>there is one and only one solution</u> $x(t)$ <u>of</u> (3.1) <u>on the interval</u> I <u>with</u>

$$x(t_0) = x_0 . \tag{3.7}$$

<u>Proof.</u> We give only the existence on closed bounded intervals at first,

postponing the uniqueness until a little later. Our proof is by successive approximations. Let J be a closed bounded interval of length L containing x_0 and let

$$x_0(t) \equiv x_0 , \qquad t \in J .$$

We then define vector functions $x_k(t)$ recursively by

$$x_{k+1}(t) = x_0 + \int_0^t [A(s)x_k(s) + f(s)]\, ds, \qquad k = 0,1,2,3,\ldots . \tag{3. 8}$$

Let $a = \max_{t \in J} \|A(t)\|$. Then for such t, using (3. 5) and (3. 6)

$$\|x_1(t) - x_0(t)\| \equiv \left\|\int_{t_0}^t [A(s)x_0(s) + f(s)]\, ds\right\|$$

$$= \left\|\int_{t_0}^t A(s)ds\, x_0 + \int_{t_0}^t f(s)ds\right\| \leq a\|x_0\| + \left\|\int_{t_0}^t f(s)ds\right\|$$

$$\leq a\|x_0\| + \sqrt{|t - t_0|}\, \|f\|_{L_n^2(J)} \quad a\|x_0\| + L^{\frac{1}{2}} \|f\|_{L_n^2(J)} \equiv c .$$

Suppose, for some k,

$$\|x_{k+1}(t) - x_k(t)\| \leq ca^k |t - t_0|^k / k! , \quad t \in J. \tag{3. 9}$$

We have shown this to be true for $k = 0$. From (3. 8),

$$\|x_{k+2}(t) - x_{k+1}(t)\| = \left\|\int_{t_0}^t A(s)(x_{k+1}(s) - x_k(s))ds\right\| \leq a\int_{t_0}^t \|x_{k+1}(s) - x_k(s)\| ds$$

$$\leq ca^{k+1} \int_{t_0}^t (|s - t_0|^k / k!) ds = ca^{k+1} |t - t_0|^{k+1} / (k+1)!$$

and, by induction, (3. 9) holds for all k. Then, for $\ell < k$, $t \in J$,

$$\|x_k(t) - x_\ell(t)\| \le \sum_{j=\ell}^{k-1} \|x_{j+1}(t) - x_j(t)\|$$

$$\le c \sum_{j=\ell}^{k-1} a^j |t - t_0|^j / j! \le c \sum_{j=\ell}^{k-1} a^j L^j / j! \quad .$$

Since $\sum_{j=0}^{\infty} a^j L^j / j!$ is a series of positive terms converging to e^{aL}, we conclude

$$\lim_{\substack{k,\ell \to \infty \\ k \ge \ell}} \|x_k(t) - x_\ell(t)\| \le \lim_{\substack{k,\ell \to \infty \\ k \ge \ell}} c \sum_{j=\ell}^{k-1} a^j L^j / j! = 0$$

and $\{x_k(t)\}$ possesses the Cauchy sequence property ([5]) uniformly on J. It follows that

$$\lim_{k \to \infty} \|x(t) - x_k(t)\| = 0, \quad \text{uniformly}, \quad t \in J,$$

for some continuous vector function $x(t)$ defined on J. From

$$x_{k+1}(t) = x_0 + \int_{t_0}^{t} [A(s) x_k(s) + f(s)] \, ds$$

and the above noted uniform convergence there follows, letting $k \to \infty$,

$$x(t) = x_0 + \int_{t_0}^{t} [A(s)x(s) + f(s)] \, ds \tag{3.10}$$

and we conclude that $x(t)$ is differentiable. Differentiating (3.10) we obtain (3.1) and the existence part of the proof is complete.

As we have already intimated, we denote the inner product of two vectors x, y in E^n interchangeably by (x, y) and $y^* x$. Here and elsewhere, if T is an $m \times n$ matrix (y is an $n \times 1$ matrix or column vector), T^* denotes its <u>adjoint</u>, i. e., conjugate transpose, an $n \times m$ matrix ([6]). Thus the formula

$$(x, y) = y^* x = \sum_{k=1}^{n} \bar{y}^k x^k$$

is consistent both with the adjoint notation and the rules for matrix multiplication.

With reference to this inner product an important relationship holds between solutions of the system

$$\dot{x} = A(t)\,x, \tag{3.11}$$

the homogeneous counterpart of (3.1), and the adjoint system

$$\dot{y} = -A^*(t)y \qquad (\text{or } \dot{y}^* = -y^*A(t)). \tag{3.12}$$

For if $x(t)$, $y(t)$ are solutions of (3.11), (3.12), respectively,

$$\frac{d}{dt}(x(t),\, y(t)) = \frac{d}{dt}(y^*(t)\,x(t)) = \dot{y}^*(t)x(t) + y^*(t)\dot{x}(t)$$
$$= -y^*(t)A(t)x(t) + y^*(t)A(t)x(t) \equiv 0$$

and we conclude that $(x(t),\, y(t))$ is constant. We can now provide the

Completion of Proof of Theorem 3.2. For the uniqueness we suppose that $\tilde{x}(t)$ is also a solution of (3.1), (3.7), in addition to the solution $x(t)$ whose existence was proved earlier. Then, by subtraction

$$\frac{d}{dt}(\tilde{x}(t) - x(t)) = A(t)(\tilde{x}(t) - x(t)), \quad t \in J.$$

Now the existence proof given earlier applies equally well to (3.12) on J. Thus we may let $y(t)$ be a solution of (3.12) on J with, for some $t_1 \in J$,

$$y(t_1) = \tilde{x}(t_1) - x(t_1).$$

Since $(\tilde{x}(t) - x(t),\, y(t))$ is constant ($\tilde{x}(t) - x(t)$ and $y(t)$ satisfy (3.11) and (3.12) respectively)

$$\|\tilde{x}(t_1) - x(t_1)\|^2 = (\tilde{x}(t_1) - x(t_1), y(t_1)) = (\tilde{x}(t_0) - x(t_0), y(t_0))$$
$$= (x_0 - x_0, y(t_0)) = 0$$

and we conclude $\tilde{x}(t_1) = x(t_1)$. Since t_1 is any point of J, $\tilde{x}(t) \equiv x(t)$, $t \in J$ and we have uniqueness.

Let J_k be an expanding sequence of closed bounded intervals with

$$\bigcup_{k=1}^{\infty} J_k = I = (\alpha, \beta) . \tag{3.13}$$

On each J_k we have the existence of a unique solution $x(t)$ of (3.1), (3.7). Uniqueness further shows $x(t)$ as defined on J_ℓ to be the restriction of $x(t)$ as defined on J_k for $k \geq \ell$. We conclude (3.1), (3.7) has a unique solution on $I = (\alpha, \beta)$. Q. E. D.

The uniqueness result, applied to the homogeneous system (3.11), has important algebraic consequences. Let $\phi_1(t), \phi_2(t), \ldots, \phi_r(t)$, $r \leq n$, be solutions of (3.11) such that $\phi_1(t_0), \phi_2(t_0), \ldots, \phi_r(t_0)$ are linearly independent vectors in E^n for some $t_0 \in (\alpha, \beta)$. Let $c_1, c_2, \ldots, c_r$ be arbitrary scalars not all zero. Then for any t_1

$$\xi(t_1) = c_1\phi_1(t_1) + c_2\phi_2(t_1) + \ldots + c_r\phi_r(t_1) \neq 0 .$$

If this were not true $\xi(t) = c_1\phi_1(t) + \ldots + c_n\phi_r(t)$ would be a solution of (3.11) vanishing at t_1 but (by linear independence of $\phi_1(t_0), \ldots, \phi_r(t_0)$) different from zero at t_0. This is impossible because $x(t) \equiv 0$ is the unique solution of (3.11) with $x(t_1) = 0$. Thus we conclude that $\phi_1(t_1), \phi_2(t_1), \ldots, \phi_r(t)$ of (3.11) are linearly independent at any $t_0 \in (\alpha, \beta)$ they are linearly independent for all $t \in (\alpha, \beta)$. Also, if $\phi_1(t_0), \ldots, \phi_r(t_0)$ are linearly dependent so that

$$\xi(t_0) = c_1\phi_1(t_0) + c_2(t_0) + \ldots + c_r\phi_r(t_0) = 0$$

then

$$\xi(t) = c_1\phi_1(t) + c_2\phi_2(t) \ldots + c_r\phi_r(t) \equiv 0, \qquad t \in (\alpha, \beta) .$$

We may therefore speak without ambiguity of linearly independent solutions or of dependent solutions of (3.11) without distinguishing whether we are referring to specific values $\phi_1(t_0), \phi_2(t_0), \ldots, \phi_r(t_0)$ or to the vector functions $\phi_1, \phi_2, \ldots, \phi_r$.

If $\phi_1, \phi_2, \ldots, \phi_n$ are n linearly independent solutions of (3.11) the $n \times n$ matrix

$$\Phi(t) = (\phi_1(t), \phi_2(t), \ldots, \phi_n(t))$$

is called a fundamental matrix solution of (3.11). It clearly satisfies the matrix differential equation

$$\dot{\Phi}(t) \equiv A(t)\,\Phi(t) \; . \tag{3.14}$$

One may verify without difficulty, using uniqueness, that the set of all fundamental matrix solutions of (3.11) and the set of all matrix functions of the form $\Phi(t)C$, where C is a nonsingular $n \times n$ matrix and $\Phi(t)$ is any particular fundamental matrix solution, coincide. We may, therefore, introduce, for $t_0 \in (\alpha, \beta)$, the fundamental matrix solution

$$\Phi(t, t_0) = \Phi(t)\,\Phi(t_0)^{-1} \; . \tag{3.15}$$

It is easy to see that

$$\Phi(t_0, t_0) = I \quad (n \times n \text{ identity matrix}) \tag{3.16}$$

and that $\Phi(t, t_0)$ is independent of the particular fundamental solution matrix $\Phi(t)$ used in (3.15). Indeed $\Phi(t, t_0)$ is the unique solution of (3.14) reducing to the identity at t_0.

Let $x_0 \in E^n$. Then

$$x(t) = \Phi(t, t_0)\,x_0 \tag{3.17}$$

is the unique solution of (3.11), (3.7). For

$$\dot{x}(t) = \frac{d}{dt}(\Phi(t, t_0)x_0) = A(t)\,\Phi(t, t_0)x_0 = A(t)x(t)$$

and

$$x(t_0) = \Phi(t_0, t_0)x_0 = x_0 \; .$$

The fact that

$$(x(t), y(t) \equiv \text{constant} \tag{3.18}$$

for solutions $x(t)$, $y(t)$ of (3.11), (3.12) respectively, generalizes immediately to show that the matrix product

$$Y^*(t)X(t) \equiv \text{constant} \ (p \times r \ \text{matrix}) \tag{3.19}$$

if $X(t)$, $Y(t)$ are $n \times r$, $n \times p$ matrix solutions, respectively, of the matrix differential equations (whose existence and uniqueness theorems follow directly from Theorem 3.2 by applying that theorem columnwise)

$$\dot{X} = A(t)X, \tag{3.20}$$

$$\dot{Y} = -A(t)^*Y \quad (\text{equiv. } \dot{Y}^* = -Y^*A(t)) \tag{3.21}$$

If $\Phi(t)$ is a fundamental solution matrix of (3.11) (equally well, (3.20))

$$\Psi(t) = (\Phi(t)^{-1})^*$$

exists since the columns of $\Phi(t)$ are independent. Moreover, $\Psi(t)$ is a fundamental matrix solution of (3.12) (or (3.21)). For, since

$$0 = \frac{d}{dt}(I) = (\Phi(t)\Phi(t)^{-1}) = \Phi(t)\frac{d}{dt}(\Phi(t)^{-1}) + \dot{\Phi}(t)\Phi(t)^{-1}$$

we have

$$\frac{d}{dt}(\Phi(t)^{-1}) = -\Phi(t)^{-1}\dot{\Phi}(t)\Phi(t)^{-1}$$

$$= -\Phi(t)^{-1}A(t)\Phi(t)\Phi(t)^{-1} = -\Phi(t)^{-1}A(t).$$

Thus $\Psi(t)^* = \Phi(t)^{-1}$ is a fundamental solution of the second equation in (3.12), (3.21). Taking the conjugate transpose, $\Psi(t)$ is a fundamental solution of (3.12) (or (3.21)). Similarly, if $\Psi(t)$ is a fundamental matrix solution of (3.12) ((3.21)), $(\Psi(t)^{-1})^*$ is a fundamental solution matrix of (3.11) ((3.20)). if (cf. (3.15)) we form the matrix $\Phi(t, \tau)$ it has the dual properties

$$\frac{\partial}{\partial t}\Phi(t,\tau) = A(t)\Phi(t,\tau), \quad \Phi(t,\tau) = I, \tag{3.22}$$

$$\frac{\partial}{\partial \tau}\Phi(t,\tau) = -\Phi(t,\tau)A(\tau), \quad \Phi(t,t) = I. \tag{3.23}$$

Moreover, it can be written either as $\Phi(t)\Phi(\tau)^{-1}$ or as $(\Psi(t)^{-1})^{*}\Psi(\tau)^{*}$ for arbitrary fundamental matrices $\Phi(t)$, $\Psi(\tau)$ of (3.11), (3.12) respectively.

We pass now to the inhomogeneous system (3.1). Let $x(t)$ be the solution of (3.1), (3.7) and consider the product $\Phi(t,\tau)x(\tau)$. We have

$$\frac{\partial}{\partial\tau}\Phi(t,\tau)x(\tau) = -\Phi(t,\tau)A(\tau)x(\tau) + \Phi(t,\tau)[A(\tau)x(\tau) + f(\tau)] = \Phi(t,\tau)f(\tau)$$

and thus

$$\Phi(t,t)x(t) - \Phi(t,t_0)x(t_0) = \int_{t_0}^{t} \Phi(t,\tau)f(\tau)\,d\tau$$

or, using (3.23),

$$x(t) = \Phi(t,t_0)x_0 + \int_{t_0}^{t} \Phi(t,\tau)f(\tau)\,d\tau . \tag{3.24}$$

The formula (3.24), which is known as the variation of parameters formula, is a fundamental result which will be used repeatedly in the sequel.

In the case of a constant coefficient system

$$\dot{x} = Ax \tag{3.25}$$

there is a fundamental matrix solution of particular importance. For the scalar case $\dot{x} = ax$, $a, x \in E^1$, we have, of course, the exponential solution

$$x(t) = e^{at} = 1 + at + \tfrac{1}{2}a^2t^2 + \ldots + \frac{1}{k!}a^kt^k + \ldots .$$

For a fixed $n \times n$ matrix A one can similarly define the series

$$e^{At} \equiv I + At + \tfrac{1}{2}A^2t^2 + \ldots + \frac{1}{k!}A^kt^k + \ldots$$

and it can be shown that the series is uniformly convergent, in each matrix entry, for t in any bounded region of E^1 (real or complex). Term by term differentiation can also be justified and we have

$$\frac{d}{dt}e^{At} = \frac{d}{dt}\left(\sum_{k=0}^{\infty} \frac{1}{k!} A^k t^k\right) = \sum_{k=1}^{\infty} \frac{1}{k!} A^k t^{k-1}$$

$$= \sum_{k=1}^{\infty} \frac{1}{(k-1)!} A^k t^{k-1} = A \sum_{k=0}^{\infty} \frac{1}{k!} A^k t^k = Ae^{At} .$$

We see readily that

$$e^{At} = I.$$

For the system (3.25), then

$$\Phi(t, \tau) = e^{A(t-\tau)}$$

and the variation of parameters formula (3.24) becomes

$$x(t) = e^{A(t-t_0)} x_0 + \int_{t_0}^{t} e^{A(t-\tau)} f(\tau)\, d\tau ,$$

giving the solution $x(t)$ of $\dot{x} = Ax + f(t)$ with $x(t_0) = x_0$.

In linear systems theory, in addition to the matrix differential equation $\dot{\Phi} = A\Phi$, satisfied by fundamental solution matrices of (3.11) - and, column for column, the same as the vector equation (3.11) - we shall have many occasions to refer to matrix differential equations of the form

$$\dot{X} = A(t)X + XB(t) + P(t), \tag{3.26}$$

where $P(t)$ is a continuous (or square integrable, for that matter) $n \times m$ matrix function, $A(t)$ and $B(t)$ are continuous $n \times n$ and $m \times m$ coefficient matrices, and $X = X(t)$ is the unknown $n \times m$ matrix solution. Corresponding to the differential equation (3.26) we may pose an initial value problem by requiring, for some t_0,

$$X(t_0) = X_0, \quad X_0 \text{ a given } n \times m \text{ matrix.} \tag{3.27}$$

The solution of (3.26), (3.27) can be described very easily. Let $\Phi(t, t_0)$, $\theta(t, t_0)$ be the fundamental $n \times n$, $m \times m$ solution matrices of (3.11) and the row vector equation $\dot{y} = yB(t)$, respectively (thus $\dot{\theta} = \theta B(t)$), with

$$\Phi(t_0, t_0) = I_n, \qquad \theta(t_0, t_0) = I_m, \tag{3.28}$$

respectively. Then

$$X(t) = \Phi(t, t_0)X_0\theta(t, t_0) + \int_{t_0}^{t} \Phi(t, s)P(s)\theta(t, s)ds . \tag{3.29}$$

To verify that (3. 29) is a solution of (3. 26), (3. 27) is easy:

$$X(t_0) = \Phi(t_0, t_0)X_0\theta(t_0, t_0) + 0 = (\text{cf. } (3.28)) = X_0$$

and, differentiating with respect to t,

$$\dot{X}(t) = A(t)\Phi(t, t_0)X_0\theta(t, t_0) + \Phi(t, t_0)X_0\theta(t, t_0)B(t)$$

$$+ \int_{t_0}^{t} [A(t)\Phi(t, s)P(s)\theta(t, s) + \Phi(t, s)P(s)\theta(t, s)B(t)]\, ds$$

$$+ \Phi(t, t)P(t)\theta(t, t) = A(t)X(t) + X(t)B(t) + P(t) .$$

That this is the unique solution of (3. 26), (3. 27) can be established, much as in Theorem 3. 1, using solutions of the "adjoint" equation

$$\dot{Y} = - A(t)^*Y - YB(t)^* . \tag{3.30}$$

The relevant inner product on the space of $n \times m$ matrices, with respect to which "adjointness" is defined, is

$$(X, Y) = \text{Tr } Y^* X,$$

where Tr denotes the trace of a matrix ([6]). As this inner product will be discussed at some length in Chapter VI we will not pursue the matter further here. Nevertheless, it is fair to say that our whole approach to linear system theory, as presented in this volume, is based on the properties of (3.26), (3. 30) and related equations.

Bibliographical Notes, Chapter I

[1] H. Goldstein: "Classical Mechanics", Addison-Wesley Publishing Co., Inc., Reading, Mass., 1950.

This remains one of the most readable treatments of classical mechanics. The equations of motion for the crane and other physical systems in this text can be derived readily after reading Goldstein's Chapter 2.

[2] E. A. Coddington and N. Levinson: "Theory of Ordinary Differential Equations", McGraw-Hill Book Co., New York, 1955.

A very accessible account of the relationship between the behavior of solutions of a nonlinear system near an equilibrium (or "critical") point may be found in Chapters 4 and 5 of

[3] F. Brauer and J. A. Nohel: "Qualitative Theory of Ordinary Differential Equations", W. A. Benjamin, Inc., 1969.

[4] F. Riesz and B. Sz.-Nagy: "Functional Analysis," F. Ungar Pub. Co., New York, 1955.

[5] W. Rudin: "Real and Complex Analysis", McGraw Hill Book Co., Inc., New York, 1966.

The basic "facts of life" in regard to linear algebra and matrix theory, as well as related material for systems of linear differential equations are ably presented in

[6] F. Brauer, J. Nohel and H. Schneider: "Linear Mathematics", W. A. Benhamin, Inc., Menlo Park, 1970.

[7] A. L. Rabenstein: "Elementary Differential Equations with Linear Algebra", Academic Press, New York, San Francisco, London, 1975.

Exercises, Chapter I

1. Develop the linearized equations of motion for the crane in the "level hold" operating mode described in (1.7) - (1.13) about the equilibrium solution

$$\psi = \psi_0, \ 0 < \psi_0 < \frac{\pi}{2}, \ \dot{\psi} = 0, \ \theta = \dot{\theta} = 0, \ u^2 = 0, \ v = 0.$$

Develop also the linearized response and observation equations, based on (1.11) - (1.13).

2. An ecological unit consists of a population of 100w wolves, 1000 r rabbits, and 10,000 v acres of vegetation on which the rabbits subsist. A very much over-simplified model for the evolution of this system might be

$$\dot{w} = \frac{\alpha(\frac{r}{w})^2 w}{1 + (\frac{r}{w})^2} - \gamma w, \qquad \alpha > \gamma > 0,$$

$$\dot{r} = \frac{\beta(\frac{v}{r})^2 r}{1 + (\frac{v}{r})^2} - \delta r - \epsilon w, \qquad \beta > \delta > 0,$$

$$\dot{v} = \theta - \psi v - \phi r .$$

For given positive values of the parameters $\alpha, \beta, \gamma, \delta, \epsilon, \theta, \psi, \phi$, determine

(i) the equilibrium points (w_0, r_0, v_0), with w_0, r_0, v_0 all > 0, in terms of the above parameters;

(ii) the linearization of the system about (w_0, r_0, v_0);

(iii) the specific equilibrium point and the specific linearized control system for the case $\alpha = \beta = \theta = 2$, $\gamma = \delta = \phi = \psi = 1$, $\epsilon = \frac{1}{4}$;

(iv) the linearization of the response equations

$$y^1 = \frac{w}{r}, \quad y^2 = \frac{r}{v},$$

both for general (w_0, r_0, v_0) and for the specific case considered in (iii).

CHAPTER II

OBSERVABILITY AND CONTROLLABILITY

1. Observability and Reconstruction

Let us consider a linear system

$$\dot{x} = A(t)x + f(t), \qquad x, f \in E^n, \tag{1.1}$$

wherein the $n \times n$ matrix $A(t)$ and the n-vector function $f(t)$ are continuous and locally square integrable (i. e. $\int_a^b \|f(t)\|^2 dt < \infty$ for each finite interval $[a, b]$), respectively, on some interval (α, β), which we assume to contain all t of interest. Along with (1.1) we suppose we have a linear observation

$$\omega = H(t)x + \hat{H}(t)f, \qquad \omega \in E^r. \tag{1.2}$$

Assuming the system (1.1) is in operation during a time interval $[t_0, t_1] \subset (\alpha, \beta)$ and that

$$x(t_0) = x_0 \in E^n,$$

we have the formula developed in I-3 (Chapter I, Section 3):

$$x(t) = \Phi(t, t_0)x_0 + \int_{t_0}^{t} \Phi(t, s) f(s) ds, \qquad t \in [t_0, t_1]. \tag{1.3}$$

The second term can be computed from knowledge of the system matrix $A(t)$ and the function f. If we suppose, as we shall here, that f is a known function - which would be the case if, for example, f had the form $f(t) = B(t)u(t)$ with $u(t)$ a control exercised by the plant operator - then in principle the term $\hat{H}(t)f$ in (1.2) and $H(t)$ times the integral in (1.3) could be subtracted from

$$\omega(t) = H(t)\Phi(t, t_0) + H(t)\int_{t_0}^{t} \Phi(t, s)f(s)ds + \hat{H}(t)f(t)$$

to yield the modified observation

$$\hat{\omega}(t) = H(t)\Phi(t, t_0)x_0 \quad . \tag{1.4}$$

Now $\Phi(t, t_0)x_0$ satisfies the homogeneous equation

$$\dot{x} = A(t)x \tag{1.5}$$

and the observation (1.4) (dropping the $\wedge$ now) has the form

$$\omega(t) = H(t)x(t) \tag{1.6}$$

Thus the question of obtaining information about (1.1) via (1.2) reduces to the same question for the corresponding homogeneous system (1.5) and the homogeneous observation (1.6).

<u>Definition 1.1.</u> <u>The system</u> (1.5), (1.6) <u>is observable (we say the pair</u> $(H(t), A(t))$ <u>is observable) on an interval</u> $[t_0, t_1]$ <u>if</u>

$$\omega(t) = H(t)x(t) \equiv 0, \qquad t \in [t_0, t_1]$$

<u>implies</u>

$$x(t) \equiv 0, \qquad t \in [t_0, t_1]$$

<u>(which, of course, is the same as saying</u> $x(t_0) = x_0 = 0$).

We want to develop conditions which, if satisfied by $A(t)$ and $H(t)$ of (1.5) and (1.6), will be sufficient for observability. To this end it will be convenient to make use of the function space $L_r^2[t_0, t_1]$ consisting, as we have noted in I-3, of r-vector functions $w(t)$ defined on $[t_0, t_1]$ with

$$\int_{t_0}^{t_1} \|w(t)\|_{E^r}^2 dt = \int_{t_0}^{t_1} w^*(t)w(t)dt < \infty .$$

We recall first of all that functions $w_1, w_2, \ldots, w_q \in L_r^2[t_0, t_1]$ (or any other linear function space) are linearly independent if the vanishing of any linear combination implies that the coefficients in that linear combination are all zero:

$$\alpha_1 w_1 + \alpha_2 w_2 + \ldots + \alpha_q w_q = 0 \Rightarrow \alpha_1 = \alpha_2 = \ldots = \alpha_q = 0 .$$

Otherwise there functions are linearly dependent.

Proposition 1. 2. Let $x_1, x_2, \ldots, x_q$ be vectors in E^n and let $x_1(t)$, $x_2(t), \ldots, x_q(t)$ be the solutions of (1. 5) on $[t_0, t_1]$ with $x_i(t_0) = x_i$, $i = 1, 2, \ldots, q$. Let observations ω_i be generated on $[t_0, t_1]$ via

$$\omega_i(t) = H(t)x_i(t), \qquad t \in [t_0, t_1] .$$

Then the "observed linear system" (1. 5), (1. 6) is observable on $[t_0, t_1]$ just in case the ω_i are linearly independent in $L_r^2[t_0, t_1]$ whenever the x_i are linearly independent in E^n.

Proof. As noted in I-3, the solutions $x_i(t)$ are linearly independent in $L_n^2[t_0, t_1]$ just in case the x_i are linearly independent in E^n. If (1. 5), (1. 6) is observable and

$$\omega(t) = \sum_{i=1}^{q} \alpha_i \omega_i(t) \equiv 0, \qquad t \in [t_0, t_1] \tag{1.7}$$

then the corresponding solution vanishes also:

$$x(t) = \sum_{i=1}^{q} \alpha_i x_i(t) \equiv 0, \qquad t \in [t_0, t_1] , \tag{1.8}$$

and, in particular

$$\sum_{i=1}^{q} \alpha_i x_i = 0 . \tag{1.9}$$

Supposing the x_i linearly independent, $\alpha_1 = \alpha_2 = \dots = \alpha_q = 0$. Since this follows from (1.7), we conclude the ω_i are linearly independent.

If on the other hand there exist linearly independent $x_1, x_2, \dots, x_q$ such that the associated observations $\omega_1(t), \omega_2(t), \dots, \omega_q(t)$ are not independent, i.e., are dependent, in $L_r^2[t_0, t_1]$, then, letting $\alpha_1, \alpha_2, \dots, \alpha_q$, not all zero, be such that

$$\omega(t) = \sum_{i=1}^{q} \alpha_i \omega_i(t) \equiv 0$$

we see that $\omega(t)$ is an identically vanishing observation on the solution

$$x(t) = \sum_{i=1}^{q} \alpha_i x_i(t)$$

which is not the zero solution of (1.5) because $x_1 = x_1(0)$, $x_2 = x_2(0), \dots$, $x_q = x_q(0)$ are linearly independent. We conclude (1.5), (1.6) is not observable. Q.E.D.

Observability may be usefully expressed in terms of the fundamental solution matrix $\Phi(t, t_0)$ of (1.3) and the observation coefficient matrix $H(t)$. We have

<u>Theorem 1.3.</u> <u>The observed linear system</u> (1.5), (1.6) <u>is observable on</u> $[t_0, t_1]$ <u>just in case the "observability Grammian" matrix</u>

$$W(t_0, t_1) \equiv \int_{t_0}^{t_1} \Phi(t, t_0)^* H(t)^* H(t) \Phi(t, t_0)\, dt$$

<u>is positive definite.</u>

<u>Proof.</u> The solution $x(t)$ of (1.5) corresponding to the initial condition $x(t_0) = x_0$ is given by

$$x(t) = \Phi(t, t_0) x_0$$

and we have, for $\omega(t) = H(t)x(t) = H(t)\Phi(t, t_0)x_0$,

$$\|\omega\|^2_{L_r^2[t_0,t_1]} = \int_{t_0}^{t_1} \omega^*(t)\omega(t)dt = x_0^* \int_{t_0}^{t_1} \Phi(t,t_0)^* H(t)^* H(t) \Phi(t,t_0) dt\, x_0 = x_0^* W(t_0, t_1) x_0,$$

a quadratic form in x_0. Clearly $W(t_0, t_1)$ is a symmetric $n \times n$ matrix. If it is positive definite then

$$\omega = 0 \Rightarrow x_0^* W(t_0, t_1) x_0 = 0 \Rightarrow x_0 = 0$$

and (1.5), (1.6) is observable on $[t_0, t_1]$. If it is not positive definite, then there is some $x_0 \neq 0$ such that $x_0^* W(t_0, t_1) x_0 = 0$. Then $x(t) = \Phi(t, t_0) x_0 \neq 0$ for $t \in [t_0, t_1]$ but $\|\omega\|^2_{L^2_r[t_0, t_1]} = 0$ so $\omega = 0$ and we conclude (1.5), (1.6) is not observable on $[t_0, t_1]$. Q. E. D.

Corollary 1.4. If (1.5), (1.6) is observable on $[t_0, t_1]$ then it is also observable on any interval $[\tilde{t}_0, \tilde{t}_1]$ such that $\tilde{t}_0 \leq t_0 < t_1 \leq \tilde{t}_1$.

Proof. Defining $\geq$ and $\leq$ for symmetric matrices in terms of the associated quadratic forms, e.g. $W \leq V$ if $x^* W x \leq x^* V x$, $x \in E^n$, clearly

$$W(\tilde{t}_0, \tilde{t}_1) = \Phi(t_0, \tilde{t}_0)^* \int_{\tilde{t}_0}^{\tilde{t}_1} \Phi(s, t_0)^* H(s)^* H(s) \Phi(s, t_0) ds\, \Phi(t_0, \tilde{t}_0) \geq \Phi(t_0, \tilde{t}_0)^* W(t_0, t_1) \Phi(t_0, \tilde{t}_0) > 0.$$

Q. E. D.

We will see in the sequel that the result of Theorem 1.3 is of considerable theoretical interest - and some practical interest. Nevertheless, in practice, determination of observability is usually carried out in terms of simple algebraic criteria.

Proposition 1.5. The observed linear system (1.5), (1.6) is observable on $[t_0, t_1]$ if there are distinct points $s_1, s_2, \ldots, s_q \in [t_0, t_1]$ such that

$$\text{rank} \begin{pmatrix} H(s_1)\Phi(s_1, t_0) \\ H(s_2)\Phi(s_2, t_0) \\ \vdots \\ H(s_q)\Phi(s_q, t_0) \end{pmatrix} \equiv \text{rank}\, V(s_1, s_2, \ldots, s_q, t_0) = n. \qquad (1.10)$$

Proof. We know from linear algebra that $V^* V > 0$ (is positive definite) just in case V has rank equal to n. Now

$$V(s_1, s_2, \ldots, s_q, t_0)^* V(s_1, s_2, \ldots, s_q, t_0) = \sum_{k=1}^{q} \Phi(s_k, t_0)^* H(s_k)^* H(s_k)\Phi(s_k, t_0).$$

Let $0 < \delta < \frac{1}{2}\min|s_k - s_\ell|$, $k \neq \ell$, and let $\pm$ be used so that the $s_k \pm \delta$ lie in $[t_0, t_1]$. Then

$$\begin{aligned} W(t_0, t_1) &= \int_{t_0}^{t_1} \Phi(s, t_0)^* H(s)^* H(s)\Phi(s, t_0)ds \\ &\geq \sum_{k=1}^{q} \left(\pm \int_{s_k}^{s_k \pm \delta} \Phi(s, t_0)^* H(s)^* H(s)\Phi(s, t_0)ds\right) \qquad (1.11) \\ &= \delta V^*(s_1, s_2, \ldots, s_q, t_0)V(s_1, s_2, \ldots, s_q, t_0) + \mathcal{O}(\delta), \ \delta \to 0. \end{aligned}$$

The term $\mathcal{O}(\delta)$ refers to a matrix function $R(\delta)$ with the property that

$$\lim_{\delta \downarrow 0} \frac{\|R(\delta)\|}{\delta} = 0 .$$

This notation will be used frequently in the sequel. (See also I-2).

If we assume the rank condition (1.10) then for δ sufficiently small,

$$\delta V^*(s_1, s_2, \ldots, s_q, t_0)V(s_1, s_2, \ldots, s_q, t_0) + \mathcal{O}(\delta) > 0 .$$

But then (1.11) implies $W(t_1, t_0) > 0$ and Theorem 2.3 shows (1.5), (1.6) to be observable on $[t_0, t_1]$. Q.E.D.

Example. Consider the second order equation

$$t^2\ddot{x} + t\dot{x} - x = 0$$

with observation $\omega = \dot{x}$. The corresponding first order system is ($x^1 = x$, $x^2 = \dot{x}$)

$$\begin{pmatrix} \dot{x}^1 \\ \dot{x}^2 \end{pmatrix} = \begin{pmatrix} 0 & 1 \\ \frac{1}{t^2} & -\frac{1}{t} \end{pmatrix} \begin{pmatrix} x^1 \\ x^2 \end{pmatrix} \qquad (1.12)$$

and

$$\omega = (0, 1)\begin{pmatrix} x^1 \\ x^2 \end{pmatrix}, \text{ i.e., } H = (0, 1). \tag{1.13}$$

It can be readily verified that

$$\Phi(t, 1) = \tfrac{1}{2}\begin{pmatrix} t + \frac{1}{t} & t - \frac{1}{t} \\ 1 - \frac{1}{t^2} & 1 + \frac{1}{t^2} \end{pmatrix},$$

$$H(t)\Phi(t, 1) = \tfrac{1}{2}\left(1 - \frac{1}{t^2},\ 1 + \frac{1}{t^2}\right).$$

Then, taking $s_1 = 1$, $s_2 = 2$,

$$\operatorname{rank}\begin{pmatrix} H(1)\Phi(1, 1) \\ H(2)\Phi(2, 1) \end{pmatrix} = \operatorname{rank}\begin{pmatrix} 0 & 1 \\ \frac{3}{8} & \frac{5}{8} \end{pmatrix} = 2$$

and the system (1.12), (1.13) is observable on $[1, 2]$.

The next proposition differs from Proposition 1.5 in that it involves only one point s but requires more differentiability of the coefficient matrices $A(t)$, $H(t)$ in (1.5), (1.6)

<u>Proposition 1.6.</u> <u>The observed linear system</u> (1.5), (1.6) <u>is observable on</u> $[t_0, t_1]$ <u>if for some positive integer</u> q <u>and point</u> $s \in [t_0, t_1]$ <u>the matrix</u> $H(t)$ <u>is</u> $q-1$ <u>times continuously differentiable at</u> s, <u>the matrix</u> $A(t)$ <u>is</u> $q-2$ <u>times differentiable at</u> s <u>and</u>

$$\operatorname{rank}\begin{pmatrix} H(s)\Phi(s, t_0) \\ \frac{d}{dt}H(t)\Phi(t, t_0)\Big|_{t=s} \\ \vdots \\ \frac{d^{q-1}}{dt^{q-1}}H(t)\Phi(t, t_0)\Big|_{t=s} \end{pmatrix} \equiv \operatorname{rank} D_q(s) = n. \tag{1.14}$$

Proof. If $\omega(t) \equiv 0$, $t \in [t_0, t_1]$ then $\frac{d^k\omega}{dt^k} \equiv 0$ there also, $k = 0, 1, \ldots, q-1$. Since $\omega(t) = H(t)\Phi(t, t_0)x_0$, the rows of (1.14) correspond to these derivatives. The equations

$$\frac{d^k\omega}{dt^k}(s) = 0, \quad k = 0, 1, 2, \ldots, q-1$$

therefore reduce to

$$D_q(s) x_0 = 0 .$$

Since $D_q(s)$ has rank n, $x_0 = 0$ and $x(t) \equiv 0$. We conclude (1.5), (1.6) is observable on $[t_0, t_1]$. Q. E. D.

Example. For the system (1.12), (1.13) the matrix $D_2(s)$ is

$$D_2(s) = \begin{pmatrix} \frac{1}{2}(1 - \frac{1}{s^2}) & \frac{1}{2}(1 + \frac{1}{s^2}) \\ \frac{1}{s^3} & -\frac{1}{s^3} \end{pmatrix} .$$

Taking $s = 1$ we have

$$D_2(1) = \begin{pmatrix} 0 & 1 \\ 1 & -1 \end{pmatrix}$$

and we again conclude (1.12), (1.13) is observable on $[t_0, t_1]$.

It should be clear that the result of Proposition 1.6 in fact applies to any interval which contains the point s and not just to a particular interval $[t_0, t_1]$.

Proposition 1.7. For the constant coefficient system

$$\dot{x} = Ax$$
$$\omega = Hx$$

we have observability on an arbitrary interval $[t_0, t_1]$, $t_0 < t_1$, if and only

<u>if for some</u> q, $0 < q \leq n$, the rank of the "observability matrix" equals n, i. e.,

$$\operatorname{rank}\begin{pmatrix} H \\ HA \\ \vdots \\ HA^{q-1} \end{pmatrix} = n . \tag{1.15}$$

<u>Proof.</u> The sufficiency of the condition (1.15) is clearly a corollary of Proposition 2.6 since

$$\frac{d^k}{dt^k}\left(He^{A(t-t_0)}\right) = HA^{k-1}e^{A(t-t_0)}$$

and hence

$$D_q(s) = \begin{pmatrix} H \\ HA \\ \vdots \\ HA^{q-1} \end{pmatrix} e^{A(s-t_0)} .$$

For the necessity we recall the Cayley-Hamilton theorem (see, e.g. [1,2]) which says that if

$$a(\lambda) = \det(\lambda I - A) = \lambda^n + a_1\lambda^{n-1} + \ldots + a_{n-1}\lambda + a_n$$

is the characteristic polynomial of A then A satisfies the matrix equation

$$A^n + a_1A^{n-1} + \ldots + a_{n-1}A + a_nI = 0 . \tag{1.16}$$

If the matrix (1.15) has rank $< n$ for all q, $0 < q \leq n$, then it certainly has rank $< n$ for $q = n$ and there is a non-zero vector $x_0 \in E^n$ such that

$$HA^k x_0 = 0, \qquad k = 0,1,2,\ldots, n-1 .$$

The equation (1.16) then shows that $HA^n = 0$. Multiplying (1.16) by $A, A^2, A^3, \ldots$ we see that $HA^k = 0$ for all non-negative integers k. Consequently

$$\omega(t) = He^{A(t-t_0)}x_0 = \sum_{k=0}^{\infty} \frac{HA^k(t-t_0)^k}{k!} \equiv 0 .$$

Thus x_0, and hence $x(t)$, is not zero while $\omega(t) \equiv 0$ and (1.5), (1.6) is not observable on any $[t_0, t_1]$. Q. E. D.

We remark that this proposition shows us that in studying constant coefficient systems it is sufficient to take $q = n$ in all cases (though affirmative results will be obtained in some cases for $q < n$). When, as in Proposition 1.7, observability, or lack of it, is independent of the particular interval, we will merely say the system is <u>observable</u>, or <u>unobservable</u>, as the case may be.

<u>Example.</u> In Figure II-1 we show, schematically, a series of n compartments separated by walls. We suppose these compartments to be filled with

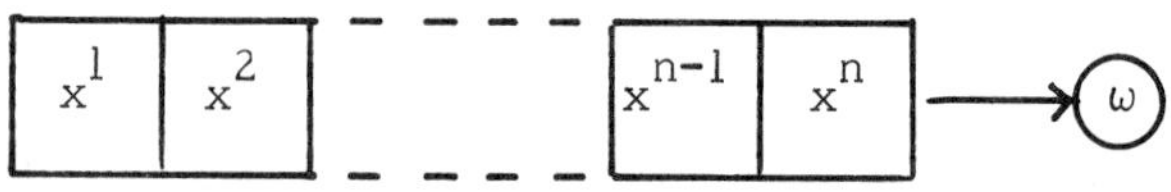

Fig. II-1. Heat Conduction System

physical substances having specific heats $\rho_1, \rho_2, \ldots, \rho_n$. The left hand walls of the 2nd, 3rd, ..., n-th compartments have heat conductivities $k_2, k_3, \ldots, k_n$; all other walls are perfect insulators. We assume the conductivities k_i, $i = 2, \ldots, n$, to be small compared with those of the substances filling the compartments and hence suppose that for each t we have essentially uniform temperatures $x^1(t)$, $x^2(t), \ldots, x^n(t)$ within the compartments. A sensor reads the temperature $\omega(t) \equiv x^n(t)$ in the n-th compartment. The system of equations governing the evolution of the temperature profile is

$$
\begin{pmatrix} x^1 \\ x^2 \\ \vdots \\ x^{i-1} \\ x^i \\ x^{i+1} \\ \vdots \\ x^n \end{pmatrix} =
$$

$$
\begin{pmatrix}
-\frac{k_2}{\rho_1} & \frac{k_2}{\rho_1} & \cdots & 0 & 0 & 0 & \cdots & 0 & 0 \\
\frac{k_2}{\rho_2} & (-\frac{k_2}{\rho_2}-\frac{k_3}{\rho_2}) & \cdots & 0 & 0 & 0 & \cdots & 0 & 0 \\
\vdots & \vdots & & \vdots & \vdots & \vdots & & \vdots & \vdots \\
0 & 0 & \cdots & (-\frac{k_{i-1}}{\rho_{i-1}}-\frac{k_i}{\rho_{i-1}}) & \frac{k_i}{\rho_{i-1}} & 0 & \cdots & 0 & 0 \\
0 & 0 & \cdots & \frac{k_i}{\rho_i} & (-\frac{k_i}{\rho_i}-\frac{k_{i+1}}{\rho_i}) & \frac{k_{i+1}}{\rho_i} & \cdots & 0 & 0 \\
0 & 0 & \cdots & 0 & \frac{k_{i+1}}{\rho_{i+1}} & (-\frac{k_{i+1}}{\rho_{i+1}}-\frac{k_{i+2}}{\rho_{i+1}}) & \cdots & 0 & 0 \\
\vdots & \vdots & & \vdots & \vdots & \vdots & & \vdots & \vdots \\
0 & 0 & \cdots & 0 & 0 & 0 & \cdots & \frac{k_n}{\rho_n} & -\frac{k_n}{\rho_n}
\end{pmatrix}
\begin{pmatrix} x^1 \\ x^2 \\ \vdots \\ x^{i-1} \\ x^i \\ x^{i+1} \\ \vdots \\ x^n \end{pmatrix}
\tag{1.17}
$$

The observation is $\omega = x^n$ so

$$H = (0, 0, \ldots, 0, 1).$$

A short computation shows that in this case (* denoting entries which have no particular significance)

$$
\begin{pmatrix} H \\ HA \\ HA^2 \\ HA^3 \\ \vdots \\ HA^{n-1} \end{pmatrix}
=
\begin{pmatrix}
0 & \cdots & 0 & 0 & 0 & 1 \\
0 & \cdots & 0 & 0 & \frac{k_n}{\rho_n} & -\frac{k_n}{\rho_n} \\
0 & \cdots & 0 & \frac{k_n}{\rho_n}\frac{k_{n-1}}{\rho_{n-1}} & * & * \\
0 & \cdots & \frac{k_n}{\rho_n}\frac{k_{n-1}}{\rho_{n-1}}\frac{k_{n-2}}{\rho_{n-2}} & * & * & * \\
\vdots & & \vdots & \vdots & \vdots & \vdots \\
\frac{k_n}{\rho_n}\frac{k_{n-1}}{\rho_{n-1}}\cdots\frac{k_2}{\rho_2} & \cdots & * & * & * & *
\end{pmatrix}
$$

and since each of the quotients $\frac{k_i}{\rho_i}$, $i = 2, 3, \ldots, n$ is non-zero we conclude the system is observable.

A slight modification of the above yields a negative result concerning observability. We consider the same system as above, modified so that the compartments form a ring as shown in Figure II-2. We suppose the

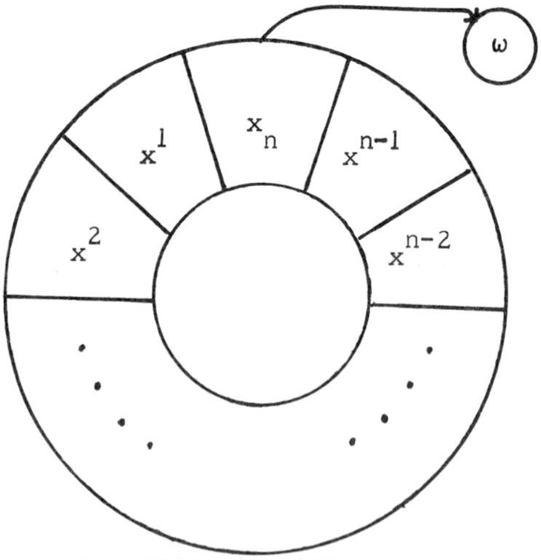

Fig. II-2. Ring System

conductivity of the wall between the 1st and n-th compartments to be k_1.

The only rows of (1.17) requiring modifications are the first and nth. They become

$$\dot{x}^1 = \left(-\frac{k_1}{\rho_1} - \frac{k_2}{\rho_2}\right)x^1 + \frac{k_2}{\rho_1}x^2 + \frac{k_1}{\rho_1}x^n$$

$$\dot{x}^n = \frac{k_1}{\rho_n}x^1 + \frac{k_n}{\rho_n}x^{n-1} + \left(-\frac{k_1}{\rho_n} - \frac{k_n}{\rho_n}\right)x^n .$$

The matrix analysis in the general case is somewhat tedious. Taking $n = 3$ and $k_1, k_2, k_3, \rho_1, \rho_2, \rho_n$ all equal to 1 we have the system

$$\begin{pmatrix} x^1 \\ x^2 \\ x^3 \end{pmatrix} = \begin{pmatrix} -2 & 1 & 1 \\ 1 & -2 & 1 \\ 1 & 1 & -2 \end{pmatrix} \begin{pmatrix} x^1 \\ x^2 \\ x^3 \end{pmatrix} \equiv Ax$$

$$\omega = Hx = (0, 0, 1) \begin{pmatrix} x^1 \\ x^2 \\ x^3 \end{pmatrix} .$$

Then

$$\begin{pmatrix} H \\ HA \\ HA^2 \end{pmatrix} = \begin{pmatrix} 0 & 0 & 1 \\ 1 & -2 & 1 \\ -3 & 6 & -3 \end{pmatrix}$$

which is singular (the rank $= 2$) and we conclude this system is not observable. Physical reasoning shows that this will always be true if

$$\rho_1 = \rho_{n-1}, \quad \rho_2 = \rho_{n-2}, \quad \rho_3 = \rho_{n-3}, \ldots \text{ etc.}$$

$$k_1 = k_n, \quad k_2 = k_{n-1}, \quad k_3 = k_{n-2}, \ldots \text{ etc.}$$

For if we start with an initial state such that

$$x_0^n = 0$$

$$x_0^1 = -x_0^{n-1}, \quad x_0^2 = -x_0^{n-2}, \quad x_0^3 = -x_0^{n-3}, \ldots \text{ etc}$$

($x_0^{n/2} = 0$ if n is an even integer) one readily verifies that these symmetry relationships continue to hold for all t, i. e.

$$x^n(t) \equiv 0$$

$$x^1(t) \equiv -x^{n-1}(t), \quad x^2(t) \equiv -x^{n-2}(t), \quad x^3(t) \equiv -x^{n-3}(t), \ldots, \text{ etc},$$

($x^{n/2}(t) \equiv 0$ if n is an even integer). Thus there are non-zero solutions for which the observation $\omega(t) \equiv x^n(t) \equiv 0$ and the system is not observable.

A natural question now arises: if a linear observed system

$$\dot{x} = A(t)x$$

$$\omega = H(t)x$$

is observable on an interval $[t_0, t_1]$, can $x_0 = x(t_0)$, the initial state for the solution on that interval, be reconstructed directly from the observation $\omega(t) = H(t)\Phi(t, t_0)x_0$?

<u>Definition 1. 8.</u> <u>The</u> $n \times r$ <u>matrix function</u> $R(t)$ <u>defined on</u> $[t_0, t_1]$ <u>is an (initial state) reconstruction kernel if and only if</u>

$$\int_{t_0}^{t_1} R(t)H(t)\,\Phi(t, t_0)dt = I \;(n \times n \text{ identity matrix}) \tag{1. 18}$$

This definition is prompted by the observation that if $R(t)$ is a reconstruction kernel and $\omega(t) = H(t)x(t) = H(t)\Phi(t, t_0)x_0$ then

$$\int_{t_0}^{t_1} R(t)\omega(t)dt = \int_{t_0}^{t_1} R(t)H(t)\,\Phi(t, t_0)dt\, x_0 = x_0 .$$

That is, the initial state $x_0 = x(t_0)$ of a solution $x(t)$ of (1. 5) can be recovered from the observation $\omega(t)$ on $[t_0, t_1]$ by forming the integral of the product $R(t)\omega(t)$ over that interval, provided (1.18) holds.

Proposition 1.9. There exists a reconstruction kernel $R(t)$ on $[t_0, t_1]$ if and only if (1.5), (1.6) is observable on $[t_0, t_1]$.

Proof. If a reconstruction kernel exists and $\omega(t) \equiv 0$ then

$$x_0 = \int_{t_0}^{t_1} R(t)\omega(t)dt = 0 .$$

Then $x(t) = 0$ and we conclude (1.5), (1.6) is observable on $[t_0, t_1]$. If, on the other hand, (1.5), (1.6) is observable on $[t_0, t_1]$ then we know from Theorem 2.3 that

$$W(t_0, t_1) = \int_{t_0}^{t_1} \Phi(t, t_0)^* H(t)^* H(t)\Phi(t, t_0)\,dt > 0 .$$

Then with

$$R_0(t) = W(t_0, t_1)^{-1}\Phi(t, t_0)^* H(t)^* \tag{1.19}$$

we have

$$\int_{t_0}^{t_1} R_0(t)H(t)\Phi(t, t_0)dt = W(t_0, t_1)^{-1}\int_{t_0}^{t_1} \Phi(t, t_0)^* H(t)^* H(t)\Phi(t, t_0)dt$$

$$= W(t_1, t_0)^{-1} W(t_1, t_0) = I$$

so that (1.19) is a reconstruction kernel. Q.E.D.

Example. Consider a mass-spring system as shown in Figure II-3 with unit mass and unit spring constant. We will suppose that the equilibrium position is given by $x = 0$. The equation of motion is

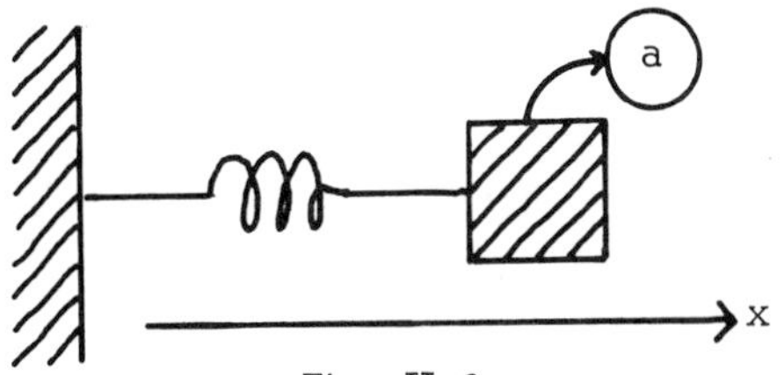

Fig. II-3
Mass-spring system

$$\ddot{x} + x = 0 \tag{1.20}$$

or, with $x^1 = x$, $x^2 = \dot{x}$,

$$\begin{pmatrix} \dot{x}^1 \\ \dot{x}^2 \end{pmatrix} = \begin{pmatrix} 0 & 1 \\ -1 & 0 \end{pmatrix} \begin{pmatrix} x^1 \\ x^2 \end{pmatrix}; \quad \left(A = \begin{pmatrix} 0 & 1 \\ -1 & 0 \end{pmatrix} \right). \tag{1.21}$$

Assume the mass includes an accelerometer a which measures $\ddot{x}$. Taking (1.20) into account we suppose we have the observation

$$\omega(t) = x^1(t) = (1, 0) \begin{pmatrix} x^1(t) \\ x^2(t) \end{pmatrix}, \quad \text{i. e.} \quad H = (1, 0).$$

Now

$$\begin{pmatrix} H \\ HA \end{pmatrix} = \begin{pmatrix} 1 & 0 \\ 0 & 1 \end{pmatrix}$$

so the system is observable. Taking $t_0 = -\pi$, $t_1 = 0$, we suppose the observation turns out to be

$$\omega(t) = x^1(t) = \tfrac{1}{2}\cos t + \tfrac{1}{2}\sin t, \quad -\pi \le t \le 0. \tag{1.22}$$

We pose the problem of computing $x^1(-\pi)$, $x^2(-\pi)$.

Now, of course, we have $x^1(-\pi) = -\frac{1}{2}$ and differentiation of (1.22) gives $x^2(-\pi) = -\frac{1}{2}\sin(-\pi) + \frac{1}{2}\cos(-\pi) = -\frac{1}{2}$. So why proceed any further? Some points need to be made. In practice $\omega(t)$ would not be available as a formula; we would only have a numerical record. One could estimate $\dot{\omega} = \dot{x}^1 = x^2$ numerically but the results would be quite unreliable because differentiation is, numerically, an unstable process. Integration, on the other hand, is a very stable process. That is why our reconstruction formula is based on integration.

A fundamental solution matrix for (1.21) is

$$\Phi(t) = \begin{pmatrix} \cos t & \sin t \\ -\sin t & \cos t \end{pmatrix}$$

so

$$\Phi(t,-\pi) = \begin{pmatrix} \cos t & \sin t \\ -\sin t & \cos t \end{pmatrix} \begin{pmatrix} -1 & 0 \\ 0 & -1 \end{pmatrix}^{-1} = \begin{pmatrix} -\cos t & -\sin t \\ \sin t & -\cos t \end{pmatrix}$$

and

$$H\Phi(t,-\pi) = (-\cos t, \quad -\sin t). \qquad (1.23)$$

Then

$$W(0,-\pi) = \int_{-\pi}^{0} \left[\begin{pmatrix} -\cos t \\ -\sin t \end{pmatrix} (-\cos t \quad -\sin t) \right] dt$$

$$= \int_{-\pi}^{0} \begin{pmatrix} \cos^2 t & \cos t \sin t \\ \cos t \sin t & \sin^2 t \end{pmatrix} dt = \frac{\pi}{2} \begin{pmatrix} 1 & 0 \\ 0 & 1 \end{pmatrix}.$$

The reconstruction formula is then (cf. (1.19), (1.23))

$$\begin{pmatrix} x^1(-\pi) \\ x^2(-\pi) \end{pmatrix} = \frac{2}{\pi} \int_{-\pi}^{0} \begin{pmatrix} -\cos t \\ -\sin t \end{pmatrix} (\tfrac{1}{2} \cos t + \tfrac{1}{2} \sin t) dt$$

$$= \frac{1}{\pi} \int_{-\pi}^{0} \begin{pmatrix} -\cos^2 t & -\cos t \sin t \\ -\sin t \cos t & -\sin^2 t \end{pmatrix} dt = \frac{1}{\pi} \begin{pmatrix} -\frac{\pi}{2} \\ -\frac{\pi}{2} \end{pmatrix} = \begin{pmatrix} -\frac{1}{2} \\ -\frac{1}{2} \end{pmatrix}.$$

Let us illustrate what we meant by stability in our earlier discussion. Suppose the observation $\omega(t)$ is "corrupted" by a "noise" term $\eta(t)$ which we shall suppose in this case to be a high frequency term $\alpha \cos(kt) + \beta \sin(kt)$, where k is a large integer. What is actually available then is

$$\tilde{\omega}(t) = \omega(t) + \eta(t) = \tfrac{1}{2} \cos t + \tfrac{1}{2} \sin t + \alpha \cos(kt) + \beta \sin(kt).$$

If one relied on $\tilde{\omega}(-\pi)$, $\dot{\tilde{\omega}}(-\pi)$ to give $x^1(-\pi)$, $x^2(-\pi)$, respectively, we would have the estimate

$$x^1(-\pi) = -\tfrac{1}{2} + \alpha(-1)^k$$

$$x^2(-\pi) = -\tfrac{1}{2} + \beta k(-1)^k$$

which, clearly, could be grossly in error. But

$$\frac{2}{\pi}\int_{-\pi}^{0} \Phi(t,-\pi)H^{*}\eta(t)dt = \frac{2}{\pi}\int_{-\pi}^{0} \begin{pmatrix} -\cos t \\ -\sin t \end{pmatrix} (\alpha \cos (kt) + \beta \sin (kt))dt$$

which, on integration, is seen to be of the order of $1/k$. Consequently, the error in estimating $x^1(-\pi)$, $x^2(-\pi)$ becomes small for large k.

We proceed now to the discussion of the reconstruction of $x_0 = x(t_0)$ from an observation $\omega(t) = H(t)x(t)$ on a solution $x(t)$ of

$$\dot{x} = A(t)x + f(t),$$

assuming $f(t)$ to be known, at least approximately, on $[t_0, t_1]$. We have seen in the beginning of this section that this can be done, at least in principle. But what we seek now is a formula analogous to the one just developed for the homogeneous case.

Let $x(t)$ be a solution of (1. 1). Then, as indicated in (1. 3), we may write

$$x(t) = \tilde{x}(t) + \hat{x}(t)$$

where $\tilde{x}(t)$ satisfies the homogeneous equation (1. 5) and

$$\tilde{x}(t_0) = x(t_0) = x_0, \quad \hat{x}(t_0) = 0.$$

We also define

$$\tilde{\omega}(t) = H(t)\tilde{x}(t), \qquad \hat{\omega}(t) = H(t)\hat{x}(t)$$

so that $\omega(t) = H(t)x(t) = \tilde{\omega}(t) + \hat{\omega}(t)$. Since $\tilde{x}(t)$ satisfies the homogeneous equation (1. 5) we have

$$\tilde{x}(t_0) = x(t_0) = x_0 = \int_{t_0}^{t_1} R(t)\tilde{\omega}(t)dt = \int_{t_0}^{t_1} R(t)[\omega(t) - \hat{\omega}(t)]\, dt .$$

But since $\hat{x}(t)$ satisfies the inhomogeneous equation (1. 1) with $\hat{x}(t_0) = 0$ we have

$$\hat{\omega}(t) = H(t)\hat{x}(t) = H(t)\int_{t_0}^{t} \Phi(t, s)f(s)ds$$

and thus

$$x_0 = \int_{t_0}^{t_1} R(t)\omega(t)dt - \int_{t_0}^{t_1} R(t)H(t)\int_{t_0}^{t} \Phi(t, s)f(s)ds\,dt .$$

Changing the order of integration in the second term we have

$$x_0 = \int_{t_0}^{t_1} R(t)\omega(t)dt - \int_{t_0}^{t_1}\left[\int_{s}^{t_1} R(t)H(t)\Phi(t, s)dt\right] f(s)ds.$$

We have therefore proved

Proposition 1.10. *If* $x(t)$ *satisfies* (1.1) *on* $[t_0, t_1]$ *and* (1.5), (1.6) *is observable on* $[t_0, t_1]$ *then* x_0 *is given by the reconstruction formula*

$$x_0 = \int_{t_0}^{t_1} [R(t)\omega(t) + Q(t)f(t)]\,dt$$

where $R(t)$ *is a reconstruction kernel as in* (1.18) *and*

$$Q(t) = -\int_{t}^{t_1} R(\tau)H(\tau)\Phi(\tau, t)\,d\tau .$$

2. Controllability and Steering Functions

A linear control system has the form

$$\dot{x} = A(t)x + B(t)u, \qquad x \in E^n, \quad u \in E^m \tag{2.1}$$

when we assume (cf. I-1) that no external disturbances are present. It is assumed that the control function u can be selected at will by the plant operator from a set of "admissible controls" in order to influence the evolution of the system state $x(t)$. In the work of this chapter the set of admissible controls on a time interval $[t_0, t_1]$ is $L_m^2[t_0, t_1]$, the Hilbert space of square integrable functions on $[t_0, t_1]$.

Definition 2.1. The linear control system (2.1) *is controllable on* $[t_0, t_1]$ *just in case, for every pair of vectors* $x_0, x_1 \in E^n$, *there is a control* $u \in L_m^2[t_0, t_1]$ *such that the solution* x *of* (2.1) *(with this control) which satisfies*

$$x(t_0) = x_0 \tag{2.2}$$

also satisfies

$$x(t_1) = x_1 . \tag{2.3}$$

When u is such that the corresponding solution x of (2.1) satisfies (2.2) and (2.3) we say that "u steers x_0 to x_1 during $[t_0, t_1]$". It is a convenient phrase, used ubiquitously in [3] for example.

Proposition 2.2. The system (2.1) *is controllable on* $[t_0, t_1]$ *if and only if for each vector* $x_1 \in E^n$ *there is a control* $u \in L_m^2[t_0, t_1]$ *which steers* 0 *to* x_1 *during* $[t_0, t_1]$.

Proof. The only if part is obvious. To establish the sufficiency it is only necessary to choose two vectors $x_0, x_1 \in E^n$ and put

$$\tilde{x}_1 = x_1 - \Phi(t_1, t_0)x_0 .$$

If u steers 0 to $\tilde{x}_1$ during $[t_0, t_1]$ then

$$\tilde{x}_1 = \int_{t_0}^{t_1} \Phi(t_1, t)u(t)\, dt .$$

But then

$$x_1 = \Phi(t, t_0)x_0 + \int_{t_0}^{t_1} \Phi(t_1, t)u(t)\, dt ,$$

showing that the solution of (2.1) for this control u satisfies both (2.2) and (2.3), i.e. steers x_0 to x_1, so (2.1) is controllable on $[t_0, t_1]$. Q.E.D.

Theorem 2.3. The system (2.1) *is controllable on* $[t_0, t_1]$ *if and only if the* "adjoint" linear observed system

$$\dot{y} = -A(t)^* y \tag{2.4}$$

$$\omega = B(t)^* y \tag{2.5}$$

is observable on $[t_0, t_1]$.

Proof. We define a linear subspace $R(t_0, t_1) \subseteq E^n$ by

$$R(t_0, t_1) = \{x_1 \in E^n \mid x_1 = \int_{t_0}^{t_1} \Phi(t_1, t)B(t)u(t)dt, \quad u \in L_m^2[t_0, t_1]\}. \tag{2.6}$$

Thus $R(t_0, t_1)$ is the subspace of states reachable from the origin using controls $u \in L_m^2[t_0, t_1]$. We will refer to $R(t_0, t_1)$ as the "reachable subspace". To investigate how large this subspace is, suppose $y_1 \in E^n$ has the property

$$y_1^* x_1 = 0, \quad x_1 \in R(t_0, t_1). \tag{2.7}$$

From (2.6) then

$$y_1^* \int_{t_0}^{t_1} \Phi(t_1, t)B(t)u(t)dt = 0, \qquad u \in L_m^2[t_0, t_1]$$

and, since $u(t)$ is an arbitrary element of $L_m^2[t_0, t_1]$, we conclude $y_1^* \Phi(t_1, t)B(t) \equiv 0$, or, transposing

$$\omega(t) \equiv B(t)^* \Phi(t_1, t)^* y_1 \equiv 0, \qquad t \in [t_0, t_1]. \tag{2.8}$$

Now $y = \Psi(t, t_1)y_1 \equiv \Phi(t_1, t)^* y_1$ is a solution of (2.4) on $[t_0, t_1]$ and $\omega(t)$ is the associated observation via (2.5). If (2.4), (2.5) is observable, (2.8) implies $y(t_0) = 0$ which in turn implies $y_1 = 0$. Thus (2.7) implies $y_1 = 0$ and we conclude that $R(t_0, t_1) = E^n$ and hence (2.1) is controllable on $[t_0, t_1]$. If (2.4), (2.5) is not observable on $[t_0, t_1]$ there is some $y_1 \equiv y(t_0) \neq 0$ such that (2.8) holds. Then, reasoning backwards, we conclude (2.7) holds for this non-zero y_1 and $R(t_0, t_1) \neq E^n$ so (2.1) is not controllable on $[t_0, t_1]$. Q. E. D.

This result is extremely important for from it we have immediately a series of propositions parallel to Theorem 1. 3 and Propositions 1. 4-1. 10 giving necessary and sufficient conditions for controllability of (2. 1) on $[t_0, t_1]$.

Theorem 2. 4. The system (2. 1) is controllable on $[t_0, t_1]$ if and only if the "controllability Grammian"

$$Z(t_0, t_1) = \int_{t_0}^{t_1} \Phi(t_1, t)B(t)B(t)^* \Phi(t_1, t)dt \tag{2. 9}$$

is positive definite.

Proof. If it is positive definite the control

$$u(t) = B(t)^* \Phi(t_1, t)^* Z(t_0, t_1)^{-1} x_1 \tag{2. 10}$$

may be defined and a short computation shows that it steers 0 to x_1 during $[t_0, t_1]$ and thus (2. 1) is controllable there. On the other hand controllability of (2. 1) on $[t_0, t_1]$ has been seen to imply observability of (2. 4), (2. 5) there. Let $\psi(t, \tau)$ be the fundamental solution of (2. 4) with $\psi(t, \tau) = I$. Theorem 1. 3 shows that if (2. 4), (2. 5) is observable on $[t_0, t_1]$ then

$$W(t_0, t_1) = \int_{t_0}^{t_1} \Psi(t, t_0)^* B(t)B(t)^* \Psi(t, t_0)dt$$

is positive definite. Now $\Psi(t, t_0)^*$ and $\Phi(t_1, t)$ satisfy

$$\frac{\partial \Psi^*(t, t_0)}{\partial t} = - \Psi(t, t_0)^* A(t), \qquad \frac{\partial \Phi(t_1, t)}{\partial t} = - \Phi(t_1, t)A(t) \quad .$$

Then $\Phi(t_1, t) = C\Psi(t, t_0)^*$ for some nonsingular $n \times n$ matrix C. Since $\Psi(t_0, t_0)^* = I$, $C = \Phi(t_1, t_0)$. Hence

$$\Phi(t_1, t) = \Phi(t_1, t_0) \Psi(t, t_0)^* \ .$$

Then

$$Z(t_0, t_1) = \Phi(t_1, t_0)W(t_0, t_1)\Phi(t_1, t_0)^*$$

and, since $W(t_0, t_1)$ is positive definite and $\Phi(t_1, t_0)$ is nonsingular, $Z(t_0, t_1)$ is positive definite. Q. E. D.

Remark. The matrix function in (2. 10), i. e. :

$$S_0(t) = B(t)^* \Phi(t_1, t)^* Z(t_0, t_1)^{-1}$$

will be referred to as a steering function since the control (2. 10) steers 0 to x_1. We will have more to say on this later.

Proposition 2. 5. *Sufficient conditions for controllability of* (2. 1) *on* $[t_0, t_1]$ *are*

(i) *the existence of points* $s_1, s_2, \ldots, s_q$ *in* $[t_0, t_1]$ *with*

$$\operatorname{rank}(\Phi(t_1, s_1)B(s_1), \quad \Phi(t_1, s_2)B(s_2), \ldots, \quad \Phi(t_1, s_q)B(s_q)) = n;$$

(ii) *(assuming appropriate differentiability of* $B(t)$, $A(t)$) *the existence of a point* $s \in [t_0, t_1]$ *with*

$$\operatorname{rank}\left(\Phi(t_1, s)B(s), \ \frac{d}{dt}\Phi(t_1, t)B(t)\Big|_{t=s}, \ldots \frac{d^{q-1}}{dt^{q-1}}\Phi(t_1, t)B(t)\Big|_{t=s}\right) = n .$$

The proof is immediate from Theorem 2. 4 and Propositions 1. 5, 1. 6. Proposition 1.7 yields the following result when combined with Theorem 2. 3.

Proposition 2. 6. *The constant coefficient control system*

$$\dot{x} = Ax + Bu \tag{2. 11}$$

is controllable (i. e., controllable on an arbitrary interval $[t_0, t_1]$) *just in case the rank of the "controllability matrix" is* n, *i. e.*,

$$\operatorname{rank}[B, AB, \ldots, A^{q-1}B] = n$$

for some integer $q \le n$.

Example. In the case of the crane linearized about a static operating condition (cf. I-2) we have

$$A = \begin{pmatrix} 0 & 1 & 0 & 0 & 0 \\ -\frac{g}{\ell_{2,0}} & 0 & 0 & 0 & 0 \\ 0 & 0 & 0 & 1 & 0 \\ 0 & 0 & 0 & 0 & 0 \\ 0 & 0 & 0 & 0 & 0 \end{pmatrix}, \quad B = \begin{pmatrix} 0 & 0 \\ 0 & -\frac{L}{\ell_{2,0}} \cos \psi_0 \\ 0 & 0 \\ 0 & 1 \\ 1 & 0 \end{pmatrix} \equiv \begin{pmatrix} 0 & 0 \\ 0 & \beta \\ 0 & 0 \\ 0 & 1 \\ 1 & 0 \end{pmatrix}$$

Taking $q = 4$ we compute $[B, AB, A^2B, A^3B]$:

$$\begin{pmatrix} 0 & 0 & 0 & \beta & 0 & 0 & 0 & \frac{-g\beta}{\ell_{2,0}} \\ 0 & \beta & 0 & 0 & 0 & \frac{-g\beta}{\ell_{2,0}} & 0 & 0 \\ 0 & 0 & 0 & 1 & 0 & 0 & 0 & 0 \\ 0 & 1 & 0 & 0 & 0 & 0 & 0 & 0 \\ 1 & 0 & 0 & 0 & 0 & 0 & 0 & 0 \end{pmatrix}$$

The first, second, fourth, sixth and eighth columns form an upper triangular 5×5 matrix when arranged in reverse order. That matrix has rank 5 just in case $\beta = -L/\ell_{2,0} \cos \psi_0 \neq 0$, i. e., just in case $\psi_0 \neq \pi/2$. We have controllability, therefore, if $\psi_0 \neq \pi/2$. Further computation will show that for $q = 5$, $[B, AB, A^2B, A^3B, A^4B]$ also has rank $= 4$ when $\psi_0 = \pi/2$ and we conclude that the system is not controllable in that case. The reason for this is that when the boom is horizontal, $\psi_0 = \pi/2$, no horizontal motion of the load can be initiated (to first order, that is) with either of the controls $u_1 = \ddot{\ell}_2$ or $u_2 = \ddot{\psi}$.

Definition 2.7. An $m \times n$ matrix function $S(t)$ with entries in $L^2_m[t_0, t_1]$ is a steering function for (2.1) on $[t_0, t_1]$ just in case

$$\int_{t_0}^{t_1} \Phi(t_1, t)B(t)S(t)dt = I, \tag{2.12}$$

$\Phi(t,\tau)$ being the fundamental solution of $\dot{x} = A(t)x$ with $\Phi(\tau,\tau) = I$.

Remark. The term is suggested by the fact that (2.12) implies that the control

$$u(t) = S(t)x_1, \qquad x_1 \in E^n,$$

steers 0 to x_1 during $[t_0, t_1]$.

Example. Consider the controlled linear harmonic oscillator

$$\ddot{x} + x = u$$

or, with $x^1 = x$, $x^2 = \dot{x}$

$$\begin{pmatrix} \dot{x}^1 \\ \dot{x}^2 \end{pmatrix} = \begin{pmatrix} 0 & 1 \\ -1 & 0 \end{pmatrix} \begin{pmatrix} x^1 \\ x^2 \end{pmatrix} + \begin{pmatrix} 0 \\ 1 \end{pmatrix} u .$$

Let us suppose that

$$\begin{pmatrix} x^1(0) \\ x^2(0) \end{pmatrix} = \begin{pmatrix} 0 \\ 0 \end{pmatrix}$$

and it is desired for some $T > 0$ to have

$$\begin{pmatrix} x^1(T) \\ x^2(T) \end{pmatrix} = \begin{pmatrix} \frac{1}{2} \\ \frac{1}{2} \end{pmatrix} .$$

Let us use the steering function suggested by (2.10) and noted earlier, i.e.,

$$S_0(t) = B(t)^* \Phi(t_1, t)^* Z(t_0, t_1)^{-1}, \tag{2.13}$$

where, since we assume a system controllable on $[t_0, t_1]$,

$$Z(t_0, t_1) = \int_{t_0}^{t_1} \Phi(t_1, t)B(t)B(t)^* \Phi(t_1, t)^* dt > 0 .$$

Applied to our example with $t_0 = 0$, $t_1 > 0$ we have

$$\Phi(T, t) = \begin{pmatrix} \cos(T-t) & \sin(T-t) \\ -\sin(T-t) & \cos(T-t) \end{pmatrix}.$$

Then, from (2.9), since $B(t) = \binom{0}{1}$,

$$Z(0, T) = \int_0^T \begin{pmatrix} \sin(T-t) \\ \cos(T-t) \end{pmatrix} (\sin(T-t), \cos(T-t))dt$$

$$= \int_0^T \begin{pmatrix} \sin^2(T-t) & \sin(T-t)\cos(T-t) \\ \sin(T-t)\cos(T-t) & \cos^2(T-t) \end{pmatrix} dt = \begin{pmatrix} \frac{T-\frac{1}{2}\sin 2T}{2} & \frac{1}{4}(1-\cos 2T) \\ \frac{1}{4}(1-\cos 2T) & \frac{T+\frac{1}{2}\sin 2T}{2} \end{pmatrix}.$$

Then

$$Z(0, T)^{-1} = \frac{4}{T^2-\frac{1}{2}(1-\cos 2T)} \begin{pmatrix} \frac{T+\frac{1}{2}\sin 2T}{2} & \frac{1}{4}(\cos 2T-1) \\ \frac{1}{4}(\cos 2T-1) & \frac{T-\frac{1}{2}\sin 2T}{2} \end{pmatrix}.$$

The desired control function is then

$$u(t) = \frac{4}{T^2-\frac{1}{2}(1-\cos 2T)} (0, 1) \begin{pmatrix} \cos(T-t) & -\sin(T-t) \\ \sin(T-t) & \cos(T-t) \end{pmatrix} \begin{pmatrix} \frac{T+\frac{1}{2}\sin 2T}{2} & \frac{1}{4}(\cos 2T-1) \\ \frac{1}{4}(\cos 2T-1) & \frac{T-\frac{1}{2}\sin 2T}{2} \end{pmatrix} \begin{pmatrix} \frac{1}{2} \\ \frac{1}{2} \end{pmatrix}$$

$$= \frac{1}{T^2-\frac{1}{2}(1-\cos 2T)} [(T-\tfrac{1}{2})(\cos(T-t) + \sin(T-t)) + \tfrac{1}{2}(\cos(T+t) - \sin(T+t))].$$

One thing which should be noted here is what happens as the control time T tends to zero. The numerator remains a uniformly bounded function but the denominator tends to zero like T^2. Thus the control function grows without bound as $T \to 0$, a circumstance which we would, of course, suspect. The control time, as is always the case for constant coefficient systems, is arbitrary but smaller control times require greater effort per unit time in order to attain the same objective.

We have no result to the effect that controls on steering functions are unique and, in fact, they are not. The next example shows the use of a steering function other than the one exhibited in (2.13). We also use this opportunity to discuss a more general control problem.

We consider a three compartment heat conduction system much the same as the one considered earlier but with some minor changes. The interior and "bottom" walls shown in Figure II.4 have unit conductivities. The upper

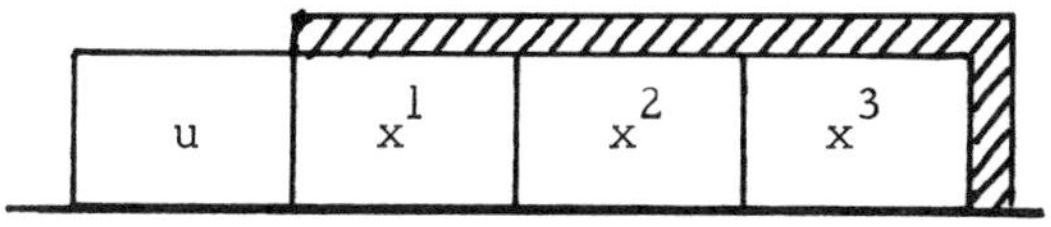

Fig. II-4. Heat conduction system

walls and the right wall of compartment 3 are insulated. The chamber marked u contains a heating/cooling device so that at any instant we can realize a given temperature u in that chamber. The material filling the shaded region below the chambers remains at constant temperature 1. The heat flow equation is then

$$\begin{pmatrix} \dot{x}^1 \\ \dot{x}^2 \\ \dot{x}^3 \end{pmatrix} = \begin{pmatrix} -3 & 1 & 0 \\ 1 & -3 & 1 \\ 0 & 1 & -2 \end{pmatrix} \begin{pmatrix} x^1 \\ x^2 \\ x^3 \end{pmatrix} + \begin{pmatrix} 1 \\ 0 \\ 0 \end{pmatrix} u + \begin{pmatrix} 1 \\ 1 \\ 1 \end{pmatrix} \tag{2.14}$$

and we have equilibrium $x^1 = x^2 = x^3 = 1$ for $u = 1$. We start with $x^1(0) = 4$, $x^2(0) = 3$, $x^3(0) = 2$ and wish to have $x^1(1) = x^2(1) = x^3(1) = 1$.

This is a slightly different problem from those discussed so far since it involves non-zero initial and terminal states and also involves a non-homogeneous vector, in this case $(1, 1, 1)^*$, in addition to the control term. However, only slight modification of our methods is necessary. If we have a system

$$\dot{x} = A(t)x + B(t)u + v \tag{2.15}$$

then

$$x(t_1) = \Phi(t_1, t_0)x_0 + \int_{t_0}^{t_1} \Phi(t_1, t)[B(t)u(t) + v(t)]\, dt .$$

If $S(t)$ is a steering function so that

$$\int_{t_0}^{t_1} \Phi(t_1, t)B(t)S(t)dt = I$$

we can achieve $x(t_1) = x_1$ by setting

$$u(t) = S(t)[x_1 - \Phi(t_1, t_0)x_0 - \int_{t_0}^{t_1} \Phi(t_1, t)\, v(t)\, dt] . \quad (2.16)$$

A particularly simple case arises when A is constant and invertible and v is constant (which is true for (2.14)). For then

$$\Phi(t_1, t) = \Phi(t_1 - t, 0) = - A^{-1} \frac{d}{dt} \Phi(t_1 - t, 0) = - \frac{d}{dt} \Phi(t_1, t)A^{-1}$$

and we have

$$u(t) = S(t)[x_1 - \Phi(t_1, t_0)x_0 + \int_{t_0}^{t_1} \frac{d}{dt} \Phi(t_1, t)dt\, A^{-1}v]$$

$$= S(t)[x_1 - \Phi(t_1, t_0)x_0 + A^{-1}v - \Phi(t_1, t_0)A^{-1}v] = S(t)[x_1 + A^{-1}v - \Phi(t_1, t_0)(x_0 + A^{-1}v)].$$

(This result can also be obtained using the change of variable $z = x + A^{-1}v$ which transforms (2.15), with A constant, into $\dot{z} = Az + B(t)u$.)

To carry this program out in our example we need $\Phi(1, t) = \Phi(1-t, 0) = e^{A(1-t)}$. Numerical computation shows the symmetric matrix in (2.14) to have approximate eigenvalues

$$\lambda_1 = -1.1980, \quad \lambda_2 = -2.5549, \quad \lambda_3 = -4.2469$$

with corresponding orthonormal eigenvectors

$$\xi_1 = \begin{pmatrix} .3280 \\ .5908 \\ .7367 \end{pmatrix}, \quad \xi_2 = \begin{pmatrix} .7370 \\ .3278 \\ -.5910 \end{pmatrix}, \quad \xi_3 = \begin{pmatrix} .5909 \\ -.7370 \\ .3279 \end{pmatrix}.$$

Then, with $X = (\xi_1, \xi_2, \xi_3)$, $X^{-1} = X^T$ and

$$\Phi(1-t, 0) = e^{A(1-t)} = X \operatorname{diag}(e^{\lambda_1 t}, e^{\lambda_2 t}, e^{\lambda_3 t})X^T . \tag{2.17}$$

Further suppose in this case that we wish to use very simple controls; controls which are constant on each of the intervals $[0, 1/3]$, $[1/3, 2/3]$, $[2/3, 1]$. Then the entries of the steering function matrix must be linear combinations of the characteristic functions $\chi_1(t)$, $\chi_2(t)$, $\chi_3(t)$ associated with these intervals. (For example, $\chi_2(t) = 1$ on $[1/3, 2/3]$, 0 everywhere else.) So

$$S(t) = (\chi_1(t), \chi_2(t), \chi_3(t))S$$

where S is a constant 3×3 matrix. The equation (2.11) for $S(t)$ yields the equation for S:

$$\int_0^1 \Phi(1,t)\begin{pmatrix} 1 \\ 0 \\ 0 \end{pmatrix} (\chi_1(t), \chi_2(t), \chi_3(t))dt)\, S = I .$$

Thus

$$S = (\int_0^{1/3} \varphi_1(1, t)dt, \int_0^{2/3} \varphi_2(1, t)dt, \int_{2/3}^{1} \varphi_3(1, t)dt)^{-1}$$

where $\Phi(1, t) = (\varphi_1(1, t), \varphi_2(1, t), \varphi_3(1, t))$. Using (2.17) these columns can be computed and integrated. The result is

$$S = \begin{pmatrix} .0390 & .0399 & .0032 \\ .0291 & .0979 & .0331 \\ .0141 & .0512 & .2465 \end{pmatrix}^{-1} = \begin{pmatrix} 36.92 & -15.92 & -1.658 \\ -11.04 & 15.75 & -1.971 \\ .1797 & -2.363 & 4.373 \end{pmatrix}$$

giving

$$S(t) = \chi_1(t)(36.92, -15.92, 1.658)$$
$$+ \chi_2(t)(-11.04, 15.75, -1.971) + \chi_3(t)(.1797, -2.363, 4.373) \tag{2.18}$$

To solve the problem originally posed we compute

$$A^{-1} = \begin{pmatrix} -.3846 & -.1538 & -.0769 \\ -.1538 & -.4615 & -.2308 \\ -.0769 & -.2308 & -.6154 \end{pmatrix}.$$

Then

$$x_1 + A^{-1}v = \begin{pmatrix} 1 \\ 1 \\ 1 \end{pmatrix} + A^{-1}\begin{pmatrix} 1 \\ 1 \\ 1 \end{pmatrix} = \begin{pmatrix} .3847 \\ .1539 \\ .0769 \end{pmatrix}$$

$$x_0 + A^{-1}v = \begin{pmatrix} 4 \\ 3 \\ 2 \end{pmatrix} + A^{-1}\begin{pmatrix} 1 \\ 1 \\ 1 \end{pmatrix} = \begin{pmatrix} 3.3847 \\ 2.1539 \\ 1.0769 \end{pmatrix}$$

$$\Phi(1,0)(x_0 + A^{-1}v) = X \operatorname{diag}(e^{\lambda_1}, e^{\lambda_2}, e^{\lambda_3})X^T(x_0 + A^{-1}v) = \begin{pmatrix} .5047 \\ .6563 \\ .6114 \end{pmatrix}.$$

The desired control function is then

$$u(t) = S(t)[x_1 + A^{-1}v - \Phi(1,0)(x_0 + A^{-1}v)]$$
$$= S(t)\begin{pmatrix} -.1199 \\ -.5024 \\ -.5345 \end{pmatrix} = \quad (\text{cf. } (2.17))$$
$$2.685\chi_1(t) - 5.536\chi_2(t) - 1.172\chi_3(t).$$

We should remark at this point that for a single control task, such as passing from a given initial point x_0 to a given terminal point x_1, it is not necessary to construct a steering function. One may select n functions

$s_1(t), s_2(t), \ldots, s_n(t)$ and postulate a control of the form

$$u(t) = \alpha^1 s_1(t) + \alpha^2 s_2(t) + \ldots + \alpha^n s_n(t). \tag{2.19}$$

Then

$$x_1 - \Phi(t_1, t_0)x_0 = \left(\int_{t_0}^{t_1} \Phi(t_1, t)B(t)s_1(t)dt, \ldots, \int_{t_0}^{t_1} \Phi(t_1, t)B(t)s_n(t)dt \right) \begin{pmatrix} \alpha^1 \\ \alpha^2 \\ \vdots \\ \alpha^n \end{pmatrix}$$

is a system of n linear equations in n unknowns which, if $s_1(t), \ldots, s_n(t)$ are appropriately chosen, will yield $\alpha^1, \alpha^2, \ldots, \alpha^n$ such that $u(t)$, as given by (2.19) will accomplish the given control objective. However, if it is desired to carry out repeated control procedures with different x_1, x_0, the matrix whose columns are the vectors $\int_{t_0}^{t_1} \Phi(t_1, t)B(t)u_i(t)dt$ should be inverted. Assuming the inverse is S, the matrix function $(s_1(t), s_2(t), \ldots, s_n(t))S$ will then be a steering function.

We have, in the foregoing, introduced the "special" steering function $S_0(t)$ and then robbed it of uniqueness by indicating that there are many other steering functions. Nevertheless $S_0(t)$ uniquely possesses certain desirable properties which will be discussed in the next section.

3. General Theory, Duality

We have hinted in Section 2 that a variety of control objectives may be formulated for a given system and it is not hard to see that the same holds true in the case of observation. Our purpose in this section is to establish a general framework for observation and control theory which applies to a variety of systems and a variety of objectives in connection with those systems. We will also see that observation and control are intimately related by certain duality relationships which we have already glimpsed in Theorems 2.3 and 2.4 and the propositions following them. Before we get into this general framework, however, we need a little more in the way of background material.

Suppose that H_1 and H_2 are Hilbert spaces, i.e., linear vector spaces in which inner products $(\ ,\ )_1$, $(\ ,\ )_2$ are defined, respectively, with associated norms

$$\|x\|_1 = [(x,x)_1]^{\frac{1}{2}}, \quad \|y\|_2 = [(y,y)_2]^{\frac{1}{2}} .$$

The term "Hilbert space" includes the assumption that H_1 and H_2 are "complete" with respect to their norms, that is, e.g., if $\{x_k\}$ is a sequence in H_1 having the Cauchy property

$$\lim_{k,\ell \to \infty} \|x_k - x_\ell\|_1 = 0$$

then there is an element $x \in H_1$ which is the limit as $k \to \infty$ of the sequence $\{x_k\}$ in the sense

$$\lim_{k \to \infty} \|x - x_k\|_1 = 0 .$$

A subspace, $\tilde{H}$ of a Hilbert space H is a subset of H which is closed under the formation of linear combinations, i.e.

$$x, \hat{x} \in \tilde{H}, \ \alpha, \hat{\alpha} \text{ scalar} \Rightarrow \alpha x + \hat{\alpha}\hat{x} \in \tilde{H} .$$

Such a subspace, or, more generally, any subset of H, is closed if every Cauchy sequence in $\tilde{H}$ converges to an element $x \in \tilde{H}$. In particular then, the whole space, H, is a closed subspace of itself.

In the applications which we make H_1, or H_2, will be either a finite dimensional Euclidean space E^n or else a space $L_n^2[t_0, t_1]$ for some interval $[t_0, t_1]$ and a positive integer n.

A linear operator F from H_1 to H_2 is a function (mapping, transformation) defined on H_1 with values in H_2 with the linearity property

$$F(\alpha x_1 + \beta x_2) = \alpha F x_1 + \beta F x_2, \quad \alpha, \beta \text{ scalar}, \ x_1, x_2 \in H_1 .$$

Such a linear operator is bounded (and we consider only bounded operators here) if there is a non-negative number M such that

$$\|Fx\|_2 \le M \|x\|_1, \quad x \in H_1 . \tag{3.1}$$

The smallest M for which (3.1) holds true is called the "norm of F" and is denoted $\|F\|$. Thus

$$\|F\| = \sup_{\substack{x_1 \in H_1 \\ \|x_1\| \neq 0}} \frac{\|Fx_1\|_2}{\|x_1\|_1} . \tag{3.2}$$

To summarize the information concerning F and its domain and range spaces we use the symbol

$$F : H_1 \to H_2$$

(read: F maps H_1 into H_2).

Given $F : H_1 \to H_2$ one may define another operator $F^* : H_2 \to H_1$ by the relationships

$$(Fx, y)_2 = (x, F^* y)_1, \quad x \in H_1, \; y \in H_2 . \tag{3.3}$$

The details of this construction may be found in any functional analysis text (see, e.g. [4], I-[4]) and it is a basic theorem that

$$\|F\| = \|F^*\|. \tag{3.4}$$

Further, if we have Hilbert spaces H_1, H_2, H_3 and

$$F_1 : H_1 \to H_2, \quad F_2 : H_2 \to H_3$$

the composition $F = F_2 F_1$ defined by $Fx = F_2(F_1 x)$ is a bounded linear operator from H_1 to H_2 with

$$\|F\| = \|F_2 F_1\| \leq \|F_2\| \, \|F_1\|. \tag{3.5}$$

The adjoint operator is

$$F^* = F_1^* F_2^* : H_3 \to H_1 . \tag{3.6}$$

When H_1 and H_2 are Euclidean spaces, E^n and E^m, so that F may be identified with an $m \times n$ matrix $F = (f_j^i)$, F^* is associated with the transposed matrix $\overline{F^T} = (\overline{f_i^j})$ if E^n, E^m are real Euclidean spaces and with $F^* = F^T = (f_i^j)$ if E^n, E^m are complex Euclidean spaces.

Definition 3.1. *A (general) linear observed system consists of a triple of Hilbert spaces* X, Y, Z *and a pair of bounded linear operators* $O : X \to Y$, $F : X \to Z$, *which we exhibit schematically as follows:*

$$\begin{array}{ccc} X & \xrightarrow{\;O\;} & Y \\ & \searrow^{F} & \\ & & Z \end{array} \tag{3.7}$$

The operator O *is called the observation operator and* F *is called the identification operator.*

Definition 3.2. *A (general) linear control system consists of a triple of Hilbert spaces* $\hat{X}, \hat{Y}, \hat{Z}$ *and a pair of bounded linear operators* C, G *such that* $C : \hat{Y} \to \hat{X}$, $G : \hat{Z} \to \hat{X}$, *which we exhibit schematically as follows:*

$$\begin{array}{ccc} \hat{X} & \xleftarrow{\;C\;} & \hat{Y} \\ & \nwarrow^{G} & \\ & & \hat{Z} \end{array} \tag{3.8}$$

The operator C *is the control operator while* G *is the objective operator.*

Before going on to define observability and controllability in this general context, we pause to relate these abstract entities to concrete observation and control situations.

Consider the linear observed system

$$\left.\begin{aligned} \dot{x} &= A(t)x \\ \omega &= H(t)x \end{aligned}\right\} \; t \in [t_0, t_1] . \tag{3.9}$$

To discuss the identification of the initial state $x_0 = x(t_0)$ by means of the observation ω we let

$$X = Z = E^n$$
$$Y = L_r^2[t_0, t_1] .$$

The initial state x_0 lies in X and it is what is to be identified. We stipulate this by taking $F = I$ (the identity): $X \to Z = X$. The observation operator is defined by

$$(Ox_0)(t) = \omega(t) = H(t)x(t) = H(t)\Phi(t, t_0)x_0, \tag{3.10}$$

carrying the initial state x_0 into the observation function $\omega \in L_r^2[t_0, t_1]$.

It might be desired, instead, to identify the final state $x_1 = x(t_1)$. If so, this is indicated by changing the identification operator from I, as above, to

$$F = \Phi(t_1, t_0)$$

for then

$$Fx_0 = \Phi(t_1, t_0)x_0 = x(t_1).$$

If it is desired to make observations only at discrete points $s_1, s_2, \ldots, s_q \in [t_0, t_1]$, as in Proposition 1.5, we would take

$$Y = E^{qr} = E^r \oplus E^r \oplus \ldots \oplus E^r, \text{ (q factors)}$$

and define

$$Ox_0 = (H(s_1)\Phi(s_1, t_0)x_0, H(s_2)\Phi(s_2, t_0)x_0, \ldots, H(s_q)\Phi(s_q, t_0)x_0).$$

One can also discuss a nonhomogeneous linear observed system

$$\dot{x} = A(t)x + B(t)u, \quad t \in [t_0, t_1], \tag{3.11}$$

$$\omega = H(t)x + J(t)u, \quad u \in L_m^2[t_0, t_1], \tag{3.12}$$

when u is a <u>known</u> input function (whether control or disturbance is immaterial here). Since u is known, it can be adjoined to the observation ω to give an augmented observation

$$\hat{\omega} = \begin{pmatrix} H(t) & J(t) \\ 0 & I \end{pmatrix} \begin{pmatrix} x \\ u \end{pmatrix}.$$

In this setting we would take

$$X = E^n \oplus L_m^2[t_0, t_1],$$

allowing the pair (x_0, u) to be an element of X. The observation operator is then $O : X \to Y = L^2_{r+m}[t_0, t_1]$:

$$(O(x_0, u))(t) = \begin{pmatrix} H(t)(\Phi(t, t_0)x_0 + \int_0^t \Phi(t, s)B(s)u(s)ds) + J(t)u(t) \\ u(t) \end{pmatrix} \tag{3.13}$$

If in this case it is desired to identify the initial state $x_0 = x(t_0)$ we would take $Z = E^n$ and define $F : X \to Z$ by

$$F(x_0, u) = x_0 = (I, 0)\begin{pmatrix} x_0 \\ u \end{pmatrix} .$$

Many other possibilities exist, some of which will be discussed later.

Passing to the general linear control system, we begin by considering

$$\dot{x} = A(t)x + B(t)u. \tag{3.14}$$

To treat the problem of steering from the origin to $x_1 \in E^n$ during the interval $[t_0, t_1]$ with a control $u \in L^2_m[t_0, t_1]$ we let

$$\hat{X} = \hat{Z} = E^n, \qquad \hat{Y} = L^2_m[t_0, t_1] \tag{3.15}$$

and we define the operators $C : \hat{Y} \to \hat{X}$, $G : \hat{Z} \to \hat{X} = \hat{Z}$ by

$$G = I, \text{ the identity on } E^n, \tag{3.16}$$

$$Cu = \int_{t_0}^{t_1} \Phi(t_1, t)B(t)u(t)\,dt . \tag{3.17}$$

If we want to steer from an arbitrary $x_0 \in E^n$ to 0 we let $\hat{X}, \hat{Y}, \hat{Z}$ be as in (3.15) but we set

$$G = -\Phi(t_1, t_0) \tag{3.18}$$

and define C as in (3.17). If we want to steer from an arbitrary $x_0 \in F^n$ to an arbitrary $x_1 \in E^n$ we take $\hat{X}, \hat{Y}$ as in (3.15) but

$$\hat{Z} = E^n \oplus E^n = E^{2n}$$

and define G by

$$G(x_0, x_1) = x_1 - \Phi(t_1, t_0)x_0 = (I, \Phi(t_1, t_0))\begin{pmatrix} x_1 \\ x_0 \end{pmatrix} \tag{3.19}$$

Other control problems can be envisioned. Suppose we consider a disturbed system

$$\dot{x} = A(t)x + B(t)u + D(t)v \tag{3.20}$$

where the disturbance $v \in L_q^2[t_0, t_1]$. If we know the disturbance v it may be possible to counteract its influence in certain respects. Specifically, suppose that a response

$$y = R(t)x, \quad y \in E^p,$$

is identified and the control problem is to determine $u \in L_m^2[t_0, t_1]$ so that, for given $x_0 = x(t_0) \in E^n$ and $v \in L_q^2[t_0, t_1]$, we have

$$y(t) = R(t)x(t) = 0, \quad t \in [t_0 + \epsilon, t_1], \tag{3.21}$$

for some fixed $\epsilon > 0$. We would then take

$$\hat{Z} = E^n \oplus L_q^2[t_0, t_1]$$

so that the known initial state x_0 and known disturbance v can be combined as an element $(x_0, v) \in \hat{Z}$, we would take $\hat{X} = L_p^2[t_0 + \epsilon, t_1]$, $\hat{Y}$ again the space $L_m^2[t_0, t_1]$. The map G would be defined by

$$(G(x_0, v))(t) = -R(t)(\Phi(t, t_0)x_0 + \int_{t_0}^{t} \Phi(t, s)D(s)v(s)ds), \quad t \in [t_0 + \epsilon, t_1], \tag{3.22}$$

while C would be defined by

$$(Cu)(t) = R(t)\int_{t_0}^{t} \Phi(t, s)B(s)u(s)ds, \quad t \in [t_0 + \epsilon, t_1]. \tag{3.23}$$

Definition 3.3. The general linear observed system (3.7) is observable if and only if, for $x \in X$,

$$Ox = 0 \Rightarrow Fx = 0. \tag{3.24}$$

Definition 3.4. The general linear control system (3.8) is controllable if and only if for each $z \in \hat{Z}$ there is at least one $y \in \hat{Y}$ such that

$$Cy = Gz \tag{3.25}$$

(that is, the range of C includes the range of G.)

In the case of the initial state identification problem for (3.9) we have observability if Ox_0 (cf. (3.10)) $= 0$ implies $x_0 = 0$, since $F = I$ here. For terminal state observability we require that $Ox_0 = 0$ implies $\Phi(t_1, t_0)x_0 = 0$. Since $\Phi(t_1, t_0)$ is nonsingular, initial and terminal state identification are equivalent here. In the case of the nonhomogeneous system (3.11), (3.12) the requirement for observability becomes $O(x_0, u) = 0$ implies $x_0 = 0$. Now (cf. (3.13)) $O(x_0, u) = 0$ if and only if $u = 0$ in $L_m^2[t_0, t_1]$ and $H(t)\Phi(t, t_0)x_0 = 0$ implies $x_0 = 0$. Since $u = 0$ implies that $\dot{x} = A(t)x$, $\omega = H(t)x$ (cf. (3.11), (3.12)) this problem is again equivalent to the earlier ones.

Passing to the control examples, we can steer (3.14) from 0 to x_1 during $[t_0, t_1]$ just in case we can find $u \in L_m^2[t_0, t_1]$ such that

$$x_1 = \int_{t_0}^{t_1} \Phi(t_1, t)B(t)u(t)\,dt ,$$

i.e. $Ix_1 = Gx_1 = Cu = \int_{t_0}^{t_1} \Phi(t_1, t)B(t)u(t)dt$ (cf. (3.16), (3.17)). To steer from x_0 to zero we need

$$0 = \Phi(t_1, t_0)x_0 + \int_{t_0}^{t_1} \Phi(t_1, t)B(t)u(t)dt$$

and this agrees with (3.25) when C is given by (3.17) and G by (3.18). The control problem corresponding to (3.19) can be interpreted similarly.

When we apply (3.25) to the case where x, u, v satisfy (3.20) and G is given by (3.22), C by (3.23), we obtain

$$R(t)[\Phi(t, t_0)x_0 + \int_{t_0}^{t} \Phi(t, s)(D(s)v(s) + B(s)u(s))ds] = 0, \quad t \in [t_0 + \epsilon, t_1],$$

which is the same as (3.21).

Definition 3.5. The general linear observed system (3.7) and the general linear control system (3.8) are dual to each other just in case

$$X = \hat{X}, \quad Y = \hat{Y}, \quad Z = \hat{Z} \tag{3.26}$$

and

$$O = C^*, \quad F = G^*. \tag{3.27}$$

We are now able to state a rather general duality theorem expressing the relationship between observability for (3.7) and controllability for (3.8). A more far-reaching result appears in [5].

Theorem 3.6. If (3.7) and (3.8) are dual systems, so that (3.26) and (3.27) obtain, then (3.7) is observable if (3.8) is controllable.

Proof. Let us suppose that the system (3.8) is controllable so that the range of C includes that of G. Then for every $z \in Z$ there is a $y \in Y$ such that $Cy = Gz$.

Let $x \in X$ be such that $Ox = C^*x = 0$. Then for every $y \in Y$

$$0 = (y, C^*x)_Y = \text{(cf. (3.3))} = (Cy, x)_X. \tag{3.28}$$

Let z be an element of Z. Then there is some y, by our controllability hypothesis, such that $Cy = Gz$. Therefore

$$(z, G^*x)_Z = (Gz, x)_X = (Cy, x)_X = 0, \tag{3.29}$$

the last equality from (3.25). But $G^*x \in Z$ so we can take $z = G^*x$ and (3.29) gives $(G^*x, G^*x)_Z = \|G^*x\|_Z^2 = 0 \Rightarrow G^*x = 0$. Thus $C^*x = 0$ implies $G^*x = 0$ and we have shown that controllability of (3.8) implies observability of (3.7) when the two are dual to each other. Q. E. D.

It is natural to enquire whether Theorem 3.6 has a converse to the effect that if (3.7) and (3.8) are dual to each other and (3.7) is observable then (3.8) is controllable. For this we need another assumption which is, however, automatically satisfied when X is finite dimensional, as it is for most applications involving ordinary differential equation control systems.

Theorem 3.7. *Let* (3.7) *and* (3.8) *be dual systems so that* (3.26) *and* (3.27) *obtain and let the range of* C *be a closed subspace of* X. *Then* (3.8) *is controllable if* (3.7) *is observable.*

Proof. For given $z \in Z$, Gz lies in X. Since the range of C is a closed subspace of X there is a point in that space which lies closest to Gz, that is, there is a $y_0 \in Y$ such that

$$\|Cy_0 - Gz\|_X = \min_{y \in Y} \|Cy - Gz\|_X .$$

(This is a standard result in linear space theory. See [3], e.g.). Let η be any point in Y and consider

$$\|C(y_0 + \alpha\eta) - Gz\|_X^2 = \|Cy_0 - Gz\|_X^2$$
$$+ \alpha(C\eta, Cy_0 - Gz)_X + \overline{\alpha}(Cy_0 - Gz, C\eta)_X + |\alpha|^2 \|C\eta\|_X^2 .$$

Taking α real and noting that

$$\frac{\partial}{\partial\alpha} \|C(y_0 + \alpha\eta) - Gz\|_X^2 \Big|_{\alpha = 0} = 0$$

by virtue of the minimality of $\|Cy_0 - Gz\|_X$, we conclude

$$(C\eta, Cy_0 - Gz)_X + (Cy_0 - Gz, C\eta)_X = 2\,\mathrm{Re}(C\eta, Cy_0 - Gz)_X = 0 ,$$

where $\mathrm{Re}\,\zeta$ denotes the real part of a complex number ζ. On the other hand with α purely imaginary we get $2\,\mathrm{Im}(C\eta, Cy_0 - Gz)_X = 0$, where $\mathrm{Im}\,\zeta$ is the imaginary part of ζ, and we have

$$(C\eta, Cy_0 - Gz)_X = 0, \quad \eta \in Y$$

whence (cf. (3.3))

$$(\eta, C^*(Cy_0 - Gz))_Y = 0, \quad \eta \in Y.$$

This being true for all $\eta \in Y$ we conclude

$$C^*(Cy_0 - Gz) = 0.$$

But then the observability condition shows that

$$G^*(Cy_0 - Gz) = 0.$$

Then

$$\begin{aligned} 0 &= (y_0, C^*(Cy_0 - Gz))_Y - (z, G^*(Cy_0 - Gz))_Z \\ &= (Cy_0, Cy_0 - Gz)_X - (Gz, Cy_0 - Gz)_X \\ &= (Cy_0 - Gz, Cy_0 - Gz)_X = \|Cy_0 - Gz\|_X^2 \end{aligned}$$

so that $Cy_0 = Gz$. Since z was any point of Z we conclude that the range of C includes that of G and (3. 8) is controllable. Q. E. D.

We remark that every subspace of a finite dimensional space X is closed so the additional assumption in Theorem 3. 7 is, indeed, satisfied automatically in that case.

We can show rather easily that Theorem 3. 7 fail, in general, without the special assumption. In the process we also indicate that rather general systems can be viewed as in the context of Definition 3. 1 (or Definition 3. 2).

Let $X = Y = Z = L^2[0,1] \equiv L_1^2[0,1]$ and let

$$\left.\begin{aligned} (Of)(t) &= \int_0^t f(s)ds \\ (Ff)(t) &= f(t) \end{aligned}\right\} t \in [0,1], \; f \in L^2[0,1]. \tag{3. 30}$$

This system is observable because

$$\int_0^t f(s)ds = 0, \; t \in [0,1] \Rightarrow f = 0 \text{ in } L^2[0,1].$$

The dual system in this case is defined by

$$\begin{aligned} (Cg)(t) &= (O^*g)(t) = \int_t^1 g(s)ds \\ (Gg)(t) &= (F^*g)(t) = g(t) \end{aligned} \qquad t \in [0,1], \; g \in L^2[0,1]. \tag{3. 31}$$

The duality follows from integration by parts:

$$(Cg, f)_{L^2[0,1]} = \int_0^1 (\int_t^1 g(s)ds)\overline{f(t)}\,dt$$

$$= [(\int_t^1 g(s)ds)(\int_0^t \overline{f(s)}ds)]_{t=0}^{t=1} + \int_0^1 g(t)(\int_0^t \overline{f(s)}\,ds)dt$$

$$= 0 + \int_0^1 g(t)(\int_0^t \overline{f(s)}ds)dt = (g, Of)_{L^2[0,1]} .$$

Even though (3. 30) is observable, (3. 31) is not controllable because the range of C, which is all functions $h(t)$ such that $h(1) = 0$ and $h(t)$ is absolutely continuous with $h\ (t) \in L^2[0,1]$, is not all of $L^2[0,1]$ while the range of G clearly is all of that space. The range of O can be shown not to be closed in $L^2[0,1]$ in this case, which explains why Theorem 3.7 does not apply.

To better relate our whole structure to observability and controllability as discussed in Sections 1 and 2, let us re-prove Theorem 2. 4. We suppose that

$$\dot{y} = - A(t)^* y \tag{3. 32}$$

$$\omega = B(t)^* y \tag{3. 33}$$

is observable on $[t_0, t_1]$, that is (cf. (3. 10))

$$(Oy_0)(t) = \omega(t) = B(t)^* \Psi(t, t_0)y_0 = 0, \quad t \in [t_0, t_1] \tag{3. 34}$$

implies that

$$y_0 = Fy_0 = 0 . \tag{3. 35}$$

Here $\Psi(t, t_0)$ is the fundamental matrix solution of (3. 32) with $\Psi(t_0, t_0) = I$. The adjoint of the observation operator $O : E^n \to L^2_m[t_0, t_1]$ is identified as follows. For $u \in L^2_m[t_0, t_1]$

$$(u, Oy_0)_{L^2_m[t_0,t_1]} = \int_{t_0}^{t_1} (u(t), B(t)^* \Psi(t, t_0)y_0)_{E^n}\, dt$$

$$= (\int_{t_0}^{t_1} \Psi(t, t_0)^* B(t)u(t)dt, y_0)_{E^n}$$

so that

$$C^*u = \int_{t_0}^{t_1} \Psi(t, t_0)^* B(t)u(t)dt .$$

The controllability equation for the dual system, i. e. (replacing x_0 by $-x_0$)

$$C^*u + F^*x_0 = 0, \quad u \in L_m^2[t_0, t_1], \quad x_0 \in E^n$$

thus becomes

$$0 = x_0 + \int_{t_0}^{t_1} \Psi(t, t_0)^* B(t)u(t)dt.$$

But (see I-3)

$$\Psi(t, t_0)^* = \Phi(t_0, t) \; (= \Phi(t, t_0)^{-1})$$

so (3. 36) is the same as

$$0 = x_0 + \int_{t_0}^{t_1} \Phi(t_0, t)B(t)u(t)dt .$$

Here $\Phi(t, \tau)$ is the fundamental matrix solution of $\dot{x} = A(t)x$ with $\Phi(\tau, \tau) = I$. Multiplying by $\Phi(t_1, t_0)$ we have

$$0 = \Phi(t_1, t_0)x_0 + \int_{t_0}^{t_1} \Phi(t_1, t)B(t)\,u(t)\,dt , \tag{3. 37}$$

which is precisely the condition for the control $u(t)$ to steer the initial state x_0 to 0 during $[t_0, t_1]$ if

$$\dot{x} = A(t)x + B(t)u . \tag{3. 38}$$

Thus the observability condition (3. 34) => (3. 35) for (3. 32), (3. 33) gives the controllability result (3. 37) for (3. 38), a result equivalent to that already obtained in Theorem 2. 4.

We are able, now, to discuss reconstruction operators and steering functions and to show the relationship between them.

Definition 3.8. *A reconstruction operator for the linear observed system* (3.7) *is a bounded operator* $R : Y \to Z$ *such that*

$$RO = F . \qquad (3.39)$$

Definition 3.9. *A steering operator for the linear control system* (3.8) *is a bounded operator* $S : \hat{Z} \to \hat{Y}$ *such that*

$$CS = G . \qquad (3.40)$$

Proposition 3.10. *If* R *is a reconstruction operator for* (3.7) *then* R^* *is a steering operator for the dual control system. Likewise, if* S *is a steering operator for* (3.8), S^* *is a reconstruction operator for the dual observed system.*

Proof. The adjoints of the operators appearing on each side of (3.39) must be equal, hence

$$O^* R^* = F^* . \qquad (3.41)$$

But if O and F are the observation and identification operators for a linear observed system, O^* and F^* are the control and objective operators for the dual linear control system and (3.41) becomes (3.40) for

$$C = O^* , \ G = F^* , \quad S = R^* .$$

Thus the adjoint of a reconstruction operator is a steering operator for the dual linear control system. The proof of the rest of the proposition is entirely similar. Q. E. D.

Definition 3.11. *A reconstruction operator,* R_0, *for the linear observed system* (3.7) *is optimal if*

$$\| R_0 \| \leq \| R \|$$

whenever R *is a reconstruction operator for* (3.7). *Similarly, a steering operator,* S_0, *for the linear control system* (3.8) *is optimal if*

$$\| S_0 \| \leq \| S \|$$

for every steering operator, S, for (3.8).

In Sections 1 and 2 of this chapter we introduced particular observation and steering operators

$$R_0 : L_r^2[t_0, t_1] \to E^n,$$

$$R_0\omega = \int_{t_0}^{t_1} W(t_0, t_1)^{-1} \Phi(t, t_0)^* H(t)^* \omega(t)\,dt , \tag{3.42}$$

$$S_0 : E^n \to L_m^2[t_0, t_1] ,$$

$$(S_0 x_1)(t) = B(t)^* \Phi(t_1, t)^* Z(t_0, t_1)^{-1} . \tag{3.43}$$

We indicated at the end of Section 2 that S_0 uniquely possesses certain "desirable properties", and one expects the same for R_0. In fact, S_0 is an optimal steering operator for the system and objective

$$\dot{x} = A(t)x + B(t)u \tag{3.44}$$

$$x(t_0) = 0, \quad x(t_1) = x_1 \in E^n, \tag{3.45}$$

as we see from the following theorem.

Theorem 3.12. The steering operator $S_0(t)$ defined by (3.43) is optimal for (3.16), (3.17). In fact it is uniquely "pointwise optimal" in the sense that, for any $x_1 \in E^n$ and any control $u \in L_m^2[t_0, t_1]$ steering 0 to x_1 during $[t_0, t_1]$

$$\|u_0\| \equiv \|S_0 x_1\| \le \|u\|$$

and equality holds if and only if $u_0 = u$ in $L_m^2[t_0, t_1]$.

Proof. We compute, $\| \;\|$ here denoting $\| \;\|_{L_m^2[t_0, t_1]}$

$$\|u\|^2 = \|u_0 + (u-u_0)\|^2 = \|u_0\|^2 + \|u-u_0\|^2 + 2\,\mathrm{Re}(u_0, u-u_0)_{L_m^2[t_0, t_1]} . \tag{3.46}$$

Now

$$(u_0, u-u_0)_{L^2_m[t_0,t_1]} = \int_{t_0}^{t_1} (u_0(t), u(t)-u_0(t))_{E^m} dt$$

$$= \int_{t_0}^{t_1} (B(t)^* \Phi(t_1, t)^* Z(t_0, t_1)^{-1} x_1 , u(t)-u_0(t))_{E^m} dt$$

$$= \int_{t_0}^{t_1} (Z(t_0, t_1)^{-1} x_1 , B(t)\Phi(t_1, t)(u(t)-u_0(t)))_{E^n} dt$$

$$(Z(t_0,t_1)^{-1}[\int_{t_0}^{t_1} \Phi(t_1, t)B(t)u(t)dt - \int_{t_0}^{t_1} \Phi(t_1,t)B(t)u_0(t)dt])_{E^n} .$$

But, since both u and u_0 steer 0 to x_1 during $[t_0, t_1]$, we have

$$\int_{t_0}^{t_1} \Phi(t_1, t)B(t)u(t)dt = \int_{t_0}^{t_1} \Phi(t_1, t)B(t)u_0(t)dt = x_1$$

and we see that

$$(u_0, u-u_0)_{L^2_m[t_0, t_1]} = 0 .$$

Then from (3.46)

$$\|u\|^2 - \|u_0\|^2 = \|u-u_0\|^2$$

and thus $\|u_0\| < \|u\|$ unless $u-u_0 = 0$ in $L^2_m[t_0, t_1]$, i.e. $u = u_0$ there.

Then, clearly, for any steering operator S and any $x_1 \in E^n$

$$\|u_0\| \equiv \|S_0 x_1\| \leq \|Sx_1\| \equiv \|u\|$$

and it follows immediately that S_0 is optimal in the sense of Definition 3.11 that S_0 is an optimal steering operator, as claimed. Q. E. D.

An optimal steering operator immediately leads to an optimal reconstruction operator. For we have seen in Proposition 3.10 that if S_0 is a steering operator then S_0^* is a reconstruction operator for the dual observed system. If R is any other reconstruction operator for the dual observed system then R^* is a steering operator for the original control system. If we then take S_0 to be optimal

$$\|S_0^*\| = \|S_0\| \leq \|R^*\| = \|R\| \tag{3.47}$$

and we conclude that $R_0 = S_0^*$ is an optimal reconstruction operator. Clearly the argument goes the other way also.

In general there is no unique optimal steering operator for (3.8) or reconstruction operator for (3.7). The operator (3.43) is, however, as we have seen, the unique pointwise optimal steering operator for (3.16), (3.17). The property of pointwise optimality is that

$$\|S_0 z\|_{\hat{Y}} \leq \|Sz\|_{\hat{Y}}, \quad z \in \hat{Z}$$

for every steering operator S, so that $y = S_0 z$ is always the least norm control, i.e., the element of least norm in $\hat{Y}$ for which $Cy = Gz$. If S_0 is pointwise optimal for (3.8), S_0^* is a reconstruction operator for the dual linear observed system which acts in the following way: for each $y \in Y$

$$S_0^* y = S_0^* \hat{\eta} \tag{3.48}$$

where η is that element of the closed subspace

$$\overline{\{y \in Y | y = Cx, \ x \in X\}} = \overline{\mathcal{R}(C)} \quad \text{(closure of the range of } C)$$

which lies closest to y, i.e.,

$$\|y - \hat{\eta}\| = \min_{\eta \in \overline{\mathcal{R}(C)}} \|y - \eta\|. \tag{3.49}$$

It is necessary to keep clearly in mind the precise nature of the dual system when comparing S and S^* (or R and R^*). We have seen that (3.43) is an optimal steering operator for (3.16), (3.17). The adjoint optimal reconstruction operator for (3.32), (3.33) is

$$R_0 \omega = \int_{t_0}^{t_1} Z(t_0, t_1)^{-1} \Phi(t_1, t) B(t) \omega(t) \, dt . \tag{3.50}$$

Now (cf. (2.9))

$$Z(t_0, t_1) = \int_{t_0}^{t_1} \Phi(t_1, t)B(t)B(t)^* \Phi(t_1, t)^* dt \qquad (3.51)$$

and, for the dual observed system (3.32), (3.33), the fundamental matrix solution $\Psi(t, \tau)$ satisfies

$$\Psi(t, t_1) = \Phi(t_1, t)^* . \qquad (3.52)$$

Combined with the fact that the observation is given by

$$\omega(t) = B(t)^* \Psi(t, t_1) y_1 \qquad (3.53)$$

if $y_1 = y(t_1)$, (3.50), (3.51), (3.52) show that

$$\begin{aligned} R_0 \omega &= \int_{t_0}^{t_1} Z(t_0, t_1)^{-1} \Phi(t_1, t)B(t)\omega(t)dt \\ &= \int_{t_0}^{t_1} Z(t_0, t_1)^{-1} \Phi(t_1, t)B(t)B(t)^* \Psi(t, t_1)dt\, y_1 \\ &= \int_{t_0}^{t_1} Z(t_0, t_1)^{-1} \Phi(t_1, t)B(t)B(t)^* \Phi(t_1, t)^* dt\, y_1 = y_1 . \end{aligned} \qquad (3.54)$$

Thus $R_0 = S_0^*$ is an optimal <u>terminal state</u> reconstruction operator. In the same way

$$(\hat{S}_0 x_0)(t) = - B(t)^* \Phi(t_1, t)^* Z(t_0, t_1)^{-1} \Phi(t_1, t_0) x_0$$

defines an optimal steering operator for (3.17), (3.18). The adjoint operator is

$$\hat{R}_0 \omega = - \int_{t_0}^{t_1} \Phi(t_1, t_0)^* Z(t_0, t_1)^{-1} \Phi(t_1, t)B(t)\,\omega(t)\,dt \qquad (3.55)$$

and, since,

$$\Phi(t_1, t_0)^* Z(t_0, t_1)^{-1} \Phi(t_1, t) = \Phi(t_1, t_0)^* Z(t_0, t_1)^{-1} \Phi(t_1, t_0) \Phi(t_0, t)$$

$$= \left(\int_{t_0}^{t_1} \Phi(t_0, t) B(t) B(t)^* \Phi(t_0, t)^* dt \right)^{-1} \Phi(t_0, t)$$

$$= \left(\int_{t_0}^{t_1} \Psi(t, t_0)^* B(t) B(t)^* \Psi(t, t_0) dt \right)^{-1} \Psi(t, t_0)^*, \qquad (3.56)$$

so, with

$$\omega(t) = B(t)^* \Psi(t, t_0) y_0 \qquad (3.57)$$

we have, combining (3.55), (3.57) and using (3.56), that

$$\hat{R}_0 \omega = y_0$$

and $\hat{R}_0$ is an (optimal, by the discussion leading to (3.47)) initial state reconstruction operator for (3.32), (3.33).

The significance of an optimal steering operator is fairly evident since it yields "small" controls. This is most strongly evident when we have the property of pointwise optimality. The significance of an optimal reconstruction operator is only slightly harder to explain. Suppose we have an observation

$$y = Cx$$

for some $x \in X$. If the observation process is corrupted in some way the actual element available is $y + \eta$, where $\eta \in Y$ represents the observation error. The reconstruction error, then, has norm

$$\|Fx - R(y+\eta)\|_Z = \|Fx - Ry - R\eta\|_Z = \|Fx - Fx - R\eta\|_Z = \|R\eta\|_Z$$

and this norm is made small by taking R_0 to be an optimal reconstruction operator.

4. Controllability of Nonlinear Systems Near an Equilibrium Point

Here we will consider systems which take the form, with $\dot{y} = \frac{dy}{dt}$,

$$\dot{y} = f(y, w, t), \quad y \in E^n, \quad w \in E^m, \tag{4.1}$$

where $f : E^{n+m+1} \to E^n$ will be assumed continuously differentiable with respect to all of its arguments in a region containing all points (y, w, t) arising in our subsequent discussion. Here y is the n-dimensional state vector and w the m-dimensional control vector. We will suppose that vectors y_0 and w_0 may be singled out so that (y_0, w_0) is an equilibrium point for all $t \in [0, \infty)$, i.e.,

$$f(y_0, w_0, t) \equiv 0 . \tag{4.2}$$

Proceeding as in I-2, the use of Taylor's formula gives us

$$(y \dot{-} y_0) = A(t)(y - y_0) + B(t)(w - w_0) + h(y, w, t) \tag{4.3}$$

where ((4.5) is stated more strongly than the corresponding relationship in I-2; it follows from continuous differentiability.)

$$A(t) = \frac{\partial f}{\partial y}(y_0, w_0, t), \quad B(t) = \frac{\partial f}{\partial w}(y_0, w_0, t) , \tag{4.4}$$

$$h(y_2, w_2, t) - h(y_1, w_1, t) = o(\|y_2 - y_1, w_2 - w_1\|), \quad (y_i, w_i) \to (y_0, w_0), \quad i = 1, 2 \tag{4.5}$$

the latter holding uniformly on any bounded t-interval in $[0, \infty)$. The linearized system (again, see I-2) is

$$\dot{x} = A(t)x + B(t)u \tag{4.6}$$

where x should be thought of as a first order approximation to $y - y_0$ and u as a first order approximation to $w - w_0$.

A great variety of existence, uniqueness, local and global behavior, etc., theorems have been developed for nonlinear systems. We refer the reader to I-[2], I-[3] and III-[2] for numerous instances of such results. We will content ourselves here with a theorem exhibiting the existence of solutions of (4.1) (equivalently, (4.3)) which correspond, via $x \sim y - y_0$, $u \sim w - w_0$, to solutions of (4.6) near the equilibrium point y_0, w_0 of (4.1).

Theorem 4.1. *Let* t_0, t_1 *with* $0 \le t_0 < t_1 < \infty$, *and* $\epsilon > 0$ *be freely chosen.*

Then there exists $\delta > 0$ such that if

$$\|x_0\|_{E^n} < \delta, \quad \|u(t)\|_{E^m} < \delta, \quad t \in [t_0, t_1],$$

then there exists a unique solution $y(t)$ of (4. 1) with

$$y(t_0) = y_0 + x_0, \quad w(t) = w_0 + u(t), \quad t \in [t_0, t_1],$$

and, for $t \in [t_0, t_1]$,

$$\|(y(t) - y_0) - x(t)\|_{E^n} < \epsilon \left(\|x_0\|_{E^n} + \sup_{t \in [t_0, t_1]} \|u(t)\|_{E^m} \right), \tag{4. 7}$$

$x(t)$ denoting the solution of (4. 6) with the indicated control function $u(t)$ and initial state $x(t) = x_0$.

Sketch of Proof. Letting $\Phi(t, s)$ denote the fundamental solution of (4. 6) with $\Phi(t, t) = I$, we have, from I-3, the variation of parameters formula

$$x(t) = \Phi(t, t_0)x_0 + \int_{t_0}^{t} \Phi(t, s)B(s)u(s)ds . \tag{4. 8}$$

Letting $\eta(t) = y(t) - y_0$ and treating the terms

$$B(t)(w(t) - w_0) + h(y(t), w(t), t) = B(t)u(t) + h(\eta(t) + y_0, u(t) + u_0, t)$$

in (4. 3) as if they corresponded to inhomogeneous terms in a linear equation, we obtain an integral equation for $\eta(t)$:

$$\eta(t) = x(t) + \int_{t_0}^{t} \Phi(t, s)h(\eta(s) + y_0, u(s) + w_0, s) ds . \tag{4. 9}$$

From (4. 5), given any $\nu > 0$, we can find $\mu > 0$ such that

$$\|y_i - y_0\|_{E^n} < \mu, \quad \|w_i - w_0\|_{E^m} < \mu, \quad i = 1, 2, \Rightarrow \|h(y_2, w_2, t) - h(y_1, w_1, t)\|$$

$$< \nu(\|y_2 - y_1\|_{E^n} + \|w_2 - w_1\|_{E^m}), \quad t \in [t_0, t_1]. \tag{4. 10}$$

Let ν be selected small enough so that

$$2\nu(t_1 - t_0) \sup_{t, s \in [t_0, t_1]} \|\Phi(t, s)\| \le \tfrac{1}{2}, \tag{4. 11}$$

and let μ be selected accordingly, satisfying, minimally, $\mu \leq 1$. As for δ, we require, first of all, that it be small enough so that, using (4.8), we can establish

$$\|x(t)\|_{E^n} \leq \frac{\mu}{2}, \quad t \in [t_0, t_1] \tag{4.12}$$

and, in addition, we require

$$\delta \leq \mu \leq 1 . \tag{4.13}$$

Now consider the family of continuous n-vector functions $\varphi(t)$ defined on $[t_0, t_1]$ such that

$$\|\varphi(t)\|_{E^n} \leq \mu, \quad t \in [t_0, t_1] . \tag{4.14}$$

For each such $\varphi(t)$, define $\psi(t)$ by

$$\psi(t) = x(t) + \int_{t_0}^{t} \Phi(t, s)h(\varphi(s) + y_0, u(s) + w_0, s)\,ds . \tag{4.15}$$

Then, using (4.10) with $y_2 = \varphi$, $y_1 \equiv y_0$, $w_2 \equiv w_0 + u$, $w_1 \equiv w_0$, $\|u(t)\|_{E^m} < \delta$,

$$\|\psi(t)\|_{E^n} < \frac{\mu}{2} + (t_1 - t_0) \sup_{t,s \in [t_0,t_1]} \|\Phi(t,s)\|\, \nu \left(\sup_{t \in [t_0,t_1]} \|\varphi(t)\|_{E^n} + \sup_{t \in [t_0,t_1]} \|u(t)\|_{E^m} \right)$$

$$\leq \frac{\mu}{2} + 2\nu\mu(t_1 - t_0) \sup_{t,s \in [t_0,t_1]} \|\Phi(t,s)\| \leq \mu , \tag{4.16}$$

the last inequality following from (4.13).

For each pair of functions $\varphi_1(t)$, $\varphi_2(t)$, satisfying (4.14), the corresponding $\psi_1(t)$, $\psi_2(t)$, computed from (4.15), satisfy

$$\|\psi_1(t) - \psi_2(t)\|_{E^n} \leq \text{(cf. (4.10) with } w_1 = w_2 = w_0 + u)$$

$$\leq (t_1 - t_0) \sup_{t,s \in [t_0,t_1]} \|\Phi(t,s)\| \sup_{t \in [t_0,t_1]} \|h(\varphi_1(t) + y_0, u(t) + w_0, t)$$

$$- h(\varphi_2(t) + y_0, u(t) + w_0, t)\|_{E^n} \leq \nu(t_1 - t_0) \sup_{t,s \in [t_0,t_1]} \|\Phi(t,s)\| \sup_{t \in [t_0,t_1]} \|\varphi_1(t) - \varphi_2(t)\|_{E^n}$$

$$\leq \text{(cf. (4.11))}\ \tfrac{1}{4} \sup \|\varphi_1(t) - \varphi_2(t)\|_{E^n} . \tag{4.17}$$

The inequalities (4.16) and (4.17) show that the map $\varphi \to \psi$ defined by (4.15) carries the set of functions (4.14) into itself and is a contraction relative to the metric $\sup_{t\in[t_0,t_1]} \|\varphi_1(t) - \varphi_2(t)\|_{E^n}$. The contraction fixed point theorem (see III-[1], e.g.) applies to show that (4.9), and hence (4.1), has a unique solution $\eta(t)$, also lying in the set of functions (4.14).

To establish the inequality (4.7) we rewrite (4.9) in the form

$$\eta(t) - x(t) = \int_{t_0}^{t} \Phi(t,s)[h(\eta(s)+y_0, u(s)+w_0, s) - h(x(s)+y_0, u(s)+w_0, s)]\,ds$$
$$+ \int_{t_0}^{t} \Phi(t,s)h(x(s)+y_0, u(s)+w_0, s)ds$$

so that, using (4.10) twice,

$$\|\eta(t) - x(t)\|_{E^n} \le \nu \max_{t,s\in[t_0,t_1]} \|\Phi(t,s)\| \int_{t_0}^{t} \|\eta(s) - x(s)\|_{E^n}\,ds$$
$$+ \nu \max_{t,s\in[t_0,t_1]} \|\Phi(t,s)\| \int_{t_0}^{t} (\|x(s)\|_{E^n} + \|u(s)\|_{E^m})ds \qquad (4.18)$$

The linear form of (4.8) enables one to obtain, very easily, a bound

$$\|x(t)\|_{E^n} \le \beta_0 \|x_0\|_{E^n} + \beta_1 \sup_{t\in[t_0,t_1]} \|u(t)\|_{E^m}$$

for constants β_0, β_1 depending only on $A(t)$, $B(t)$ and the interval $[t_0, t_1]$. Using this in (4.18) we obtain an inequality of the form

$$\|\eta(t) - x(t)\|_{E^n} \le \nu(\gamma_0 \int_{t_0}^{t_1} \|\eta(s) - x(s)\|_{E^n} ds\,(t - t_0)[\gamma_1 \|x_0\|_{E^n}$$
$$+ \gamma_2 \sup_{t\in[t_0,t_1]} \|u(t)\|_{E^m}])$$

from which (see I-[3]) the Gronwall inequality gives (since $x(t_0) = \eta(t_0) = x_0$)

$$\|\eta(t) - x(t)\|_{E^n} \le \int_{t_0}^{t} e^{\nu\gamma_0 s} ds[\nu\gamma_1 \|x_0\|_{E^n} + \gamma_2 \sup_{t\in[t_0,t_1]} \|u(t)\|_{E^m}] .$$

Then, since $\eta(t) = y(t) - y_0$, (4.7) is satisfied if we take ν small enough (which, of course, entails possible further restriction on μ and δ as introduced earlier). Q. E. D.

The foregoing theorem shows that on bounded intervals, and in a small neighborhood of the equilibrium point y_0, w_0, the trajectories of the nonlinear system (4. 1) closely follow those of the linearized system (4. 6). This provides at least some justification for replacing (4. 1) by (4. 6) in control analysis, provided we have reason to believe that our system will remain in an appropriately small neighborhood of y_0, w_0. We will have more to say on this in IV- 5.

We are in a position, now, to study the control problem for (4. 1) at least in a small neighborhood of $\mathring{y}_0, w_0$. To avoid inessential complication we will confine our treatment to the "reachability problem" of steering a solution of (4. 1) from the equilibrium point y_0 to a nearby point y_1 during the time interval $[t_0, t_1]$. In Exercise 16 below we suggest that the reader extend our work to treat the problem of steering from a point y_1 near y_0 to a second point y_2, also near y_0, during the interval $[t_0, t_1]$. In each case a control $w(t)$ with $w(t)$ near w_0 for $t_0 \leq t \leq t_1$ is constructed with an application of the implicit function theorem ([6]). Our result presented here, and more general results, appear in [7] .

In Theorem 2. 4 we constructed a control, steering (4. 6) from 0 to a point $x_1 \in E^n$ during $[t_0, t_1]$, in the form

$$u(t) = B^* \Phi(t_1, t)^* \xi, \quad t \in [t_0, t_1], \quad \xi \in E^n, \tag{4.19}$$

with ξ in the form (cf. (2. 10))

$$\xi = Z(t_0, t_1)^{-1} x_1 . \tag{4.20}$$

This establishes a nonsingular linear transformation

$$x_1 = Z(t_0, t_1) \xi , \tag{4.21}$$

with inverse (4. 20), between control parameters ξ and terminal states x_1 in the case of the linear system (4. 6).

We write (4. 1) in the form (4. 3) and use the variation of parameters formula, as before, to obtain (4. 9). Expressing $x(t)$ in the form (4. 8), with $x_0 = 0$ since $y(t_0) = y_0$, and again letting $\eta(t) = y(t) - y_0$, $u(t) = w(t) - w_0$, we have, for $t = t_1$,

$$\eta(t_1) = \int_{t_0}^{t_1} \Phi(t_1, t)B(t)u(t)dt + \int_{t_0}^{t_1} \Phi(t_1, t)h(\eta(t) + y_0, u(t) + w_0, t)\,dt \,. \tag{4.22}$$

Let us substitute (4. 19), with ξ undetermined as yet, into (4. 22) to yield

$$\eta(t_1, \xi) = \int_{t_0}^{t_1} \Phi(t_1, t)B(t)B(t)^* \Phi(t_1, t)^* dt\, \xi$$
$$+ \int_{t_0}^{t_1} \Phi(t_1, t)h(\eta(t, \xi) + y_0, B(t)^* \Phi(t_1, t)^* \xi + w_0, t)dt \tag{4.23}$$

where

$$y(t, \xi) = \eta(t, \xi) + y_0 \tag{4.24}$$

is the solution of (4. 1) with $u(t) = w(t) - w_0$ and $y(0, \xi) = y_0$. Using the various regularity theorems available for differential equations (see, e. g. I- [2], [3]) one can establish that the function $F : E^n \to E^n$ given by

$$F(\xi) = \eta(t_1, \xi) \tag{4.25}$$

s continuously differentiable, at least for ξ near 0. We compute, using the definition, (2.9), of $Z(t_0, t_1)$

$$\frac{\partial F}{\partial \xi} = \frac{\partial \eta(t_1, \xi)}{\partial \xi} = Z(t_0, t_1)$$
$$+ \int_{t_0}^{t_1} \Phi(t_1, t)[\frac{\partial h}{\partial y}(\eta(t, \xi) + y_0, B(t)^* \Phi(t_1, t)^* \xi + w_0, t)\frac{\partial \eta(t, \xi)}{d\xi}$$
$$+ \frac{\partial h}{\partial w}(\eta(t, \xi) + y_0, B(t)^* \Phi(t_1, t)^* \xi + w_0, t)B(t)^* \Phi(t_1, t)^*]\, dt \,.$$

Now, as $\xi \to 0$,

$$x(t, \xi) = \int_{t_0}^{t} \Phi(t, s)B(s)B(s)^* \Phi(t_1, s)^* ds\, \xi$$

tends to zero uniformly for $t \in [t_0, t_1]$ and Theorem 4. 1 shows that the same must be true for $\eta(t, \xi)$. Theorem 4. 1 also shows that

$$\left.\frac{\partial\eta(t,\xi)}{\partial\xi}\right|_{\xi=0} = \left.\frac{\partial x(t,\xi)}{\partial\xi}\right|_{\xi=0} = \int_{t_0}^{t} \Phi(t,s)B(s)B(s)^*\Phi(t_1,s)^* ds$$

and only a little more work, not carried out here for the sake of brevity, shows $\frac{\partial\eta(t,\xi)}{\partial\xi}$ to be continuous near $\xi = 0$. It follows that

$$\frac{\partial F}{\partial\xi}(0) = Z(t_0,t_1) + \int_{t_0}^{t_1} \Big[\frac{\partial h}{\partial y}(y_0,w_0,t)\int_{t_0}^{t} \Phi(t,s)B(s)B(s)^*\Phi(t,s)^* ds$$
$$+ \frac{\partial h}{\partial w}(y_0,w_0,t)B(t)^*\Phi(t_1,t)^*\Big] dt = Z(t_0,t_1),$$

the latter because (4.4), (4.5) imply

$$\frac{\partial h}{\partial y}(y_0,w_0,t) \equiv 0, \quad \frac{\partial h}{\partial w}(y_0,w_0,t) \equiv 0 .$$

Since $\frac{\partial F}{\partial\xi}(0) = Z(t_0,t_1)$, controllability of (4.6) during $[t_0,t_1]$ implies $Z(t_0,t_1)$ is nonsingular and the implicit function theorem ([6]) shows that $\eta = F(\xi)$ maps a neighborhood of $\xi = 0$ in E^n onto a neighborhood of $\eta = 0$ in E^n. In particular, if $y_1 - y_0$ is sufficiently small, there is a unique ξ_1 near 0 such that $F(\xi_1) = y_1 - y_0$. Thus the control

$$u_1(t) = B(t)^*\Phi(t_1,t)^*\xi_1$$

yields

$$y(t_1) = \eta(t_1,\xi_1) + y_0 = y_1$$

so that $u_1(t)$ steers (4.1) from y_0 to y_1 during $[t_0,t_1]$ and we have proved

Theorem 4.2. If the linearized system (4.6) is controllable during $[t_0,t_1]$ then the nonlinear system (4.1) is locally controllable near y_0, w_0 during $[t_0,t_1]$ in the sense that, given y_1 sufficiently close to y_0, there is a control $w(t) = u(t) + w_0$, with $\|u(t)\|_{E^m}$ small on $[t_0,t_1]$, steering a solution $y(t)$ of (4.1) from y_0 to y_1 during $[t_0,t_1]$.

Bibliographical Notes, Chapter II

[1] F. R. Gantmacher: "The Theory of Matrices", Vols. I, II, Chelsea Pub. Co., New York, 1959.

The two volume work by Gantmacher can be fairly described as a classic treatment of linear algebra. The Hamilton-Cayley theorem appears on p. 83 of Vol. I. The first volume includes all material to which we will make reference in this book. An extremely accessible account of most aspects of linear algebra needed here is provided in

[2] J. N. Franklin: "Matrix Theory", Prentice Hall, Inc., Englewood Cliffs, N. J., 1968.

[3] E. B. Lee and L. Markus: "Foundations of Optimal Control Theory", John Wiley and Sons, Inc., New York, 1967.

[4] G. Bachman and L. Narici: "Functional Analysis", Academic Press, New York, 1966.

Chapter 20 of Bachman and Narici contains a very fine treatment of adjoint operators and duality notions in general. A treatment restricted to the case of bounded operators is given in Chapter V of reference I-[4].

The notion of duality between an abstract linear control system and an associated abstract linear observed system is further developed in

[5] S. Dolecki and D. L. Russell: "A general theory of observation and control", SIAM J. on Control and Optimization, Vol. 15 (1977), pp. 185-220.

[6] L. M. Graves: "The Theory of Functions of Real Variables", McGraw-Hill Book Co., New York, Toronto, London, 1956.

[7] D. L. Lukes: "Global controllability of nonlinear systems", SIAM J. Control 10 (1972), pp. 112-116.

Exercises, Chapter II

1. Let $\dot{x} = Ax$, $\omega = Hx$, denote the heat conduction system (1.17) with $n = 3$, $H = (0,0,1)$, $k_2 = k_3 = \rho_1 = \rho_2 = \rho_3 = 1$. Compute the reconstruction kernel $R_0(t)$ as in (1.19) for the interval $[t_0, t_1] = [0, 1]$ and verify (1.18) directly.

2. Let τ be an arbitrary point in the interval $[t_0, t_1]$. How should the definition (1.18) of a reconstruction kernel be modified so as to yield a "τ-reconstruction kernel", $R(t, \tau)$ such that

$$x(\tau) = \int_{t_0}^{t_1} R(t, \tau)\omega(t)\,dt \,?$$

What is the corresponding analog of the formula (1.19), i.e., what is $R_0(t, \tau)$?

3. How should the formula for $Q(t)$ in Proposition 1.10 be modified if we wish to reconstruct $x(\tau)$ from $\omega = Hx$ when $\dot{x} = A(t)x + f(t)$? Let the resulting matrix be $Q(t, \tau)$. Verify directly that if

$$x(\tau) = \int_{t_0}^{t_1} [\,R(t, \tau)\omega(t) + Q(t, \tau)f(t)]\,dt$$

then

$$\frac{dx(\tau)}{d\tau} = A(\tau)x(\tau) + f(\tau)$$

when f is a continuous n-vector function on $[t_0, t_1]$.

4. Consider the constant coefficient linear control system with scalar control

$$\dot{x} = Ax + bu, \quad x, b \in E^n, \quad u \in E^1.$$

Show that a necessary condition for controllability is that the minimal polynomial of A (the monic polynomial q of least degree for which $q(A) = 0$) coincides with the characteristic polynomial.

5. Show that the system of Exercise 4 is controllable if and only if there is a nonsingular $n \times n$ matrix P such that

$$P^{-1}AP = \Lambda, \qquad P^{-1}b = \hat{b} ,$$

where Λ and $\hat{b}$ have the block structure

$$\Lambda = \begin{pmatrix} \Lambda_1 & 0 & \cdots & 0 \\ 0 & \Lambda_2 & \cdots & 0 \\ \vdots & \vdots & \ddots & \vdots \\ 0 & 0 & \cdots & \Lambda_r \end{pmatrix}, \quad \hat{b} = \begin{pmatrix} \hat{b}_1 \\ \hat{b}_2 \\ \vdots \\ \hat{b}_r \end{pmatrix},$$

for some positive integer r, the Λ_k have the form

$$\Lambda_k = \begin{pmatrix} \lambda_k & 1 & 0 & \cdots & 0 & 0 \\ 0 & \lambda_k & 1 & \cdots & 0 & 0 \\ \vdots & \vdots & \vdots & & \vdots & \vdots \\ 0 & 0 & 0 & \cdots & \lambda_k & 1 \\ 0 & 0 & 0 & \cdots & 0 & \lambda_k \end{pmatrix},$$

(all entries of the first superdiagonal row are 1's)

for distinct λ_k, $k = 1, 2, \ldots, r$, and the $\hat{b}_k$ have the form

$$\hat{b}_k = \begin{pmatrix} \beta_k^1 \\ \beta_k^2 \\ \vdots \\ \beta_k^{\mu_k} \end{pmatrix} \qquad (\mu_1 + \mu_2 + \ldots + \mu_r = 1)$$

with $\beta_k^{\mu_k} \neq 0$, $k = 1, 2, \ldots, r$.

6. Let $\dot{x} = A(t)x + B(t)u$ be a control system, $x \in E^n$, $u \in E^m$, $t \in [t_0, t_1]$, and let $y = \Gamma x$ for some $r \times n$ matrix Γ of rank r. We define the system to be Γ-controllable on $[t_0, t_1]$ if for every $y \in E^r$ there is a control $u \in L_m^2[t_0, t_1]$ steering the system from $x(t_0) = x_0$ to $x(t_1) = x_1$, x_1 being such that $\Gamma x_1 = y$. The system is controllable,

of course, if it is Γ-controllable with $\Gamma = I$ on E^n. Develop a criterion for Γ-controllability which extends Theorem 2.4 and a similar extension of Proposition 2.6 for the case where A and B are constant. Are there any $r \times n$ matrices Γ with $r < n$ such that Γ-controllability implies controllability (for given, constant A, B)?

7. Obtain the linearized control system corresponding to the "level hold" crane (cf. (1.9), Chapter I) in the case $L_1 = \sqrt{2}$, $\ell_2 = 1$ (i.e. $y_0^3 = 0$) $\theta_0 = \dot{\theta}_0 = 0$, $\psi_0 = \frac{\pi}{4}$, $\dot{\psi}_0 = 0$, $\ddot{\psi} = u^2(t)$ the (scalar control). For which $r \times 4$ matrices Γ-controllable (on an arbitrary interval $[0, T]$) in the sense of Exercise 6 above?

8. Let $s_i(t)$, $i = 1, 2, \ldots, n$, be functions in $L_n^2[t_0, t_1]$. Under what conditions does there exist a steering function for $\dot{x} = A(t)x + B(t)u$ on $[t_0, t_1]$ of the form

$$S(t) = (s_1(t),\ s_2(t), \ldots, s_n(t))S,$$

for some nonsingular $n \times n$ matrix S?

9. Set the "τ-observation" problem of Exercise 2 in the general framework of Definition 3.1, i.e. define X, Y, Z, O, F. Find the dual "τ-control" problem, compute an appropriate τ-steering function $S(t, \tau)$ and relate $S(t, \tau)$ to $R(t, \tau)$ in Exercise 2.

10. Set the "Γ-controllability" problem of Exercise 6 in the general framework of Definition 3.2 and find the corresponding dual observation problem.

11. Define and interpret the dual linear observed system for the control system (3.22), (3.23). When controllable, does (3.22), (3.23) have a bounded steering operator? What about a reconstruction operator for the dual linear observed system?

12. Consider the ecological system of Chapter I, Exercise 2. Using the linearized system derived in part (ii) of that problem, determine conditions on the parameters $\alpha, \gamma, \beta, \delta, \epsilon, \theta, \psi, \phi$ such that measurement of the rabbit population, $r(t)$, on an interval $0 \le t \le T$ suffices to

determine the complete state of the system.

13. Again referring to the system of Chapter I, Exercise 2, the θ parameter in the third equation can be thought of as representing the "carrying capacity" of the land, a combination of available area, soil fertility, etc. Presumably this can be varied by use of fertilizer or by fencing off various areas or growing crops not usable by rabbits. Taking $\theta = \theta(t)$, repeat part (ii) of the indicated exercise, obtaining a linearized control system with control $u(t) = \theta(t) - \theta_0$ and analyze the controllability of the system.

14. Suppose one desires to steer a solution of system (4.1) from y_0 to a point $y_1 = y_0 + \eta_1$, near y_0, during $[t_0, t_1]$. This can be done with a control of the form (4.19) if we can solve (cf. (4.23), (4.24))

$$\eta_1 = Z(t_0,t_1)\xi + \int_{t_0}^{t_1} \Phi(t_1, t)h(\eta(t, \xi) + y_0, \; B(t)^* \Phi(t_1, t)^* \xi + w_0, t)dt$$

for the unknown vector $\xi \in E^n$. Show that for η_1 sufficiently small the unique solution of this equation near the origin in E^n can be approximated as closely as desired by use of the iterative technique based on the equations

$$Z(t_0, t_1)\xi_{k-1} = \eta_1 - \int_{t_0}^{t_1} \Phi(t_1, t)h(\eta(t, \xi_k) + y_0, \; B(t)^* \Phi(t_1, t)^* \xi_k + w_0, t)dt,$$

$$k = 0, 1, 2, \ldots, \quad \xi_0 = 0.$$

15. Use the method of Exercise 14, using a suitable numerical integration technique (e.g., the trapezoidal rule on Simpson's rule) to approximate the integrals on the right hand side and the solution $y(t, \xi_k) = \eta(t, \xi_k) + y_0$ of (4.1), to obtain an approximation to a control function for the ecological system described in Exercise 13 (and earlier exercises). Take the control to be $\theta(t) =$ (cf. I-Exercise 2, part (iii)) $= 2 + u(t)$, take $y_0 = (w_0, r_0, v_0)$ as determined in I-Exercise 2, part (iii), and take

$$y_1 = (w_0 + .1, \; r_0 + .1, \; v_0 + .1).$$

16. Generalize Theorem 4.2 to the case where both the initial point, y_1, and the terminal point, y_2, are vectors near, but not necessarily coinciding with, y_0. Hint: Let $y_1 = y_0 + \eta_1$, $y_2 = y_0 + \eta_2$. The initial state for (4.24) becomes $y(0, \xi) = y_0 + \eta_1$ and (4.23) is replaced by an equation of the form (cf. (4.25))

$$\eta_2 = \hat{F}(\xi, \eta_1).$$

Linearize this equation about the point $\eta_2 = \eta_1 = \xi = 0$, analyze the solvability of the linearized equation and then use the implicit function theorem.

CHAPTER III

ALGEBRAIC THEORY OF LINEAR CONSTANT COEFFICIENT CONTROL SYSTEMS

1. Asymptotic Stability and Liapounov's Matrix Equation

Although we now have the rather satisfying result that the control problem

$$\dot{x} = A(t)x + B(t)u$$

$$x(t_0) = x_0 ,$$

$$x(t_1) = x_1 ,$$

has a solution in $L_m^2[t_0, t_1]$ if an appropriate controllability condition is met, it turns out that our success has more didactic than practical interest. True, we have a fairly simple formula for a steering function and have seen that very simple ones indeed can be developed (such as in the case of the heat conduction system of II-2, for example. But such controls are not commonly used in practice. Among the reasons why this is the case we may list the following:

(i) Complexity of the hardware required to determine precisely the initial state $x(t_0)$, keep track of the current time, t, and precisely implement the control $u(t) = S(t)x_1$;

(ii) A feeling (which can be strongly supported by mathematical and engineering arguments) that such methods are inflexible and rarely meet the elementary requirements of smoothness and reliability of control action;

(iii) Difficulties in "translating" the theory into the "frequency domain" language classically used by control practitioners who have developed criteria for "desirable" control performance in terms of that language.

Even as we write these lines some of these factors are undergoing fundamental change. Certainly the on-going revolution in the control engineering curriculum will affect (iii) and the rapid development and proliferation of special purpose "mini-computers" must eventually meet some of the objections raised in (i). Whether (ii) is fundamental or just a corollary of (i) and (iii) is hard to say.

In this chapter we will be concerned exclusively with a linear constant coefficient control system

$$\dot{x} = Ax + Bu, \quad x \in E^n, \ u \in E^m. \tag{1.1}$$

Many of the concepts which we will discuss, however, are readily extendable to more complicated systems as we will see in a later chapter.

Definition 1.1. The linear homogeneous system

$$\dot{x} = Ax, \quad x \in E^n \tag{1.2}$$

is

(i) stable, if every solution $x(t)$ of (1.2) remains bounded as $t \to \infty$;

(ii) unstable if (i) is not true;

(iii) asymptotically stable if every solution of (1.2) satisfies

$$\lim_{t\to\infty} \|x(t)\| = 0;$$

(iv) completely unstable if every solution of (1.2) satisfies

$$\lim_{t\to\infty} \|x(t)\| = \infty. \tag{1.3}$$

The following theorem relates (i) - (iv) of Definition 1.1 to basic invariants of the matrix A appearing in (1.2).

Theorem 1. 2. Let the matrix A *of* (1. 2) *be reduced via similarity to the block diagonal form*

$$\Lambda = \begin{pmatrix} \lambda_1 I_1 + N_1 & 0 & \cdots & 0 \\ 0 & \lambda_2 I_2 + N_2 & \cdots & 0 \\ \vdots & \vdots & \ddots & \vdots \\ 0 & 0 & \cdots & \lambda_\nu I_\nu + N_\nu \end{pmatrix}, \tag{1. 4}$$

wherein $\lambda_1, \lambda_2, \ldots, \lambda_\nu$ *are the distinct eigenvalues of* A *with multiplicities*

$$\mu_1 + \mu_2 + \ldots + \mu_\nu = n,$$

I_k *is the* $\mu_k \times \mu_k$ *identity matrix and* N_k *is a* $\mu_k \times \mu_k$ *matrix whose only non-zero entries lie strictly above the main diagonal. Then*

(i) *the system* (1. 2) *is stable if and only if for each* $k = 1, 2, \ldots, \nu$, *either*

(a) $\mathrm{Re}(\lambda_k) < 0$

or

(b) $\mathrm{Re}(\lambda_k) = 0$ *and* $N_k = 0$ *(which is true automatically, of course, if* $\mu_k = 1$);

(ii) *the system* (1. 2) *is asymptotically stable if and only if* $\mathrm{Re}(\lambda_k) < 0$ *for* $k = 1, 2, \ldots, \nu$;

(iii) *the system* (1. 3) *is completely unstable if and only if for each* $k = 1, 2, \ldots, \nu$ *we have*

$$\mathrm{Re}(\lambda_k) > 0 .$$

We will presently give a constructive procedure for finding (1. 4), i. e. for finding a non-singular $n \times n$ matrix P such that

$$P^{-1} A P = \Lambda . \tag{1. 5}$$

For the moment we assume this can be done and proceed with the

Proof of Theorem 1. 2. We recall that the fundamental matrix solution of (1. 2), i. e.

$$\Phi(t,0) \equiv e^{At} ,$$

has the power series representation

$$e^{At} = \sum_{k=0}^{\infty} \frac{t^k}{k!} A^k ,$$

convergent for all $t \in (-\infty, \infty)$ (and, in fact, for all complex t). From (1. 5) we have

$$P^{-1}e^{At}P = \sum_{k=0}^{\infty} \frac{t^k}{k!} P^{-1}A^kP = \sum_{k=0}^{\infty} \frac{t^k}{k!}(P^{-1}AP)^k = \sum_{k=0}^{\infty} \frac{t^k}{k!} \Lambda^k = e^{\Lambda t} . \tag{1. 6}$$

Now, from (1. 4),

$$e^{\Lambda t} = \sum_{k=0}^{\infty} \frac{t^k}{k!}\Lambda^k = \sum_{k=0}^{\infty} \frac{t^k}{k!} \begin{pmatrix} (\lambda_1 I_1+N_1)^k & 0 & \cdots & 0 \\ 0 & (\lambda_2 I_2+N_2)^k & \cdots & 0 \\ \vdots & \vdots & & \vdots \\ 0 & 0 & \cdots & (\lambda_\nu I_\nu+N_\nu)^k \end{pmatrix}$$

$$= \begin{pmatrix} e^{(\lambda_1 I_1+N_1)t} & 0 & \cdots & 0 \\ 0 & e^{(\lambda_2 I_2+N_2)t} & \cdots & 0 \\ \vdots & \vdots & \ddots & \vdots \\ 0 & 0 & \cdots & e^{(\lambda_\nu I_\nu+N_\nu)t} \end{pmatrix} . \tag{1. 7}$$

For each $k = 1, 2, \ldots, \nu$ the diagonal matrix $\lambda_k I_k$ commutes with N_k, so that (see I, Ex. 4)

$$e^{(\lambda_k I_k+N_k)t} = e^{(\lambda_k I_k)t} e^{N_k t} = e^{\lambda_k t} e^{N_k t} .$$

The matrix N_k, being a $\mu_k \times \mu_k$ strictly superdiagonal matrix, is nilpotent with index p_k, i. e.,

$$(N_k)^{p_k} = 0 \tag{1.8}$$

for some least integer p_k, $0 < p_k \leq \mu_k$. Thus

$$e^{N_k t} = \sum_{j=0}^{\infty} \frac{t^j}{j!} (N_k)^j = \sum_{j=0}^{p_k-1} \frac{t^j}{j!} (N_k)^j$$

is a matrix polynomial in t and

$$e^{(\lambda_k I_k + N_k)t} = e^{\lambda_k t} \left(\sum_{j=0}^{p_k-1} \frac{t^j}{j!} (N_k)^j \right) . \tag{1.9}$$

We ask the reader to verify the easy result that (1.2) is stable if and only if the matrix e^{At} is bounded for $t \in [0,\infty)$. Then we use (1.6) and (1.7) to see that this is true if and only if each of the matrices (1.9) is likewise bounded for $t \in [0,\infty)$. If $\operatorname{Re}(\lambda_k) < 0$ we have, for $t \geq 0$

$$\left\| e^{\lambda_k t} \sum_{j=0}^{p_k-1} \frac{t^j}{j!} (N_k)^j \right\| \leq M \sum_{j=0}^{p_k-1} t^j e^{\operatorname{Re}(\lambda_k)t} ,$$

provided $M \geq \max_{j=0,\ldots,p_k-1} \{\frac{1}{j!} \|N_k\|^j\}$ and we conclude that if $\operatorname{Re}(\lambda_k) < 0$,

$$\lim_{t \to +\infty} \left\| e^{(\lambda_k I_k + N_k)t} \right\| = 0 . \tag{1.10}$$

From (1.10) it immediately follows that

$$\left\| e^{(\lambda_k I_k + N_k)t} \right\| \leq B, \quad t \in [0,\infty), \quad \operatorname{Re}(\lambda_k) < 0 \tag{1.11}$$

for some positive B.

If $\operatorname{Re}(\lambda_k) = 0$, then

$$\left\| e^{(\lambda_k I_k + N_k)t} \right\| = \left\| e^{i(\operatorname{Im}(\lambda_k))t} \sum_{j=0}^{p_k-1} \frac{t^j}{j!} (N_k)^j \right\| = \left\| \sum_{j=0}^{p_k-1} \frac{t^j}{j!} (N_k)^j \right\| . \tag{1.12}$$

Using this together with (1.15) (to be established below) we see that

$$\{\|e^{(\lambda_k I_k + N_k)t}\| \leq B,\ t \in [0,\infty),\ \mathrm{Re}(\lambda_k) = 0\} \Leftrightarrow \{N_k = 0\}. \tag{1.13}$$

If $\mathrm{Re}(\lambda_k) > 0$ then

$$\|e^{(\lambda_k I_k + N_k)t}\| = e^{\mathrm{Re}(\lambda_k)t} \left\| \sum_{j=0}^{p_k-1} \frac{t^j}{j!} (N_k)^j \right\|$$

tends to ∞ as $t \to \infty$ if we can establish that $\|\Sigma_{j=0}^{p_k-1} \frac{t^j}{j!} (N_k)^j\|$ is bounded away from zero for large t. Since p_k is the least integer for which (1.8) is true,

$$(N_k)^{p_k-1} \neq 0$$

(note that $p_k - 1 = 0$ if $N_k = 0$) and

$$\lim_{t\to\infty} \left(\frac{1}{t^{p_k-1}} \left\| \sum_{j=0}^{p_k-1} \frac{t^j}{j!} (N_k)^j \right\|\right) = \lim_{t\to\infty} \left(\left\| \sum_{j=0}^{p_k-1} \frac{t^{j-p_k+1}}{j!} (N_k)^j \right\|\right) = \frac{1}{(p_k-1)!} \|(N_k)^{p_k-1}\| \neq 0 \tag{1.14}$$

and we conclude that there is a positive number b such that, for sufficiently large t,

$$\left\| \sum_{j=0}^{p_k-1} \frac{t^j}{j!} (N_k)^j \right\| \geq b \frac{t^{p_k-1}}{(p_k-1)!} \|(N_k)^{p_k-1}\| . \tag{1.15}$$

Thus

$$\lim_{t\to+\infty} \|e^{(\lambda_k I_k + N_k)t}\| = \infty, \quad \mathrm{Re}(\lambda_k) > 0 . \tag{1.16}$$

Part (i) of our theorem follows immediately from (1.11), (1.13) and (1.16). Part (ii) follows from (1.10) and (1.16) and the additional observation that if $\mathrm{Re}(\lambda_k) = 0$ we have (1.13) and hence (from (1.15)) $\lim_{t\to+\infty} \|e^{(\lambda_k I_k + N_k)t}\| \neq 0$. Part (iii) is a direct consequence of the observation that e^{At} is completely unstable if and only if e^{-At} is asymptotically stable. Q. E. D.

We remark that whenever the system $\dot{x} = Ax$ takes the form

$$\begin{pmatrix} \dot{x}^1 \\ x^2 \\ \vdots \\ x^{n-1} \\ x^n \end{pmatrix} = \begin{pmatrix} 0 & 1 & \cdots & 0 & 0 \\ 0 & 0 & \cdots & 0 & 0 \\ \vdots & \vdots & & \vdots & \vdots \\ 0 & 0 & \cdots & 0 & 1 \\ -a^n & -a^{n-1} & \cdots & -a^2 & -a^1 \end{pmatrix} \begin{pmatrix} x^1 \\ x^2 \\ \vdots \\ x^{n-1} \\ x^n \end{pmatrix},$$

as it does when we begin with the scalar linear differential equation

$$x^{(n)} + a_1 x^{(n-1)} + \dots + a_{n-1}\dot{x} + a_n x = 0$$

and set $x^1 = x$, $x^2 = \dot{x}$, etc., then all of the matrices N_k associated with eigenvalues λ_k of multiplicity $\mu_k > 1$ are non-zero. Indeed, in such cases μ_k is the smallest integer for which $N_k^{\mu_k} = 0$. In such systems, therefore, stability obtains if and only if all eigenvalues have non-negative real part and those with zero real part are simple.

Before going on to further material related to stability and/or asymptotic stability we show, briefly, how the block diagonal form (1.4) of the matrix A may be obtained in practice. Let the (not necessarily distinct) eigenvalues of A be $\lambda_1, \lambda_2, \dots, \lambda_n$. Let φ_1 be an eigenvector corresponding to λ_1. We first assume that the first component, φ_1^1, of φ_1 is not zero. If so we let

$$P_1 = (\varphi_1, e_2, e_3, \dots, e_n),$$

where $e_1, e_2, \dots, e_n$ are the columns of the $n \times n$ identity matrix I. The matrix P_1 is nonsingular and

$$\Lambda_1 = P_1^{-1} A P_1 = \begin{pmatrix} \lambda_1 & a^1 \\ 0 & A_1 \end{pmatrix}$$

where a^1 is an $n-1$ dimensional row vector and A_1 is an $(n-1) \times (n-1)$ matrix with eigenvalues $\lambda_2, \lambda_3, \dots, \lambda_n$. If $\varphi_1^1 = 0$, we select $\varphi_1^j \neq 0$ (which can be done since φ_1 is not the zero vector) and we put

$$P_1 = (e_j e_2 \dots e_{j-1}, e_1, e_{j+1}, \dots e_n)(\varphi_1, e_2, e_3, \dots, e_n)$$

to achieve the same result.

Since A_1 is a diagonal block of Λ_1, the eigenvalues of A_1 are the remaining eigenvalues $\lambda_2, \ldots, \lambda_n$ of A. We then repeat the above procedure with A_1 and the eigenvalue λ_2 instead of A and λ_1. A non-singular matrix $\hat{P}_2$ of dimension $(n-1) \times (n-1)$ is found so that

$$\hat{P}_2^{-1} A_1 \hat{P}_2 = \begin{pmatrix} \lambda_2 & a^2 \\ 0 & A_2 \end{pmatrix}$$

and then with

$$P_2 = \begin{pmatrix} 1 & 0 \\ 0 & \hat{P}_2 \end{pmatrix}$$

we have

$$\Lambda_2 = P_2^{-1} \Lambda_1 P_2 = P_2^{-1} P_1^{-1} A P_1 P_2 = \begin{pmatrix} \lambda_1 & * & * \\ 0 & \lambda_2 & a^2 \\ 0 & 0 & A_2 \end{pmatrix},$$

the elements represented by $**$ being the appropriate components of the row vector $a^1 P_2$.

This process is carried out repeatedly for $n-1$ steps to give

$$\tilde{P}^{-1} A \tilde{P} = P_{n-1}^{-1} P_{n-2}^{-1} \cdots P_2^{-1} P_1^{-1} A P_1 P_2 \cdots P_{n-2} P_{n-1} = \tilde{\Lambda}$$

where $\tilde{\Lambda}$ is triangular. Assuming that the distinct eigenvalues of A are $\lambda_1, \lambda_{1+\mu_1}, \lambda_{1+\mu_1+\mu_2}, \ldots, \lambda_{1+\mu_1+\ldots+\mu_{\nu-1}}$ with

$$\lambda_{1+\mu_1+\ldots+\mu_k+\ell} = \lambda_{1+\mu_1+\ldots+\mu_k}, \quad \begin{cases} \ell = 1, 2, \ldots, \mu_{k+1}-1, \\ k = 1, 2, \ldots, \nu-1, \end{cases}$$

we rename them:

$$\lambda_1 = \lambda_1, \lambda_2 = \lambda_{1+\mu_1}, \lambda_3 = \lambda_{1+\mu_1+\mu_2}, \ldots, \lambda_\nu = \lambda_{1+\mu_1+\mu_2+\ldots+\mu_{\nu-1}}$$

and we have

$$\tilde{\Lambda} = \begin{pmatrix} \lambda_1 I_1 + N_1 & \tilde{A}_2^1 & \cdots & \tilde{A}_\nu^1 \\ 0 & \lambda_2 I_2 + N_2 & \cdots & \tilde{A}_\nu^2 \\ \vdots & \vdots & \ddots & \vdots \\ 0 & 0 & \cdots & \lambda_\nu I_\nu + N_\nu \end{pmatrix}$$

where $N_1, N_2, \ldots, N_\nu$ are as described in Theorem 1.2 and $\tilde{A}_j^i$ is a matrix of dimension $\mu_i \times \mu_j$, $i = 1, 2, \ldots, \nu-1$, $j = i+1, \ldots, \nu$.

We now let

$$Q = \begin{pmatrix} I_1 & Q_2^1 & \cdots & Q_\nu^1 \\ 0 & I_2 & \cdots & Q_\nu^2 \\ \vdots & \vdots & & \vdots \\ 0 & 0 & \cdots & I_\nu \end{pmatrix}$$

and require that, with Λ as in (1.4),

$$\Lambda = Q^{-1}\tilde{\Lambda}Q ,$$

or equivalently, since Q is nonsingular

$$Q\Lambda - \tilde{\Lambda}Q = 0 . \tag{1.17}$$

Carrying out the indicated multiplications we find that (1.17) is automatically true for blocks on or below the main diagonal. For the blocks above the main diagonal we obtain the equation

$$Q_j^i(\lambda_j I_j + N_j) - (\lambda_i I_i + N_i)Q_j^i - \tilde{A}_j^i - \sum_{\ell=i+1}^{j-1} \tilde{A}_\ell^i Q_j^\ell = 0, \quad i = 1, 2, \ldots, \nu-1, \; j = i+1, \ldots, \nu .$$

(The last sum is empty for $j = i+1$). These equations can be solved in the order

$$\binom{i}{j} = \binom{1}{2}, \binom{2}{3}, \ldots, \binom{\nu-1}{\nu}, \binom{1}{3}, \binom{2}{4}, \ldots, \binom{\nu-2}{\nu}, \ldots, \binom{\nu-1}{\nu}$$

(i.e. one superdiagonal row at a time, moving upwards) provided we have the following lemma, which will also prove useful in later work.

Lemma 1.3. Consider the linear matrix equation

$$\tilde{Q}(\lambda_\sigma I_\sigma + N_\sigma) - (\lambda_\rho I_\rho + N_\rho)\tilde{Q} = R \tag{1.18}$$

wherein $\tilde{Q}$ and R have dimension $\mu_\rho \times \mu_\sigma$, I_σ and I_ρ are the identity matrices of dimension $\mu_\sigma \times \mu_\sigma$, $\mu_\rho \times \mu_\rho$, respectively, N_σ and N_ρ are nilpotent with

$$N_\sigma^{\mu_\sigma} = 0, \quad N_\rho^{\mu_\rho} = 0 \tag{1.19}$$

and λ_σ, λ_ρ are distinct complex numbers. Then (1.18) has the unique solution (for given R)

$$\tilde{Q} = \sum_{k=0}^{\mu_\rho+\mu_\sigma-2} \frac{1}{(\lambda_\sigma-\lambda_\rho)^{k+1}} \left(\sum_{\ell=0}^{k} (-1)^\ell \binom{k}{\ell} N_\rho^{k-\ell} R N_\sigma^\ell\right). \tag{1.20}$$

Proof. We rewirte (1.18) as

$$\tilde{Q} = \frac{1}{\lambda_\sigma-\lambda_\rho} N_\rho \tilde{Q} - \tilde{Q}N_\sigma + \frac{1}{\lambda_\sigma-\lambda_\rho} R \equiv M_\rho\tilde{Q} - \tilde{Q}M_\sigma + S.$$

We repeatedly substitute this equation into itself (cf. "fixed point" iteration [1]):

$$\begin{aligned}\tilde{Q} &= M_\rho(M_\rho\tilde{Q} - \tilde{Q}M_\sigma + S) - (M_\rho\tilde{Q} - \tilde{Q}M_\sigma + S)M_\sigma + S\\ &= M_\rho^2\tilde{Q} - 2M_\rho\tilde{Q}M_\sigma + \tilde{Q}M_\sigma^2 + M_\rho S - SM_\sigma + S\\ &= M_\rho^2(M_\rho\tilde{Q} - \tilde{Q}M_\sigma + S) - 2M_\rho(M_\rho\tilde{Q} - \tilde{Q}M_\sigma + S)M_\sigma\\ &\quad + (M_\rho\tilde{Q} - \tilde{Q}M_\sigma + S)M_\sigma^2 + M_\rho S - SM_\sigma + S\\ &= M_\rho^3\tilde{Q} - 3M_\rho^2\tilde{Q}M_\sigma + 3M_\rho\tilde{Q}M_\sigma^2 - \tilde{Q}M_\sigma^3\\ &\quad + M_\rho^2 S - 2M_\rho SM_\sigma + SM_\sigma^2 + M_\rho S - SM_\sigma + S = \ldots\end{aligned}$$

and so on. Eventually, carrying out this process $\mu_\rho + \mu_\sigma - 1$ times we find that

$$\tilde{Q} = \sum_{k=0}^{\mu_\rho+\mu_\sigma-1} (-1)^k \binom{\mu_\rho+\mu_\sigma-1}{k} M_\rho^{\mu_\rho+\mu_\sigma-1-k}\tilde{Q}M_\sigma^k + \sum_{k=0}^{\mu_\rho+\mu_\sigma-2}\left(\sum_{\ell=0}^{k}(-1)^\ell\binom{k}{\ell}M_\rho^{k-\ell}SM_\sigma^\ell\right). \tag{1.21}$$

Because $\mu_\rho + \mu_\sigma - 1 - k$ and k total to $\mu_\rho + \mu_\sigma - 1$, either

$$\mu_\rho + \mu_\sigma - 1 \geq \mu_\rho \quad \text{or} \quad k \geq \mu_\sigma$$

and since (1.19) shows $M_\sigma^{\mu_\sigma} = 0$, $M_\rho^{\mu_\rho} = 0$, we conclude that the first sum in (1.21) is zero. The remaining equation is the same as (1.20). Q. E. D.

We remark that the most efficient way to actually solve (1.18) is to set $\tilde{Q}_0 = 0$ and successively compute

$$\tilde{Q}_k = \frac{1}{\lambda_\sigma - \lambda_\rho}(N_\rho \tilde{Q}_{k-1} - \tilde{Q}_{k-1} N_\sigma) + \frac{1}{\lambda_\sigma - \lambda_\rho} R, \quad k = 1, 2, \ldots, \mu_\rho + \mu_\sigma - 1.$$

Then $\tilde{Q} = \tilde{Q}_{\mu_\rho + \mu_\sigma - 1}$

Corollary 1.4. *Let A and B be $n \times n$ and $m \times m$ matrices, respectively. Then the matrix equation*

$$AX + XB = C$$

has a unique $n \times m$ matrix solution X for each given $n \times m$ matrix C if and only if A and -B have no common eigenvalues.

Proof. See Exercise 4 at the end of this chapter.

It is possible to check for asymptotic stability of (1.2) by applying the Routh-Hurwitz criterion (II-[1], Vol. II) to the characteristic polynomial, $p(\lambda)$, of A without actually computing the eigenvalues of A (i.e., zeros of $p(\lambda)$). We do not present this material here as our primary interest lies in the method of A. A. Liapounov [2, 3] which is related to the energy conservation (or dissipation) laws which apply to many physical systems. For a linear homogeneous constant coefficient system

$$\dot{x} = Ax, \quad x \in E^n, \tag{1.22}$$

the Liapounov criterion is expressed in terms of the Liapounov matrix equation

$$A^* X + XA + W = 0 \tag{1.23}$$

and its $n \times n$ solution matrices X. Here W and X are symmetric, with the understanding, now and subsequently, that this means hermitian when the entities involved a complex.

<u>Theorem 1.5.</u> <u>Suppose</u> W <u>is an</u> $n \times n$ <u>symmetric matrix with</u>

$$W \geq H^* H, \qquad (H, A) \text{ observable}. \tag{1.24}$$

<u>Then</u>: (i) <u>if the system</u> (1.22) <u>is asymptotically stable the equation</u> (1.23) <u>has a unique solution</u> X <u>and</u> X <u>is symmetric and positive definite</u>; (ii) <u>if</u> (1.23) <u>has a symmetric positive definite solution</u> X <u>then</u> (1.22) <u>is asymptotically stable.</u>

<u>Proof.</u> Suppose first that (1.22) is asymptotically stable, i.e. all eigenvalues of A have negative real parts. Let A be reduced to the block diagonal canonical form (1.4), i.e. let P, non-singular, be such that

$$P^{-1}AP = \Lambda = \begin{pmatrix} \lambda_1 I_1 + N_1 & 0 & \cdots & 0 \\ 0 & \lambda_2 I_2 + N_2 & \cdots & 0 \\ \vdots & \vdots & & \vdots \\ 0 & 0 & \cdots & \lambda_\nu I_\nu + N_\nu \end{pmatrix} \tag{1.25}$$

Let

$$\gamma = - \min_{k=1,2,\ldots,\nu} \{\operatorname{Re}(\lambda_k)\}.$$

Since $e^{N_k t}$ is a matrix polynomial in t there is a positive number M_k such that

$$\|e^{N_k t}\| \leq M_k e^{\gamma t/2}, \quad t \geq 0, \quad k = 1, 2, \ldots, \nu.$$

Then

$$\|e^{At}\| \leq \|P\| \, \|P^{-1}\| \, \|e^{\Lambda t}\| \leq \|P\| \, \|P^{-1}\| \max_{k=1,2,\ldots,\nu} \|e^{(\lambda_k I_k + N_k)t}\|$$

$$\leq \|P\| \, \|P^{-1}\| e^{-\gamma/2t} \max_{k=1,2,\ldots,\nu} \{M_k\} \equiv M e^{-\gamma t/2}, \quad t \geq 0. \tag{1.26}$$

Define the matrix

$$X = \int_0^\infty e^{A^* t} W e^{At} dt.$$

The inequality (1. 26) guarantees that the integral converges. Now

$$A^* X + XA = \int_0^\infty [A^* e^{A^* t} We^{At} + e^{A^* t} We^{At} A] dt$$

$$= \int_0^\infty [A^* e^{A^* t} We^{At} + e^{A^* t} W Ae^{At}] dt$$

$$= \int_0^\infty \frac{d}{dt}[e^{A^* t} We^{At}] dt = \lim_{T \to \infty} (e^{A^* T} We^{AT} - W)$$

$$= \text{(again using (1. 26))} = - W.$$

Thus X is a solution of (1. 23). It is clear that X is symmetric and non-negative. To show that X is positive we note that for any finite $T > 0$ (cf. (1. 24))

$$X = \int_0^\infty e^{A^* t} W e^{At} dt \geq \int_0^T e^{A^* t} H^* He^{At} dt. \tag{1. 27}$$

But the last integral is just the matrix $W(0, T)$ as defined in II - 1 for the linear observed system

$$\dot{x} = Ax$$

$$\omega = Hx .$$

Since this system is assumed observable, $W(0, T)$ is positive and we conclude from (1. 27) that X is positive.

To show that X is the unique solution of (1. 23) we use (1. 25) to obtain

$$P^* A^* (P^*)^{-1} P^* XP + P^* XPP^{-1} AP + P^* WP = 0 ,$$

i. e.,

$$\Lambda^* \tilde{X} + \tilde{X}\Lambda + \tilde{W} = 0 , \quad \tilde{X} = P^* XP, \quad \tilde{W} = P^* W P . \tag{1. 28}$$

Letting $\tilde{X}$ and $\tilde{W}$ be subdivided into blocks $\tilde{X}^\rho_\sigma$, $\tilde{W}^\rho_\sigma$, $\rho, \sigma = 1,2,\dots, \nu$, compatible with the block structure of Λ, we find that (1. 28) gives

$$(\overline{\lambda}_\rho I_\rho + N_\rho^*)\tilde{X}_\sigma^\rho + \tilde{X}_\sigma^\rho(\lambda_\sigma I_\sigma + N_\sigma) + \tilde{W}_\sigma^\rho = 0 .$$

Since (1.2) is asymptotically stable, $\overline{\lambda}_\rho + \lambda_\sigma \neq 0$ and a trivial modification of Lemma 1.3 gives

$$\tilde{X}_\sigma^\rho = -\sum_{k=0}^{\mu_\rho+\mu_\sigma-2} \frac{1}{\overline{\lambda}_\rho + \lambda_\sigma}\left(\sum_{\ell=0}^{k}\binom{k}{\ell}(N_\rho^*)^{k-\ell}\tilde{W}_\sigma^\rho N_\sigma^\ell\right), \quad \rho,\sigma = 1,2,\ldots,\nu ,$$

showing that $\tilde{W}$ uniquely determines $\tilde{X}$ and hence that W uniqely determines X in (1.23).

Now suppose that for some W satisfying (1.24) there is a symmetric positive solution X for (1.23). Let $x_0 \in E^n$ and let $x(t)$ be the solution of (1.22) for $t \geq 0$. We compute

$$\frac{d}{dt}(x(t)^* X x(t)) = x(t)^* A^* X x(t) + x(t)^* X A(t)$$

$$= -x(t)^* W x(t) \leq -x(t)^* H^* H x(t) \leq 0 . \qquad (1.29)$$

Let $T > 0$ be fixed. Then for any positive integer k

$$x((k-1)T)^* X x((k-1)T) - x(kT)^* X x(kT) = \int_{(k-1)T}^{kT} x(t)^* H^* H x(t) dt$$

$$= \int_{(k-1)T}^{kT} x((k-1)T)^* e^{A^*(t-(k-1)T)} H^* H e^{A(t-(k-1)T)} x((k-1)T) dt$$

$$= x((k-1)T)^* \left[\int_0^T e^{A^* t} H^* H e^{At} dt\right] x((k-1)T)$$

$$= x((k-1)T)^* W(0, T) x((k-1)T) . \qquad (1.30)$$

Since (H, A) is observable W(0, T), as defined in II-1, is positive definite and for some positive number θ

$$W(0, T) \geq \theta X. \qquad (1.31)$$

Since (1.29) shows $x(t)^* X x(t)$ to be non-increasing we must have $0 < \theta < 1$ and (1.30), (1.31) yield

$$x(kT)^* X x(kT) \le (1-\theta)\, x((k-1)T)^* X x((k-1)T).$$

This being true for all positive integers k we conclude

$$\lim_{k\to\infty} x(kT)^* X x(kT) = 0. \tag{1.32}$$

Using again the fact that $x(t)^* X x(t)$ is non-increasing together with the positivity of X, (1.32) implies

$$\lim_{t\to\infty} \|x(t)\| = 0.$$

Since this is true for each $x_0 \in E^n$ we conclude that (1.22) is asymptotically stable. Q.E.D.

As an example of application of this theorem we consider the system shown in Figure III.1, consisting of two mass-spring systems coupled by friction. The equations of motion can be taken to be

$$m_1\ddot{x} + \gamma(\dot{x} - \dot{y}) + k_1 x = 0$$

$$m_2\ddot{y} + \gamma(\dot{y} - \dot{x}) + k_2 y = 0$$

or, in first order form with $x^1 = x$, $x^2 = \dot{x}$, $x^3 = y$, $x^4 = \dot{y}$,

$$\begin{pmatrix} \dot{x}^1 \\ x^2 \\ x^3 \\ x^4 \end{pmatrix} = \begin{pmatrix} 0 & 1 & 0 & 0 \\ -\frac{k_1}{m_1} & -\frac{\gamma}{m_1} & 0 & \frac{\gamma}{m_1} \\ 0 & 0 & 0 & 1 \\ 0 & \frac{\gamma}{m_2} & -\frac{k_2}{m_2} & -\frac{\gamma}{m_2} \end{pmatrix} \begin{pmatrix} x^1 \\ x^2 \\ x^3 \\ x^4 \end{pmatrix} \equiv \begin{pmatrix} x^1 \\ x^2 \\ x^3 \\ x^4 \end{pmatrix} \tag{1.33}$$

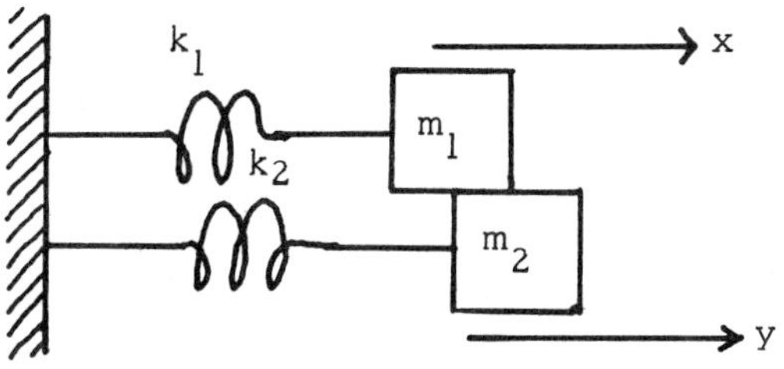

Fig. III-1

We take

$$X = \frac{1}{2}\begin{pmatrix} k_1 & 0 & 0 & 0 \\ 0 & m_1 & 0 & 0 \\ 0 & 0 & k_2 & 0 \\ 0 & 0 & 0 & m_2 \end{pmatrix},$$

inspired by the fact that $x^* X x$ is the system energy, as ordinarily defined. Then

$$A^* X + XA = \begin{pmatrix} 0 & 0 & 0 & 0 \\ 0 & -\gamma & 0 & \gamma \\ 0 & 0 & 0 & 0 \\ 0 & \gamma & 0 & -\gamma \end{pmatrix},$$

from which we see that

$$A^* X + XA + W = 0$$

with

$$W = H^* H, \quad H = (0, \gamma^{\frac{1}{2}}, 0, -\gamma^{\frac{1}{2}}). \tag{1.34}$$

If the pair (H, A) were not observable we should have a solution of (1.33) for which

$$\dot{x}(t) - \dot{y}(t) = x^2(t) - x^4(t) \equiv 0 .$$

Then (since $\frac{\gamma}{m_1}(x^2(t) - x^4(t)) \equiv \frac{\gamma}{m_2}(x^2(t) - x^4(t)) \equiv 0$)

$$0 = \ddot{x}(t) - \ddot{y}(t) = \dot{x}^2(t) - \dot{x}^4(t) = -\frac{k_1}{m_1}x^1(t) + \frac{k_2}{m_2}x^3(t) = \frac{k_1}{m_1}(x^3(t) - x^1(t)) + x^3(t)\left(\frac{k_2}{m_2} - \frac{k_1}{m_1}\right). \tag{1.35}$$

Since $\dot{x}^3(t) - \dot{x}^1(t) = x^2(t) - x^4(t) \equiv 0$, $x^3(t) - x^1(t)$ is constant. If

$$\frac{k_2}{m_2} - \frac{k_1}{m_1} \neq 0 \tag{1.36}$$

(1.35) implies $x^3(t)$ is constant and hence $x^1(t)$ is constant also. Then

$x^2(t) \equiv 0,\ x^4(t) \equiv 0$ and since we then have

$$x^1(t) = -\frac{m_1}{k_1}\dot{x}^2(t), \quad x^3(t) = -\frac{m_2}{k_2}\dot{x}^4(t)$$

we conclude the solution is identically zero. Thus (H, A) is observable if (1.36) is true and we conclude (1.33) is asymptotically stable if (1.36) is true. If on the other hand

$$\frac{k_2}{m_2} - \frac{k_1}{m_1} = 0$$

any nonzero solution $\begin{pmatrix} x^1(t) \\ x^2(t) \end{pmatrix}$ of

$$\begin{pmatrix} \dot{x}^1 \\ x^2 \end{pmatrix} = \begin{pmatrix} 0 & 1 \\ -\frac{k_1}{m_1} & 0 \end{pmatrix} \begin{pmatrix} x^1 \\ x^2 \end{pmatrix}$$

is also a nonzero solution of

$$\begin{pmatrix} \dot{x}^3 \\ x^4 \end{pmatrix} = \begin{pmatrix} 0 & 1 \\ -\frac{k_2}{m_2} & 0 \end{pmatrix} \begin{pmatrix} x^3 \\ x^4 \end{pmatrix}$$

and putting $x^3(t) \equiv x^1(t),\ x^4(t) \equiv x^2(t)$ we obtain a non-zero solution of (1.33) lying for all t in the null space of H as given by (1.34). Thus in this case (H, A) is not observable and it is easily verified that the solution thus constructed does not tend to 0 as $t \to \infty$.

It may fairly be asked at this point why so much emphasis is placed on asymptotically stable systems. There are many reasons; perhaps the following is as important as any. Ordinarily a linear homogeneous system (1.22) represents the linearization of some operating system about an equilibrium point, represented by $x = 0$ in (1.22, which is, of course, an equilibrium point for (1.22) itself. Maintaining $x = 0$ corresponds to keeping the original system at equilibrium, which often amounts to keeping it in a "desirable" operating configuration. But almost all systems are subject to unwanted "outside disturbances", represented in the linearized system by an additive inhomogeneous term:

$$\dot{x} = Ax + v(t). \tag{1.37}$$

The effect of the disturbance v will be to displace the system from the equilibrium at $x = 0$. A small displacement in itself is often not too serious but can become a matter of concern if, for any reason, it has a tendency to grow. The definition of asymptotic stability (cf. Def. 1.1-3) shows that if the disturbance ceases at some time t_1, so that $x(t)$ satisfies $\dot{x} = Ax$ for $t \geq t_1$, then $\lim_{t \to \infty} x(t) = 0$, so that the system will return to equilibrium. But even more important is the so-called BIBO (bounded input-bounded output) property possessed by (1.37) when $\dot{x} = Ax$ is asymptotically stable. This property states that if the disturbance v is bounded, i.e., there exists some $v_0 \geq 0$ such that

$$\|v(t)\| \leq v_0, \qquad t \geq 0, \tag{1.38}$$

then there is a constant $M > 0$, depending only on A, such that

$$\|x(t)\| \leq Mv_0, \qquad t \geq 0 \tag{1.39}$$

if $x(0) = 0$. (If $x(0) = x_0$, (1.38) is replaced by $\|x(t)\| \leq M_0 \|x_0\| + M_1 v_0$). We make this precise in the following theorem, wherein we introduce for the first time the term "stability matrix" to refer to a matrix A for which $\dot{x} = Ax$ is asymptotically stable, i.e. A has only eigenvalues with negative real parts. (The term "Hurwitzian matrix" is also used.)

Theorem 1.6. *We consider the linear inhomogeneous system* (1.37). *There is an* $M > 0$, *depending only on* A, *such that* (1.39) *holds whenever* $x(0) = 0$ *and* v *obeys* (1.38), *if and only if* A *is a stability matrix.*

Proof. For the sufficiency of the condition we let X be the unique symmetric positive definite solution of

$$A^* X + XA + I = 0. \tag{1.40}$$

We let x satisfy (1.37) with v satisfying (1.38) and compute (cf. (1.29))

$$\begin{aligned}\frac{d}{dt}x(t)^* X\, x(t) &= (Ax(t) + v(t))^* X\, x(t) + x(t)^* X(Ax(t) + v(t)) \\ &= x(t)^*(A^*X + XA)x(t) + v(t)^* X x(t) + x(t)^* X\, v(t) \\ &= (\text{using } (1.40)) - \|x(t)\|^2 + v(t)^* X x(t) + x(t)^* X\, v(t) \\ &\leq - \|x(t)\|^2 + 2\,\|X\|\,\|v(t)\|\,\|x(t)\| \\ &\leq - \|x(t)\|^2 + 2\,\|X\|\, v_0\, \|x(t)\|\end{aligned}$$

and we conclude that $\frac{d}{dt}x(t)^* X\, x(t) < 0$ if

$$\|x(t)\| > 2\,\|X\|\, v_0 .$$

Since X is positive definite by Theorem 1.5, we can find $\xi > 0$ such that

$$x(t)^* X\, x(t) > \xi \Rightarrow \|x(t)\| > 2\,\|X\|\, v_0 ,$$

indeed we need only take $\xi = 4\lambda_n \|X\|^2 v_0^2$ where λ_n is the largest eigenvalue of X. It follows that $x(t)^* X x(t)$ is decreasing whenever it is greater than ξ. Hence if $x(0) = 0$, $x(t)^* X\, x(t)$ can never exceed ξ. Thus

$$x(t)^* X\, x(t) \leq \xi, \qquad t \geq 0,$$

which gives

$$\lambda_1 \|x(t)\|^2 \leq x(t)^* X\, x(t) \leq 4\lambda_n \|X\|^2 v_0^2 ,$$

or

$$\|x(t)\| \leq 2\left(\frac{\lambda_n}{\lambda_1}\right)^{\frac{1}{2}} \|X\|\, v_0 \equiv M v_0 ,$$

where λ_1 is the smallest eigenvalue of A. Since X is determined by A and λ_1, λ_n are eigenvalues of A, M depends only on A.

For the necessity, suppose A has an eigenvalue λ with $\text{Re}(\lambda) \geq 0$. Let φ be an associated eigenvector of unit norm. If $\text{Re}(\lambda) > 0$, let

$$v(t) = v_0 \varphi .$$

The solution of (1.37) with $x(0) = 0$ is

$$x(t) = \frac{v_0}{\lambda} e^{\lambda t} \varphi$$

and $\|x(t)\| \to \infty$ as $t \to \infty$. If $\mathrm{Re}\,(\lambda) = 0$ we have $\lambda = i\nu$ as an eigenvalue. Take

$$v(t) = v_0 e^{i\nu t} \varphi$$

and we find that

$$x(t) = \int_0^t e^{i\nu (t-s)} e^{i\nu s} ds\, v_0 \varphi = v_0 t e^{i\nu t} \varphi$$

and again $\|x(t)\| \to \infty$ as $t \to \infty$. Q. E. D.

2. Stabilization via Linear Feedback Control

Let us consider a linear control system

$$\dot{x} = Ax + Bu, \quad x \in E^n, \quad u \in E^m. \tag{2.1}$$

If this represents the linearization of some operating plant about a "desired" equilibrium state, that state is here represented by $x = 0$, $u = 0$. Now it may well be that the uncontrolled $(u = 0)$ homogeneous system

$$\dot{x} = Ax$$

fails to be asymptotically stable. One of the tasks of the control analyst is to use the control u in such a way as to remedy this situation. Because of ease of both implementation and analysis, the traditionally favored means for accomplishing this objective is the use of a linear feedback relation

$$u = Kx, \tag{2.2}$$

wherein the control $u(t)$ is determined as a linear function of the current system state $x(t)$. The problem now becomes that of choosing the $m \times n$ "feedback matrix", K, in such a way that the modified homogeneous system realized by substituting (2.2) into (2.1), i. e.

$$\dot{x} = (A + BK)x \tag{2.3}$$

is such that $A + BK$ has only eigenvalues with negative real parts.

We pause to introduce some terminology standard in control literature. The system (2.1) is called an "open loop" system while the modified system (2.3) is called a "closed loop" system. The terminology arises from the associated block diagrams which are widely used in this area (though not in this book). If we assume $x(0) = 0$, the solution $x(t)$ of (2.1) is determined by $u(t)$. We can represent this schematically by

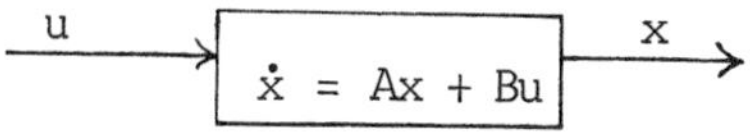

This loop is "open". A control $u(t)$ is determined externally and, via (2.1), it gives rise to $x(t)$. If, on the other hand, $u(t)$ is generated from $x(t)$ by means of (2.2), the above diagram is modified to

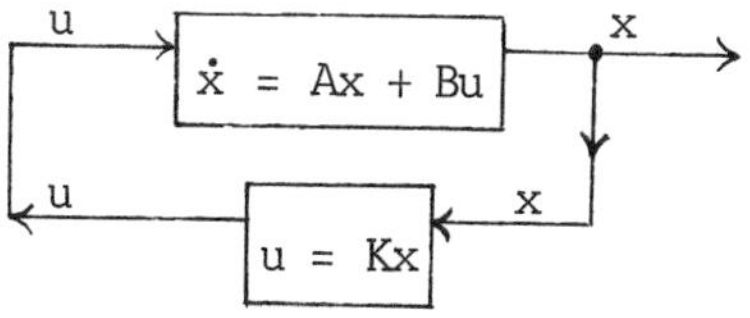

,

i.e., it is "closed" in that the control is now internally generated within the system rather than determined externally.

One of the most basic results in the control theory of constant coefficient linear systems is that "controllability implies stabilizability", the latter being the property of (2.1) which admits the possibility of selecting K so that $A + BK$ is a stability matrix (see Definition 2.3 below). We present here a proof of this result due (independently, it appears) to Lukes [5] and Kleinman [4].

Theorem 2.1. Let the system (2.1) be controllable, so that the adjoint system

$$\dot{x} = -A^*x \tag{2.4}$$

$$\omega = B^*x \tag{2.5}$$

is observable and the corresponding observability matrix

$$W_T \equiv W(0, T) = \int_0^T e^{-At}BB^*e^{-A^*t}\,dt \tag{2.6}$$

is positive for $T > 0$. Then for each $T > 0$ the linear feedback control law

$$u = -B^* W_T^{-1} x \equiv K_T x \tag{2.7}$$

stabilizes (2.1), i.e. $A - BB^* W_T^{-1}$ is a stability matrix, having only eigenvalues with negative real parts.

Proof. We compute

$$AW_T + W_T A^* = \int_0^T [Ae^{-At} BB^* e^{-A^* t} + e^{-At} BB^* e^{-A^* t} A^*]\, dt$$

$$= -\int_0^T \frac{d}{dt} [e^{-At} BB^* e^{-A^* t}]\, dt = -e^{-AT} BB^* e^{-A^* T} + BB^* . \tag{2.8}$$

Since W_T is positive and symmetric for $T > 0$ we can modify (2.8) as follows:

$$(A - BB^* W_T^{-1}) W_T + W_T (A - BB^* W_T^{-1})^* + BB^* + e^{-AT} BB^* e^{-A^* T} = 0 .$$

Now

$$BB^* + e^{-AT} BB^* e^{-A^* T} \geq BB^*$$

and, since W_T is positive, we see that $(A - BB^* W_T^{-1})^*$, and hence $A - BB^* W_T^{-1}$ itself, is shown by Theorem 1.5 to be a stability matrix, provided we can show that the controllability of the pair (A, B) implies that of $(A - BB^* W_T^{-1}, B)$ (and hence the observability of the pair $(B^*, (A - BB^* W_T^{-1})^*)$). This is important enough in itself to deserve presentation in a separate lemma below. So we regard the proof of Theorem 2.1 to be complete, pending that result. Q. E. D.

In the control of a plant modelled by a system (2.2) the control u is frequently decomposed into two parts:

$$u = \tilde{u} + \hat{u} .$$

The $\tilde{u}$ component is determined automatically via a control law (2.2) (with the use of an appropriate servomechanical system) so that (2.3) becomes

$$\dot{x} = (A + BK)x + B\hat{u} . \tag{2.9}$$

The component $\hat{u}$ is reserved for the use of the plant operator to enable him to cause the system to achieve various objectives. The automatic component $\tilde{u}$, determined via $\tilde{u} = Kx$, will have been engineered in such a way as to make the task of the human (or, perhaps, electronic) operator less demanding than it otherwise might have been. Clearly it is of first importance that the control system (2.9) should remain controllable.

Lemma 2.2. The pair $(A + BK, B)$ is controllable if and only if the pair (A, B) is controllable.

Proof. We merely note that for an integer $q \geq 0$,

$$(B, (A+BK)B, (A+BK)^2B, \ldots, (A+BK)^qB) = (B, AB, A^2B, \ldots, A^qB)\begin{pmatrix} I & KB & KAB+(KB)^2 & \cdots & * \\ 0 & I & KB & \cdots & * \\ 0 & 0 & I & & * \\ \vdots & \vdots & \vdots & \ddots & \vdots \\ 0 & 0 & 0 & \cdots & I \end{pmatrix}$$

and thus the matrix on the left has the same rank as does $(B, AB, A^2B, \ldots, A^qB)$. Appealing to Proposition 2.6 of Chapter II we have the result. Q. E. D.

We provide a very simple example of the application of Theorem 2.1. With suitable units the linearized equations of motion for an inverted

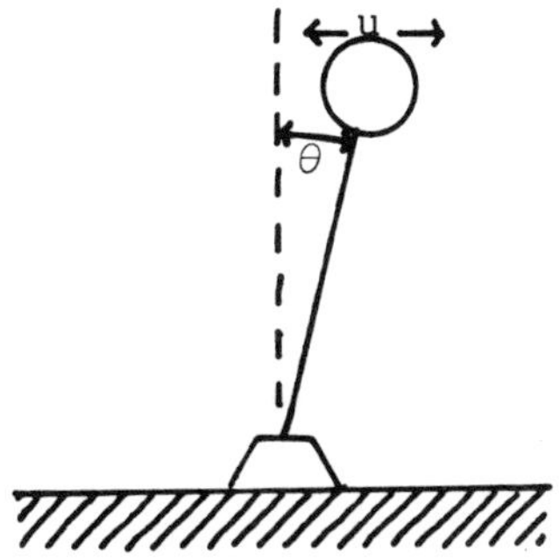

Fig. III-2. Inverted Pendulum

pendulum, as shown in Figure III-2, take the form

$$\ddot{\theta} - \theta = u$$

or, in first order form with $\theta^1 = \theta$, $\theta^2 = \dot{\theta}$,

$$\begin{pmatrix} \dot{\theta}^1 \\ \dot{\theta}^2 \end{pmatrix} = \begin{pmatrix} 0 & 1 \\ 1 & 0 \end{pmatrix} \begin{pmatrix} \theta^1 \\ \theta^2 \end{pmatrix} + \begin{pmatrix} 0 \\ 1 \end{pmatrix} u \,. \tag{2.10}$$

We have

$$A = \begin{pmatrix} 0 & 1 \\ 1 & 0 \end{pmatrix}, \quad B = \begin{pmatrix} 0 \\ 1 \end{pmatrix}, \quad e^{-At} = \frac{1}{2} \begin{pmatrix} e^t + e^{-t}, & -e^t + e^{-t} \\ -e^t + e^{-t}, & e^t + e^{-t} \end{pmatrix},$$

we set $T = 1$ and compute the matrix of (2.6)

$$W_1 = \frac{1}{4} \int_0^1 \begin{pmatrix} -e^t + e^{-t} \\ e^t + e^{-t} \end{pmatrix} (-e^t + e^{-t}, \; e^t + e^{-t}) dt$$

$$= \frac{1}{4} \int_0^1 \begin{pmatrix} e^{2t} + e^{-2t} - 2 & -e^{2t} + e^{2t} \\ -e^{2t} + e^{-2t} & e^{2t} + e^{-2t} + 2 \end{pmatrix} dt$$

$$= \frac{1}{8} \left[\begin{pmatrix} e^{2t} - e^{-2t} - 4t & -e^{2t} - e^{-2t} \\ -e^{2t} - e^{-2t} & e^{2t} - e^{-2t} + 4t \end{pmatrix} \right]_0^1$$

$$= \frac{1}{8} \begin{pmatrix} e^2 - e^{-2} - 4 & -e^2 - e^{-2} + 2 \\ -e^2 - e^2 + 2 & e^2 - e^{-2} + 4 \end{pmatrix} = \begin{pmatrix} .4067 & -.6906 \\ -.6906 & 1.4067 \end{pmatrix}.$$

Then

$$W(0,1)^{-1} = \begin{pmatrix} 14.780 & 7.2559 \\ 7.2559 & 4.2731 \end{pmatrix}.$$

Consequently, a stabilizing feedback matrix is

$$K_1 = -B^* W_1^{-1} = -(0,1) \begin{pmatrix} 14.780 & 7.2559 \\ 7.2559 & 4.2731 \end{pmatrix} = (-7.2559, \; -4.2731)\,.$$

With use of the control law

$$u = -B^* W_1^{-1} \begin{pmatrix} \theta_1 \\ \theta_2 \end{pmatrix} = -7.2559\,\theta^1 - 4.2731\,\theta^2$$

we obtain the closed loop system with matrix $A - BB^* W_1^{-1}$:

$$\begin{pmatrix} \dot{\theta}_1 \\ \dot{\theta}_2 \end{pmatrix} = \begin{pmatrix} 0 & 1 \\ -6.2559 & -4.2731 \end{pmatrix} \begin{pmatrix} \theta^1 \\ \theta^2 \end{pmatrix} \tag{2.11}$$

Whereas the original matrix A in (2.10) had eigenvalues ± 1, those of the matrix in (2.11), are $-2.1366 \pm 1.3004\,i$.

For the record we state

Definition 2.3. *The linear constant coefficient control system* (2.1) *is stabilizable if there exists an* $m \times n$ *matrix* K *such that* $A + BK$ *is a stability matrix.*

Theorem 2.1 shows that every controllable system is stabilizable. The converse is not true. A system

$$\dot{x} = Ax + 0u$$

with A a stability matrix at the outset is stabilizable (any K will do) but clearly not controllable. Nevertheless, a slightly stronger form of stabilizability does imply controllability as we will see later.

The foregoing stabilization method has the disadvantage that it is necessary to compute the matrix function $e^{-At}B$ (note that this is ordinarily less difficult than computing e^{-At}) and then integrate $e^{-At}BB^*e^{-A^*t}$ from 0 to T to obtain W_T. Then W_T must be inverted. These operations would often need to be done numerically. A more straightforward approach is due to Bass [6]. It relies first of all on

Lemma 2.3. *The pair* (A, B) *is controllable if and only if* $(A - \lambda I, B)$ *is controllable for arbitrary scalar* λ.

Proof. We merely note that

$$(B, (A-\lambda I)B, (A-\lambda I)^2B, \ldots, (A-\lambda I)^qB)$$

$$= (B, AB, A^2B, \ldots, A^qB)\begin{pmatrix} I & -\lambda & \lambda^2 & \cdots & (-1)^q\lambda^q \\ 0 & I & -2\lambda & \cdots & (-1)^{q-1}q\lambda^{q-1} \\ 0 & 0 & I & \cdots & (-1)^{q-2}\binom{q}{q-2}\lambda^{q-2} \\ \vdots & \vdots & \vdots & \ddots & \vdots \\ 0 & 0 & 0 & \cdots & I \end{pmatrix}$$

and draw the conclusion from the same argument as is used in Lemma 2. 2. Q. E. D.

The indicated method proceeds as follows. Let $\lambda \geq 0$ be chosen large enough so that $-(A + \lambda I)$ is a stability matrix. This is assured, for example, if $\lambda > \max_i (\Sigma_{j=1}^n |a_j^i|)$ or if $\lambda > \max_j (\Sigma_{i=1}^n |a_j^i|)$ (see II-[2]). Then we consider the following equation for Z:

$$(A + \lambda I)Z + Z(A + \lambda I)^* = BB^*, \tag{2.12}$$

or, equivalently

$$-(A + \lambda I)Z - Z(A + \lambda I)^* + BB^* = 0 .$$

Combining Lemma 2. 3 with the result of Theorem 1. 5, we conclude that there is a unique positive definite symmetric solution Z, for $(A + \lambda I, B)$ controllable implies $(B^*, -(A + \lambda I)^*)$ observable. Rewriting (2. 12) in the form

$$(A + \lambda I - BB^*Z^{-1})Z + Z(A + \lambda I - BB^*Z^{-1})^* + BB^* = 0$$

and using both Lemma 2. 2 and Lemma 2. 3 with Theorem 1. 5 we conclude that $(A + \lambda I - BB^*Z^{-1})^*$, and hence $A + \lambda I - BB^*Z^{-1}$ is a stability matrix. But then, since $\lambda \geq 0$, $A - BB^*Z^{-1}$ must also be a stability matrix.

An important feature of this method is that it allows us to specify a minimal distance between the eigenvalues of the closed loop matrix $A - BB^*Z^{-1}$, which lie in the left half plane, and the imaginary axis. That distance is $\geq \lambda$ since the eigenvalues of $A + \lambda I - BB^*Z^{-1}$ already lie in the left half plane and those of $A - BB^*Z^{-1}$ lie λ units to the left of those of $A + \lambda I - BB^*Z^{-1}$. (See Exercise 11 for a comparable modification of Theorem 2.1.)

Example. Consider again the inverted pendulum problem. We stabilize via Bass' method. The matrix A is $\begin{pmatrix} 0 & 1 \\ 1 & 0 \end{pmatrix}$ for which $\max_i (\Sigma_{j=1}^{i} |a_j^i|) = \max_j (\Sigma_{i=1}^{i} |a_j^i|) = 1$. So we take $\lambda = 2$. Then equation (2.12) becomes, with $Z = \begin{pmatrix} x & y \\ y & z \end{pmatrix}$,

$$\begin{pmatrix} 2 & 1 \\ 1 & 2 \end{pmatrix}\begin{pmatrix} x & y \\ y & z \end{pmatrix} + \begin{pmatrix} x & y \\ y & z \end{pmatrix}\begin{pmatrix} 2 & 1 \\ 1 & 2 \end{pmatrix} = \begin{pmatrix} 0 & 0 \\ 0 & 1 \end{pmatrix}$$

giving

$$\begin{aligned} 4x + 2y \qquad &= 0 \\ x + 4y + \ z &= 0 \\ 2y + 4z &= 1 \ . \end{aligned}$$

The solution is

$$Z = \begin{pmatrix} \frac{1}{24} & -\frac{7}{84} \\ -\frac{7}{84} & \frac{7}{24} \end{pmatrix}$$

and we compute

$$Z^{-1} = \begin{pmatrix} 56 & 16 \\ 16 & 8 \end{pmatrix}.$$

Then

$$A - BB^*Z^{-1} = \begin{pmatrix} 0 & 1 \\ 1 & 0 \end{pmatrix} - \begin{pmatrix} 0 & 0 \\ 0 & 1 \end{pmatrix}\begin{pmatrix} 56 & 16 \\ 16 & 8 \end{pmatrix} = \begin{pmatrix} 0 & 1 \\ -15 & -8 \end{pmatrix}.$$

The eigenvalues of this closed loop matrix are the roots of

$$0 = \lambda^2 + 8\lambda + 15 = (\lambda + 3)(\lambda + 5)$$

and are thus equal to -3 and -5.

While the above method and the method of Theorem 2.1 both readily yield a stabilizing feedback matrix K when $\dot{x} = Ax + Bu$ is controllable, neither is widely used for ultimate control synthesis because they are rather inflexible. There are other requirements which must be met in control system design beside stability. Nevertheless they are useful, particularly as "starting methods" for more realistic design techniques which we will study later.

We have noted that stabilizability does not imply controllability. With just a little more than stabilizability, however, we do obtain controllability.

Theorem 2.4. Suppose there are $m \times n$ matrices K^+, K^- such that $A+BK^+$ is a stability matrix and $-(A + BK^-)$ is a stability matrix (i.e., all eigenvalues of $A + BK^-$ have positive real parts.) Then the system (2.1) is controllable.

Proof. Since $A + BK^+$ and $-(A + BK^-)$ are both stability matrices we can find $T > 0$ such that for some $\gamma > 0$, $0 \leq \gamma < 1$,

$$\|e^{(A+BK^+)T}\| \leq \gamma, \quad \|e^{-(A+BK-)T}\| \leq \gamma. \tag{2.13}$$

Let $\tilde{x}_0 \in E^n$ and let $x^+(t)$ be the solution of

$$\dot{x}^+(t) = (A + BK^+)x^+(t), \quad x^+(0) = \tilde{x}_0,$$

for $0 \leq t \leq T$. Then set

$$x^+(T) = \hat{x},$$

and let $x^-(t)$ be the solution of

$$\dot{x}^-(t) = (A+BK^-)x^-(t), \quad x^-(t) = -\hat{x},$$

for $0 \leq t \leq T$. Then let

$$x(t) = x^+(t) + x^-(t), \quad u(t) = K^+x^+(t) + K^-x^-(t)$$

on the same interval. It is easy to see that

$$\dot{x}(t) = Ax(t) + Bu(t), \quad 0 \leq t \leq T, \tag{2.14}$$

$$x(0) = (I - e^{-(A+BK^-)T}e^{(A+BK^+)T})\tilde{x}_0, \tag{2.15}$$

$$x(T) = x^+(T) + x^-(T) = \hat{x} - \hat{x} = 0.$$

Thus the control $u(t)$ steers the system (2.1) from the initial state (2.15) to the final state 0. Since the condition (2.13) guarantees that the matrix in (2.15) is invertible, we may solve the equation (2.15) for $\tilde{x}_0$, given any $x(0) \in E^n$. It follows that any initial state $x(0) \in E^n$ can be steered to zero at time T. Following remarks made in II-2 we conclude that the system (2.1) is controllable. Q. E. D.

A slightly more careful treatment allows us to obtain a steering function from this result. In fact, since

$$x^+(t) = e^{(A+BK^+)t}\tilde{x}_0$$

$$x^-(t) = -e^{(A+BK^-)(t-T)}\hat{x} = -e^{(A+BK^-)(t-T)}e^{(A+BK^+)T}\tilde{x}_0$$

the formula for the control $u(t)$ is

$$u(t) = [K^+e^{(A+BK^+)t} - K^-e^{(A+BK^-)(t-T)}e^{(A+BK^+)T}]\tilde{x}_0$$

$$= [K^+e^{(A+BK^+)t} - K^-e^{(A+BK^-)(t-T)}e^{(A+BK^+)T}]$$

$$\times [I - e^{-(A+BK^-)T}e^{(A+BK^+)T}]^{-1}x_0 \equiv S(K^+, K^-)(t)x_0 .$$

Since, from the variation of parameters formula,

$$0 = e^{AT}x_0 + \int_0^T e^{A(T-t)}Bu(t)dt,$$

we conclude that $S(K^+, K^-)(t)$ must satisfy

$$\int_0^T e^{A(T-t)}B\,S(K^+, K^-)(t)dt + e^{AT} = 0$$

and hence is a null steering function for our system. This relationship may be verified algebraically and, in fact, continues to be valid as long as the matrix $I - e^{-(A+BK^-)T}e^{(A+BK^+)T}$ is invertible, leading to

<u>Proposition 2.5.</u> <u>The linear constant coefficient control system</u>

$$\dot{x} = Ax + Bu$$

<u>is controllable if and only if there exist</u> $m \times n$ <u>feedback matrices</u> K^+, K^- <u>for which the matrix</u>

$$I - e^{-(A + BK^-)T} e^{(A+BK^+)T}$$

<u>is invertible.</u>

Theorem 2.4 yields, almost immediately, an important result on the equivalence of stabilizability and controllability for certain second order systems.

<u>Proposition 2.6.</u> <u>Consider the linear second order control system</u>

$$\ddot{x} + CH\dot{x} + Dx = Cu, \quad x \in E^n, \; u \in E^m, \quad C_{n\times m}, \; H_{m\times n}, \; D_{n\times n} \; . \tag{2.16}$$

<u>Such a system is controllable if and only if it is stabilizable.</u>

<u>Proof.</u> The system (2.16) has the form (2.1) with x replaced by $\binom{x}{\dot{x}}$, $A = \begin{pmatrix} 0 & I \\ CH & D \end{pmatrix}$, $B = \binom{0}{C}$. Suppose the feedback law

$$u = K_1 x + K_2 \dot{x} \quad (K_1, K_2 \quad m \times n) \tag{2.17}$$

stabilizes (2.16) for $t \to -\infty$. The closed loop system is

$$\ddot{x} + C(H - K_1)\dot{x} + (D - CK_1)x \; . \tag{2.18}$$

If we let $t = -\tau$, (2.16) becomes

$$\frac{d^2x}{d\tau^2} - CH\frac{dx}{d\tau} + Dx = Cu \; . \tag{2.19}$$

Then setting

$$u = K_1 x + (K_2 - 2H)\frac{dx}{d\tau} = K_1 x - (K_2 - 2H)\dot{x} \tag{2.20}$$

(2.19) becomes

$$\frac{d^2x}{d\tau^2} + C(H - K_2)\frac{dx}{d\tau} + (D - CK_1)x = 0$$

which, comparing with (2.18), has solutions tending to 0 as $\tau \to \infty$, i.e., as $t \to -\infty$. Thus (2.17) stabilizes (2.16) in the positive direction, (2.20) in the negative direction, and controllability follows from Theorem 2.4. We already know that controllability implies stabilizability. Q. E. D.

3. Control Canonical Form, Eigenvalue Specification

We have seen in the foregoing section that controllability of the constant coefficient linear control system

$$\dot{x} = Ax + Bu, \quad x \in E^n, \; u \in E^m, \tag{3.1}$$

is strongly related to the degree to which it is possible to move the eigenvalues of the matrix $A + BK$ by varying the $m \times n$ matrix K. We will carry this relationship further now - showing that (3.1) is controllable if and only if, by appropriate choice of K, any desired set of eigenvalues $\lambda_1, \lambda_2, \ldots, \lambda_n$ may be realized for the matrix $A + BK$. The "if" part is already established - it is an immediate consequence of Theorem 2.4. What we have to show then is that if (3.1) is controllable this "eigenvalue specification property" obtains. To do so, we introduce a transformation which reduces (3.1) to what is called the "control canonical form".

One of the best known facts about linear differential equations is that a scalar n-th order system

$$\frac{d^n x}{dt^n} + a_1 \frac{d^{n-1} x}{dt^{n-1}} + \ldots + a_{n-1} \frac{dx}{dt} + a_n x = u(t)$$

can be equivalently expressed as a first order n-dimensional system

$$\frac{d}{dt}\begin{pmatrix} \zeta^1 \\ \zeta^2 \\ \vdots \\ \zeta^{n-1} \\ \zeta^n \end{pmatrix} = \begin{pmatrix} 0 & 1 & \cdots & 0 & 0 \\ 0 & 0 & \cdots & 0 & 0 \\ \vdots & \vdots & & \vdots & \vdots \\ 0 & 0 & \cdots & 0 & \\ -a_n & -a_{n-1} & \cdots & -a_2 & -a_1 \end{pmatrix}\begin{pmatrix} \zeta^1 \\ \zeta^2 \\ \vdots \\ \zeta^{n-1} \\ \zeta^n \end{pmatrix} + \begin{pmatrix} 0 \\ 0 \\ \vdots \\ 0 \\ 1 \end{pmatrix} u(t) \tag{3.2}$$

by setting $\zeta^1 = x$, $\zeta^2 = \frac{dx}{dt}$, $\ldots$, $\zeta^n = \frac{d^{n-1}x}{dt^{n-1}}$. Only slightly less well known is the fact that, while not every system (3.1), with $m = 1$, can be reduced to a system of the form (3.2) - hence one equivalent to a scalar n-th order equation - "almost all" systems can be. The condition is precisely that the system be controllable. (We remark that a homogeneous system $\dot{x} = Ax$ can be reduced to the form (3.2) (with $u = 0$) if and only if the minimal polynomial of A is equal to the characteristic polynomial of A, which, in turn, is equivalent to the existence of some vector b for which (A, b) is a controllable pair.)

It will be seen that that the theory extends to arbitrary m(i.e. B need not be a vector, b) but we will begin with the scalar control case.

Theorem 3.1. Consider the linear control system with scalar control,

$$\dot{x} = Ax + bu,$$

and suppose (A, b) to be a controllable pair, i.e.,

$$C = (A^{n-1}b, A^{n-2}b, \ldots, Ab, b) \tag{3.3}$$

is a nonsingular matrix. Then there is a nonsingular coordinate transformation

$$x = Q\zeta$$

such that the resulting system

$$\dot{\zeta} = Q^{-1}AQ\zeta + Q^{-1}bu \equiv \hat{A}\zeta + \hat{b}u$$

has the form (3.2).

Proof. The matrix Q is formed in two steps. We first let

$$x = C\tilde{\zeta}$$

where C is the matrix (3.3). There results the system

$$\dot{\tilde{\zeta}} = C^{-1}AC\tilde{\zeta} + C^{-1}bu \equiv \tilde{A}\tilde{\zeta} + \tilde{b}u. \tag{3.4}$$

From (3.3) we see that

$$C^{-1}(A^{n-1}b, A^{n-2}b, \ldots, Ab, b) = I = (e_1, e_2, \ldots, e_n), \tag{3.5}$$

the e_j indicating the columns of the $n \times n$ identity matrix. Then certainly

$$C^{-1}b = \tilde{b} = e_n \,. \tag{3.6}$$

Also

$$AC = (A^n b,\ A^{n-1}b, \dots, A^2 b,\ Ab)$$

from which we see that

$$C^{-1}AC = (C^{-1}A^n b,\ e_1, e_2, \dots, e_{n-1}). \tag{3.7}$$

The Cayley-Hamilton theorem says that

$$A^n + a_1 A^{n-1} + \dots + a_{n-1}A + a_n I = 0$$

where

$$a(\lambda) = \lambda^n + a_1\lambda^{n-1} + \dots + a_{n-1}\lambda + a_n$$

is the characteristic polynomial of the matrix A. Hence (cf. (3.5))

$$C^{-1}A^n b = C^{-1}[-a_1 A^{n-1}b - a_2 A^{n-2}b - \dots - a_{n-1}Ab - a_n b]$$

$$= -a_1 e_1 - a_2 e_2 - \dots - a_{n-1}e_{n-1} - a_n e_n = \begin{pmatrix} -a_1 \\ -a_2 \\ \vdots \\ -a_{n-1} \\ -a_n \end{pmatrix}.$$

Putting this together with (3.6), (3.7) we see that the system (3.4) has the form

$$\frac{d}{dt}\begin{pmatrix} \tilde{\zeta}_1 \\ \tilde{\zeta}_2 \\ \vdots \\ \tilde{\zeta}_{n-1} \\ \tilde{\zeta}_n \end{pmatrix} = \begin{pmatrix} -a_1 & 1 & \dots & 0 & 0 \\ -a_2 & 0 & \dots & 0 & 0 \\ \vdots & \vdots & & \vdots & \vdots \\ -a_{n-1} & 0 & \dots & 0 & 1 \\ -a_n & 0 & \dots & 0 & 0 \end{pmatrix}\begin{pmatrix} \tilde{\zeta}^1 \\ \tilde{\zeta}^2 \\ \vdots \\ \tilde{\zeta}^{n-1} \\ \tilde{\zeta}^n \end{pmatrix} + \begin{pmatrix} 0 \\ 0 \\ \vdots \\ 0 \\ 1 \end{pmatrix} u \,. \tag{3.8}$$

This is tantalizingly close to the form that we want and it has its own significance. But a further transformation is required to actually realize the form (3.2). This transformation is

$$\tilde{\zeta} = \Phi\zeta \equiv \begin{pmatrix} 1 & 0 & 0 & \cdots & 0 & 0 \\ a_1 & 1 & 0 & \cdots & 0 & 0 \\ a_2 & a_1 & 1 & \cdots & 0 & 0 \\ \vdots & \vdots & \vdots & & \vdots & \vdots \\ a_{n-2} & a_{n-3} & a_{n-4} & \cdots & 1 & 0 \\ a_{n-1} & a_{n-2} & a_{n-3} & \cdots & a_1 & 1 \end{pmatrix} \begin{pmatrix} \zeta^1 \\ \zeta^2 \\ \zeta^3 \\ \vdots \\ \zeta^{n-1} \\ \zeta^n \end{pmatrix} \tag{3.9}$$

To see that this does the job we want it to we note that with $\tilde{A}$ denoting the matrix in (3.4), (3.8) we have

$$\tilde{A}\Phi = \begin{pmatrix} 0 & 1 & 0 & \cdots & 0 & 0 \\ 0 & a_1 & 1 & \cdots & 0 & 0 \\ 0 & a_2 & a_1 & \cdots & 0 & 0 \\ \vdots & \vdots & \vdots & & \vdots & \vdots \\ 0 & a_{n-2} & a_{n-3} & \cdots & a_1 & 1 \\ -a_n & 0 & 0 & \cdots & 0 & 0 \end{pmatrix}$$

$$= (-a_n\varphi_n, \varphi_1 - a_{n-1}\varphi_n, \varphi_2 - a_{n-2}\varphi_n, \ldots, \varphi_{n-2} - a_2\varphi_n, \varphi_{n-1} - a_1\varphi_n),$$

where we have used the notation

$$\Phi = (\varphi_1, \varphi_2, \ldots, \varphi_{n-1}, \varphi_n)$$

to indicate the columns of Φ. Then

$$\Phi^{-1}\tilde{A}\Phi = (-a_n e_n, e_1 - a_{n-1}e_n, e_2 - a_{n-2}e_n, \ldots, e_{n-2}, e_{n-1} - a_1 e_n)$$

which is just the matrix $\hat{A}$ in (3.2). Clearly

$$\hat{b} = \Phi^{-1} e_n = e_n \tag{3.10}$$

so the transformation (3.9) does indeed carry (3.4), (3.8) into (3.2). Then the matrix

$$Q = C\Phi$$

has the property required. Q. E. D.

<u>Corollary 3.2.</u> <u>Let the system</u> (3.1) <u>be controllable. Then the closed-loop system</u>

$$\dot{x} = (A + bk^*)x \tag{3.11}$$

<u>has the eigenvalue placement property: given any n-tuple</u> $\lambda_1, \lambda_2, \ldots, \lambda_n$ <u>of complex numbers (repetitions permitted) there is one and only one row vector</u> $k^* = (k_1, k_2, \ldots, k_n)$ <u>such that the matrix</u> $A + bk^*$ <u>has the eigenvalues</u> $\lambda_1, \lambda_2, \ldots, \lambda_n$.

<u>Proof.</u> We have seen that the transformation $x = Q\zeta$ carries (3.1) into (3.2). The feedback law $u = k^*x$ is likewise carried over into

$$u = k^*Q\zeta = \hat{k}^*\zeta ,$$

where

$$\hat{k}^* = (\hat{k}_1, \hat{k}_2, \ldots, \hat{k}_n)$$

is again an n dimensional row vector. Thus (3.11) becomes

$$\dot{\zeta} = (\hat{A} + \hat{b}\hat{k}^*)\zeta ,$$

where now, in view of (3.10) and the form (3.2) for $\hat{A}$

$$(\hat{A} + \hat{b}\hat{k}^*) = \begin{pmatrix} 0 & 1 & \cdots & 0 & 0 \\ 0 & 0 & \cdots & 0 & 0 \\ \vdots & \vdots & & \vdots & \vdots \\ 0 & 0 & \cdots & 0 & 1 \\ -a_n+\hat{k}_1 & -a_{n-1}+\hat{k}_2 & \cdots & -a_2+\hat{k}_{n-1} & -a_1+\hat{k}_n \end{pmatrix} .$$

The characteristic polynomial of the matrix $\hat{A} + \hat{b}\hat{k}^*$, and hence of $A + bk^* = Q(\hat{A} + \hat{b}\hat{k}^*)Q^{-1}$, is

$$a(\lambda) = \lambda^n + (a_n - \hat{k}_1)\lambda^{n-1} + (a_{n-1} - \hat{k}_2)\lambda^{n-2} + \dots + (a_2 - \hat{k}_{n-1})\lambda + (a_1 - \hat{k}_n).$$

There is a one to one correspondence between sets of coefficients, and hence zeros, and values of $\hat{k}_1, \hat{k}_2, \dots, \hat{k}_{n-1}, \hat{k}_n$. Since the correspondence between k and $\hat{k}$ is also one to one it follows that each choice of eigenvalues $\lambda_1, \lambda_2, \dots, \lambda_n$ for $A + bk^*$ corresponds to exactly one row vector k^* and vice versa. Q. E. D.

We illustrate the use of the above technique with the following example. We consider a stiff rod hinged to a movable platform to which a horizontal force u may be applied. We let θ denote the angle which the rod makes with the vertical axis. We assume the lower end of the rod is fixed relative to the platform but free to rotate about the "hinge" attaching the rod to the platform. Assuming the platform to have unit mass, the rod to have unit length and uniform mass density 1, we compute

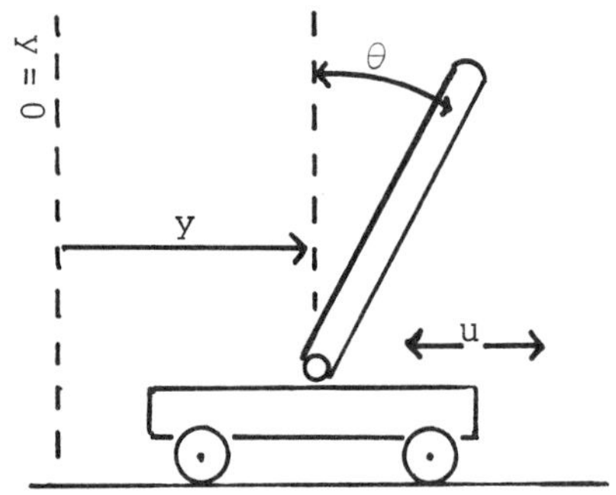

Fig. III-3. Rod hinged to movable platform

$$\text{Kinetic energy}: \; K(y, \dot{y}, \theta, \dot{\theta}) = \tfrac{1}{2}(\dot{y})^2 + \int_0^1 \tfrac{1}{2}[(s \sin\theta\dot{\theta})^2 + (\dot{y} + s\cos\theta\dot{\theta})^2]\, ds$$

$$= (\dot{y})^2 + \frac{1}{6}(\dot{\theta})^2 + \tfrac{1}{2}\cos\theta\,\dot{y}\,\dot{\theta}$$

$$\text{Potential energy}: \; V(y, \dot{y}, \theta, \dot{\theta}) = \int_0^1 s(\cos\theta - 1)ds = \tfrac{1}{2}(\cos\theta - 1).$$

With applied force u we have the work-energy relationship

$$\frac{d}{dt}((\dot{y})^2 + \frac{1}{6}(\dot{\theta})^2 + \frac{1}{2}\cos\theta\,\dot{y}\,\dot{\theta} + \frac{1}{2}(\cos\theta - 1)) = \dot{y}\,u(t)$$

or

$$2\dot{y}\,\ddot{y} + \frac{1}{3}\dot{\theta}\ddot{\theta} + \frac{1}{2}\cos\theta\,\dot{y}\,\ddot{\theta} + \frac{1}{2}\cos\ddot{y}\,\dot{\theta} - \frac{1}{2}\sin\theta\,\dot{y}\,(\dot{\theta})^2 - \frac{1}{2}\sin\theta\,\dot{\theta} = \dot{y}\,u(t)\,.$$

Hence

$$\dot{y}\,(2\ddot{y} + \frac{1}{2}\cos\theta\ddot{\theta} - \frac{1}{2}\sin\theta(\dot{\theta})^2 - u(t)) + \dot{\theta}\,(\frac{1}{3}\ddot{\theta} + \frac{1}{2}\cos\theta\,\ddot{y} - \frac{1}{2}\sin\theta) = 0.$$

Setting the coefficients of $\dot{y}$, $\dot{\theta}$ equal to zero:

$$2\ddot{y} + \frac{1}{2}\cos\theta\,\ddot{\theta} = \frac{1}{2}\sin\theta(\dot{\theta})^2 + u$$

$$\frac{1}{2}\cos\theta\ddot{y} + \frac{1}{3}\ddot{\theta} = \frac{1}{2}\sin\theta\,.$$

The linearized dynamics about $\theta = \dot{\theta} = 0$ are

$$2\ddot{y} + \frac{1}{2}\ddot{\theta} = u$$

$$\frac{1}{2}\ddot{y} + \frac{1}{3}\ddot{\theta} = \frac{1}{2}\theta\,,$$

and solving for $\ddot{y}$, $\ddot{\theta}$ we have

$$\ddot{y} = -\frac{3}{5}\theta + \frac{4}{5}u\,,$$

$$\ddot{\theta} = \frac{12}{5}\theta - \frac{6}{5}u\,.$$

Assuming that y itself is of no interest, we put $x^1 = \dot{y}$, $x^2 = \theta$, $x^3 = \dot{\theta}$ and have

$$\dot{\begin{pmatrix} x^1 \\ x^2 \\ x^3 \end{pmatrix}} = \begin{pmatrix} 0 & -\frac{3}{5} & 0 \\ 0 & 0 & 1 \\ 0 & \frac{12}{5} & 0 \end{pmatrix} \begin{pmatrix} x^1 \\ x^2 \\ x^3 \end{pmatrix} + \begin{pmatrix} \frac{4}{5} \\ 0 \\ -\frac{6}{5} \end{pmatrix} u\,. \tag{3.12}$$

It is easy to check that this system is controllable and hence, by Corollary

3.2, we may place the eigenvalues where we will by linear feedback

$$u = (k_1, k_2, k_3)\begin{pmatrix} y^1 \\ y^2 \\ y^3 \end{pmatrix} = k_1 y^1 + k_2 y^2 + k_3 y^3 .$$

Let us select as the desired eigenvalues

$$\lambda_1 = -1, \quad \lambda_2 = -1+i, \quad \lambda_2 = -1-i . \tag{3.13}$$

We compute the matrix

$$C = (A^2 b,\ Ab,\ b) = \begin{pmatrix} \frac{18}{25} & 0 & \frac{4}{5} \\ 0 & -\frac{6}{5} & 0 \\ -\frac{72}{25} & 0 & -\frac{6}{5} \end{pmatrix} .$$

Then

$$C^{-1} = \begin{pmatrix} -\frac{5}{6} & 0 & -\frac{5}{9} \\ 0 & -\frac{5}{6} & 0 \\ 2 & 0 & \frac{1}{2} \end{pmatrix}$$

and we check that

$$C^{-1}b = \begin{pmatrix} -\frac{5}{6} & 0 & -\frac{5}{9} \\ 0 & -\frac{5}{6} & 0 \\ 2 & 0 & \frac{1}{2} \end{pmatrix}\begin{pmatrix} \frac{4}{5} \\ 0 \\ -\frac{6}{5} \end{pmatrix} = \begin{pmatrix} 0 \\ 0 \\ 1 \end{pmatrix} \equiv \tilde{b} .$$

$$C^{-1}AC = C^{-1}(A^3 b, A^2 b, Ab) = \begin{pmatrix} -\frac{5}{6} & 0 & -\frac{5}{9} \\ 0 & -\frac{5}{6} & 0 \\ 2 & 0 & \frac{1}{2} \end{pmatrix}\begin{pmatrix} 0 & \frac{18}{25} & 0 \\ -\frac{72}{25} & 0 & -\frac{6}{5} \\ 0 & -\frac{72}{25} & 0 \end{pmatrix} = \begin{pmatrix} 0 & 1 & 0 \\ \frac{12}{5} & 0 & 1 \\ 0 & 0 & 0 \end{pmatrix} \equiv \tilde{A}. \tag{3.14}$$

The matrix Φ is then

$$\Phi = \begin{pmatrix} 1 & 0 & 0 \\ 0 & 1 & 0 \\ -\frac{12}{5} & 0 & 1 \end{pmatrix}$$

with inverse

$$\Phi^{-1} = \begin{pmatrix} 1 & 0 & 0 \\ 0 & 1 & 0 \\ \frac{12}{5} & 0 & 1 \end{pmatrix} .$$

We check easily that (cf. (3.14))

$$\Phi^{-1}\tilde{A}\Phi = \begin{pmatrix} 0 & 1 & 0 \\ 0 & 0 & 1 \\ 0 & \frac{12}{5} & 0 \end{pmatrix} \equiv A .$$

The control canonical form is thus

$$\begin{pmatrix} \dot{\zeta}^1 \\ \dot{\zeta}^2 \\ \dot{\zeta}^3 \end{pmatrix} = \begin{pmatrix} 0 & 1 & 0 \\ 0 & 0 & 1 \\ 0 & \frac{12}{5} & 0 \end{pmatrix} \begin{pmatrix} \zeta^1 \\ \zeta^2 \\ \zeta^3 \end{pmatrix} + \begin{pmatrix} 0 \\ 0 \\ 1 \end{pmatrix} u .$$

The characteristic polynomial is $\lambda^3 - \frac{12}{5}\lambda$, indicating eigenvalues

$$\lambda_1 = 0, \quad \lambda_2 = 1.549\ldots, \quad \lambda_3 = -1.549\ldots .$$

The polynomial having the zeros (3.13) is

$$p(\lambda) = (\lambda+1)(\lambda+1-i)(\lambda+1+i) = \lambda^3 + 3\lambda^2 + 4\lambda + 2 .$$

With feedback

$$u = \hat{k}_1\zeta^1 + \hat{k}_2\zeta^2 + \hat{k}_3\zeta^3$$

the matrix of the closed loop system becomes

$$\begin{pmatrix} 0 & 1 & 0 \\ 0 & 0 & 1 \\ \hat{k}_1 & \hat{k}_2+\frac{12}{5} & \hat{k}_3 \end{pmatrix}.$$

To get the desired eigenvalues we need

$$\hat{k}_1 = -2, \quad \hat{k}_2 + \frac{12}{5} = -4, \quad \hat{k}_3 = -3,$$

giving

$$\hat{k}^* = (\hat{k}_1, \hat{k}_2, \hat{k}_3) = (-2, \ -\frac{32}{5}, \ -3).$$

In terms of the original system (3.12) the desired feedback matrix is

$$k^* = \hat{k}^* \Phi^{-1} C^{-1}$$

$$= (-2, -\frac{32}{5}, -3) \begin{pmatrix} 1 & 0 & 0 \\ 0 & 1 & 0 \\ \frac{12}{5} & 0 & 1 \end{pmatrix} \begin{pmatrix} -\frac{5}{6} & 0 & -\frac{5}{9} \\ 0 & -\frac{5}{6} & 0 \\ 2 & 0 & \frac{1}{2} \end{pmatrix} = (\frac{15}{9}, \frac{48}{9}, \frac{65}{18}). \quad (3.15)$$

The resulting closed loop system is then

$$\begin{pmatrix} \dot{x}^1 \\ \dot{x}^2 \\ \dot{x}^3 \end{pmatrix} = \begin{pmatrix} 0 & -\frac{3}{5} & 0 \\ 0 & 0 & 1 \\ 0 & \frac{12}{5} & 0 \end{pmatrix} + \begin{pmatrix} \frac{4}{5} \\ 0 \\ -\frac{6}{5} \end{pmatrix} (\frac{15}{9}, \frac{48}{9}, \frac{65}{18}) \begin{pmatrix} x^1 \\ x^2 \\ x^3 \end{pmatrix}$$

$$= \begin{pmatrix} \frac{4}{3} & \frac{11}{3} & \frac{26}{9} \\ 0 & 0 & 1 \\ -2 & -4 & -\frac{13}{3} \end{pmatrix} \begin{pmatrix} x^1 \\ x^2 \\ x^3 \end{pmatrix}.$$

One can easily verify that this system matrix has the desired eigenvalues -1, $-1+i$, $-1-i$. Thus, assuming that $\dot{y}$, θ, $\dot{\theta}$ can all be measured at each instant t and the control u given by (cf. (3.15))

$$u(t) = \frac{15}{9}\dot{y} + \frac{48}{9}\theta + \frac{65}{18}\dot{\theta}$$

actually implemented in some way, the rod will be balanced with the vertical stationary position as an asymptotically stable equilibrium.

We are now ready to extend the concept of control canonical form to controllable systems (3.1) with $m > 1$. To avoid the possibility that one might have $Bu(t) \equiv 0$ for $u(t) \not\equiv 0$ we shall suppose that the columns $b_1, b_2, \ldots, b_m$ of B are linearly independent. We then arrange the columns of the matrix

$$C = (A^{n-1}B,\ A^{n-2}B, \ldots, AB, B) \tag{3.16}$$

in an array

$$\begin{matrix} A^{n-1}b_1 & A^{n-1}b_2 & \cdots & A^{n-1}b_m \\ A^{n-2}b_1 & A^{n-2}b_2 & \cdots & A^{n-2}b_m \\ \vdots & \vdots & & \vdots \\ Ab_1 & Ab_2 & \cdots & Ab_m \\ b_1 & b_2 & \cdots & b_m \end{matrix} \quad . \tag{3.17}$$

The first step in the development of the control canonical form involves the formation of a nonsingular $n \times n$ matrix Ψ whose columns are a subset of the vectors in (3.17). Our linear independence assumption on $b_1, b_2, \ldots, b_m$ implies $m \le n$. If $m = n$ we just take

$$\Psi = (b_1, b_2, \ldots, b_m).$$

The more interesting and more usual situation is that $m < n$. We then proceed as follows. We examine the vectors in the array (3.17), going from bottom to top, left to right, one row at a time, i.e., in the order $b_1, b_2, \ldots, b_m,\ Ab_1, Ab_2, \ldots, Ab_m, \ldots, A^{n-1}b_1, A^{n-1}b_2, \ldots, A^{n-1}b_m$. Whenever one of these vectors is linearly dependent on vectors appearing earlier in the sequence it is discarded. When we are all done we may assume, reordering the b_j if necessary, that the vectors remaining are

$$\begin{array}{ccccc} A^{\mu_1-1}b_1 & & & & \\ \vdots & & A^{\mu_j-1}b_j & & \\ & & & & A^{\mu_m-1}b_m \\ \vdots & \cdots & \vdots & \cdots & \vdots \\ Ab_1 & & Ab_j & & Ab_m \\ b_1 & & b_j & & b_m \end{array} , \tag{3.18}$$

where, without loss of generality

$$\mu_1 \geq \mu_2 \geq \cdots \geq \mu_m$$

and, since rank $C = n$ ((3.1) is assumed controllable)

$$\mu_1 + \mu_2 + \cdots + \mu_m = n .$$

For each j such that $\mu_j < \mu_1$ we must now carry out a transformation of of the control space whose significance is not easy to appreciate until later. Since the j-th column terminates with $A^{\mu_j-1}b_j$, there is a linear dependence relationship

$$A^{\mu_j}b_j + \sum_{k=1}^{j-1} \alpha_j^k A^{\mu_j}b_k + \sum_{k=1}^{m} \sum_{\ell=0}^{\min\{\mu_j,\mu_k\}-1} \alpha_{j,\ell}^k A^{\ell} b_k = 0 \tag{3.19}$$

($\alpha_j^k = 0$ if $\mu_j = \mu_k$). We then set

$$u = M_j u_j \quad (u_j \text{ is a vector here, not a component}) \tag{3.20}$$

where

$$M_j = \begin{pmatrix} 1 & \cdots & 0 & \alpha_j^1 & 0 & \cdots & 0 \\ \vdots & & \vdots & \vdots & \vdots & & \vdots \\ 0 & \cdots & 1 & \alpha_j^{j-1} & 0 & \cdots & 0 \\ 0 & \cdots & 0 & 1 & 0 & \cdots & 0 \\ 0 & \cdots & 0 & 0 & 1 & \cdots & 0 \\ \vdots & & \vdots & \vdots & \vdots & & \vdots \\ 0 & \cdots & 0 & 0 & 0 & \cdots & 1 \end{pmatrix} \tag{3.21}$$

The result is a new system

$$\dot{x} = Ax + B_j u_j$$

wherein the $b_{j,k}$, $k = 1, 2, \ldots, m$, of B_j coincide with those of B except for $b_{j,j}$ which is now

$$b_{j,j} = b_j + \sum_{k=1}^{j-1} \alpha_j^k b_k .$$

The array (3.18) is left as is except for the j-th column, where b_j is replaced by $b_{j,j}$. The relationship (3.19) now becomes

$$A^{\mu_j} b_{j,j} + \sum_{k=1}^{m} \sum_{\ell=0}^{\min\{\mu_j, \mu_k\}-1} \alpha_{j,j,\ell}^k A^\ell b_k = 0$$

for certain coefficients $\alpha_{j,j,\ell}^k$.

Transformations of this type are performed, one after the other, for each j such that $\mu_j < \mu_1$, say for $j = J, J+1, \ldots, m$. Denoting the composite change of control variable by

$$u = \underline{M}\underline{u} \equiv M_J M_{J+1} \cdots M_m \underline{u} \tag{3.22}$$

we have finally the system

$$\dot{x} = Ax + \underline{B}\underline{u} . \tag{3.23}$$

For each j such that $\mu_j < \mu_1$ we now have

$$A^{\mu_j}\underline{b}_j + \sum_{k=1}^{m} \sum_{\ell=0}^{\min\{\mu_j,\mu_k\}-1} a^k_{j,\ell} A^\ell b_k = 0 \qquad (3.24)$$

i.e. $A^{\mu_j}\underline{b}_j$ can be expressed as a linear combination of vectors $\underline{b}_j$ in rows below the μ_j-th row when the $A^\ell \underline{b}_k$ are arranged as are the $A^\ell b_k$ in (3.18)

We now apply to the system (3.23) the transformation

$$x = \Psi\tilde{\zeta} , \qquad (3.25)$$

where the columns of Ψ are the linearly independent vectors $A^\ell \underline{b}_k$, $k = 1, 2, \dots, m$, $\ell = \mu_k - 1, \mu_k - 2, \dots, 1$. The result is

$$\dot{\tilde{\zeta}} = \Psi^{-1}A\Psi\tilde{\zeta} + \Psi^{-1}\underline{B}\underline{u} \equiv \tilde{A}\tilde{\zeta} + \tilde{B}\underline{u} . \qquad (3.26)$$

Since the columns of $\underline{B}$ constitute the μ_1-th, $(\mu_1+\mu_2)$-th, ... , nth columns of Ψ we see immediately that

$$\tilde{B} = \Psi^{-1}\underline{B} = (e_{\mu_1}, e_{\mu_1+\mu_2}, \dots, e_n). \qquad (3.27)$$

On the other hand from the form of Ψ we see that

$$A\Psi = (A^{\mu_1}\underline{b}_1, \dots, A^2\underline{b}_1, A\underline{b}_1, A^{\mu_2}\underline{b}_2, \dots, A^2\underline{b}_2, A\underline{b}_2, \dots, A^{\mu_m}\underline{b}_m, \dots, A^2\underline{b}_m, A\underline{b}_m)$$

from which it is clear that

$$\Psi^{-1}A\Psi = (\Psi^{-1}A^{\mu_1}b_1, e_1, \dots, e_{\mu_1-1}, \Psi^{-1}A^{\mu_2}b_2, e_{\mu_1+1}, \dots, e_{\mu_1+\mu_2-1}, \dots, \Psi^{-1}A^{\mu_m}b_m, e_{\mu_1+\mu_2+\dots\mu_{m-1}+1}, \dots, e_{n-1})$$

To treat the columns $\Psi^{-1}A^{\mu_j}b_j$ we use (3.24):

$$\begin{aligned}\Psi^{-1}A^{\mu_j}b_j &= -\sum_{k=1}^{m} \sum_{\ell=0}^{\min\{\mu_j,\mu_k\}-1} a^k_{j,\ell}\Psi^{-1}A^\ell b_k \\ &= -\sum_{k=1}^{m} \sum_{\ell=0}^{\min\{\mu_j,\mu_k\}-1} a^k_{j,\ell}\, e_{\mu_1+\dots+\mu_k-\ell} .\end{aligned} \qquad (3.28)$$

Thus $\tilde{A}$ has the block arrangement

$$\tilde{A} = \begin{pmatrix} \tilde{A}_1^1 & \tilde{A}_2^1 & \cdots & \tilde{A}_m^1 \\ \tilde{A}_1^2 & \tilde{A}_2^2 & \cdots & \tilde{A}_m^2 \\ \vdots & \vdots & & \vdots \\ \tilde{A}_1^m & \tilde{A}_1^m & \cdots & \tilde{A}_m^m \end{pmatrix} . \tag{3.29}$$

The diagonal blocks have a structure

$$\tilde{A}_k^k = \begin{pmatrix} -a_{k,\mu_{k-1}}^k & 1 & \cdots & 0 \\ \vdots & \vdots & & \vdots \\ -a_{k,1}^k & 0 & \cdots & 1 \\ -a_{k,0}^k & 0 & \cdots & 0 \end{pmatrix} . \tag{3.30}$$

In the case of off-diagonal blocks the form depends on whether $\min\{\mu_k, \mu_j\} = \mu_k$ or whether it equals μ_j. If the latter is the case (which is true for $k < j$) then

$$\tilde{A}_j^k = \begin{pmatrix} 0 & 0 & \cdots & 0 \\ \vdots & \vdots & & \vdots \\ 0 & 0 & \cdots & 0 \\ -a_{j,\mu_j-1}^k & 0 & \cdots & 0 \\ \vdots & \vdots & & \vdots \\ -a_{j,1}^k & 0 & \cdots & 0 \\ -a_{j,0}^k & 0 & \cdots & 0 \end{pmatrix} . \quad \text{(first } \mu_k - \mu_j \text{ rows)} \tag{3.31}$$

If the former is the case (which is true for $k > j$) then

$$\tilde{A}_j^k = \begin{pmatrix} -a^k_{j,\mu_k-1} & 0 & \cdots & 0 \\ \vdots & \vdots & & \vdots \\ -a^k_{j,1} & 0 & \cdots & 0 \\ -a^k_{j,0} & 0 & \cdots & 0 \end{pmatrix} . \tag{3.32}$$

The important point is that the number of non-zero entries in the first column cannot exceed the number of columns in $\tilde{A}_j^k$, which is μ_j. The transformation (3.22) was carried out to ensure that this would be the case.

The system (3.26) with $\tilde{A}$, $\tilde{B}$ as described in (3.29), (3.30), (3.31), (3.32) and (3.27) is said to be in control normal form. In the special case $m = 1$ this is just (3.4) and the matrix Ψ constructed above is just the matrix C shown in (3.3).

The matrix Φ of the transformation which carries the control normal form into the control canonical form also has a structure rather similar to the scalar case. Indeed

$$\Phi = \begin{pmatrix} \Phi_1^1 & \Phi_2^1 & \cdots & \Phi_m^1 \\ \Phi_1^2 & \Phi_2^2 & \cdots & \Phi_m^2 \\ \vdots & \vdots & & \vdots \\ \Phi_1^m & \Phi_2^m & \cdots & \Phi_m^m \end{pmatrix} . \tag{3.33}$$

Each diagonal block has the structure

$$\Phi_k^k = \begin{pmatrix} 1 & 0 & 0 & \cdots & 0 \\ a^k_{k,\mu_k} & 1 & 0 & \cdots & 0 \\ a^k_{k,\mu_k-2} & a^k_{k,\mu_k} & 1 & \cdots & 0 \\ \vdots & \vdots & \vdots & & \vdots \\ a^k_{k,1} & a^k_{k,2} & a^k_{k,3} & \cdots & 1 \end{pmatrix} \tag{3.34}$$

Those off-diagonal blocks for which $k \leq j$ have the structure

$$\Phi_j^k = \left(\begin{array}{ccccc} 0 & 0 & \cdots & 0 & 0 \\ \vdots & \vdots & & \vdots & \vdots \\ 0 & 0 & \cdots & 0 & 0 \\ a^k_{j,\mu_j-1} & 0 & \cdots & 0 & 0 \\ \vdots & \vdots & \ddots & \vdots & \vdots \\ a^k_{j,2} & a^k_{j,3} & \cdots & 0 & 0 \\ a^k_{j,1} & a^k_{j,2} & \cdots & a^k_{j,\mu_j-1} & 0 \end{array}\right) \begin{array}{l} \left.\begin{array}{l} \\ \\ \\ \end{array}\right\} \mu_k-\mu_j+1 \text{ rows} \\ \\ \\ \\ \\ \end{array} \qquad (3.35)$$

while those for which $j \leq k$ have the structure

$$\Phi_j^k = \left(\begin{array}{cccccccc} 0 & 0 & \cdots & 0 & \overbrace{0 & \cdots & 0}^{\mu_j-\mu_k+1 \text{ columns}} \\ a^k_{j,\mu_k-1} & 0 & \cdots & 0 & 0 & \cdots & 0 \\ \vdots & \vdots & \ddots & \vdots & \vdots & & \vdots \\ a^k_{j,2} & a^k_{j,3} & \cdots & 0 & 0 & \cdots & 0 \\ a^k_{j,1} & a^k_{j,2} & \cdots & a^k_{j,\mu_k-1} & 0 & \cdots & 0 \end{array}\right) . \qquad (3.36)$$

Aside from the diagonal row structure of the (possibly) non-zero entries, the most important thing to notice is that the last column of each diagonal block is all zeros except the "1" in the last position while the last column of each off-diagonal block is all zeros. It was to achieve this structure of the off-diagonal blocks of Φ, corresponding to the special structure already noted for the off-diagonal blocks of $\tilde{A}$, that we made the transformation (3.22) in the control space. As a result of this we can say that Φ has the structure

$$\Phi = (\phi_1 \cdots \phi_{\mu_1}, \phi_{\mu_1+1}, \cdots, \phi_{\mu_1+\mu_2}, \cdots, \phi_{\mu_1+\cdots+\mu_{n-1}+1}, \cdots, \phi_n) \qquad (3.37)$$

with

$$\phi_{\mu_1} = e_{\mu_1}, \quad \phi_{\mu_1+\mu_2} = e_{\mu_1+\mu_2}, \cdots, \phi_n = e_n. \qquad (3.38)$$

Thus the m columns of $\tilde{B}$ correspond to the columns (3.37) of Φ and from this it follows that the transformation

$$\tilde{\zeta} = \Phi\zeta$$

which carries (3.26) into

$$\dot{\zeta} = \Phi^{-1}\tilde{A}\Phi\zeta + \Phi^{-1}\tilde{B}\underline{u} \equiv \hat{A}\zeta + \hat{B}\underline{u} \qquad (3.39)$$

leaves $\tilde{B}$ invariant:

$$\hat{B} = \Phi^{-1}\tilde{B} = (e_{\mu_1}, e_{\mu_1+\mu_2}, \cdots, e_n). \qquad (3.40)$$

It is not easy to appreciate the role played by the structure of Φ as it related to that of $\tilde{A}$. For clarity we consider here only the case $m = 2$ and we shall suppose that $\mu_1 > \mu_2$. We may then represent $\tilde{A}$ in the form (changing notation slightly)

$$\tilde{A} = \begin{pmatrix}
-a_1 & 1 & 0 & \cdots & 0 & 0 & 0 & 0 & \cdots & 0 \\
-a_2 & 0 & 1 & \cdots & 0 & 0 & \vdots & \vdots & & \vdots \\
 & & & & & & 0 & 0 & \cdots & 0 \\
\vdots & \vdots & \vdots & \ddots & \vdots & \vdots & -b_1 & 0 & \cdots & 0 \\
-a_{\mu_1-1} & 0 & 0 & \cdots & 0 & 1 & \vdots & \vdots & & \vdots \\
-a_{\mu_1} & 0 & 0 & \cdots & 0 & 0 & -b_{\mu_2} & 0 & \cdots & 0 \\
\hline
-c_1 & 0 & 0 & \cdots & 0 & 0 & -d_1 & 1 & 0 & \cdots & 0 & 0 \\
 & & & & & & -d_2 & 0 & 1 & \cdots & 0 & 0 \\
\vdots & \vdots & \vdots & & \vdots & \vdots & \vdots & \vdots & \vdots & & \vdots & \vdots \\
 & & & & & & -d_{\mu_2-1} & 0 & 0 & \cdots & 0 & 1 \\
-c_{\mu_2} & 0 & 0 & \cdots & 0 & 0 & -d_{\mu_2} & 0 & 0 & \cdots & 0 & 0
\end{pmatrix} \qquad (3.41)$$

$$\left.\begin{array}{l}\\ \\ \end{array}\right\} \begin{array}{l}\mu_1-\mu_2 \\ \text{rows}\end{array}$$

Then Φ has the corresponding form

$$
\Phi = \left(\begin{array}{cccccc|cccccc}
1 & 0 & 0 & \cdots & & 0 & 0 & 0 & 0 & \cdots & 0 & 0 \\
 & & & & & & \vdots & \vdots & \vdots & & \vdots & \vdots \\
a_1 & 1 & 0 & \cdots & & 0 & 0 & 0 & 0 & \cdots & 0 & 0 \\
a_2 & a_1 & 1 & \cdots & & 0 & b_1 & 0 & 0 & \cdots & 0 & 0 \\
 & & & & & & b_2 & b_1 & 0 & \cdots & 0 & 0 \\
\vdots & \vdots & \vdots & & & \vdots & \vdots & \vdots & \vdots & & \vdots & \vdots \\
a_{\mu_1-1} & a_{\mu_1-2} & a_{\mu_1-3} & \cdots & & 1 & b_{\mu_2-1} & b_{\mu_2-2} & b_{\mu_2-3} & \cdots & b_1 & 0 \\
\hline
0 & 0 & 0 & \cdots & 0 & 0 \cdots 0 & 1 & 0 & 0 & \cdots & & 0 \\
c_1 & 0 & 0 & \cdots & 0 & 0 \cdots 0 & d_1 & 1 & 0 & \cdots & & 0 \\
c_2 & c_1 & 0 & \cdots & 0 & 0 \cdots 0 & d_2 & d_1 & 1 & \cdots & & 0 \\
\vdots & \vdots & & & \vdots & \vdots \quad \vdots & \vdots & \vdots & \vdots & & & \vdots \\
 & & & & 0 & 0 \cdots 0 & & & & & & \\
c_{\mu_2-1} & c_{\mu_2-2} & c_{\mu_2-3} & \cdots & c_1 & \underbrace{0 \cdots 0}_{\substack{\mu_1-\mu_2+1 \\ \text{columns}}} & d_{\mu_2-1} & d_{\mu_2-2} & d_{\mu_2-3} & \cdots & & 1
\end{array}\right)
\begin{array}{l}
\left.\vphantom{\begin{array}{c}0\\ \vdots\\ 0\end{array}}\right\}\mu_1-\mu_2+1 \text{ rows} \\
\vphantom{\begin{array}{c}0\\0\\ \vdots\\0\\0\\0\\0\\ \vdots\\0\\0\end{array}}
\end{array}
. \quad (3.42)
$$

Again using the notation (3. 37) to indicate the columns of Φ, careful checking shows that

$$
\begin{aligned}
\tilde{A}\Phi = (&-a_{\mu_1} e_{\mu_1} - c_{\mu_2} e_n, \ \phi_1 - a_{\mu_1-1} e_{\mu_1} - c_{\mu_2-1} e_n, \\
&\ldots, \phi_{\mu_1-1} - a_1 e_{\mu_1} - c_1 e_n, \ -d_{\mu_2} e_n - b_{\mu_2} e_{\mu_1}, \\
&\phi_{\mu_1+1} - d_{\mu_2-1} e_n - b_{\mu_2-1} e_{\mu_1}, \ldots, \phi_{n-1} - d_1 e_n - b_1 e_{\mu_1})
\end{aligned}
$$

and from this we see that

$$
\begin{aligned}
\Phi^{-1}\tilde{A}\Phi = (&-a_{\mu_1} e_{\mu_1} - c_{\mu_2} e_n, \ e_1 - a_{\mu_1-1} e_{\mu_1} - c_{\mu_2-1} e_n, \\
&\ldots, e_{\mu_1-1} - a_1 e_{\mu_1} - c_1 e_n, \ -d_{\mu_2} e_n - b_{\mu_2} e_{\mu_1}, \\
&e_{\mu_1+1} - d_{\mu_2-1} e_n - b_{\mu_2-1} e_{\mu_1}, \ldots, e_{n-1} - d_1 e_n - b_1 e_{\mu_1}) .
\end{aligned}
$$

Thus, in this case with $m = 2$ the final system is

$$\dot{\zeta} = \hat{A}\zeta + \hat{B}\underline{u}$$

with

$$\hat{A} = \begin{pmatrix} 0 & 1 & \cdots & 0 & 0 & 0 & \cdots & 0 \\ \vdots & \vdots & & \vdots & \vdots & \vdots & & \vdots \\ 0 & 0 & \cdots & 1 & 0 & 0 & \cdots & 0 \\ -a_{\mu_1} & -a_{\mu_1-1} & \cdots & -a_1 & -b_{\mu_2} & -b_{\mu_2-1} & \cdots & -b_1 \\ 0 & 0 & \cdots & 0 & 0 & 1 & \cdots & 0 \\ \vdots & \vdots & & \vdots & \vdots & \vdots & & \vdots \\ 0 & 0 & \cdots & 0 & 0 & 0 & \cdots & 1 \\ -c_{\mu_2} & -c_{\mu_2-1} & \cdots & 0 & -d_{\mu_2} & -d_{\mu_2-1} & \cdots & -d_1 \end{pmatrix} \tag{3.43}$$

(In the lower left hand block, $-c_1$ occurs in the μ_2-th column and there are zeros in the (μ_2+1)-st through μ_1-th columns.) and

$$\hat{B} = \begin{pmatrix} 0 & 0 \\ \vdots & \vdots \\ 0 & 0 \\ 1 & 0 \\ 0 & 0 \\ \vdots & \vdots \\ 0 & 0 \\ 0 & 1 \end{pmatrix} \begin{matrix} \left.\begin{matrix} \\ \\ \\ \\ \end{matrix}\right\} \mu_1 \text{ rows} \\ \left.\begin{matrix} \\ \\ \\ \\ \end{matrix}\right\} \mu_2 \text{ rows} \end{matrix} \tag{3.44}$$

It may be verified, though painfully, that this structure persists in the general case of a m-dimensional control. Each diagonal block of $\hat{A}$ has the form

$$\hat{A}_k^k = \begin{pmatrix} 0 & 1 & \cdots & 0 \\ \vdots & \vdots & \cdots & \vdots \\ 0 & 0 & \cdots & 1 \\ -a_{k,0}^k & -a_{k,1}^k & \cdots & -a_{k,\mu_k-1}^k \end{pmatrix} \tag{3.45}$$

those off diagonal blocks $\hat{A}_j^k$ for which $k > j$ have the structure

$$\hat{A}_j^k = \begin{pmatrix} 0 & 0 & \cdots & 0 & 0 & & 0 \\ \vdots & \vdots & & \vdots & \vdots & & \vdots \\ 0 & 0 & \cdots & 0 & 0 & \cdots & 0 \\ -a_{j,0}^k & -a_{j,1}^k & \cdots & -a_{j,\mu_k-1}^k & 0 & \cdots & 0 \end{pmatrix} \tag{3.46}$$

and those for which $k < j$ look like

$$\hat{A}_j^k = \begin{pmatrix} 0 & 0 & \cdots & 0 \\ \vdots & \vdots & & \vdots \\ 0 & 0 & \cdots & 0 \\ -a_{j,0}^k & -a_{j,1}^k & \cdots & -a_{j,\mu_j-1}^k \end{pmatrix} . \tag{3.47}$$

The structure of $\hat{B}$ has already been given. We now state

Theorem 3.3. *An arbitrary controllable system*

$$\dot{x} = Ax + Bu \tag{3.48}$$

can be transformed, via nonsingular transformations

$$x = Q\zeta, \qquad u = M\underline{u}$$

into control canonical form

$$\dot{\zeta} = \hat{A}\zeta + \hat{B}\underline{u} \tag{3.49}$$

wherein $\hat{A}$ *has the structure exhibited in* (3.45) - (3.47) *and* $\hat{B}$ *that of* (3.40). (*In contrast to the case* $m = 1$, *this canonical form is not, in* in general, unique.) As a consequence, any desired set of n complex eigenvalues $\lambda_1, \lambda_2, \ldots, \lambda_n$ (including multiplicities) may be realized for the closed loop system

$$\dot{x} = (A + BK)x$$

by appropriate linear feedback

$$u = Kx$$

in the original control system

$$\dot{x} = Ax + Bu .$$

Proof. The first part of the theorem has already been proved with $Q = \Psi\Phi$ (cf. (3.25), (3.37)). The non-uniqueness follows from the fact that the choice of a maximal linearly independent subset of the vectors in (3.18) is not, in general, unique. The eigenvalue placement result follows from the observation that, except for the 1's in superdiagonal rows of diagonal blocks of $\hat{A}$, all (possibly) non-zero entries of $\hat{A}$ lie in the same row as a 1 in $\hat{B}$. Thus if the $\mu_1 + \mu_2 + \ldots + \mu_k$-th row of A is $\alpha_1, \alpha_2, \ldots, \alpha_n$, by setting

$$\underline{u}^k = (\beta_1 - \alpha_1)\zeta^1 + (\beta_2 - \alpha_2)\zeta^2 + \ldots + (\beta_n - \alpha_n)\zeta^n$$

one can change this row to any other set $\beta_1, \beta_2, \ldots, \beta_n$. This is true for all such rows, $k = 1, 2, \ldots, m$. In particular, then, one can reduce all off-diagonal blocks of $\hat{A} + \hat{B}\hat{K}$ to zero and place any desired coefficients in the last row of each diagonal block, thus achieving any desired characteristic polynomial, and hence any desired eigenvalues, for $\hat{A} + \hat{B}\hat{K}$. The feedback law

$$\underline{u} = \hat{K}\zeta$$

translates to

$$u = M^{-1}\hat{K}\Phi^{-1}\Psi^{-1}x \equiv Kx$$

in terms of the original coordinates. Q. E. D.

The reduction to control canonical form permits an essentially complete treatment of a number of questions for controllable systems. We wish to make some remarks now concerning the general control system

$$\dot{x} = Ax + Bu, \quad x \in E^n, \quad u \in E^m . \tag{3.50}$$

For this study we do not assume the system is necessarily controllable. We will assume, however, that the matrix

$$B = (b_1, b_2, \ldots, b_m)$$

has linearly independent columns, precluding the possibility that one might have $Bu(t) \equiv 0$ for some $u(t)$ which does not vanish identically.

Definition 3.4. The controllable subspace for the system (3.50) is the subspace $\mathcal{C}(A, B) \subseteq E^n$ which is spanned by the columns of the matrix

$$C = (A^{n-1}B, A^{n-2}B, \ldots, AB, B) . \tag{3.51}$$

The name for this subspace is justified by

Proposition 3.5. The control problem

$$x(0) = x_0, \quad x(T) = x_1 \tag{3.52}$$

for the system (3.50) is solvable if and only if

$$x_1 - e^{AT}x_0 \in \mathcal{C}(A, B) . \tag{3.53}$$

Proof. The control problem (3.52) is solvable if and only if there is a control $u \in L_m^2[0, T]$ such that

$$x_1 - e^{AT}x_0 = \int_0^T e^{A(T-t)}Bu(t)\,dt . \tag{3.54}$$

Since the Cayley-Hamilton theorem shows that, for each t, the matrix

$$e^{A(T-t)}B = \sum_{k=0}^{\infty} \frac{(T-t)^k}{k!} A^k B$$

can be expressed as a linear combination of the submatrices $A^{n-1}B, \ldots, AB, B$

of C, the integral in (3.54) is necessarily expressible as a linear combination of the columns of C and we conclude that $x_1 - e^{AT}x_0$ must lie in $\mathscr{C}(A, B)$ if there is to be $u \in L_m^2[0, T]$ for which (3.54) holds.

To establish the sufficiency of the condition we assume (3.53) and we let ξ be a vector in E^n. Suppose that

$$\xi^* \int_0^T e^{A(T-t)} Bu(t)\, dt = 0$$

for all $u \in L_m^2[0, T]$. Then clearly

$$\xi^* e^{A(T-t)}B \equiv 0, \quad t \in [0, T]$$

and, evaluating this function and its derivatives at $t = T$ we find

$$\xi^* B = \xi^* AB = \dots = \xi^* A^{n-1} B = 0$$

and we conclude that $\xi^* \in \mathscr{C}(A, B)^{\perp}$. It follows that the range of the integral expression $\int_0^T e^{A(T-t)}Bu(t)dt$ is all of $\mathscr{C}(A, B)$ and thus (3.52) can be solved whenever $x_1 - e^{AT}x_0$ lies in this space. Q. E. D.

<u>Proposition 3.6.</u> <u>$\mathscr{C}(A, B)$ is an invariant subspace of the matrix</u> A.

<u>Proof.</u> If $x \in \mathscr{C}(A, B)$ we have

$$x = A^{n-1}Bu_1 + A^{n-2}Bu_2 + \dots + ABu_{n-1} + Bu_n$$

for some m-vectors $u_1, u_2, \dots, u_m$, by definition of $\mathscr{C}(A, B)$. Then

$$\begin{aligned} Ax &= A^n Bu_1 + A^{n-1}Bu_2 + \dots + A^2 Bu_{n-1} + ABu_n \\ &= \text{(again invoking the Cayley-Hamilton theorem)} \\ &= A^{n-1}B(u_2 - a_1u_1) + A^{n-2}B(u_3 - a_2u_1) \\ &\quad + \dots + AB(u_n - a_{n-1}u_1) + B(-a_nu_n) \end{aligned}$$

and hence continues to lie in $\mathscr{C}(A, B)$. Q. E. D.

<u>Theorem 3.7.</u> <u>Let</u> $\mathscr{C}(A, B)$ <u>have dimension</u> $\nu \le n$ <u>and let</u> P <u>be any</u>

nonsingular matrix such that the vectors in its first ν rows, $p_1, p_2, \ldots, p_\nu$ form a basis for $\mathscr{C}(A, B)$. (One need only let $p_1, p_2, \ldots, p_\nu$ be a maximal linearly independent subset of the columns of the matrix C, for example.) Then the change of variable

$$x = Py \tag{3.55}$$

carries (3.50) into

$$\dot{y} = \tilde{A}y + \tilde{B}u \tag{3.56}$$

which has the decomposition (A_1 being a $\nu \times \nu$ matrix and y^1 a ν vector)

$$\begin{pmatrix} \dot{y}^1 \\ \dot{y}^2 \end{pmatrix} = \begin{pmatrix} A_1 & A_3 \\ 0 & A_2 \end{pmatrix} \begin{pmatrix} y^1 \\ y^2 \end{pmatrix} + \begin{pmatrix} B_1 \\ 0 \end{pmatrix} u \tag{3.57}$$

Moreover, the ν-dimensional system

$$\dot{y}^1 = A_1 y^1 + B_1 u \tag{3.58}$$

is controllable.

Proof. Let $P = (P_1, P_2)$, P_1 being the $n \times \nu$ matrix whose columns $p_1, p_2, \ldots, p_\nu$ form a basis for $\mathscr{C}(A, B)$. The transformation (3.55) yields (3.56) with

$$\tilde{A} = P^{-1}AP, \quad \tilde{B} = P^{-1}B.$$

Writing $P^{-1} = \begin{pmatrix} Q_1 \\ Q_2 \end{pmatrix}$ we have

$$\begin{pmatrix} Q_1P_1 & Q_1P_2 \\ Q_2P_1 & Q_2P_2 \end{pmatrix} = \begin{pmatrix} I_\nu & 0 \\ 0 & I_{n-\nu} \end{pmatrix}.$$

Since the columns of B lie in $\mathscr{C}(A, B)$ we can write

$$B = P_1 B_1$$

where B_1 is some $\nu \times \nu$ matrix. Since $\mathcal{C}(A, B)$ is invariant under A and the columns of P_1 lie in $\mathcal{C}(A, B)$, the columns of AP_1 lie in $\mathcal{C}(A, B)$ and we can write

$$AP_1 = P_1 A_1$$

where, again, A_1 is a $\nu \times \nu$ matrix. Then

$$\tilde{B} = P^{-1}B = \begin{pmatrix} Q_1 \\ Q_2 \end{pmatrix} P_1 B_1 = \begin{pmatrix} Q_1 P_1 B_1 \\ Q_2 P_1 B_1 \end{pmatrix} = \begin{pmatrix} B_1 \\ 0 \end{pmatrix},$$

$$\tilde{A} = P^{-1}AP = \begin{pmatrix} Q_1 \\ Q_2 \end{pmatrix} (P_1 A_1, AP_2) = \begin{pmatrix} Q_1 P_1 A_1 & Q_1 AP_2 \\ Q_2 P_1 A_1 & Q_2 AP_2 \end{pmatrix} = \begin{pmatrix} A_1 & Q_1 AP_2 \\ 0 & Q_2 AP_2 \end{pmatrix} \equiv \begin{pmatrix} A_1 & A_3 \\ 0 & A_2 \end{pmatrix}.$$

To show that the ν-dimensional system (3. 58) is controllable we note that $\dim \mathcal{C}(A, B) = \operatorname{rank}(A^{n-1}B, A^{n-2}B, \ldots, AB, B) = \nu = \operatorname{rank}(\tilde{A}^{n-1}\tilde{B}, \tilde{A}^{n-2}\tilde{B}, \ldots, \tilde{A}\tilde{B}, \tilde{B})$. But

$$(\tilde{A}^{n-1}\tilde{B}, \tilde{A}^{n-2}\tilde{B}, \ldots, \tilde{A}\tilde{B}, \tilde{B}) = \left(\begin{pmatrix} (A_1)^{n-1}B_1 \\ 0 \end{pmatrix}, \begin{pmatrix} (A_1)^{n-2}B_1 \\ 0 \end{pmatrix} \ldots, \begin{pmatrix} A_1 B_1 \\ 0 \end{pmatrix}, \begin{pmatrix} B_1 \\ 0 \end{pmatrix} \right)$$

from which we conclude, again using the Cayley-Hamilton theorem, that

$$\operatorname{rank}(A_1^{\nu-1}B_1, A_1^{\nu-2}B_1, \ldots, A_1 B_1, B_1) = \operatorname{rank}(A_1^{n-1}B_1, A_1^{n-2}B_1, \ldots, A_1 B_1, B_1) = \nu$$

and the ν-dimensional system (3. 58) is controllable. Q. E. D.

We note in passing that the complementary $n-\nu$ dimensional system

$$\dot{y}_2 = A_2 y_2 + 0u$$

is "completely uncontrollable". This system is void, of course, if the original system (3. 50) is controllable.

It is clear that the eigenvalues of A_2 constitute $n-\nu$ eigenvalues of A which persist as eigenvalues of $A + BK$ for any feedback matrix K. Combined with our results on the stabilizability of controllable systems we have

<u>Proposition 3.8.</u> <u>The linear control system</u> (3.50), <u>i.e.</u>,

$$\dot{x} = Ax + Bu,$$

<u>is stabilizable if and only if, after reduction to the form</u> (3.57), A_2 <u>is a stability matrix.</u>

<u>Proof.</u> Since (3.50) is related to (3.57) by a similarity transformation, it is clear that (3.50) is stabilizable if and only if (3.57) is. Setting

$$u = Ky = K_1 y^1 + K_2 y^2$$

in (3.57), that equation becomes

$$\begin{pmatrix} \dot{y}^1 \\ \dot{y}^2 \end{pmatrix} = \begin{pmatrix} A_1 + B_1 K_1 & A_3 + B_1 K_2 \\ 0 & A_2 \end{pmatrix} \begin{pmatrix} y^1 \\ y^2 \end{pmatrix} . \tag{3.59}$$

The eigenvalues of the matrix in (3.59) consist of eigenvalues of $A_1 + B_1 K_1$ together with eigenvalues of A_2. The former can be assigned negative real parts because the pair (A_1, B_1) is controllable. The eigenvalues of A_2 cannot be changed by feedback and the proposition follows immediately.

<u>Example.</u> Suppose a scalar control u is used to influence two identical mass-spring systems (cf. (1.33) with $\gamma = 0$). The control system is

$$\begin{pmatrix} \dot{x}^1 \\ \dot{x}^2 \\ \dot{\xi}^1 \\ \dot{\xi}^2 \end{pmatrix} = \begin{pmatrix} 0 & 1 & 0 & 0 \\ -1 & 0 & 0 & 0 \\ 0 & 0 & 0 & 1 \\ 0 & 0 & -1 & 0 \end{pmatrix} \begin{pmatrix} x^1 \\ x^2 \\ \xi^1 \\ \xi^2 \end{pmatrix} + \begin{pmatrix} 0 \\ \alpha \\ 0 \\ \beta \end{pmatrix} u \tag{3.60}$$

with α, β non-zero. The vectors spanning $\mathscr{C}(A, B)$ are

$$\begin{pmatrix} 0 \\ \alpha \\ 0 \\ \beta \end{pmatrix}, \begin{pmatrix} \alpha \\ 0 \\ \beta \\ 0 \end{pmatrix}$$

as formation of C (cf. (3. 51)) readily shows. Let

$$P = \begin{pmatrix} 0 & \alpha & 0 & 0 \\ \alpha & 0 & 0 & 0 \\ 0 & \beta & 1 & 0 \\ \beta & 0 & 0 & 1 \end{pmatrix}.$$

Then

$$P^{-1} \begin{pmatrix} 0 \\ \alpha \\ 0 \\ \beta \end{pmatrix} = \begin{pmatrix} 1 \\ 0 \\ 0 \\ 0 \end{pmatrix}$$

and

$$P^{-1}AP = P^{-1} \begin{pmatrix} \alpha & 0 & 0 & 0 \\ 0 & -\alpha & 0 & 0 \\ \beta & 0 & 0 & 1 \\ 0 & -\beta & -1 & 0 \end{pmatrix} = \begin{pmatrix} 0 & -1 & 0 & 0 \\ 1 & 0 & 0 & 0 \\ 0 & 0 & 0 & 1 \\ 0 & 0 & -1 & 0 \end{pmatrix}$$

so that in this case the system (3. 57) is

$$\begin{pmatrix} \dot{y}^1 \\ y^2 \\ \eta^1 \\ \eta^2 \end{pmatrix} = \begin{pmatrix} 0 & -1 & 0 & 0 \\ 1 & 0 & 0 & 0 \\ 0 & 0 & 0 & 1 \\ 0 & 0 & -1 & 0 \end{pmatrix} \begin{pmatrix} y^1 \\ y^2 \\ \eta^1 \\ \eta^2 \end{pmatrix} + \begin{pmatrix} 1 \\ 0 \\ 0 \\ 0 \end{pmatrix} u .$$

Note that the pair $\begin{pmatrix} 0 & -1 \\ 1 & 0 \end{pmatrix}$, $\begin{pmatrix} 1 \\ 0 \end{pmatrix}$ is controllable. The matrix $\begin{pmatrix} 0 & 1 \\ -1 & 0 \end{pmatrix}$ in the lower right hand corner has eigenvalues $\pm i$ and we conclude that the system (3. 60) is not stabilizable.

4. Stabilization with Restricted Feedback, Observation Canonical Form, Observer Theory

In the discussion of the previous section we have seen that a stabilizable system

$$\dot{x} = Ax + Bu \tag{4.1}$$

can be stabilized by linear feedback

$$u = Kx \tag{4.2}$$

and, in fact, the eigenvalues of the closed-loop matrix $A + BK$ can be chosen at will if we make the slightly stronger assumption that (4.1) is controllable. This result might seem to settle the issue of stabilization once and for all, but, actually, it does not. A practical objection is the following. As we have anticipated by the introduction of the term "observation", the complete system state, x, is not always available. One normally has an observation $\omega = Hx + Ju$ which, after the known control u is accounted for, may be replaced by

$$\omega = Hx, \quad \omega \in E^r, \quad x \in E^n, \tag{4.3}$$

with $r < n$. We have seen in Section 1 that the state $x(t)$ can be recovered by integrating the product of $\omega(t)$ with a reconstruction kernel $R(t)$ over an interval $[t-T, t]$ and, at least in principle, such a technique could be used to supply the state $x(t)$ needed for formation of u as indicated in (4.2). In practice this procedure is ordinarily considered too complicated, primarily because one does not wish to include in the system the computational hardware which would be necessary to compute the integral in question. Thus one must be content with feedback which uses only the observation (4.3), which amounts to requiring that K have the form

$$K = K_1 H, \quad K_1 \ m \times r, \tag{4.4}$$

or resort to other measures. The "other measures" we will discuss presently. Let us begin by considering feedback (4.2) with K of the form (4.4).

One might suspect that the closed loop system, in this case

$$\dot{x} = (A + BK_1 H)x$$

could be stabilized if (A, B) were controllable and (H, A) observable, but, in fact, this is not necessarily the case. For the controlled mass spring system

$$\ddot{x} + x = u,$$

or

$$\begin{pmatrix} \dot{x}^1 \\ \dot{x}^2 \end{pmatrix} = \begin{pmatrix} 0 & 1 \\ -1 & 0 \end{pmatrix} \begin{pmatrix} x^1 \\ x^2 \end{pmatrix} + \begin{pmatrix} 0 \\ 1 \end{pmatrix} u \tag{4.5}$$

with observation

$$\omega = x = x^1 = (1, 0) \begin{pmatrix} x^1 \\ x^2 \end{pmatrix},$$

the general restricted feedback control law is

$$u = k_1 (1, 0) \begin{pmatrix} x^1 \\ x^2 \end{pmatrix} \tag{4.6}$$

where $k_1 = K_1$ is scalar. The closed loop system is

$$\begin{pmatrix} \dot{x}^1 \\ \dot{x}^2 \end{pmatrix} = \begin{pmatrix} 0 & 1 \\ -1 + k_1 & 0 \end{pmatrix} \begin{pmatrix} x^1 \\ x^2 \end{pmatrix}.$$

The trace of the matrix here is zero for any k_1, and since the trace is the sum of the eigenvalues, it is clearly impossible for both of those eigenvalues to have negative real part. Thus stabilization of (4.5) with restricted feedback (4.6) is impossible. One easily sees that $\left(\begin{pmatrix} 0 & 1 \\ -1 & 0 \end{pmatrix}, \begin{pmatrix} 0 \\ 1 \end{pmatrix}\right)$ is controllable and $\left((1, 0), \begin{pmatrix} 0 & 1 \\ -1 & 0 \end{pmatrix}\right)$ is observable.

If we take, in place of (4.6),

$$\omega = \dot{x} = x^2 = (0, 1)\begin{pmatrix} x^1 \\ x^2 \end{pmatrix}$$

so that feedback

$$u = k_1 (0, 1)\begin{pmatrix} x^1 \\ x^2 \end{pmatrix}$$

is allowed, it turns out that (4. 5) is stabilizable in this case. Thus it is difficult to see the outline of any general theory. Some computational and theoretical techniques have been developed but they are not wholly satisfactory. (See [7], [8].)

What is frequently done in practice is the following. One ignores the feedback restriction (4. 4) and determines an $m \times n$ feedback matrix K such that $u = Kx$ yields $\dot{x} = (A + BK)x$ with $A + BK$ a stability matrix. Then one constructs (electronically, usually) a second "observer" or "state estimator" system

$$\dot{y} = (A + BK)y + L(\omega - Hy) = (A + BK - LH)y + L\omega , \tag{4.7}$$

for some "feed-forward" matrix L of dimension $n \times r$. Here ω is the observation $\omega(t) = Hx(t)$ continuously available from the ongoing process (4. 1), (4. 3). It is fed into the system (4. 7) as an external input, modified by the matrix L. Now (4. 7), being a "laboratory" system, is assumed completely accessible, i. e., we assume that all components of its state, y, are available to us. This being assumed, one generates u in (4. 1) by

$$u = Ky \tag{4.8}$$

instead of $u = Kx$, which is, in general, inadmissible if only the observation $\omega = Hx$ is available to us. Then (4. 1), (4. 7), (4. 8) constitute a coupled system

$$\dot{x} = Ax + BKy \tag{4.9}$$

$$\dot{y} = (A + BK - LH)y + LHx , \tag{4.10}$$

or, in partitioned matrix notation,

$$\begin{pmatrix} \dot{x} \\ y \end{pmatrix} = \begin{pmatrix} A & BK \\ LH & A + BK - LH \end{pmatrix} \begin{pmatrix} x \\ y \end{pmatrix} . \qquad (4.11)$$

We now need

Definition 4.1. The linear observed system

$$\dot{x} = Ax, \qquad x \in E^n$$

$$\omega = Hx, \qquad \omega \in E^r,$$

is detectable (alternatively, (H, A) is a detectable pair) if and only if there exists an $n \times r$ feed-forward matrix L such that $A - LH$ has only eigenvalues with negative real parts; i.e., is a stability matrix.

Proposition 4.2. The pair (H, A) is detectable if and only if the pair $(A^*, -H^*)$ is stabilizable.

Proof. We have

$$(A - LH)^* = A^* + (-H^*)L^* .$$

If $(A^*, -H^*)$ is stabilizable an $r \times n$ feedback matrix L^* can be found so that $A^* + (-H^*)L^*$ is a stability matrix. But then $A - LH$ is also a stability matrix. Conversely, if L can be found so that $A - LH$ is a stability matrix, then $A^* + (-H^*)L^*$ is a stability matrix. Q. E. D.

Theorem 4.3. If (A, B) is stabilizable and (H, A) is detectable, then there exist matrices K, L, $m \times n$ and $n \times r$, respectively, such that the matrix in (4.11), i.e.,

$$\begin{pmatrix} A & BK \\ LH & A + BK - LH \end{pmatrix} \qquad (4.12)$$

is a stability matrix.

Proof. Let K and L be chosen so that $A + BK$ and $A - LH$ are stability matrices. The eigenvalues of (4.12) coincide with the eigenvalues of

$$\begin{pmatrix} I_n & 0 \\ I_n & -I_n \end{pmatrix}^{-1} \begin{pmatrix} A & BK \\ LH & A+BK-LH \end{pmatrix} \begin{pmatrix} I_n & 0 \\ I_n & -I_n \end{pmatrix} = \begin{pmatrix} I_n & 0 \\ I_n & -I_n \end{pmatrix} \begin{pmatrix} A+BK & -BK \\ A+BK & -A-BK+LH \end{pmatrix}$$

$$= \begin{pmatrix} A+BK & -BK \\ 0 & A-LH \end{pmatrix} .$$

Since the eigenvalues of the matrix on the right hand side of this matrix are those, jointly, of $A + BK$ and $A - LH$, the theorem follows. Q. E. D.

As an example, consider the controlled harmonic oscillator

$$\begin{pmatrix} \dot{x}^1 \\ \dot{x}^2 \end{pmatrix} = \begin{pmatrix} 0 & 1 \\ -1 & 0 \end{pmatrix} \begin{pmatrix} x^1 \\ x^2 \end{pmatrix} + \begin{pmatrix} 0 \\ u \end{pmatrix} .$$

The feedback relation

$$u = -2x^2 = (0, -2) \begin{pmatrix} x^1 \\ x^2 \end{pmatrix}$$

yields a closed loop system with -1 as an eigenvalue of multiplicity 2. But if we assume, as in (4.6), that only x^1 can be directly observed, this is an inadmissible feedback relation.

We introduce the observer system (cf. (4.10))

$$\begin{pmatrix} \dot{y}^1 \\ \dot{y}^2 \end{pmatrix} = \left[\begin{pmatrix} 0 & 1 \\ -1 & -2 \end{pmatrix} - \begin{pmatrix} \ell^1 \\ \ell^2 \end{pmatrix} (1, 0) \right] \begin{pmatrix} y^1 \\ y^2 \end{pmatrix} + \begin{pmatrix} \ell^1 \\ \ell^2 \end{pmatrix} (1, 0) \begin{pmatrix} x^1 \\ x^2 \end{pmatrix} .$$

The matrix $A - LH$ in this case is

$$\begin{pmatrix} 0 & 1 \\ -1 & 0 \end{pmatrix} - \begin{pmatrix} \ell^1 \\ \ell^2 \end{pmatrix} (1, 0) = \begin{pmatrix} -\ell^1 & 1 \\ -1-\ell^2 & 0 \end{pmatrix}$$

and is stable, for example, if we take

$$\ell^1 = 4, \quad \ell^2 = 3 .$$

Indeed $A - LH$ has -2 as an eigenvalue of multiplicity 2. Then taking

$$u = (0, -2)\begin{pmatrix} y^1 \\ y^2 \end{pmatrix} = -2y^2$$

we have the composite system

$$\begin{pmatrix} \dot{x}^1 \\ \dot{x}^2 \\ \dot{y}^1 \\ \dot{y}^2 \end{pmatrix} = \begin{pmatrix} 0 & 1 & 0 & 0 \\ -1 & 0 & 0 & -2 \\ 4 & 0 & -4 & 1 \\ 3 & 0 & -4 & -2 \end{pmatrix}\begin{pmatrix} x^1 \\ x^2 \\ y^1 \\ y^2 \end{pmatrix}. \tag{4.13}$$

The characteristic polynomial is easily seen to be

$$p(\lambda) = \lambda^4 + 6\lambda^3 + 13\lambda^2 + 12\lambda + 4$$

with zeros -1, -2, each of multiplicity 2, in agreement with the theory.

The observer system (4.10), i.e.,

$$\dot{y} = (A + BK)y + L(\omega - Hy)$$

is important for reasons which go beyond the problem of stabilization with restricted observation. We will discuss these questions in later sections.

Having seen that the eigenvalues of the composite system (4.11) are those of $A + BK$ together with those of $A - LH$, and knowing from Theorem 3.3 that if (A, B) is controllable the eigenvalues of $A + BK$ can be placed at will by appropriate choice of the $m \times n$ feedback matrix K, it is natural to ask if all eigenvalues of (4.12) can be placed at will if, in addition, (H, A) is observable, which reduces to the question of the placement of the eigenvalues of $A - LH$ by appropriate selection of the $n \times r$ "feedforward" matrix L. It is quite immediate that this can be done if the pair (H, A) is observable, for then $(A^*, -H^*)$ is controllable and, given any desired eigenvalues $\mu_1, \mu_2, \ldots, \mu_n$, we can find an $r \times n$ matrix L^* such that $A^* - H^*L^*$ has the eigenvalues $\bar{\mu}_1, \bar{\mu}_2, \ldots, \bar{\mu}_n$. Then $A - LH$ has the eigenvalues $\mu_1, \mu_2, \ldots, \mu_n$.

This result can also be obtained by reducing the linear observed system

$$\dot{y} = Ay \tag{4.14}$$

$$\omega = Hy \tag{4.15}$$

to "observation canonical form". This process is just the transposition of the corresponding process for the control canonical form and we discuss it explicitly only for the case of a scalar observation

$$\omega = h^* y,$$

leaving the case of a higher dimensional observation as an exercise. Assuming (4.14), (4.15) observable, the "observability matrix"

$$O = O(h^*, A) = \begin{pmatrix} h^* A^{n-1} \\ \vdots \\ h^* A \\ h^* \end{pmatrix} \tag{4.16}$$

is nonsingular and one may consider the transformation

$$y = O^{-1}\tilde{\eta},$$

which carries (4.14), (4.15) into

$$\dot{\tilde{\eta}} = OAO^{-1}\tilde{\eta} \equiv \tilde{\mathcal{A}}\,\tilde{\eta} \tag{4.17}$$

$$\omega = h^* O^{-1}\tilde{\eta}. \tag{4.18}$$

Since h^* is the last row of O, we have

$$h^* O^{-1} = e_n^* \quad \text{(last row of } n \times n \text{ identity matrix)},$$

so that (4.18) amounts to

$$\omega = e_n^* \tilde{\eta} \quad (= \tilde{\eta}^n). \tag{4.19}$$

On the other hand

$$OA = \begin{pmatrix} h^*A^n \\ \vdots \\ h^*A^2 \\ h^*A \end{pmatrix} .$$

Comparing with (4.16), we see that

$$OAO^{-1} = \begin{pmatrix} h^*A^nO^{-1} \\ e_1^* \\ \vdots \\ e_{n-1}^* \end{pmatrix} .$$

Again using the Cayley-Hamilton theorem and supposing the characteristic polynomial of A to be

$$a(\mu) = \mu^n + a_1\mu^{n-1} + \dots + a_{n-1}\mu + a_n ,$$

we have

$$h^*A^nO^{-1} = h^*(-a_1A^{n-1} \dots - a_{n-1}A - a_nI)O^{-1}$$

$$= - a_1e_1^* \dots - a_{n-1}e_{n-1}^* - a_ne_n^* = - (a_1, \dots, a_{n-1}, a_n).$$

Thus the system (4.17), (4.18) reduces to (4.19) and

$$\begin{pmatrix} \dot{\tilde{\eta}}^1 \\ \dot{\tilde{\eta}}^2 \\ \vdots \\ \dot{\tilde{\eta}}^{n-1} \\ \dot{\tilde{\eta}}^n \end{pmatrix} = \begin{pmatrix} -a_1 & -a_2 & \dots & -a_{n-1} & -a_n \\ 1 & 0 & \dots & 0 & 0 \\ \vdots & \vdots & & \vdots & \vdots \\ 0 & 0 & \dots & 0 & 0 \\ 0 & 0 & \dots & 0 & 0 \end{pmatrix} \begin{pmatrix} \tilde{\eta}^1 \\ \tilde{\eta}^2 \\ \vdots \\ \tilde{\eta}^{n-1} \\ \tilde{\eta}^n \end{pmatrix} . \tag{4.20}$$

Next we introduce the matrix

$$\Theta = \begin{pmatrix} 1 & a_1 & \cdots & a_{n-2} & a_{n-1} \\ 0 & 1 & \cdots & a_{n-3} & a_{n-2} \\ \vdots & \vdots & & \vdots & \vdots \\ 0 & 0 & \cdots & 1 & a_1 \\ 0 & 0 & \cdots & 0 & 1 \end{pmatrix}$$

and carry out the transformation

$$\tilde{\eta} = \Theta^{-1}\eta \,.$$

We find immediately that

$$\omega = e_n^* \Theta^{-1} \eta = e_n^* \eta$$

since Θ is again an upper triangular matrix with "1's" on the main diagonal. Now, from the form of $\tilde{\mathcal{A}}$ (cf. (4.17), (4.20)) we have

$$\Theta \tilde{A} = \begin{pmatrix} -a_n e_n^* \\ \theta^1 - a_{n-1} e_n^* \\ \vdots \\ \theta^{n-2} - a_2 e_n^* \\ \theta^{n-1} - a_1 e_n \end{pmatrix},$$

where θ^k, $k = 1, 2, \ldots, n$, are the rows of θ. Then

$$\Theta\tilde{A}\Theta^{-1} = \begin{pmatrix} -a_n e_n^* \\ e_1^* - a_{n-1} e_n^* \\ \vdots \\ e_{n-2}^* - a_2 e_n \\ e_{n-1}^* - a_1 e_n \end{pmatrix} = \begin{pmatrix} 0 & 0 & \cdots & 0 & -a_n \\ 1 & 0 & \cdots & 0 & -a_{n-1} \\ \vdots & \vdots & & \vdots & \vdots \\ 0 & 0 & \cdots & 0 & -a_2 \\ 0 & 0 & \cdots & 1 & -a_1 \end{pmatrix} \equiv \hat{\mathcal{A}} \,. \quad (4.21)$$

Thus we have proved

Theorem 4.4. There is a nonsingular matrix $P(= O^{-1}\Theta^{-1})$ such that the change of variable

$$y = P\eta$$

carries the linear observed system (4.14), (4.15) into

$$\dot{\eta} = \hat{\mathcal{A}}\eta$$

$$\omega = e_n^* \eta$$

with $\hat{\mathcal{A}}$ as shown in (4.20). Thus, given any complex numbers $\mu_1, \mu_2, \ldots, \mu_n$, an $n \times 1$ matrix (vector) ℓ can be found such that $A - \ell h^*$ has these numbers as eigenvalues.

Proof. The first part has already been established. For the eigenvalue placement property we note that

$$\hat{\mathcal{A}} - \hat{\ell} e_n^* = \begin{pmatrix} 0 & 0 & \cdots & 0 & -a_n - \hat{\ell}^1 \\ 1 & 0 & \cdots & 0 & -a_{n-1} - \hat{\ell}^2 \\ \vdots & \vdots & & \vdots & \vdots \\ 0 & 0 & \cdots & 0 & -a_2 - \hat{\ell}^{n-1} \\ 0 & 0 & \cdots & 1 & -a_1 - \hat{\ell}^n \end{pmatrix}$$

has the characteristic polynomial

$$\mu^n + (a_1 + \hat{\ell}^n)\mu^{n-1} + \ldots + (a_{n-1} + \hat{\ell}^2)\mu + (a_n + \hat{\ell}^1).$$

Since its coefficients can be chosen at will, so can its zeros, which are $\mu_1, \mu_2, \ldots, \mu_n$. The same is then true of

$$P(\hat{\mathcal{A}} - \hat{\ell} e_n^*)P^{-1} = A - P\hat{\ell} h^* \equiv A - \ell h^*$$

and the proof is complete.

A comparable result is true for the case of a multi-dimensional observation. (See Exercise 8 at the end of this chapter.)

The analysis of partially observable systems proceeds in much the same way, again transposing the developments for partially controllable systems. We consider (4.14), (4.15) again, i.e.,

$$\dot{y} = Ay$$

$$\omega = Hy$$

and we suppose the rows of H to be linearly independent. We define (H, A) to be the subspace of E^{n*} spanned by the rows of

$$O(H,A) = \begin{pmatrix} HA^{n-1} \\ HA^{n-2} \\ \vdots \\ HA \\ H \end{pmatrix} .$$

We suppose that the dimension of $O(H,A) = \rho$ and we let F be a $\rho \times n$ matrix whose rows, selected from those of $O(H,A)$ and including the rows of H, form a basis for $O(H,A)$. Then we let

$$Q = \begin{pmatrix} F \\ G \end{pmatrix} ,$$

G being an $(n-\rho) \times n$ matrix such that Q is nonsingular. One may verify that if φ is a row vector in $O(H,A)$ then φA also lies in this space. It follows, therefore, that

$$FA = JF \tag{4.22}$$

for some $(n-\rho) \times (n-\rho)$ matrix J.

Now we let

$$y = Q^{-1}w,$$

thereby changing (4.14), (4.15) to

$$\dot{w} = \begin{pmatrix} F \\ G \end{pmatrix} A(\Phi, \Gamma)w \tag{4.23}$$

$$w = H(\Phi, \Gamma)w \tag{4.24}$$

where Φ is $n \times \rho$, Γ is $n \times (n-\rho)$ and $Q^{-1} = (\Phi, \Gamma)$. Using (4.22) we see that

$$\begin{pmatrix} F \\ G \end{pmatrix} A(\Phi, \Gamma) = \begin{pmatrix} FA\Phi & FA\Gamma \\ GA\Phi & GA\Gamma \end{pmatrix} = \begin{pmatrix} JF\Phi & JF\Gamma \\ GA\Phi & GA\Gamma \end{pmatrix} = \begin{pmatrix} \mathcal{A}_1^1 & 0 \\ \mathcal{A}_1^2 & \mathcal{A}_2^2 \end{pmatrix} \tag{4.25}$$

and, since the rows of H are included in those of F, (4.24) takes the form

$$\omega = (\mathcal{H}_1, 0)\begin{pmatrix} w^1 \\ w^2 \end{pmatrix} . \tag{4.26}$$

From (4.25), the transformed system (4.23) is

$$\begin{pmatrix} \dot{w}^1 \\ w^2 \end{pmatrix} = \begin{pmatrix} \mathcal{A}_1^1 & 0 \\ \mathcal{A}_1^2 & \mathcal{A}_2^2 \end{pmatrix}\begin{pmatrix} w^1 \\ w^2 \end{pmatrix} . \tag{4.27}$$

An argument similar to that given earlier for control systems shows that the pair $(\mathcal{H}_1, \mathcal{A}_1^1)$ is observable. On the other hand, any initial state

$$\begin{pmatrix} w^1(0) \\ w^2(0) \end{pmatrix} = \begin{pmatrix} 0 \\ w_0^2 \end{pmatrix}$$

leads to a solution of the form $\begin{pmatrix} 0 \\ w^2(t) \end{pmatrix}$ for which the observation

$$w(t) = (\mathcal{H}_1, 0)\begin{pmatrix} 0 \\ w^2(t) \end{pmatrix} \equiv 0 .$$

<u>Proposition 4.5.</u> <u>The linear observed system</u> (4.14), (4.15) <u>is detectable if and only if</u>

$$\lim_{t \to \infty} \|\omega(t)\| = 0$$

$$\Longrightarrow \lim_{t \to \infty} \|x(t)\| = 0 .$$

This follows from Definition 4.1, Theorem 4.4, the form (4.27) and the next proposition.

Proposition 4.6. The linear observed system (4.14), (4.15) is detectable if and only if, in the form (4.23), all of the eigenvalues of the matrix $\mathcal{A}_2^2$ have negative real parts.

One can also see that the eigenvalues of $\mathcal{A}_2^2$ are precisely those eigenvalues of A which persist as eigenvalues of $A - LH$ for every choice of L.

The observer structure discussed in Theorem 3.11 and the preceding material is by no means the only way in which the system (4.1) can be stabilized using the information available via the observation (4.3). An extensive "theory of observers" has been developed by Luenberger [9] and others. One indication that a more general theory is required arises from the fact that r components of the state, i.e., the components of Hx, are already available. It seems superfluous, then, to have to use an n-dimensional observer system (4.7) to generate $n-r$ complementary state components. Luenberger's theory is particularly addressed to this question of the construction of "reduced order" observers. The theory is based on the observation that almost any system

$$\dot{z} = Fz + Gx, \quad z \in E^{\nu}, \tag{4.28}$$

wherein F is a stability matrix and the "forcing term", Gx, arises from the system

$$\dot{x} = Ax, \tag{4.29}$$

can serve as an observer for (4.29) in the sense that $z(t)$ tends to a linear function of $x(t)$. This is contained in

Lemma 4.7. If there is a $\nu \times n$ matrix T which solves the matrix equation

$$FT - TA + G = 0 \tag{4.30}$$

then

$$z(t) = Tx(t) + e^{Ft}(z(0) - Tx(0)). \tag{4.31}$$

Proof. Using (4.29), (4.30), (4.31) we have

$$\frac{d}{dt}(z(t) - Tx(t)) = Fz(t) + Gx(t) - TAx(t)$$

$$= Fz(t) + Gx(t) - FTx(t) - Gx(t) = F(z(t) - Tx(t)).$$

Hence $z(t) - Tx(t)$ satisfies the linear homogeneous differential equation in E^{ν} with coefficient matrix F and it follows that

$$z(t) - Tx(t) = e^{Ft}(z(0) - Tx(0))$$

which is the same as (4. 31). Q. E. D.

We can extend this result very easily to systems

$$\dot{x} = Ax + Bu \tag{4. 32}$$

just by adding a corresponding forcing term to the system (4. 28), viz.:

$$\dot{z} = Fz + Gx + TBu . \tag{4. 33}$$

If x, z satisfy (4. 32), (4. 33), respectively, then $z(t)$, $Tx(t)$ will still satisfy (4. 31). Since $u(t)$ is presumably known we assume that we can form the term $TBu(t)$ and insert it in (4. 33).

Suppose all of the eigenvalues of F lie to the left of those of A. We may then expect the term $e^{Ft}(z(0) - Tx(0))$ to tend to zero more rapidly than $Tx(t)$ so that, in this sense, we have the asymptotic relationship

$$z(t) \cong Tx(t), \qquad t \to \infty.$$

This enables us, in the limit, to infer certain information about $x(t)$ from $z(t)$. In fact suppose the matrix

$$\begin{pmatrix} H \\ T \end{pmatrix} \tag{4. 34}$$

is invertible. Then $\begin{pmatrix} H \\ T \end{pmatrix}^{-1} \begin{pmatrix} H \\ T \end{pmatrix} x = x$, or

$$\begin{pmatrix} H \\ T \end{pmatrix}^{-1} \begin{pmatrix} \omega(t) \\ z(t) - e^{Ft}(z(0) - Tx(0)) \end{pmatrix} = x(t)$$

whence

$$\begin{pmatrix} H \\ T \end{pmatrix}^{-1} \begin{pmatrix} \omega(t) \\ z(t) \end{pmatrix} \cong x(t), \qquad t \to \infty .$$

Theorem 4.8. Let T be a matrix of dimension (n-r) × n satisfying (4.30) for some (n-r) × (n-r) stability matrix F and suppose that the matrix (4.34) is invertible. Let K be an m × n feedback matrix such that A + BK is a stability matrix. Then, writing

$$\begin{pmatrix} H \\ T \end{pmatrix}^{-1} = (M_1, M_2)$$

and setting

$$u = KM_1\omega + KM_2 z = KM_1 Hx + KM_2 z \qquad (4.35)$$

in (4.32) and (4.33), we have the composite "original plant-observer" system

$$\begin{pmatrix} \dot{x} \\ z \end{pmatrix} = \begin{pmatrix} A + BKM_1 H & BKM_2 \\ G + TBKM_1 H & F + TBKM_2 \end{pmatrix} \begin{pmatrix} x \\ z \end{pmatrix} \qquad (4.36)$$

and this system is asymptotically stable, indeed the eigenvalues of this (2n-r) × (2n-r) dimensional system are those of A + BK together with those of F.

Proof. We replace z by a new variable

$$\zeta = z - Tx ,$$

the "error" between the "observation" z and Tx. This corresponds to the nonsingular change of variable

$$\begin{pmatrix} x \\ \zeta \end{pmatrix} = \begin{pmatrix} I_n & 0 \\ -T & I_{n-r} \end{pmatrix} \begin{pmatrix} x \\ z \end{pmatrix} .$$

In the x, ζ coordinates we have

$$\dot{x} = (A + BKM_1H)x + (BKM_2)(\zeta + Tx)$$

$$= [A + BK(M_1, M_2)\binom{H}{T}]\, x + BKM_2\zeta = (A + BK)x + BKM_2\zeta$$

$$\dot{\zeta} = (G + TBKM_1H)x + (F + TBKM_2)z - T(A+BKM_1H)x - TBKM_2z$$

$$= (G - TA)x + Fz = (G-TA)x + F(\zeta + Tx) = (G-TA+FT)x + F\zeta = F\zeta,$$

i. e.

$$\begin{pmatrix} \dot{x} \\ \dot{\zeta} \end{pmatrix} = \begin{pmatrix} A+BK & BKM_2 \\ 0 & F \end{pmatrix}\begin{pmatrix} x \\ \zeta \end{pmatrix},$$

from which the result follows. Q. E. D.

Naturally, the most convenient observer is the one which arises for $T = I$ in which case $\omega = Hx$ does not need to be used directly. We can hope for this only when $\nu = n$. If $T = I$ then (4. 30) reads $F - A + G = 0$ so that

$$F = A - G$$

and the observer system (4. 33) becomes

$$\dot{z} = (A - G)z + Gx + Bu.$$

If the only information on $x(t)$ which is directly available is the observation

$$\omega = Hx \tag{4. 37}$$

then, necessarily, G must take the form $G = LHx$ and we have

$$\dot{z} = (A - LH)z + LHx + Bu. \tag{4. 38}$$

If in (4. 32) and we now set

$$u = Kz$$

we obtain the system studied in Theorem 3. 11 and the preceding material.

Our main interest lies, however, with the possibility of constructing an observer system of reduced dimension, namely, dimension $n-r$ if ω has dimension r, and still such that the composite system consisting of the original system and the observer can be assigned arbitrary eigenvalues.

We consider therefore the linear observed system (4. 32), (4. 33), assuming that the rows of H are linearly independent. An easy change of variables then allows us to assume that H has the form

$$H = (I_r, \ 0) \tag{4.39}$$

and introducing the partitioning

$$x = \binom{y}{w}, \quad y \in E^r, \quad w \in E^{n-r}$$

we now have

$$w = y,$$

and (4. 32) becomes

$$\begin{pmatrix} \dot{y} \\ w \end{pmatrix} = \begin{pmatrix} A_1^1 & A_2^1 \\ A_1^2 & A_2^2 \end{pmatrix} \begin{pmatrix} y \\ w \end{pmatrix} + \begin{pmatrix} B^1 \\ B^2 \end{pmatrix} u \ . \tag{4.40}$$

The basic idea is the following. The vector y is available and u is assumed known. If we differentiate and use (4. 40) we have

$$\dot{y} = A_1^1 y + A_2^1 w + B^1 u \ ,$$

which now gives us

$$A_2^1 w = \dot{y} - A_1^1 y - B^1 u \ .$$

One then attempts to construct an ordinary "identity observer", as described above, for the state component w, viewing A_2^1 as an observation on w. This prompts

Lemma 4. 9. The pair $((I_r, 0), A)$ *is observable if and only if* (A_2^1, A_2^2) *is observable.*

Proof. Suppose there were a non-zero solution w(t) of

$$\dot{w} = A_2^2 w \tag{4.41}$$

for which

$$A_2^1 w(t) \equiv 0.$$

Then $\binom{0}{w(t)}$ (in which $y(t) \equiv 0$) is a non-zero solution of

$$\begin{pmatrix} \dot{y} \\ w \end{pmatrix} = \begin{pmatrix} A_1^1 & A_2^1 \\ A_1^2 & A_2^2 \end{pmatrix} \begin{pmatrix} y \\ w \end{pmatrix} \tag{4.42}$$

and we conclude $((I_r, 0), A)$ is not observable. We conclude that if $((I_r, 0), A)$ is observable, then (A_2^1, A_2^2) is observable.

Suppose there were a non-zero solution of (4.42) of the form $\binom{0}{w(t)}$, i.e., a solution for which the observation $y(t) \equiv 0$. Then $w(t)$ is a non-zero solution of (4.41) for which

$$A_2^1 w(t) = \dot{y}(t) - A_1^1 y(t) \equiv 0$$

and we would conclude that (A_2^1, A_2^2) is not observable. Thus the observability of (A_2^1, A_2^2) implies that of $((I_r, 0), A)$. Q. E. D.

To actually construct the observer we initially define it in the form, suggested by our earlier observations,

$$\begin{aligned} \dot{\hat{w}} = {} & (A_2^2 - LA_2^1) w(t) + A_1^2 y(t) + B^2 u(t) \\ & + L(\dot{y}(t) - A_1^1 y(t)) - LB^1 u(t), \end{aligned}$$

selecting the $(n-r) \times r$ matrix L so that $A_2^2 - LA_2^1$ is a stability matrix - indeed, from Lemma 3.16 the eigenvalues of $A_2^2 - LA_2^1$ can be selected at will. Now defining

$$z(t) = \hat{w}(t) - Ly(t)$$

we have the final form of the observer:

$$\dot{z} = (A_2^2 - LA_2^1)z + (A_2^2 - LA_2^1)Ly + (A_1^2 - LA_1^1)y + (B^2 - LB^1)u(t) \tag{4.43}$$

This corresponds to

$$F = A_2^2 - LA_2^1 \tag{4.44}$$

$$G = ((A_2^2 - LA_2^1)L + (A_1^2 - LA_1^1), \; 0) . \tag{4.45}$$

Since the term multiplying $u(t)$ should be TB, and it is actually $-LB^1 + B^2$, we are led to take

$$T = (-L, I_{n-r}). \tag{4.46}$$

Then (4.43) becomes

$$\dot{z} = Fz + Gx + TBu$$

as in (4.33). The matrix $F = A_2^2 - LA_2^1$ has already been made stable by choice of L. We have

$$FT - TA = (A_2^2 - LA_2^1)(-L, I_{n-r}) - (-L, I_{n-r})\begin{pmatrix} A_1^1 & A_2^1 \\ A_1^2 & A_2^2 \end{pmatrix}$$

$$= (-(A_2^2 - LA_2^1)L, \; A_2^2 - LA_2^1) + (LA_1^1 - A_1^2, \; LA_2^1 - A_2^2)$$

$$= (-(A_2^2 - LA_2^1)L - (A_1^2 - LA_1^1), \; 0) = -G. \tag{4.47}$$

Thus all of the assumptions of Theorem 4.8 are verified and the construction of a reduced order observer with arbitrary eigenvalues has been shown to be feasible. This result is summarized in

Theorem 4.10. If the linear observed system

$$\dot{x} = Ax, \quad x \in E^n$$

$$\omega = Hx, \quad \omega \in E^r$$

is observable, then there exists a reduced $(n-r)$-th order observer (4.33) with F, G, T as described above, and the $2n-r$ eigenvalues of the composite "original plant-observer" system (4.36) can be chosen at will.

We give a simple example of the application of this method, along the same lines as our original example for the observer of full dimension. We again consider the controlled linear harmonic oscillator

$$\begin{pmatrix} \dot{x}^1 \\ \dot{x}^2 \end{pmatrix} = \begin{pmatrix} 0 & 1 \\ -1 & 0 \end{pmatrix} \begin{pmatrix} x^1 \\ x^2 \end{pmatrix} + \begin{pmatrix} 0 \\ 1 \end{pmatrix} u \tag{4.48}$$

and assume that by measuring $\dot{x}^2$ $(= \ddot{x})$ we can obtain

$$\omega = x^1 = (1, 0) \begin{pmatrix} x^1 \\ x^2 \end{pmatrix} = H \begin{pmatrix} x^1 \\ x^2 \end{pmatrix} . \tag{4.49}$$

One can realize -1 as a double eigenvalue of the closed loop system by setting

$$u = -2x^2 , \tag{4.50}$$

but this is inadmissible if (4.49) is the only information available. Let us set as our goal the construction of a reduced dimension observer (dimension 1 in this case, of course) such that the combined "original plant-observer" system has -1 as a triple eigenvalue.

The system (4.48), (4.49) is already in the form corresponding to (4.39), (4.40). We thus have, with $y = x^1$, $w = x^2$

$$\omega = y$$

$$\begin{pmatrix} \dot{y} \\ \dot{w} \end{pmatrix} = \begin{pmatrix} 0 & 1 \\ -1 & 0 \end{pmatrix} \begin{pmatrix} y \\ w \end{pmatrix} + \begin{pmatrix} 0 \\ 1 \end{pmatrix} u ,$$

and (cf. (4.40)) we have

$$A_1^1 = 0, \; A_2^1 = 1, \; A_1^2 = -1, \; A_2^2 = 0, \; B^1 = 0, \; B^2 = 1.$$

The matrix $A_2^2 - LA_2^1$ of (4.43) is just

$$0 - L \cdot 1 = -L$$

and it has (trivially) the eigenvalue -1 if we take

$$L = 1.$$

The equation (4.43) is then

$$\dot{z} = -z - 2y + u \ . \tag{4.51}$$

Continuing to (4.44), (4.45), (4.46) we have

$$F = -1, \quad G = (-2, 0), \quad T = (-1, 1)$$

and since

$$FT - TA = (-1)(-1, 1) - (-1, 1)\begin{pmatrix} 0 & 1 \\ -1 & 0 \end{pmatrix} = (2, 0) = -G ,$$

equation (4.47) is verified. Referring back to Theorem 4.8, we have

$$\begin{pmatrix} H \\ T \end{pmatrix} = \begin{pmatrix} 0 & 0 \\ -1 & 1 \end{pmatrix}, \quad \begin{pmatrix} H \\ T \end{pmatrix}^{-1} = \begin{pmatrix} 1 & 0 \\ 1 & 1 \end{pmatrix}$$

and thus

$$M_1 = \begin{pmatrix} 1 \\ 1 \end{pmatrix}, \quad M_2 = \begin{pmatrix} 0 \\ 1 \end{pmatrix}.$$

Accordingly, the feedback relation determining u is (cf. (4.35), (4.50))

$$u = (0, -2)\begin{pmatrix} 1 \\ 1 \end{pmatrix} y + (0, -2)\begin{pmatrix} 0 \\ 1 \end{pmatrix} z = -2y - 2z \ . \tag{4.52}$$

Substituting (4.52) into (4.48) (going back to x^1, x^2 now) and (4.51) we have the composite third order system

$$\begin{pmatrix} \dot{x}^1 \\ \dot{x}^2 \\ \dot{z} \end{pmatrix} = \begin{pmatrix} 0 & 1 & 0 \\ -3 & 0 & -2 \\ -4 & 0 & -3 \end{pmatrix}\begin{pmatrix} x^1 \\ x^2 \\ z \end{pmatrix}. \tag{4.53}$$

The characteristic polynomial is easily seen to be $\lambda^3 + 3\lambda^2 + 3\lambda + 1$, so (4.53) has -1 as an eigenvalue of multiplicity 3.

5. Transfer Matrices and Transfer Functions

The use of the matrix T, defined by (4.30), in Lemma 4.7 and Theorem 4.8 to show the observer system state to be, asymptotically, a linear transformation of the system state x, constitutes an example of a technique which has very wide applicability in systems theory. The basic scenario is a partially coupled system

$$\begin{aligned} \dot{x} &= Ax + By, \\ \dot{y} &= Cy, \end{aligned} \quad \text{i. e.,} \quad \begin{pmatrix} x \\ y \end{pmatrix}^{\bullet} = \begin{pmatrix} A & B \\ 0 & C \end{pmatrix} \begin{pmatrix} x \\ y \end{pmatrix}, \quad \begin{aligned} x &\in E^n, \\ y &\in E^p, \end{aligned} \tag{5. 1}$$

wherein the evolution of the vector $x(t)$ is influenced by that of $y(t)$ - but not vice versa. The first question which might occur to one might very well be that of decoupling: can (5. 1) be reduced to two systems of dimension n and p which do not influence each other, i. e.

$$\begin{pmatrix} \xi \\ \eta \end{pmatrix}^{\bullet} = \begin{pmatrix} D & 0 \\ 0 & E \end{pmatrix} \begin{pmatrix} \xi \\ \eta \end{pmatrix} ? \tag{5. 2}$$

If we set

$$\begin{pmatrix} x \\ y \end{pmatrix} = \begin{pmatrix} I & T \\ 0 & I \end{pmatrix} \begin{pmatrix} \xi \\ \eta \end{pmatrix}, \quad \text{i. e.,} \quad \begin{aligned} \eta &= y \\ x = \xi + T\eta &= \xi + Ty, \end{aligned}$$

the result is that (5. 1) becomes

$$\begin{pmatrix} \xi \\ y \end{pmatrix}^{\bullet} = \begin{pmatrix} I & T \\ 0 & I \end{pmatrix} \begin{pmatrix} A & B \\ 0 & C \end{pmatrix} \begin{pmatrix} I & T \\ 0 & I \end{pmatrix} \begin{pmatrix} \xi \\ y \end{pmatrix} = \begin{pmatrix} A & AT - TC + B \\ 0 & C \end{pmatrix} \begin{pmatrix} \xi \\ y \end{pmatrix}.$$

From Section 1, Corollary 1. 4, we can solve

$$AT - TC + B = 0 \tag{5. 3}$$

if A and C have no common eigenvalues. This gives us (5. 2) (with $\eta = y$), $D = A$, $E = C$. We may justifiably refer to T, therefore, as the "decoupling matrix" whenever this condition on the eigenvalues of A and C obtains.

Now suppose the eigenvalues of A lie in a half-plane

$$\mathrm{Re}\,(\lambda) \le a, \tag{5. 4}$$

where a is a real number, and the eigenvalues of C lie in another half plane

$$\mathrm{Re}\,(\mu) \geq c > a \ . \tag{5.5}$$

One may see from block diagonalization of A and C (see Section 1 again) that, for positive numbers M_0, M_1,

$$\|\xi(t)\|_{E^n} \leq M_0 e^{at} \|\xi(0)\|_{E^n}, \quad \|y(t)\|_{E^p} \geq M_1 e^{ct} \|y(0)\|_{E^p}, \tag{5.6}$$

so that, assuming $y(0) \neq 0$, in the relation

$$x(t) = \xi(t) + Ty(t) \tag{5.7}$$

the term $Ty(t)$ will dominate $\xi(t)$ for large t - unless T were so curiously configured that

$$Ty(t) \equiv 0 \ . \tag{5.8}$$

This can be ruled out by supposing the pair (A, B) controllable. For, if we assume (5.8) and differentiate repeatedly, using (5.3) in the form

$$TC = AT + B,$$

we obtain

$$\begin{pmatrix} B \\ AB \\ \vdots \\ A^{n-1}B \end{pmatrix} y(t) \equiv 0$$

and the controllability rank condition implies $y(t) \equiv 0$, whence $y(0) = 0$, contrary to our assumption. Assuming this controllability condition then, (5.6), (5.7) together give the asymptotic relation

$$x(t) \simeq Ty(t), \quad t \to \infty, \quad \|y(0)\| \neq 0 \ . \tag{5.9}$$

($\simeq$: asymptotically equal to). In this sense, then, T indicates the manner in which the behavior of $y(t)$ is transferred over into corresponding asymptotic behavior of $x(t)$. For this reason we will refer to T as the transfer matrix for the triple (A, B, C) whenever the eigenvalue separation conditions

(5.4), (5.5) obtain. If an output or response

$$z = Fx, \quad (F \ q \times n), \tag{5.10}$$

(ordinarily $q < n$) is selected for some intrinsic interest, then

$$z(t) \simeq FTy(t) \tag{5.11}$$

and FT is referred to as the transfer matrix for (F, A, B, C).

In classical control theory one considers, at least initially, a scalar input, scalar output system

$$\dot{x} = Ax + by(t), \qquad x \in E^n, \quad y \in E^1, \tag{5.12}$$

$$z = f^* x, \qquad z \in E^1, \tag{5.13}$$

one assumes an eigenvalue condition (5.4) for A, usually with a negative, and one studies scalar exponential inputs

$$y(t) = e^{st}. \tag{5.14}$$

As long as s is not an eigenvalue of A it may easily be seen that (5.12), (5.14) has the particular solution

$$x_s(t) = (sI - A)^{-1} b e^{st} . \tag{5.15}$$

The general solution is of the form

$$x(t) = e^{At} x + x_s(t), \qquad x \in E^n,$$

and (5.15) is the dominant term if $\text{Re}(s) > a$. Under these circumstances

$$z(t) = f^* x(t) \simeq f^* (sI - A)^{-1} b e^{st} \equiv \mathcal{T}(s) e^{st} . \tag{5.16}$$

This so called "transfer function",

$$\mathcal{T}(s) = f^* (sI - A)^{-1} b = q(s)/p(s), \tag{5.17}$$

is a rational function of the complex variable s whose denominator, $p(s)$, is the characteristic polynomial of A and whose numerator, $q(s)$, is a polynomial in s of degree $\leq n-1$. The term "transfer function" indicates its role in transferring the exponential input e^{st} over into the corresponding dominant (for $\mathrm{Re}(s) > a$) exponential term in the output $z(t)$. It has importance beyond this fact for the following reason. If $y \in L^2[0,\infty)$, it has the Laplace transform (see [11], [12], e.g.)

$$\hat{y}(s) = \int_0^\infty e^{-st} y(t)\,dt,$$

an analytic function of the complex variable s for $\mathrm{Re}(s) > 0$. Solving (5.12) by the variation of parameters formula and forming the output (5.13), we have

$$z(t) = f^* e^{At} x(0) + f^* \int_0^t e^{A(t-\tau)} b y(\tau)\,d\tau . \tag{5.18}$$

This output has the Laplace transform

$$\begin{aligned}
\hat{z}(s) &= \int_0^\infty e^{-st} [f^* e^{At} x(0) + f^* \int_0^t e^{A(t-\tau)} b y(\tau)\,d\tau]\,dt \\
&= f^*(sI - A)^{-1} x(0) + f^* \int_0^\infty \int_0^t e^{-st} e^{A(t-\tau)} b y(\tau)\,d\tau\,dt \\
&= \text{(using the change of variable } t = \sigma + \tau) \\
&= f^*(sI - A)^{-1} x(0) + f^* \int_0^\infty \int_0^\infty e^{-s(\sigma+\tau)} e^{A\sigma} b y(\tau)\,d\tau\,d\sigma \\
&= f^*(sI - A)^{-1} x(0) + f^* \int_0^\infty \int_0^\infty e^{(-sI+A)\sigma}\,d\sigma\, b e^{-s\tau} y(\tau)\,d\tau \\
&= f^*(sI - A)^{-1} x(0) + f^*(sI - A)^{-1} \int_0^\infty e^{-s\tau} y(\tau)\,d\tau ,
\end{aligned} \tag{5.19}$$

provided $\mathrm{Re}(s) > a$. For $x(0) = 0$ we have, from (5.19),

$$\hat{z}(s) = \mathcal{T}(s)\hat{y}(s) . \tag{5.20}$$

The term $f^* e^{AT} x(0)$ in (5.18) is called the transient term and will ordinarily, for large t, be small relative to the second term if a in (5.4) is negative. Thus the asymptotic relationship of the output $z(t)$ to the input $y(t)$, obtained by omitting the first term in (5.18), carries over, via the Laplace transform, to (5.20), the transform of the asymptotic output being equal to the transform of the $L^2[0,\infty)$ input multiplied by the transfer function $\mathcal{T}(s)$.

The transfer matrix FT, defined by (5.3), (5.11) generalizes $\mathcal{T}(s)$. For (5.12), (5.14) can equally well be replaced by

$$\begin{aligned} \dot{x} &= Ax + by, \\ \dot{y} &= \phantom{Ax+{}} sy \end{aligned} \qquad \text{i.e.,} \quad \begin{pmatrix} \dot{x} \\ \dot{y} \end{pmatrix} = \begin{pmatrix} A & b \\ 0 & s \end{pmatrix} \begin{pmatrix} x \\ y \end{pmatrix}$$

and then (5.3) becomes

$$AT - Ts + b = 0,$$

with T now an n-dimensional column vector. Then

$$T = (sI - A)^{-1}b, \quad s \text{ not an eigenvalue of } A,$$

and, replacing F by f^*,

$$FT = f^*(sI - A)^{-1}b = \mathcal{T}(s).$$

The generalization lies in the fact that FT ($= FT(C)$) is defined for $p \times p$ matrices C while $\mathcal{T}(s)$ is defined for complex numbers s - the two are identical when C is 1×1.

The transfer matrices FT may be used in much the same way as transfer functions are used in classical control theory. (See II-[3], e.g.). We present an example. The series of systems

$$\begin{aligned} \dot{y} &= Cy \\ \dot{x}_1 &= A_1x_1 + B_1y \\ \dot{x}_2 &= A_2x_2 + B_2x_1 \\ &\vdots \\ \dot{x}_r &= A_rx_r + B_rx_{r-1} \\ z &= F_rx_r \end{aligned} \tag{5.21}$$

form what is called a "cascaded" system, each feeding into the next. With

$$x = \begin{pmatrix} x_r \\ x_{r-1} \\ x_{r-2} \\ \vdots \\ x_2 \\ x_1 \end{pmatrix}, \quad A = \begin{pmatrix} A_r & B_r & 0 & \dots & 0 & 0 \\ 0 & A_{r-1} & B_{r-1} & \dots & 0 & 0 \\ 0 & 0 & A_{r-2} & \dots & 0 & 0 \\ \vdots & \vdots & \vdots & & \vdots & \vdots \\ 0 & 0 & 0 & \dots & A_2 & B_2 \\ 0 & 0 & 0 & \dots & 0 & A_1 \end{pmatrix}, \tag{5.22}$$

$$B = \begin{pmatrix} 0 \\ 0 \\ 0 \\ \vdots \\ 0 \\ B_1 \end{pmatrix}, \quad F = (F_r, 0, 0, \dots, 0, 0) \tag{5.23}$$

the transfer matrix is, from (5.3), (5.11),

$$FT, \quad AT - TC + B = 0 .$$

<u>Proposition 5.1.</u> <u>The transfer matrix</u> FT <u>for the cascaded system</u> (5.21), i.e., (cf. (5.22), (5.23))

$$\begin{pmatrix} \dot{x} \\ y \end{pmatrix} = \begin{pmatrix} A & B \\ 0 & C \end{pmatrix} \begin{pmatrix} x \\ y \end{pmatrix}, \tag{5.24}$$

<u>can be computed recursively with use of the formulae</u>

$$FT = F_r T_r, \tag{5.25}$$

$$\begin{aligned} &A_r T_r - T_r C + B_r T_{r-1} = 0, \\ &A_{r-1} T_{r-1} - T_{r-1} C + B_{r-1} T_{r-2} = 0 \\ &\vdots \\ &A_2 T_2 - T_2 C + B_2 T_1 = 0, \\ &A_1 T_1 - T_1 C + B_1 = 0 . \end{aligned} \tag{5.26}$$

Proof. With

$$T = \begin{pmatrix} T_r \\ T_{r-1} \\ \vdots \\ T_2 \\ T_1 \end{pmatrix},$$

and F as given by (5.23), (5.25) is immediate and (5.26) follows from

$$\begin{pmatrix} A_r & B_r & 0 & \cdots & 0 & 0 \\ 0 & A_{r-1} & B_{r-1} & \cdots & 0 & 0 \\ 0 & 0 & A_{r-2} & \cdots & 0 & 0 \\ \vdots & \vdots & \vdots & & \vdots & \vdots \\ 0 & 0 & 0 & \cdots & A_2 & B_2 \\ 0 & 0 & 0 & \cdots & 0 & A_1 \end{pmatrix} \begin{pmatrix} T_r \\ T_{r-1} \\ T_{r-2} \\ \vdots \\ T_2 \\ T_1 \end{pmatrix} - \begin{pmatrix} T_r \\ T_{r-1} \\ T_{r-2} \\ \vdots \\ T_2 \\ T_1 \end{pmatrix} C + \begin{pmatrix} 0 \\ 0 \\ 0 \\ \vdots \\ 0 \\ B_1 \end{pmatrix} = 0,$$

which is just the expanded form of (5.3). Q.E.D.

If the solution of the linear matrix equation

$$MX + XC + D = 0,$$

with M and C having no common eigenvalues, is denoted in operator form by

$$X = \mathcal{T}_{M,C}(D),$$

our result can be expressed in terms of composition of such operators:

$$FT = F_r \mathcal{T}_{A_r, C}(B_r \mathcal{T}_{A_{r-1}, C}(B_{r-1} \cdots B_2 \mathcal{T}_{A_1, C}(B_1)\ldots)). \qquad (5.27)$$

For C = s, scalar, and for cascaded single input, single output systems

$$z(t) = f_r^* x_r(t),$$

$$\dot{x}_r = A_r x_r + b_r f_{r-1}^* x_{r-1},$$

$$\dot{x}_{r-1} = A_{r-1} x_{r-1} + b_{r-1} f_{r-2}^* x_{r-2},$$

$$\vdots$$

$$\dot{x}_2 = A_2 x_2 + b_2 f_1^* x_1,$$

$$\dot{x}_1 = A_1 x_1 + b_1 y(t),$$

$$y(t) = e^{st},$$

the formula (5.27) reduces to the familiar result to the effect that the transfer function $\mathcal{T}(s)$ for the cascaded system is the ordinary product of the transfer functions $\mathcal{T}_k(s)$, $k = 1, 2, \ldots, r$, for the systems

$$\eta_k(t) = e^{st}, \quad \dot{x}_k = A_k x_k + b_k \eta_k, \quad \zeta_k = f_k^* x_k,$$

that is,

$$\mathcal{T}(s) = \mathcal{T}_r(s)\mathcal{T}_{r-1}(s)\ldots\mathcal{T}_2(s)\mathcal{T}_1(s),$$

where

$$\mathcal{T}_k(s) = f_k^*(sI - A_k)^{-1} b_k.$$

One of the basic themes of this book is the use of Liapounov equations

$$A^* X + XA + W = 0,$$

linear matrix equations in general,

$$AX + XC + W = 0,$$

and transfer matrices, as defined herein, to analyze the properties of multiple input - multiple output linear control systems and as an aid to the design of such systems to meet various performance criteria. This theme will pervade the next four chapters.

In IV-3 we will encounter what amounts to a generalization of T as described by (5. 3), which we view aa a decoupling matrix now. If we have a system

$$\begin{pmatrix} \dot{x} \\ y \end{pmatrix} = \begin{pmatrix} A & B \\ D & C \end{pmatrix} \begin{pmatrix} x \\ y \end{pmatrix}, \tag{5.28}$$

in which x and y both influence each other, and if it is desired to decouple x from y, we still set

$$\begin{pmatrix} x \\ y \end{pmatrix} = \begin{pmatrix} I & T \\ 0 & I \end{pmatrix} \begin{pmatrix} \xi \\ \eta \end{pmatrix}$$

to obtain

$$\begin{pmatrix} \dot{\xi} \\ \eta \end{pmatrix} = \begin{pmatrix} I & -T \\ 0 & I \end{pmatrix} \begin{pmatrix} A & B \\ D & C \end{pmatrix} \begin{pmatrix} I & T \\ 0 & I \end{pmatrix} \begin{pmatrix} \xi \\ \eta \end{pmatrix} = \begin{pmatrix} A-TD & AT-TC-TDT+B \\ D & C+DT \end{pmatrix} \begin{pmatrix} \xi \\ \eta \end{pmatrix}. \tag{5.29}$$

To decouple x from y, we need to solve

$$AT - TC - TDT + B = 0, \tag{5.30}$$

a quadratic matrix equation in T. If this eqation can be solved, and, under special circumstancès we will obtain an affirmative result in this direction in IV- 4, then (5. 29) becomes

$$\begin{pmatrix} \dot{\xi} \\ \eta \end{pmatrix} = \begin{pmatrix} A-TD & 0 \\ D & C+DT \end{pmatrix} \begin{pmatrix} \xi \\ \eta \end{pmatrix}. \tag{5.31}$$

Bibliographical Notes, Chapter III

[1] J. M. Ortega and W. C. Rheinboldt: "Iterative Solution of Nonlinear Equations in Several Variables", Academic Press, New York, 1970.

[2] J. LaSalle and S. Lefschetz: "Stability by Liapounov's Direct Method, with Applications", Academic Press, New York, 1961.

The above treatment by LaSalle and Lefschetz is recommended as an introduction to the Liapounov theory.

[3] A. A. Liapounov: "Problème général de la stabilité du mouvement", photo-reproduced as Annals of Mathematics Study No. 17 (Princeton University Press, Princeton, New Jersey) from the 1907 French translation of Liapounov's 1892 paper in Russian.

[4] D. L. Lukes: "Stabilizability and optimal control", Funkcialaj Ekvacioj, 11 (1968), pp. 39-50.

[5] D. Kleinman: "An easy way to stabilize a linear constant system", IEEE Trans. Auto. Control, AC-15 (1970), p. 692.

[6] R. W. Bass: "Lecture Notes on Control Synthesis and Optimization", presented at NASA Langley Research Center, August 1961.

The problem of eigenvalue assignment using feedback based on a lower dimensional observation $y = Hx$, $y \in E^p$, $p < n$, leading to a closed loop system $\dot{x} = (A + BKH)x$, has been studied by a number of authors. We list two contributions here. Further references may be found in [8].

[7] E. J. Davison and S. H. Wang: "On pole assignment in linear multivariable systems using output feedback", IEEE Trans. Automat. Contr., AC-20 (1975), pp. 516-518.

[8] H. Kimura: "A further result on the problem of pole assignment by output feedback", Ibid., AC-22(1977), pp. 458-463.

[9] D. G. Luenberger" "An introduction to observers", Ibid., AC-16 (1971), pp. 596-602.

[10] B. Gopinath: "On the control of linear multiple input-output systems", Bell Syst. Tech. Jour., March 1971.

The Laplace transform is not extensively used in this book since we are emphasizing the approach based on the Liapounov theory and linear matrix equations generally. Nevertheless, a basic understanding of this important tool is essential to an understanding of classical control theory and most control problems wherein frequency considerations play an important role. The elements of the theory of the Laplace transform appear in I - [6], while the following classic treatises provide a more complete background.

[11] D. V. Widder: "The Laplace Transform", Princeton University Press, Princeton, 1946.

[12] G. Doetsch: "Guide to Applications of Laplace Transforms", Van Nostrand, London, 1967.

Exercises, Chapter III

1. Consider the crane system linearized about $\psi = 45^\circ$, $\dot{\psi} = 0$, $\theta = \dot{\theta} = 0$, $\ell_2 = 1$, $v = 0$. Supposing $g = 1$, $L_1 = \sqrt{2}$ we have

$$\begin{pmatrix} x^1 \\ x^2 \\ x^3 \\ x^4 \\ x^5 \end{pmatrix} = \begin{pmatrix} 0 & 1 & 0 & 0 & 0 \\ -1 & 0 & 0 & 0 & 0 \\ 0 & 0 & 0 & 1 & 0 \\ 0 & 0 & 0 & 0 & 0 \\ 0 & 0 & 0 & 0 & 0 \end{pmatrix} \begin{pmatrix} x^1 \\ x^2 \\ x^3 \\ x^4 \\ x^5 \end{pmatrix} + \begin{pmatrix} 0 & 0 \\ 0 & -1 \\ 0 & 0 \\ 0 & 1 \\ 1 & 0 \end{pmatrix} \begin{pmatrix} u^1 \\ u^2 \end{pmatrix}$$

An approximate "level hold" operating condition is obtained by choosing u^1 in such a way that, with

$$y^3 = -L_1 \sin \psi_0 \, x^3 - x^5 = -x^3 - x^5$$

representing the first order variation in the height of the load, we have

$$\dot{y}^3 = -x^4 - u^1 = 0, \quad \text{i.e.} \quad u^1 = -x^4 .$$

Show that setting $u^1 = -x^4$ results in a four dimensional system with control u^2, state x^1, x^2, x^3, x^4, and

(a) analyze the stability properties of this four dimensional system;

(b) analyze the controllability of the system.

(c) Use Bass' method to stabilize the system in such a way that all eigenvalues have real part ≤ -1.

2. (a) Suppose we have a control system with scalar control:

$$\dot{x} = Ax + bu$$

wherein, for distinct λ_j, $j = 1, 2, \ldots, r$,

$$A = \begin{pmatrix} A_1 & 0 & \ldots & 0 \\ 0 & A_2 & \ldots & 0 \\ \vdots & \vdots & & \vdots \\ 0 & 0 & \ldots & A_r \end{pmatrix}, \quad A_j = \begin{pmatrix} \lambda_j & e_j^1 & 0 & \ldots & 0 & 0 \\ 0 & \lambda_j & e_j^2 & \ldots & 0 & 0 \\ \vdots & \vdots & \vdots & & \vdots & \vdots \\ 0 & 0 & 0 & \ldots & \lambda_j & e_j^{\mu_j - 1} \\ 0 & 0 & 0 & \ldots & 0 & \lambda_j \end{pmatrix},$$

e_j^k being either zero or one, and

$$b = \begin{pmatrix} b_1 \\ b_2 \\ \vdots \\ b_r \end{pmatrix}, \quad b_j = \begin{pmatrix} b_j^1 \\ b_j^2 \\ \vdots \\ b_j^{\mu_j} \end{pmatrix}.$$

Show that such a system is controllable if and only if

(i) all $e_j^k = 1$, and

(ii) all $b_j^{\mu_j} \neq 0$.

(b) Suppose in the five dimensional crane system of Exercise 1 we impose a constant $\alpha u^1 + \beta u^2 = 0$. Show that if α and β are not both zero, the system is not controllable with the remaining control capacity.

3. Obtain an alternative reduction to control canonical form for a controllable system with scalar control by showing that the system $\dot{x} = Ax + bu$ and the control canonical form (3. 2) can both be reduced to a system of the form exhibited in 2(a) above, with

$$b_j^1 = b_j^2 = \ldots = b_j^{\mu_j - 1} = 0, \quad b_j^{\mu_j} = 1, \quad j = 1, 2, \ldots, r.$$

4. Prove Corollary 1. 4. Hint: See the proof of Theorem 1. 5. What does this say about the solvability of equation (4. 30) in general? Does this result play any role in the proof of Theorem 4. 10?

5. Let $\dot{x} = Ax + Bu$ with A antihermitian, i. e.,

$$A^* = -A.$$

Show that $A - BB^*$ and $-A - BB^*$ are both stability matrices if and only if (A, B) is controllable. Then consider the <u>second order</u> systems in E^n

$$\ddot{x} \pm DD^*\dot{x} + Cx \begin{cases} C & n \times n, \text{ symmetric, positive definite,} \\ D & n \times m, \end{cases}$$

and show that all solutions tend to 0 as $t \to \pm\infty$, respectively, if and only if (C, D) is controllable. Hint: Use Theorem 2.4.

6. Construct a control canonical form for the five dimensional linearized crane equation, with two dimensional control, shown in problem 1. Then determine a feedback control law

$$\begin{pmatrix} u^1 \\ u^2 \end{pmatrix} = K \begin{pmatrix} x^1 \\ x^2 \\ x^3 \\ x^4 \\ x^5 \end{pmatrix}, \quad K \quad 2 \times 5,$$

so that the closed loop system has the eigenvalues -1, $-1 \pm i$, $-1 \pm 2i$.

7. (a) For the linear harmonic oscillator

$$\begin{pmatrix} \dot{x}^1 \\ x^2 \end{pmatrix} = \begin{pmatrix} 0 & 1 \\ -1 & 0 \end{pmatrix} \begin{pmatrix} x^1 \\ x^2 \end{pmatrix} + \begin{pmatrix} 0 \\ 1 \end{pmatrix} u$$

with observation

$$\omega = x^1 = (1, 0) \begin{pmatrix} x^1 \\ x^2 \end{pmatrix}$$

construct a combined "original plant-observer" system of total dimension 4 with the eigenvalues of $A + BK$ being -1, -2 and those of $A - LH$ being -3, -4.

(b) Construct a reduced dimension observer so that the combined three dimensional system has eigenvalues -1, -2, -3.

8. State and prove the theorem dual to Theorem 3.3 for linear observed systems

$$\dot{x} = Ax, \quad \omega = Hx, \quad \omega \in E^r,$$

with (H, A) observable. (See Theorem 4.4 for $r = 1$.)

9.* Consider the five dimensional crane equation of Exercise 1 again. Suppose we can observe only

$$\theta(=x^1),\quad \psi(=x^3),\quad \ell_2(=x^5).$$

Construct a two-dimensional reduced order observer so that the combined crane-observer system has eigenvalues

$$-1,\ -1\pm i,\ -1\pm 2i,\ -2\pm i.$$

10.* With an appropriate routine for numerical solution of systems of differential equations (or exact solution if you prefer), generate solutions of

$$\begin{pmatrix} \dot{x}^1 \\ \dot{x}^2 \end{pmatrix} = \begin{pmatrix} 0 & 1 \\ -1 & -2 \end{pmatrix}\begin{pmatrix} x^1 \\ x^2 \end{pmatrix},\quad \begin{pmatrix} x^1(0) \\ x^2(0) \end{pmatrix} = \begin{pmatrix} 1 \\ 0 \end{pmatrix},$$

and of (4.13), (4.53)

$$\begin{pmatrix} x^1(0) \\ x^2(0) \\ y^1(0) \\ y^2(0) \end{pmatrix} = \begin{pmatrix} 1 \\ 0 \\ 0 \\ 0 \end{pmatrix},\quad \begin{pmatrix} x^1(0) \\ x^2(0) \\ z(0) \end{pmatrix} = \begin{pmatrix} 1 \\ 0 \\ 0 \end{pmatrix},$$

respectively. Compare the results, insofar as x^1, x^2 are concerned, graphically, for $0 \le t \le 5$ (or any other suitable interval). How would you improve the performance of the reduced order observer? Hint: see discussion preceding Theorem 4.8.

11. Kleinman's method, described in Section 2, for stabilization of $\dot{x} = Ax + Bu$, can also be modified so that the closed loop eigenvalues lie to the left of a given line $\mathrm{Re}(z) = -\lambda$, $\lambda > 0$. The matrix W_T is replaced by

$$W_{T,\lambda} = \int_0^T e^{-(A+\lambda I)t} BB^* e^{-(A+\lambda I)^* t}\, dt.$$

Carry out the details and show that when the eigenvalues of $A + \lambda I$ lie in the right half plane $\mathrm{Re}(z) > 0$ then Bass' method is a limiting form of Kleinman's method obtained by letting $T \to \infty$.

12. The reduction to block triangular form

$$P^{-1}AP = \Lambda$$

described in Section 1, can be used to define f(A) whenever f(A) is an analytic function expressible by the convergent power series

$$f(\lambda) = \sum_{k=0}^{\infty} f_k \lambda^k .$$

One sets

$$f(A) = \sum_{k=0}^{\infty} f_k A^k .$$

Show that the right hand side here has the form

$$P\left(\sum_{k=0}^{\infty} f_k \Lambda^k\right) P^{-1}$$

and hence converges if and only if the series $\sum_{k=0}^{\infty} f_k \lambda^k$ converges whenever λ is an eigenvalue of A. Show that under these circumstances the "spectral mapping" theorem holds, i. e., the eigenvalues of f(A) have the form $f(\lambda)$ where λ is an eigenvalue of A.

Remark: Just as $f(\lambda)$ can often be extended beyond the original disc of convergence by analytic continuation, the same can be done for f(A) when the eigenvalues of A lie outside that disc.

13. Obtain a generalization of formula (5. 20). Let S be an $m \times m$ matrix whose eigenvalues have positive real parts. Given a $p \times m$ matrix function Y(t) whose entries, $y_j^i(t)$, lie in $L^2[0,\infty)$, define the matrix Laplace transform of Y by

$$(\mathcal{L}Y)(S) = \int_0^{\infty} Y(t)e^{-St}\,dt .$$

Let $\Phi(t)$, $n \times m$, solve

$$\dot{\Phi} = A\Phi + Y(t), \qquad \Phi(0) = 0,$$

where A, $n \times n$, is a stability matrix and now Y(t) is $n \times m$ with entries in $L^2[0,\infty)$. Show that $\Phi(t)$ also has entries in $L^2[0,\infty)$ and

$$(\mathcal{L}\Phi)(S) = \text{(cf. (5. 27))} = \mathcal{T}_{A,S}((\mathcal{L}Y)(S)).$$

14. (i) Suppose a collection of systems is connected as indicated by the following diagram:

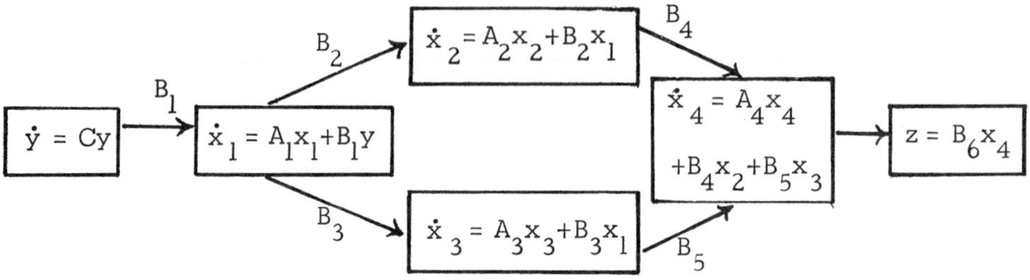

How is the transfer matrix for the composite system related to the transfer matrices for (A_1, B_1, C), (A_2, B_2, C), (A_3, B_3, C), (A_4, B_4, C), (A_4, B_5, C)? (See definition in the paragraph before (5.10).)

(ii) Essentially the same question, but for the following collection of systems:

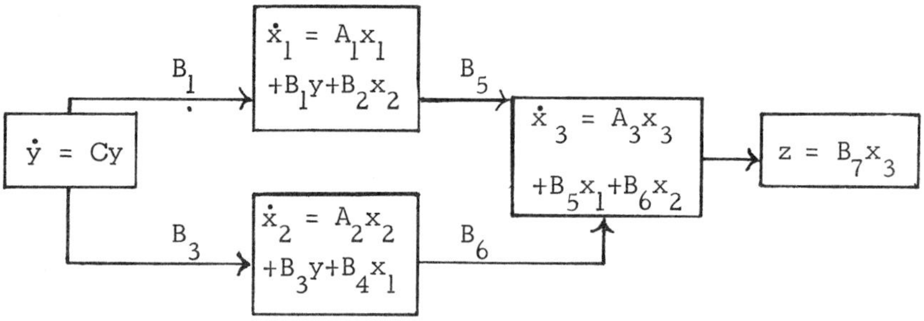

CHAPTER IV

OPTIMAL CONTROL WITH QUADRATIC PERFORMANCE CRITERIA AND RELATED TOPICS

1. Quadratic Performance Criteria

We have seen in Chapter III that linear feedback control laws

$$u = Kx$$

applied to a controllable, or stabilizable, linear control system

$$\dot{x} = Ax + Bu, \tag{1.1}$$

have the capability of producing widely varying dynamic behavior in the closed loop system

$$\dot{x} = (A + BK)x. \tag{1.2}$$

But something is missing from our analysis so far. We have not adequately addressed ourselves to the question of precisely what type of behavior we wish to realize in (1.2), except that ordinarily we will wish to have asymptotic stability. We are faced in practice with the problem of describing and classifying dynamic behavior in terms more comprehensive than is permitted by the very coarse division into "stable", "asymptotically stable", and "unstable" categories. Beyond that, even if we do develop some satisfactory method for describing dynamic behavior, we have the further problem of determining feedback matrices K for which (1.2) exhibits a given type of dynamic behavior. Both of these objectives are treated in the present chapter through the introduction of a quadratic cost functional and the problem of minimization of such a cost functional by appropriate control action.

The Liapounov approach to the study of asymptotic stability provides a clue as to what might be useful in describing dynamic behavior generally. Let us recall the role which the Liapounov equation

$$A^*X + XA + W = 0 \tag{1.3}$$

plays. Assuming X a symmetric positive definite solution for some symmetric non-negative W as detailed in III-1, the fact that A must then be a stability matrix is deduced from the fact that

$$\lim_{t \to \infty} x(t)^* X\, x(t) = 0$$

when $x(t)$ is a solution of $\dot{x} = Ax$. We can think of the function

$$\xi(x) = (x^* X x)^{\frac{1}{2}} \tag{1.4}$$

as defining a norm, or distance from the origin, 0, of the vector x with respect to a coordinate system defined by the matrix X. The qualitative property of asymptotic stability is associated with the fact that this norm tends to zero as $t \to \infty$. We can use this function in other ways, however. We can define a quantitative descriptor associated with solutions of $\dot{x} = Ax$ on an interval $[0, T]$, namely

$$\int_0^T x^*(t)\, W x(t)\, dt \;, \tag{1.5}$$

representing the cumulative sum of values of the square of the "distance" $\omega(x) = (x^* W x)^{\frac{1}{2}}$ on $[0, T]$. As in III-1, differentiation and use of (1.3) gives

$$\int_0^T x^*(t) W x(t)\, dt = x^*(0) X x(0) - x^*(T) X x(T) \,.$$

When we have asymptotic stability, $x(T) \to 0$ and the integral converges as $T \to \infty$ so that

$$\int_0^\infty x^*(t) W x(t) dt = x^*(0) X x(0) \,. \tag{1.6}$$

Thus the value of the quadratic form $x^* X x$ at the initial state $x(0)$ yields the integral on the left of (1.6), where we assume, of course, that $x(t)$ is the solution of $\dot{x} = Ax$ which corresponds to this initial state. It is not hard to see that (1.6) can provide a useful quantitative descriptor for various types of dynamic behavior. One can select any linear output

$$y = Fx, \tag{1.7}$$

such that (F, A) is observable. Then, with

$$W = F^*F,$$

$$\int_0^\infty \|y(t)\|^2 dt = \int_0^\infty x^*(t)Wx(t)dt = x(0)^* Xx(0), \tag{1.8}$$

where X and W are related by (1.3). By choosing the components of y to be outputs or responses which we wish to closely monitor or are for some other reason, significant, (1.8) becomes a numerical quantity of some importance and usefulness. If, instead of $\dot{x} = Ax$, we have the closed loop system (1.2), then (1.3) becomes

$$(A + BK)^* X(K) + X(K)(A + BK) + W = 0 \tag{1.9}$$

and the expression on the right hand side of (1.8) is replaced by $x(0)^* X(K)x(0)$. The goal of keeping the output (1.7) "small", as measured by (1.8), can then be translated into the problem of selecting K in (1.9) so that the solution matrix $X(K)$ is appropriately small.

In pursuing the objective just outlined one must not lose sight of the fact that control is not free. Implementation of a control u in (1.1) involves expenditure of energy, stress on the operating plant, etc. In order to take the cost of control into account it is convenient to again employ a quadratic expression

$$\int_0^T u(t)^* \tilde{U} u(t)\, dt, \tag{1.10}$$

$\tilde{U}$ being some $m \times m$ positive definite weighting matrix. When $u(t) = Kx(t)$ obtains, (1.10) becomes

$$\int_0^T x(t)^* K^* \tilde{U} K x(t)\, dt$$

and when $A + BK$ is a stability matrix, we may consider the integral

$$\int_0^{\infty} x(t)^* K^* \tilde{U} K x(t) dt = x(0)^* \tilde{X}(K) x(0), \tag{1.11}$$

where now

$$(A+BK)^* \tilde{X}(K) + \tilde{X}(K)(A+BK) + K^* \tilde{U} K = 0. \tag{1.12}$$

The complete control analysis must then include consideration of the way in which both of the matrices, $X(K)$ and $\tilde{X}(K)$, depend on the feedback matrix K.

Commonly the "response cost" (1.8) and the control cost (1.10) are combined in a single expression. In fact, applications often dictate a slightly more complex formulation. As noted in I-2, the response may well be control dependent:

$$y = Fx + Gu, \quad y \in E^q.$$

Letting $\tilde{W}$ be a positive definite symmetric $q \times q$ "weighting" matrix, we consider the weighted response cost

$$\int_0^T y(t)^* \tilde{W} y(t) dt = \int_0^T [x(t)^* F^* \tilde{W} F x(t) + x(t)^* F^* \tilde{W} G u(t)$$

$$+ u(t)^* G^* \tilde{W} F x(t) + u(t)^* G^* \tilde{W} G u(t)] \, dt.$$

If we add to this a control cost of the form (1.10) we have the quadratic integral

$$\int_0^T [x(t)^* F^* \tilde{W} F x(t) + x(t)^* F^* \tilde{W} G u(t) + u(t)^* G^* \tilde{W} F x(t) + u(t)^* (\tilde{U} + G^* \tilde{W} G) u(t)] \, dt$$

$$= \int_0^T [x(t)^* W x(t) + x(t)^* R u(t) + u(t)^* R^* x(t) + u(t)^* U u(t)] \, dt, \tag{1.13}$$

where

$$W = F^* \tilde{W} F \quad R = F^* \tilde{W} G, \quad U = \tilde{U} + G^* \tilde{W} G.$$

When $u = Kx$ and $A + BK$ is a stability matrix, and we let $T = \infty$, the cost (1.13) becomes, using an argument like that which lead to (1.8),

$$\int_0^\infty x(t)^*[W + RK + K^*R^* + K^*UK]\,x(t)dt = x(0)^*\hat{X}(K)x(0), \tag{1.14}$$

where

$$(A + BK)^*\hat{X}(K) + \hat{X}(K)(A + BK) + W + RK + K^*R^* + K^*UK = 0. \tag{1.15}$$

The objective, again in rough terms, is to make the matrix $X(K)$ small by appropriate selection of K. The solution of this problem will be presented in Section 3 in a slightly more general context.

As an example we consider the controlled harmonic oscillator once more:

$$\ddot{x} + x = u, \quad \text{or} \quad \begin{pmatrix} \dot{x}^1 \\ \dot{x}^2 \end{pmatrix} = \begin{pmatrix} 0 & 1 \\ -1 & 0 \end{pmatrix}\begin{pmatrix} x^1 \\ x^2 \end{pmatrix} + \begin{pmatrix} 0 \\ 1 \end{pmatrix} u. \tag{1.16}$$

This time let us think of $x = x^1$ as representing the vertical position of a load of mass 1 borne by a wheel and spring of spring constant 1 - perhaps the simplest possible conveyance involving a suspension system. Unevenness in the "road surface" results in deviations of $x(t)$ from its equilibrium position $x = 0$. The control u represents a force in the vertical direction

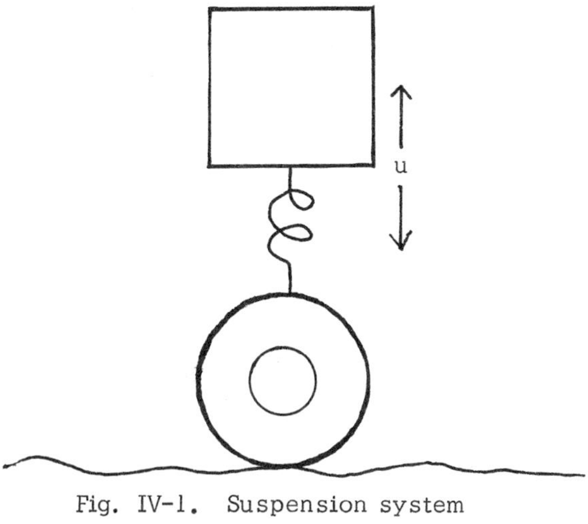

Fig. IV-1. Suspension system

which may be applied for various ends.

Let us identify as the responses which we wish to monitor the quantities

$$\text{position} \equiv y^1 \equiv x^1$$

$$\text{velocity} \equiv y^2 \equiv x^2$$

$$\text{acceleration} \quad y^3 \equiv \dot{x}^2 = -x^1 + u.$$

Then the response vector is

$$y = Fx + Gu, \quad F = \begin{pmatrix} 1 & 0 \\ 0 & 1 \\ -1 & 0 \end{pmatrix}, \quad G = \begin{pmatrix} 0 \\ 0 \\ 1 \end{pmatrix} . \tag{1.17}$$

The position and velocity enter into the definition of the system energy $E = \frac{1}{2}((x^1)^2 + (x^2)^2)$. The acceleration, $\dot{x}^2$, is important if, for example, the load represents a human passenger or a crate of eggs. Whether it is large or small affects what is sometimes called the "ride quality". We can balance the two considerations against each other with a weighting matrix

$$\tilde{W} = \begin{pmatrix} \alpha & 0 & 0 \\ 0 & \alpha & 0 \\ 0 & 0 & \beta \end{pmatrix}, \quad \alpha > 0, \ \beta > 0. \tag{1.18}$$

Including a term δu^2 to reflect control cost we have the quadratic integral

$$\int_0^T [y(t)^* \tilde{W} y(t) + \delta u(t)^2] \, dt \tag{1.19}$$

which can easily be seen to take the form (1.13) with

$$W = F^* \tilde{W} F = \begin{pmatrix} \alpha+\beta & 0 \\ 0 & \alpha \end{pmatrix} \tag{1.20}$$

$$R = F^* \tilde{W} G = \begin{pmatrix} -\beta \\ 0 \end{pmatrix} \tag{1.21}$$

$$U = \delta + G^* \tilde{W} G = \delta + \beta . \tag{1.22}$$

By varying α, β and δ, within this framework, varied emphases on system energy, ride quality and control cost can be brought into play. We will return to this example in Section 3 of this chapter.

With the foregoing as motivation, we proceed now to a detailed study of linear systems with quadratic cost functionals and the problem of choosing the control so as to minimize such a cost functional.

2. Optimal Control on a Finite Time Interval

Let us suppose that the expression

$$y(t) = F(t)x + G(t)u, \quad y \in E^q, \quad x \in E^n, \quad u \in E^m,$$

represents a (possibly) time varying response vector which one wishes to keep small by appropriate selection of the control function u appearing in the system

$$\dot{x} = A(t)x + B(t)u, \tag{2.1}$$

defined for $t_0 \leq t \leq t_1$, $t_0 < t_1$. As noted in Section 1, this leads us to consideration of the quadratic "cost" integral

$$\int_{t_0}^{t_1} (F(t)x(t) + G(t)u(t))^* \tilde{W}(t)(F(t)x(t) + G(t)u(t))\,dt \tag{2.2}$$

where $\tilde{W}(t)$ is a continuous, symmetric positive definite $q \times q$ matrix - the output weighting matrix - again defined for $t_0 \leq t \leq t_1$. We adjoin to (2.2) the "control cost"

$$\int_{t_0}^{t_1} u(t)^* \tilde{U}(t)u(t)\,dt \tag{2.3}$$

where $\tilde{U}(t)$ is a continuous, positive definite symmetric $m \times m$ matrix. Additionally, it may be desirable to place some weight on the final state $x(t_1)$, which leads to a further expression

$$x(t_1)^* V x(t_1) \tag{2.4}$$

wherein we shall suppose that V is a symmetric non-negative $n \times n$ matrix.

Supposing $x(t)$ to be a solution of (2.1) corresponding to a given initial state

$$x(t_0) = x_0 \in E^n \tag{2.5}$$

and a control

$$u \in L_m^2[t_0, t_1]$$

we combine (2.2), (2.3) and (2.4) above into a single cost functional

$J(x_0, u, t_0, t_1)$ ($\equiv J(u)$ when x_0, t_0, t_1 are understood)

$$= \int_{t_0}^{t_1} (x(t)^*, u(t)^*) \begin{pmatrix} F(t)^* \tilde{W}(t) F(t) & F(t)^* \tilde{W}(t) G(t) \\ G(t)^* \tilde{W}(t) F(t) & \tilde{U}(t) + G(t)^* \tilde{W}(t) G(t) \end{pmatrix} \begin{pmatrix} x(t) \\ u(t) \end{pmatrix} dt$$

$$+ x(t_1)^* V x(t_1) \equiv \text{(cf. (1.13))}$$

$$\equiv \int_{t_0}^{t_1} (x(t)^*, u(t)^*) \begin{pmatrix} W(t) & R(t) \\ R(t)^* & U(t) \end{pmatrix} \begin{pmatrix} x(t) \\ u(t) \end{pmatrix} dt + x(t_1)^* V x(t_1). \tag{2.6}$$

The conditions laid down so far imply

$$U(t) = \tilde{U}(t) + G(t)^* \tilde{W}(t) G(t) > 0, \quad t \in [t_0, t_1];$$

and

$$\begin{pmatrix} W(t) & R(t) \\ R(t)^* & U(t) \end{pmatrix} \geq 0, \quad t \in [t_0, t_1].$$

Throughout this chapter we will assume that all matrices occurring are real matrices and we consider only real controls u and real solutions of (2.1). The theory is, however, readily extended to the complex case, see e.g. Exercise 2 below.

We now pose our

Optimization Problem. Find, if possible, a control $\hat{u} \in L_m^2[t_0, t_1]$ such that for all $u \in L_m^2[t_0, t_1]$

$$J(\hat{u}) \le J(u), \qquad u \in L_m^2[t_0, t_1] .$$

The existence and uniqueness of the optimal control u will arise out of later developments which in turn depend on the characterization of u presented in

Theorem 2.1. *Let $\hat{u} \in L_m^2[t_0, t_1]$ and let $\hat{x}(t)$ be the corresponding solution of* (2.1), (2.5) *on* $[t_0, t_1]$. *Let $\psi(t)$ be the solution on $[t_0, t_1]$ of*

$$\dot{\psi} = -A(t)^*\psi - W(t)\hat{x}(t) - R(t)\hat{u}(t) \tag{2.7}$$

$$\psi(t_1) = V\hat{x}(t_1) . \tag{2.8}$$

Then $J(x_0, t_0, t_1, \hat{u}) \le J(x_0, t_0, t_1, u)$ *for all* $u \in L_m^2[t_0, t_1]$ *if and only if*

$$\hat{u}(t) = -U(t)^{-1}[B^*(t)\psi(t) + R(t)^*\hat{x}(t)] \quad \text{in } L_m^2[t_0, t_1] . \tag{2.9}$$

Proof. Let $u_1 \in L_m^2[t_0, t_1]$ and consider, for real ϵ, $J(\hat{u} + \epsilon u_1)$. Letting $x_1(t)$ satisfy (2.1) with $u = u_1$ and $x_1(t_0) = 0$ we have (omitting the "t" argument in the integrand to save space)

$$\begin{aligned} J(\hat{u} + \epsilon u_1) &= \int_{t_0}^{t_1} [(\hat{x} + \epsilon x_1)^* W(\hat{x} + \epsilon x_1) + 2(\hat{x} + \epsilon x_1)^* R(\hat{u} + \epsilon u_1) \\ &\quad + (\hat{u} + \epsilon u_1)^* U(\hat{u} + \epsilon u_1)]\, dt + (\hat{x}(t_1) + \epsilon x_1(t_1))^* V(\hat{x}(t_1) + \epsilon x_1(t_1)) \\ &= \int_{t_0}^{t_1} [\hat{x}^* W \hat{x} + 2\hat{x}^* R\hat{u} + \hat{u}^* U \hat{u}]\, dt + \hat{x}(t_1)^* V \hat{x}(t_1) \\ &\quad + \epsilon^2 \left(\int_{t_0}^{t_1} [x_1^* W x_1 + 2x_1^* R u_1 + u_1^* U u_1]\, dt + x_1(t_1)^* V x_1(t_1) \right) \\ &\quad + 2\epsilon \int_{t_0}^{t_1} [\hat{x}^* W x_1 + \hat{x}^* R u_1 + x_1^* R \hat{u} + \hat{u}^* U u_1]\, dt + \hat{x}(t_1)^* V x_1(t_1). \end{aligned} \tag{2.10}$$

The coefficient of ϵ^2 here is $\frac{d^2}{d\epsilon^2} J(\hat{u} + \epsilon u_1)\Big|_{\epsilon = 0} = J(0, t_0, t_1, u_1) \ge 0$, and the first integral (coefficient of "1") is $J(\hat{u})$. Thus $J(\hat{u}+\epsilon u_1) \geq J(\hat{u})$ for all real ϵ just in case $\frac{d}{d\epsilon} J(\hat{u}+\epsilon u_1)\Big|_{\epsilon=0} = 0$, which is just

$$\int_{t_0}^{t_1} [\hat{x}^* W x_1 + \hat{x}^* R u_1 + x_1^* R\hat{u} + \hat{u}^* U u_1]\, dt + \hat{x}(t_1)^* V x_1(t_1) = 0. \tag{2.11}$$

Now let $\psi(t)$ satisfy (2.7), (2.8) and consider the equation (recall $x_1(t_0) = 0$)

$$\begin{aligned} 0 = \psi(t_0)^* x_1(t_0) &= \psi(t_1)^* x_1(t_1) - \int_{t_0}^{t_1} \frac{d}{dt}(\psi(t)^* x_1(t))dt \\ &= \hat{x}(t_1)^* V x_1(t_1) - \int_{t_0}^{t_1} [(-\psi^*(t)A - \hat{x}(t)^* W - \hat{u}(t)^* R^*)x_1 + \psi(t)^*(Ax_1(t) + Bu_1(t))]\, dt \\ &= \hat{x}(t_1)^* V x_1(t_1) + \int_{t_0}^{t_1} [\hat{x}^* W x_1 + \hat{x}^* R u_1 + x_1^* R\hat{u} + \hat{u}^* U u_1]\, dt \\ &\quad - \int_{t_0}^{t_1} (\psi^* B + \hat{x}^* R + \hat{u}^* U) u_1\, dt . \end{aligned} \tag{2.12}$$

Comparing (2.11) and (2.12) we see that $J(\hat{u}+\epsilon u_1) \geq J(\hat{u})$ for all real ϵ just in case

$$\int_{t_0}^{t_1} (\psi^* B + \hat{x}^* R + \hat{u}^* U) u_1\, dt = 0. \tag{2.13}$$

Now $J(u) \geq J(\hat{u})$ for all $u \in L_m^2[t_0, t_1]$ just in case $J(\hat{u}+\epsilon u_1) \geq J(\hat{u})$ for all real ϵ and all $u_1 \in L_m^2[t_0, t_1]$, which we have seen to be true just in case (2.13) holds for all $u_1 \in L_m^2[t_0, t_1]$. But then, taking

$$u_1(t) = B^* \psi + R^* \hat{x} + U\hat{u},$$

(2.13) gives

$$\int_{t_0}^{t} \| U(t)\hat{u}(t) + B^*(t)\psi(t) + R^*(t)\hat{x}(t) \|_{E^m}^2\, dt = 0$$

from which (2.9) follows immediately. Q. E. D.

Corollary 2.2. If there is a control $\hat{u}$ for which $J(\hat{u}) \leq J(u)$, $u \in L_m^2[t_0, t_1]$ (and hence such that (2.9) is satisfied with $\psi(t)$ as defined by (2.7), (2.8)) then

$$\psi(t_0)^* \hat{x}(t_0) = J(\hat{u}) \equiv J(x_0, t_0, t_1, \hat{u}). \tag{2.14}$$

Proof. Repeating the calculations in (2.12), but with $x(t)$, $u(t)$ replacing $\hat{x}_1(t)$, $\hat{u}_1(t)$, we have

$$\begin{aligned}\psi(t_0)^* \hat{x}(t_0) &= \hat{x}(t_1)^* V \hat{x}(t_1) + \int_{t_0}^{t_1} [\hat{x} W \hat{x} + 2\hat{x}^* Ru + \hat{u}^* U u]\, dt \\ &+ \int_{t_0}^{t_1} [(\psi^* B + \hat{x}^* R + \hat{u}^* U]\, dt .\end{aligned} \tag{2.15}$$

But the fact that $\hat{u}$ satisfies (2.9) shows that the last integral in (2.15) is zero. Then comparing (2.6) and (2.15) the identity (2.14) follows immediately. Q.E.D.

Remark. The initial instant t_0 may be replaced by any $\tau \in [t_0, t_1]$ and the above argument repeated on $[\tau, t_1]$ to give

$$\psi(\tau)^* \hat{x}(\tau) = J(\hat{x}(\tau), \tau, t_1, \hat{u}), \qquad t_0 \leq \tau \leq t_1 . \tag{2.16}$$

This identity will prove to be of considerable importance.

As a result of Theorem 2.1, an optimal control $\hat{u}$, if such exists, is characterized in terms of a two point boundary value problem consisting of the equation (2.7) and the boundary condition (2.8) given at $t = t_1$, the equation (2.1) and the boundary condition (2.5) given at $t = t_0$ and, finally, the equation (2.9) which gives $\hat{u}$ in terms of ψ and $\hat{x}$. If we substitute (2.9) into (2.1) and (2.7) we have

$$\dot{\hat{x}} = [A(t) - B(t)U(t)^{-1}R(t)^*]\hat{x} - B(t)U(t)^{-1}B(t)^*\psi, \tag{2.17}$$

$$\hat{x}(t_0) = x_0, \tag{2.18}$$

$$\dot{\psi} = [-A(t)^* + R(t)U(t)^{-1}B(t)^*]\psi + [R(t)U(t)^{-1}R(t)^* - W(t)]\hat{x} \tag{2.19}$$

$$\psi(t_1) = V\hat{x}(t_1) \tag{2.20}$$

It is not easy to solve this two point boundary problem as it stands. An ingenious method of solution, however, was developed by R. E. Kalman in 1960 ([1]). The method is related to the Hamilton-Jacobi theory of the calculus of variations ([9]) and to the dynamic programming method of R. Bellman and others ([10]). Perhaps the best way to motivate it here is to ask the reader to observe that the equations (2.17) - (2.20) are all linear and homogeneous and that everything ultimately depends on the given initial state vector x_0. One could guess, then, that the optimal value of the cost $J(x_0, t_0, t_1, u)$ is a quadratic form in x_0:

$$J(x_0, t_0, t_1, \hat{u}) = x_0^* Q(t_0) x_0, \quad Q(t_0)\ n \times n, \text{ symmetric.} \tag{2.21}$$

Comparing (1.21) with (1.14) leads one to conjecture also that

$$\psi(t_0) = Q(t_0)x_0. \tag{2.22}$$

Now let us carry this a little further. If $\hat{x}(t)$, $\hat{u}(t)$ are optimal on $[t_0, t_1]$ it is not hard to convince one's self of the "principle of optimality", as Bellman has called it; namely, that on any smaller interval $[\tau, t_1]$, $\hat{u}$ must solve the problem of minimizing (2.6) with t_0 replaced by τ, subject to (2.1) and the initial condition $x(\tau) = \hat{x}(\tau)$. (More on this in Section 5.) One then obtains (2.17) - (2.20) all over again on $[\tau, t_1]$ and, reasoning as we did to get (2.21), (2.22) we conjecture that for $\tau \in [t_0, t_1]$

$$J(\hat{x}(\tau), \tau, t_1, \hat{u}) = \hat{x}(\tau)^* Q(\tau)\hat{x}(\tau), \tag{2.23}$$

$$\psi(\tau) = Q(\tau)\hat{x}(\tau), \tag{2.24}$$

for some $n \times n$ symmetric matrix function $Q(\tau)$. Since this should be true for all $\tau \in [t_0, t_1)$ and since we have (2.20), one is led to expect that

$$Q(t_1) = V. \tag{2.25}$$

Moreover, since $\psi(t)$ and $\hat{x}(t)$ are differentiable on $[t_0, t_1]$, one expects

Q(t) to be differentiable there. Assuming this for the present, we carry out the following computation:

$$\frac{d}{dt}(\hat{x}(t)^*Q(t)\hat{x}(t)) = \hat{x}(t)^*\dot{Q}(t)\hat{x}(t) + \dot{\hat{x}}(t)^*Q(t)\hat{x}(t)$$

$$+ \hat{x}(t)^*Q(t)\dot{\hat{x}}(t) = (\text{using } (1.17)) = \hat{x}(t)^*\dot{Q}(t)\hat{x}(t)$$

$$+ (\hat{x}(t)^*[A(t)^* - R(t)U(t)^{-1}B(t)^*] - \psi(t)^*B(t)U(t)^{-1}B(t)^*)Q(t)\hat{x}(t)$$

$$+ \hat{x}(t)^*Q(t)([A(t) - B(t)U(t)^{-1}R(t)^*]\hat{x}(t) - B(t)U(t)^{-1}B(t)^*\psi(t))$$

$$= (\text{using } (1.24))\, \hat{x}(t)^*(\dot{Q}(t) + [A(t)^* - R(t)U(t)^{-1}B(t)^*$$

$$- Q(t)B(t)U(t)^{-1}B(t)^*]\, Q(t)$$

$$+ Q(t)[A(t) - B(t)U(t)^{-1}R(t)^* - B(t)U(t)^{-1}B(t)^*Q(t)])\hat{x}(t). \qquad (2.26)$$

But now, applying Corollary 2.2 together with (2.23) and performing the same differentiation we have

$$\frac{d}{dt}(\hat{x}(t)^*Q(t)\hat{x}(t) = \frac{d}{dt}(\psi(t)\hat{x}(t)) = \frac{d}{dt}J(\hat{x}(t), t, t_1, \hat{u})$$

$$= \frac{d}{dt}\int_t^{t_1}(\hat{x}(s)^*\hat{u}(s)^*)\begin{pmatrix} W(s) & R(s) \\ R(s)^* & U(s) \end{pmatrix}\begin{pmatrix} \hat{x}(s) \\ \hat{u}(s) \end{pmatrix} ds + \hat{x}(t_1)Gx(t_1)$$

$$= -(\hat{x}(t)^*, \hat{u}(t)^*)\begin{pmatrix} W(t) & R(t) \\ R(t)^* & U(t) \end{pmatrix}\begin{pmatrix} \hat{x}(t) \\ \hat{u}(t) \end{pmatrix}$$

$$= (\text{from } (1.9) \text{ and } (1.24))$$

$$= -\hat{x}(t)^*(I_n, -[R(t)+Q(t)B(t)]U(t)^{-1})\begin{pmatrix} W(t) & R(t) \\ R(t)^* & U(t) \end{pmatrix}\begin{pmatrix} I_n \\ -U(t)^{-1}[R(t)^* + B(t)^*Q(t)] \end{pmatrix}\hat{x}(t)$$

$$= (\text{carrying out the indicated matrix operations})$$

$$= \hat{x}(t)^*[-W(t) + R(t)U(t)^{-1}R(t)^* - Q(t)B(t)U(t)^{-1}B(t)^*Q(t)]\,\hat{x}(t)\,. \qquad (2.27)$$

Comparing (2. 26) and (2. 27), we see that the two are satisfied simultaneously, for all $\hat{x}(t)$, if $Q(t)$ satisfies the "matrix Riccati differential equations"

$$\dot{Q}(t) + A(t)^* Q(t) + Q(t)A(t) + W(t) - [Q(t)B(t) + R(t)]U(t)^{-1}[B(t)^* Q(t) + R(t)^*] = 0. \tag{2.28}$$

If one retraces one's steps here it is possible to see quite readily that if $Q(t)$ satisfies the differential equation (2. 28) on an interval $[\tau, t_1]$ and the boundary condition (2. 25) at t_1, then a solution of the two point boundary value problem (2. 17), (2. 19), (2. 20) with an initial condition

$$\hat{x}(\tau) = x_\tau \tag{2.29}$$

can be constructed step-by-step in the following manner.

(i) Let $\hat{x}(t)$ be the solution of (cf. (2. 17))

$$\dot{\hat{x}}(t) = (A(t) - B(t)U(t)^{-1}[R(t)^* + B(t)^* Q(t)])\hat{x}(t) \tag{2.30}$$

on $[\tau, t_1]$ with $\hat{x}(\tau)$ given by (2. 29). This solution can be obtained readily because $Q(t)$ is assumed to be a known matrix function on $[\tau, t]$. The equation (2. 30) arises from setting

$$\hat{u}(t) = - U(t)^{-1}[R(t)^* + B(t)^* Q(t)]\, \hat{x}(t) \tag{2.31}$$

in the control system (2. 1).

(ii) Let

$$\psi(t) = Q(t)\hat{x}(t), \qquad t \in [\tau, t_1]. \tag{2.32}$$

From (1. 25) we then have

$$\psi(t_1) = Q(t_1)\hat{x}(t_1) = V\hat{x}(t_1) \tag{2.33}$$

and we can compute

$$\dot{\psi}(t) = \frac{d}{dt}(Q(t)\hat{x}(t)) = \dot{Q}(t)\hat{x}(t) + Q(t)\dot{\hat{x}}(t)$$

$$= \text{(from (2.28) and (2.30))}$$

$$(-A(t)^*Q(t)-Q(t)A(t)-W(t)+[Q(t)B(t)+R(t)]\,U(t)^{-1}[R(t)^*+B(t)^*Q(t)]\,)\hat{x}(t)$$

$$+ Q(t)(A(t)-B(t)U(t)^{-1}[R(t)^* + B(t)^*Q(t)]\,)\hat{x}(t)$$

$$= (-A(t)^*Q(t)-W(t)+R(t)U(t)^{-1}[R(t)^*+B(t)^*Q(t)]\,)\hat{x}(t)$$

$$= (-A(t)^* - R(t)U(t)^{-1}B(t)^*)\psi(t) + (R(t)U(t)^{-1}R(t)^*-W(t))\hat{x}(t)$$

which is the same as (2.19).

If we now follow the proof of Theorem 2.1 but replace x_0 by x_τ and the interval $[t_0, t_1]$ by $[\tau, t_1]$, we can see that we have proved

__Proposition 2.3.__ _If_ $Q(t)$ _solves_ (2.25), (2.28) _on_ $[\tau, t_1]$ _then the control_ $\hat{u}(t)$ _generated by use of the feedback law_ (2.31) _in the equation_ (2.1) _with_ $\hat{x}(\tau) = x_\tau$ _has the property_

$$J(x_\tau, \tau, t_1, \hat{u}) \le J(x_\tau, \tau, t_1, u), \qquad u \in L_m^2[\tau, t_1].$$

Since the general existence and uniqueness theory for ordinary differential equations guarantees a solution $Q(t)$ of (2.28), corresponding to the terminal condition (2.25), on some (possibly small) interval $[\tau, t_1]$, Proposition 2.3 shows that the problem of minimization of $J(x_\tau, \tau, t_1, u)$ does, indeed have a solution for every $x_\tau \in E^n$ if $t_1 - \tau$ is sufficiently small. However, since the differential equation (2.28) is nonlinear, the existence and uniqueness of $Q(t)$ on the interval $[t_0, t_1]$, t_0 arbitrary, cannot be inferred without further analysis. (See [5] for background.)

The general existence and uniqueness theory ([5]) does tell us that $Q(t)$ exists on some maximal interval $I \subseteq [t_0, t_1]$ with $t_1 \in I$ and, moreover, if $I \ne [t_0, t_1]$, then I necessarily has the form $(\alpha, t_1]$ with

$$\limsup_{t \downarrow \alpha} \|Q(t)\| = \infty .$$

Thus if one can demonstrate _a priori_ that there is a positive number M such that necessarily

$$\|Q(t)\| \leq M, \qquad t \in [\tau, t_1] \tag{2.34}$$

for any subinterval $[\tau, t_1] \subseteq [t_0, t_1]$ whereon $Q(t)$ satisfies (2.25), (2.28), we can infer that I must, in fact, be $[t_0, t_1]$. We obtain this *a priori* bound in

Proposition 2.4. *There is a positive number* M *depending only on* V *and the continuous matrix functions* $A(t)$, $B(t)$, $W(t)$, $R(t)$, $U(t)$ *as defined on* $[t_0, t_1]$ *such that* (2.34) *holds on any subinterval* $[\tau, t_1] \subseteq [t_0, t_1]$, *provided* $Q(t)$ *satisfies* (2.25) *and the differential equation* (2.28) *on* $[\tau, t_1]$. *As a consequence there is a unique solution* $Q(t)$ *of* (2.25), (2.28) *on the interval* $[t_0, t_1]$.

Proof. Let x_τ be an arbitrary vector in E^n and let $\hat{u}$ be generated on $[\tau, t_1]$ as described in Proposition 2.3. Then $\hat{u}$ minimizes $J(x_\tau, \tau, t_1, u)$, $u \in L_m^2[\tau, t_1]$ and, in particular, yields a smaller cost than does the control $u(t) \equiv 0$, i.e.,

$$J(x_\tau, \tau, t_1, \hat{u}) \leq J(x_\tau, \tau, t_1, 0).$$

Now the solution $\xi(t)$ of (2.1) which corresponds to the control $u(t) \equiv 0$ satisfies

$$\dot{\xi} = A(t)\xi, \qquad t \in [\tau, t_1]$$

and, letting $\|A(t)\| \leq M_0$, $t \in [t_0, t_1]$, we see (since then $\|\dot{\xi}\| \leq M_0 \|\xi\|$) that with $\xi(\tau) = x_\tau$ we have

$$\|\xi(t)\| \leq e^{M_0(t-\tau)} \|x_\tau\| \leq e^{M_0(t_1-t_0)} \|x_\tau\| \equiv M_1 \|x_\tau\|, \quad t \in [\tau, t_1].$$

Then

$$\begin{aligned} J(x_\tau, \tau, t_1, 0) &= \int_\tau^{t_1} \xi(t)^* W(t) \xi(t) dt + \xi(t_1)^* V \xi(t_1) \\ &\leq [(t_1 - \tau) M_2 (M_1)^2 + M_3 (M_1)^2] \|x_\tau\|^2 \\ &\leq [(t_1 - t_0) M_2 + M_3] M_1^2 \|x_\tau\|^2 \equiv M \|x_\tau\|^2, \end{aligned}$$

where $\|W(t)\| \leq M_2$, $t \in [t_0, t_1]$ and $\|V\| = M_3$.

Since the solution $\hat{x}(t)$ corresponding to $\hat{u}(t)$ with $\hat{x}(\tau) = x_\tau$ and $\psi(t) = Q(t)\hat{x}(t)$ together satisfy (2.17), (2.19), (2.20) on $[\tau, t_1]$, Corollary 2.2 and the subsequent remark give

$$x_\tau^* Q(\tau)x_\tau = J(x_\tau, \tau, t_1, \hat{u}) \leq J(x_\tau, \tau, t_1, 0) \leq M\|x_\tau\|^2$$

from which, taking the symmetry of $Q(\tau)$ into account, we have (2.34) for $t = \tau$. Since the same argument holds for any $\hat{\tau} \in [\tau, t_1]$, the inequality (2.34) follows and the proof of the proposition is complete. Q. E. D.

We are now able to prove

<u>Theorem 2.5.</u> <u>Let</u> $Q(t)$ <u>be the unique solution of</u> (2.25), (2.28) <u>on</u> $[t_0, t_1]$, <u>as demonstrated in Proposition</u> 2.4. <u>Then</u> (<u>cf.</u> (2.6)) <u>the problem</u>

$$\min_{u \in L_m^2[t_0, t_1]} J(x_0, t_0, t_1, u), \quad x_0 \in E^n \tag{2.35}$$

<u>has exactly one solution</u> $\hat{u}(t)$ <u>which is obtained as outlined in</u> (i), (ii) <u>above</u> (cf. (2.30), (2.31) <u>in particular) with</u> $\tau = t_0$. <u>Moreover, as anticipated in</u> (2.21),

$$J(x_0, t_0, t_1, \hat{u}) = x(t_0)^* Q(t_0) x(t_0). \tag{2.36}$$

<u>Proof.</u> That $\hat{u}$ as thus obtained minimizes $J(x_0, t_0, t_1, u)$ follows from Theorem 2.1 and Proposition 2.3 specialized to $\tau = t_0$. The only remaining question, therefore, is uniqueness. Conceivably there could be another optimal control $\tilde{u}(t)$ associated with a solution of the two point boundary value problem (2.17) - (2.20) other than the one obtained by the process (i), (ii) (equation (2.30) - (2.33)) outlined above. To show that this cannot be the case we let $\tilde{x}(t)$ satisfy $\dot{\tilde{x}} = A(t)\tilde{x} + B(t)\tilde{u}(t)$, $\tilde{x}(t_0) = x_0$ and compute

$$\frac{d}{dt}(\tilde{x}(t)^* Q(t)\tilde{x}(t)) = (A(t)\tilde{x}(t)+B(t)\tilde{u}(t))^* Q(t)\tilde{x}(t)$$

$$- \tilde{x}(t)^*[A(t)^* Q(t)+Q(t)A(t)+W(t)-(Q(t)B(t)+R(t))U(t)^{-1}(R(t)^*+B(t)^*Q(t))]\tilde{x}(t)$$

$$+ \tilde{x}(t)^* Q(t)(A(t)\tilde{x}(t) + B(t)\tilde{u}(t))$$

$$= -\tilde{x}(t)^* W(t)\tilde{x}(t) + 2\tilde{u}(t)^* B(t)^* Q(t)\tilde{x}(t)$$

$$+ \tilde{x}(t)^*(Q(t)B(t)+R(t))U(t)^{-1}(R(t)^* + B(t)^* Q(t))\tilde{x}$$

$$= -\tilde{x}(t)^* W(t)\tilde{x}(t) - 2\tilde{u}(t)^* R(t)^* \tilde{x}(t) - \tilde{u}(t)^* U(t)\tilde{u}(t)$$

$$+ 2\tilde{u}(t)^* B(t)^* Q(t)\tilde{x}(t) + 2\tilde{u}(t)^* R(t)\tilde{x}(t) + \tilde{u}(t)^* U(t)\tilde{u}(t)$$

$$+ \tilde{x}(t)^*(Q(t)B(t) + R(t))U(t)^{-1}(R(t)^* + B(t)^* Q(t))\tilde{x}(t).$$

Collecting the last four terms here and integrating from $t = t_0$ to $t = t_1$ we have

$$J(x_0, t_0, t_1, u) = x_0^* Q(t_0)x_0 = \tilde{x}(t_1)^* Q(t_1)\tilde{x}(t_1)$$

$$- \int_{t_0}^{t_1} \frac{d}{dt}(\tilde{x}(t)^* Q(t)\tilde{x}(t))dt$$

$$= \tilde{x}(t_1)^* G\tilde{x}(t_1) + \int_{t_0}^{t_1} [\tilde{x}(t)^* W(t)\tilde{x}(t) + 2\tilde{u}(t)^* R(t)^* \tilde{x}(t) + \tilde{u}(t)^* U(t)\tilde{u}(t)]\, dt$$

$$- \int_{t_0}^{t_1} (\tilde{u}(t) + U(t)^{-1}[B(t)^* Q(t)+R(t)^*]\tilde{x}(t))U(t)(\tilde{u}(t)+U(t)^{-1}[B(t)^* Q(t)+R(t)^*]\tilde{x}(t))dt$$

$$= J(x_0, t_0, t_1, \tilde{u}) - \int_{t_0}^{t_1} \|U(t)^{\frac{1}{2}}(\tilde{u}(t)+U(t)^{-1}[B(t)^* Q(t)+R(t)^*]\tilde{x}(t))\|^2\, dt,$$

where (see II-[1],[2]) $U(t)^{\frac{1}{2}}$ is the positive definite square root of the symmetric positive definite matrix $U(t)$. Thus

$$J(x_0, t_0, t_1, \hat{u}) \le J(x_0, t_0, t_1, \tilde{u}) \tag{2.37}$$

(which again confirms the minimality of $\hat{u}$) and, since $U(t)^{\frac{1}{2}}$ is positive definite, equality holds if and only if

$$\tilde{u}(t) = -U(t)^{-1}[B(t)^* Q(t) + R(t)^*]\tilde{x}(t) \quad \text{in} \quad L_m^2[t_0, t_1]. \tag{2.38}$$

But this is the same feedback law which generates $\hat{u}(t)$ so (2. 38) implies $\tilde{u}(t) = \hat{u}(t)$, $\tilde{x}(t) = \hat{x}(t)$ and we conclude that equality holds in (2. 37) if and only if $\tilde{u}(t) = \hat{u}(t)$ in $L_m^2[t_0, t_1]$, thus completing the proof. Q. E. D.

The following example, involving a one dimensional system, is not representative of the complexity encountered in higher dimensional applications. Nevertheless it does illustrate how the various steps of the procedure outlined above are carried out in order to effect a solution.

We consider a two compartment heat-transfer system, as shown in Figure IV-2. The two compartments are isolated from the external environment

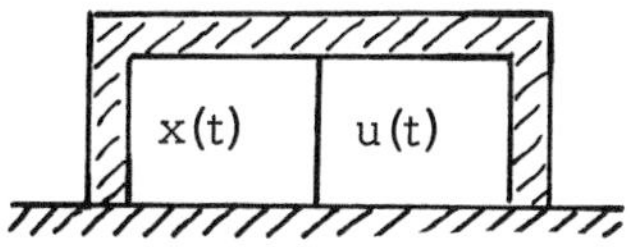

Fig. IV-2. Two compartment heat transfer system

and separated by a wall having unit conductivity. The left compartment has unit specific heat and the temperature is designated by $x(t)$. The right compartment is the control compartment, its temperature, $u(t)$, can be determined at will in order to influence $x(t)$. The differential equation governing $x(t)$ is

$$\dot{x} = -x + u \quad (A = -1, \quad B = 1).$$

Let us select as the responses of interest the quantities x and $x - u$, the latter we may suppose to be of interest since an excessive temperature difference between the two compartments might damage the wall separating them. Thus

$$y = \begin{pmatrix} y^1 \\ y^2 \end{pmatrix} = \begin{pmatrix} 1 \\ 1 \end{pmatrix} x + \begin{pmatrix} 0 \\ -1 \end{pmatrix} u \quad \begin{pmatrix} F = \begin{pmatrix} 1 \\ 1 \end{pmatrix} \\ G = \begin{pmatrix} 0 \\ -1 \end{pmatrix} \end{pmatrix}.$$

We take

$$\tilde{W} = \begin{pmatrix} \alpha & 0 \\ 0 & \beta \end{pmatrix}$$

and

$$\tilde{U} = \gamma ,$$

$$V = \delta \ ,$$

$\alpha, \beta, \gamma, \delta$ all being non-negative. We take $t_1 = 0$, $t_2 = 1$. Our objective then, given an initial temperature

$$x(0) = x_0 ,$$

is to select $u \in L^2[0,1]$ so as to minimize the cost functional

$$J(x_0, 0, 1, u) = \int_0^1 \left[(x, x-u) \begin{pmatrix} \alpha & 0 \\ 0 & \beta \end{pmatrix} \begin{pmatrix} x \\ x-u \end{pmatrix} + \gamma u^2 \right] dt$$

$$+ \delta x(1)^2 = \int_0^1 (x, u) \begin{pmatrix} \alpha+\beta & -\beta \\ -\beta & \beta+\gamma \end{pmatrix} \begin{pmatrix} x \\ u \end{pmatrix} dt + \delta x(1)^2 ,$$

i. e.,

$$W = \alpha + \beta ,$$

$$R = - \beta ,$$

$$U = \beta + \gamma .$$

In this case the Riccati differential equation (2. 28) reduces to the scalar ordinary differential equation

$$\dot{q}(t) - 2q(t) + \alpha + \beta - (\beta + \gamma)^{-1} (q(t) - \beta)^2 = 0 \qquad (2.39)$$

and the terminal condition is

$$q(1) = \delta \equiv V. \qquad (2.40)$$

The differential equation (2. 39) can be solved by the separation of variables technique (see [6], e. g.) to give the solution corresponding to (2. 40). The equation (2. 39) can be written

$$a \frac{dq}{dt} = q^2 + bq + c$$

with

$$a = \beta + \gamma, \quad b = 2\gamma, \quad c = -\beta\gamma - \alpha(\beta + \gamma).$$

Then, for $s \leq 1$

$$\frac{1-s}{a} = \frac{1}{a} \int_s^1 dt = \int_{q(s)}^{q(1)=\delta} \frac{1}{a \frac{dq}{dt}} dq$$

$$= \int_{q(s)}^{\delta} \frac{dq}{q^2 + bq + c} = \frac{1}{r_1 - r_2} \int_{q(s)}^{\delta} \left[\frac{1}{q - r_1} - \frac{1}{q - r_2}\right] dq \qquad (2.41)$$

where r_1, r_2 are the (assumed distinct) roots of $q^2 + bq + c = 0$, i.e.,

$$r_1 = -\gamma + \sqrt{\gamma^2 + \beta\gamma + \alpha(\beta + \gamma)},$$

$$r_2 = -\gamma - \sqrt{\gamma^2 + \beta\gamma + \alpha(\beta + \gamma)}.$$

Carrying out the indicated integration and taking the exponential of both sides of (2.41) we have

$$\exp\left(\frac{r_1 - r_2}{a}(1-s)\right) = \exp\left[\log\left(\frac{\delta - r_1}{q(s) - r_1}\right) - \log\left(\frac{\delta - r_2}{q(s) - r_2}\right)\right] = \frac{\delta - r_1}{\delta - r_2} \frac{q(s) - r_2}{q(s) - r_1}$$

which can be solved to give

$$q(s) = \frac{(\delta - r_2) r_1 \exp\left(\frac{r_1 - r_2}{a}(1-s)\right) - (\delta - r_1) r_2}{(\delta - r_2) \exp\left(\frac{r_1 - r_2}{a}(1-s)\right) - (\delta - r_1)}. \qquad (2.42)$$

That $q(1) = \delta$ is readily checked.

The optimal control on $0 \leq t \leq 1$ is then generated by the feedback law (cf. (2.31))

$$\hat{u}(t) = -\frac{1}{\beta + \gamma}(q(t) - \beta)\hat{x}(t)$$

and the optimal system state trajectory satisfies

$$\hat{x}(0) = x_0$$

$$\dot{\hat{x}}(t) = -\hat{x}(t) + \hat{u}(t) = -(1 + \frac{1}{\beta+\gamma}(q(t)-\beta))\hat{x}(t) .$$

Then

$$\hat{x}(t) = x_0 \exp(-\int_0^t (1 + \frac{1}{\beta+\gamma}(q(s)-\beta))ds),$$

$$\hat{u}(t) = -\frac{x_0}{\beta+\gamma}(q(t)-\beta)\exp(-\int_0^t (1 + \frac{1}{\beta+\gamma}(q(s)-\beta))ds),$$

$q(s)$ being given by (2.42). We have, then, an explicit solution of the problem, lacking only the computation of the integral

$$\int_0^t (1 + \frac{1}{\beta+\gamma}(q(s)-\beta))ds = t(1 - \frac{\beta}{\beta+\gamma}) + \frac{1}{\beta+\gamma}\int_0^t q(s)ds . \qquad (2.43)$$

This can be done analytically, using the change of variable

$$\frac{r_1 - r_2}{a}(1 - s) = \log \sigma .$$

In Exercise [3] at the end of this chapter the reader is invited to obtain analytic expressions for $\hat{x}(t)$, $\hat{u}(t)$ in terms of elementary functions and to study the way in which the optimal solution depends on the parameters $\alpha, \beta, \gamma, \delta$.

3. Optimal Control of Constant Coefficient Systems on an Infinite Time Interval

We specialize now to control systems

$$\dot{x} = Ax + Bu, \quad x \in E^n, \ u \in E^m, \qquad (3.1)$$

wherein A and B are constant $n \times n$ and $n \times m$ matrices, respectively. Throughout this section we will assume that the pair (A, B) is stabilizable. We again consider a cost functional of the form (2.6), but now with constant matrices F, G, $\tilde{W}$, $\tilde{U}$, (Hence constant W, R and U) and we shall take V, the weighting matrix on the terminal state, to be zero. Since the coefficients are now independent of t we may, without loss of generality, take the time interval involved to be $[0,T]$ and thus consider the cost

$$J(x_0, T, u) \equiv \int_0^T (x(t)^*, u(t)^*) \begin{pmatrix} W & R \\ R^* & U \end{pmatrix} \begin{pmatrix} x(t) \\ u(t) \end{pmatrix} dt . \tag{3.2}$$

Our intention is to study a similar cost functional on the infinite interval, i. e.

$$J(x_0, u) \equiv \int_0^\infty (x(t)^*, u(t)^*) \begin{pmatrix} W & R \\ R^* & U \end{pmatrix} \begin{pmatrix} x(t) \\ u(t) \end{pmatrix} dt , \tag{3.3}$$

by taking account of what happens as we let $T \to \infty$ in (3. 2). The cost (3. 3) is defined whenever $u(t) \in L_m^2[0,\infty)$ and $x(t) \in L_n^2[0,\infty)$. The stabilizability condition is used to guarantee that the set of controls and associated trajectories for which (3. 3) is finite is not an empty set.

Proposition 3. 1. The set of control and trajectory pairs $u(t)$, $x(t)$ for which (3. 3) is finite is non-empty for each $x_0 \in E^n$ if (A, B) is a stabilizable pair.

Proof. Since (A, B) is stabilizable there is an $m \times n$ feedback matrix K such that $A + BK$ is a stability matrix. Then

$$\|e^{(A+BK)t}\| \le Me^{-\gamma t}, \quad t \ge 0 \tag{3.4}$$

for some positive numbers M, γ. Then for $x_0 \in E^n$ and $x(t) = e^{(A+BK)t}$, $u(t) = Ke^{(A+BK)t}$, $x(t)$ and $u(t)$ together satisfy (3. 1) and, since

$$\|x(t)\| \le Me^{-\gamma t}\|x_0\|, \quad \|u(t)\| \le M\|K\|e^{-\gamma t}\|x_0\| \tag{3.5}$$

we have $x(t) \in L_n^2[0,\infty)$, $u(t) \in L_m^2[0,\infty)$ and the cost (3. 3) is, indeed, finite. Q. E. D.

Having established this proposition, it is now reasonable to ask if for each $x_0 \in E^n$ there is a control $\hat{u}(t) \in L_m^2[0,\infty)$ and corresponding solution $\hat{x}(t)$ of (3. 1), satisfying $\hat{x}(t_0) = x_0$, such that $\hat{u}, \hat{x}$ minimize (3. 3) when compared to other controls u, x satisfying (3. 1) and

$$x(t_0) = x_0 \tag{3.6}$$

which yield a finite value for the cost (3. 3). To study this question we return to the Riccati matrix differential equation

$$\dot{Q} + A^*Q + QA + W - (QB + R)U^{-1}(R^* + B^*Q) = 0 \tag{3.7}$$

on a finite interval $[0, T]$, now with the terminal condition (since $V = 0$)

$$Q(T) = 0. \tag{3.8}$$

Since our attention is now focussed on what happens as T varies, in particular, what happens as $T \to \infty$, we distinguish the solution of (3. 7), (3. 8) by the subscript T; viz.: $Q_T(t)$. The fact that (3. 7) is a constant coefficient system shows that the value of $Q_T(t)$ depends only on the difference between T and t, so that for $T_2 > T_1$

$$Q_{T_2}(t + T_2 - T_1) = Q_{T_1}(t). \tag{3.9}$$

Now let $x_0 \in E^n$ be given and let $u_T(t)$ denote the optimal control in $L^2_m[0, T]$ for the cost (3. 2) and let $x_T(t)$ denote the corresponding optimal solution of (3. 1), (3. 6). From Theorem 2. 5 we know that these two are related by

$$u_T(t) = -U^{-1}[B^*Q_T(t) + R^*]x_T(t). \tag{3.10}$$

<u>Proposition 3. 2.</u> <u>If</u> $T_2 \geq T_1$ <u>then</u>

$$Q_{T_2}(0) \geq Q_{T_1}(0).$$

<u>Equivalently, for fixed</u> t, $Q_T(t)$ <u>is a non-increasing (in the sense of quadratic forms) matrix function of</u> t <u>and for fixed</u> T <u>it is a non-decreasing function of</u> T.

<u>Proof.</u> From (2. 36) we know, for $x_0 \in E^n$, that

$$x_0^* Q_{T_1}(0)x_0 = J(x_0, T_1, u_{T_1}) \leq \text{(by optimality)}$$

$$\leq J(x_0, T_1, u_{T_2}) = \int_0^{T_1} (x_{T_2}(t)^*, u_{T_2}(t)^*) \begin{pmatrix} W & R \\ R^* & U \end{pmatrix} \begin{pmatrix} x_{T_2}(t) \\ u_{T_2}(t) \end{pmatrix} dt$$

$$\leq \left(\text{since} \begin{pmatrix} W & R \\ R^* & U \end{pmatrix} \geq 0\right) \leq \int_0^{T_2} (x_{T_2}(t)^*, u_{T_2}(t)^*) \begin{pmatrix} W & R \\ R^* & U \end{pmatrix} \begin{pmatrix} x_{T_2}(t) \\ u_{T_2}(t) \end{pmatrix} dt$$

$$= J(x_0, T_2, u_{T_2}) = x_0^* Q_{T_2}(0)x_0, \tag{3.11}$$

where $u_{T_2}(t) \equiv -U^{-1}[B^* Q_{T_2}(t) + R^*]x_{T_2}(t)$ generates the optimal control for $J(x_0, T_2, u)$. Since this is true for every $x_0 \in E^n$, $Q_{T_2}(0) \geq Q_{T_1}(0)$. Q. E. D.

<u>Proposition 3. 3.</u> <u>If the pair</u> (A, B) <u>is stabilizable there is a positive number</u> q <u>such that</u>

$$Q_T(0) \leq qI, \qquad 0 \leq T < \infty .$$

<u>Proof.</u> Let K be an $m \times n$ matrix such that $A + BK$ is a stability matrix. Then we have the estimates (3. 5) of Proposition 3. 1 and, for $x_0 \in E^n$

$$x_0^* Q_T(0)x_0 = \int_0^T (x_T(t)^*, u_T(t)^*) \begin{pmatrix} W & R \\ R^* & U \end{pmatrix} \begin{pmatrix} x_T(t) \\ u_T(t) \end{pmatrix} dt$$

$$\leq \int_0^T M^2(\|W\| + 2\|R\|\|K\| + \|U\|\|K\|^2)e^{-2\gamma t} dt \|x_0\|^2$$

$$\leq \frac{M^2(\|W\|+2\|R\|\|K\|+\|U\|\|K\|^2)}{2\gamma} \|x_0\|^2 \equiv q\|x_0\|^2 \tag{3.12}$$

and since the estimate (3. 12) is independent of T, the proposition follows. Q. E. D.

Our next step is to prove that there is a matrix Q such that $\lim_{T \to \infty} Q_T(0) = Q$. In order to do this we need

<u>Lemma 3. 4.</u> <u>Let</u> $\{S_k\}$ <u>be a sequence of real</u> $n \times n$ <u>symmetric matrices</u>

<u>such that</u> $\{S_k\}$ <u>is monotone and bounded; without loss of generality,</u> $\{S_k\}$ <u>is increasing:</u>

$$S_1 \leq S_2 \leq \cdots \leq S_k \leq S_{k+1} \leq \cdots ,$$

<u>and</u> $\{S_k\}$ <u>is bounded above: there is some real</u> s <u>such that</u>

$$S_k \leq sI, \quad k = 1, 2, 3, \ldots .$$

Then there is a real $n \times n$ symmetric matrix S such that

$$\lim_{k \to \infty} S_k = S.$$

<u>Remark.</u> One may equally well replace the discrete index k by a continuous parameter T and obtain the same result for S_T.

<u>Proof of Lemma 3.4.</u> Let e_i be the vector appearing in the i-th column of the $n \times n$ identity matrix. Then

$$(S_k)_i^i = e_i^* S_k e_i \leq e_i^* S_{k+1} e_i = (S_{k+1})_i^i \leq s$$

and the usual convergence theorem for monotone sequences of real numbers ([7]) gives

$$\lim_{k \to \infty} (S_k)_i^i = S_i^i, \qquad i = 1, 2, \ldots, n,$$

for some numbers S_i^i. Thus the diagonal entries of the S_k converge. Since for $k > \ell$

$$|(S_k)_j^i - (S_\ell)_j^i| = |e_i^*(S_k - S_\ell)e_j| \leq \tfrac{1}{2}(e_i^*(S_k - S_\ell)e_i + e_j^*(S_k - S_\ell)e_j)$$

$$= \tfrac{1}{2}((S_k)_i^i - (S_\ell)_i^i + (S_k)_j^j - (S_\ell)_j^j) \qquad (3.12)$$

we conclude that $\{(S_k)_j^i\} = \{(S_k)_i^j\}$ is Cauchy and hence, for some S_j^i,

$$\lim_{k \to \infty} (S_k)_j^i = \lim_{k \to \infty} (S_k)_i^j = S_j^i = S_i^j .$$

Thus the off-diagonal elements converge also. Q. E. D. (The inequality in (3.12) is obtained by noting that

$$(e_i \pm e_j)^*(S_k - S_\ell)(e_i \pm e_j) \geq 0, \qquad k > \ell.)$$

Lemma 3.4 with Propositions 3.2, 3.3 now leads us to

Theorem 3.5. If the pair (A, B) is stabilizable we have

$$\lim_{T \to \infty} Q_T(0) = Q \geq 0 ,$$

where the symmetric $n \times n$ matrix Q satisfies the quadratic matrix equation

$$A^* Q + QA + W - (QB + R)U^{-1}(R^* + B^* Q) = 0 . \tag{3.13}$$

Proof. We have seen in Propositions 3.2, 3.3 that $Q_T(0)$, as a matrix function of T, is nondecreasing and bounded above when (A, B) is stabilizable. Applying Lemma 3.4 to $Q_T(0)$ we have

$$\lim_{T \to \infty} Q_T(0) = Q \tag{3.14}$$

for some symmetric matrix Q. Since $Q_T(0) \geq 0$ for all T we have $Q \geq 0$. From (3.14) we have

$$\lim_{T \to \infty} [A^* Q_T(0) + Q_T(0)A + W - (Q_T(0)B + R)U^{-1}(R^* + B^* Q_T(0))]$$
$$= A^* Q + QA + W - (QB + R)U^{-1}(R^* + B^* Q) \tag{3.15}$$

from which we conclude, using (3.7), that

$$\lim_{T \to \infty} \dot{Q}_T(0) \text{ exists.}$$

If we fix T and let t vary

$$\lim_{t \to -\infty} Q_T(t) = Q, \quad \lim_{t \to -\infty} \dot{Q}_T(t) = \lim_{T \to \infty} \dot{Q}_T(0) .$$

But the first limit cannot exist unless the second limit is zero. Hence

$$\lim_{T \to \infty} \dot{Q}_T(0) = 0$$

and then, since (3.7) gives

$$\dot{Q}_T(0) + A^* Q_T(0) + Q_T(0)A + W - (Q_T(0)B + R)U^{-1}(R^* + B^* Q_T(0)) = 0$$

for all T, (3.15) reduces to (3.13). Q. E. D.

We have seen that on any finite interval $[0,T]$ the cost $J(x_0, T, u)$ is minimized by implementing the time varying linear feedback law

$$u_T(t) = -U^{-1}[B^* Q_T(t) + R^*]\, x_T(t). \tag{3.16}$$

As we let $T \to \infty$, for each fixed t the matrix in (3.16) tends to the matrix $-U^{-1}[B^* Q + R^*]$ and we are led to consider the limiting feedback relationship

$$\hat{u}(t) = -U^{-1}[B^* Q + R^*]\, \hat{x}(t) \tag{3.17}$$

which, substituted into (3.1), gives the closed loop system

$$\dot{\hat{x}} = (A - BU^{-1}[B^* Q + R^*])\, \hat{x}. \tag{3.18}$$

We will see that implementation of this feedback law yields the unique control $\hat{u}$ minimizing $J(x_0, u)$ (cf. (3.3)). We establish this in several steps. First of all, we recall that our cost functional is originally given in the form (cf. (3.6))

$$J(x_0, u) = \int_0^\infty [(Fx(t) + Gu(t))^* \tilde{W} (Fx(t) + Gu(t)) + u(t)^* \tilde{U} u(t)]\, dt \tag{3.19}$$

with $\tilde{W}$ and $\tilde{U}$ symmetric positive definite $q \times q$ and $m \times m$ matrices, respectively.

<u>Proposition 3.6.</u> <u>Let</u> (A, B) <u>be stabilizable and</u> (F, A) <u>observable. Then for each</u> $T > 0$, $Q_T(0)$ <u>is positive definite and, since</u> $Q_T(0)$ <u>increases with</u> T, $Q = \lim_{T \to \infty} Q_T(0)$ <u>is likewise positive definite.</u>

<u>Proof.</u> We have

$$x_0^* Q_T(0) x_0 = J(x_0, T, u_T) = \int_0^T [(Fx_T(t) + Gu_T(t))^* \tilde{W} (Fx_T(t) + Gu_T(t))$$

$$+ u_T(t)^* \tilde{U} u_T(t)]\, dt. \tag{3.20}$$

If $u_T \neq 0$ in $L_m^2[0,T]$ then $J(x_0, T, u_T) > 0$ since the first term in the integrand of (3.19) is ≥ 0 and $\tilde{U}$ is positive definite. On the other hand, if $u_T = 0$ then

$$\dot{x}_T(t) = Ax_T(t), \quad t \in [0,T] .$$

If $x_0 \neq 0$, observability of (F, A) implies $Fx_T(t) \neq 0$ in $L_q^2[0,T]$. Then (3.18) reads

$$J(x_0, T, u_T) = \int_0^T (Fx_T(t))^* \tilde{W} Fx_T(t)dt > 0 .$$

Thus, in any event, $J(x_0, T, u_T) > 0$ when $x_0 \neq 0$ and the first equation in (3.11) shows $Q_T(0) > 0$. As noted, $Q_T(0)$ is nondecreasing as T increases and, since $Q = \lim_{T\to\infty} Q_T(0)$, we conclude $Q > 0$. Q. E. D.

Proposition 3.7. If (A, B) is stabilizable and (F, A) is observable, then the matrix $A - BU^{-1}[B^*Q + R^*]$ in (3.18), obtained by using the feedback law (3.17) in $\dot{x} = Ax + Bu$ is a stability matrix, the cost $J(x_0, \hat{u})$ resulting from use of this feedback law is finite for each $x_0 \in E^n$ and

$$J(x_0, \hat{u}) = x_0^* Q x_0, \quad x_0 \in E^n \tag{3.21}$$

where Q is the solution (3.14) of (3.13) constructed above.

Proof. Rearranging the quadratic matrix equation (3.13) we have

$$(A-BU^{-1}[B^*Q+R^*])^*Q + Q(A-BU^{-1}[B^*Q+R^*]) + W$$

$$+ (QB + R)U^{-1}B^*Q + QBU^{-1}[B^*Q+R^*] - (QB+R)U^{-1}(B^*Q+R^*) = 0$$

or

$$(A-BU^{-1}[B^*Q+R^*])^*Q + Q(A-BU^{-1}[B^*Q+R^*] + W - RU^{-1}R^* + QBU^{-1}B^*Q = 0. \tag{3.22}$$

Comparing with the computations performed in (2.27), we see that this corresponds to

$$(A - BU^{-1}[B^*Q+R^*])^*Q + Q(A - BU^{-1}[B^*Q + R^*])$$

$$+ (I, -[QB+R]U^{-1}) \begin{pmatrix} W & R \\ R^* & U \end{pmatrix} \begin{pmatrix} I \\ -U^{-1}[B^*Q + R^*] \end{pmatrix} = 0 \tag{3.23}$$

which in turn may be rewritten in terms of the original $\tilde{W}$, $\tilde{U}$ as

$$(A-BU^{-1}[B^*Q+R^*])^*Q+Q(A-BU^{-1}[B^*Q+R^*])$$
$$+(F-GU^{-1}[B^*Q+R^*])^*\tilde{W}(F-GU^{-1}[B^*Q+R^*])$$
$$+(QB+R)U^{-1}\tilde{U}U^{-1}(B^*Q+R^*) = 0 . \tag{3.24}$$

If we now let $\hat{x}(t)$ satisfy (3.18) and compute $\frac{d}{dt}\hat{x}(t)^*Q\hat{x}(t)$ to find, after integrating from 0 to τ, that

$$\hat{x}(0)^*Q\hat{x}(0)-\hat{x}(\tau)^*Q\hat{x}(\tau)$$
$$=\int_0^\tau [\hat{x}(t)^*(F-GU^{-1}[B^*Q+R^*])^*\tilde{W}(F-GU^{-1}[B^*Q+R^*])\hat{x}(t)$$
$$+\hat{x}(t)^*(QB+R)U^{-1}\tilde{U}U^{-1}(B^*Q+R^*)\hat{x}(t)]\,dt . \tag{3.25}$$

Now the integral here is positive for every initial state $x(0) = x_0 \in E^n$. Since U and $\tilde{U}$ are both positive definite and $\tilde{W}$ is positive definite, the integral is positive if

$$\hat{u}(t) = -U^{-1}[B^*Q+R^*]\hat{x}(t) \neq 0 \text{ in } L_m^2[0,\tau] .$$

If, on the other hand, $-U^{-1}[B^*Q+R^*]\hat{x}(t) = 0$ in $L_m^2[0,\tau]$ then $\hat{x}(t)$ satisfies

$$\dot{\hat{x}}(t) = A\hat{x}(t)$$

and (3.25) reads

$$x(0)^*Qx(0)-x(\tau)^*Qx(\tau) = \int_0^\tau \hat{x}(t)^*F^*\tilde{W}F\hat{x}(t)\,dt > 0 \tag{3.26}$$

because (F, A) is observable and $\tilde{W} > 0$. Combining these and using $\hat{x}(t) = e^{(A-BU^{-1}[B^*Q+R^*])t}x_0$, we see that with

$$P_\tau = \int_0^\tau e^{(A-BU^{-1}[B^*Q+R^*])^*t}(((F-GU^{-1}[B^*Q+R^*])^*\tilde{W}(F-GU^{-1}[B^*Q+R^*])$$
$$+(QB+R)U^{-1}\tilde{U}U^{-1}(B^*Q+R^*))\,e^{(A-BU^{-1}[B^*Q+R^*])t}\,dt \tag{3.27}$$

we have

$$\hat{x}(0)^* Q\hat{x}(0) - \hat{x}(\tau)^* Q\hat{x}(\tau) = \hat{x}(0)^* P_\tau \hat{x}(0) > 0, \quad \hat{x}(0) \neq 0. \tag{3.28}$$

We may then proceed as in Theorem 2.4 to show

$$\lim_{t\to\infty} \hat{x}(t) = 0$$

for each initial state $\hat{x}(0) = x_0 \in E^n$. Hence the matrix $A - BU^{-1}[B^*Q+R^*]$ is a stability matrix.

Now letting $\tau \to \infty$ in (3.25) we will have

$$\lim_{\tau\to\infty} \hat{x}(\tau)^* Q\hat{x}(\tau) = 0$$

and the integral converges since $A - BU^{-1}[B^*Q+R^*]$ stable implies we have some estimate $\|\hat{x}(t)\| \leq Me^{-\gamma t}\|\hat{x}(0)\|$, $M, \gamma > 0$. Thus, if $\hat{x}(0) = x_0$, as $\tau \to \infty$ (3.25) becomes

$$x_0^* Qx_0 = \int_0^\infty \hat{x}(t)^* (F-GU^{-1}[B^*Q+R^*])^* \tilde{W}(F-GU^{-1}[B^*Q+R^*])\hat{x}(t)$$

$$+ \hat{x}(t)^* (QB+R)U^{-1}\tilde{U}U^{-1}(B^*Q+R^*)\hat{x}(t)dt$$

$$= \int_0^\infty [(F\hat{x}(t)+G\hat{u}(t))^* \tilde{W}(F\hat{x}(t)+G\hat{u}(t))+\hat{u}(t)^* U\hat{u}(t)]dt$$

$$= J(x_0, \hat{u}). \quad \text{Q.E.D.} \tag{3.29}$$

Finally, we have

<u>Theorem 3.8.</u> <u>If</u> (A, B) <u>is stabilizable and</u> (F, A) <u>is observable the control</u> $\hat{u}(t)$ <u>generated by the feedback law</u> (3.17) <u>is the unique solution of the problem</u>

$$\min_{u\in L^2[0,\infty)} J(x_0, u). \tag{3.30}$$

<u>Proof.</u> The proof of optimality goes fairly quickly. Let u be any control for which $J(x_0, u)$ is finite and let $0 < T_1 < T_2$. Then

$$J(x_0, u_{T_2}, T_1) \leq J(x_0, T_2, u_{T_2}) \leq J(x_0, T_2, u),$$

the first inequality following from $T_1 < T_2$ and the non-negativity of the integrand defining the cost $J(x_0, T, u)$, the second following from the optimality of u_{T_2} relative to the cost $J(x_0, T_2, u)$. Now let $T_2 \to \infty$. On $[0, T_1]$ $x_{T_2}(t)$ satisfies

$$\dot{x}_{T_2}(t) = (A - BU^{-1}[B^* Q_{T_2}(t) + R^*])x_{T_2}(t)$$

$$x_{T_2}(0) = x_0 .$$

Since $Q_{T_2}(t) = Q_{T_2 - t}(0)$ and, for $t \in [0, T_1]$

$$Q_{T_2 - T_1}(0) \leq Q_{T_2 - t}(0) \leq Q$$

we conclude from (3.14) that $Q_{T_2}(t)$ converges uniformly to Q for $t \in [0, T_1]$ and from this it is not hard to show using (2.30), (3.18) (with $Q(t) \equiv Q_{T_2}(t)$ in (2.30)) that

$$\lim_{T_2 \to \infty} x_{T_2}(t) = \hat{x}(t), \text{ uniformly for } t \in [0, T_1], \tag{3.31}$$

where, as earlier, $\hat{x}(t)$ is the solution of (3.18) with $\hat{x}(0) = x_0$. Then (cf. (3.10), (3.17))

$$\lim_{T_2 \to \infty} u_{T_2}(t) = \lim_{T_2 \to \infty} -U^{-1}[B^* Q_{T_2}(t) + R^*] x_{T_2}(t)$$

$$= -U^{-1}[B^* Q + R^*] x(t), \text{ uniformly for } t \in [0, T_1], \tag{3.32}$$

and the formula (2.36) for $J(x_0, T, u)(t_0 = 0,\ t_1 = T)$ together with the fact that $J(x_0, u)$ is finite implies

$$J(x_0, T_1, \hat{u}) = \lim_{T_2 \to \infty} J(x_0, T_1, u_{T_2}) \leq \lim_{T_2 \to \infty} J(x_0, T_2, u) = J(x_0, u). \tag{3.33}$$

Now let $T_1 \to \infty$ in (3.33). Because $\hat{u}(t)$, $\hat{x}(t)$ decay exponentially as $t \to \infty$ we have

$$J(x_0, \hat{u}) = \lim_{T_1 \to \infty} J(x_0, T_1, \hat{u}) \leq J(x_0, u)$$

and we see that $\hat{u}$ does indeed solve the minimization problem (3.30).

For the proof that $\hat{u}$ is the unique control minimizing $J(x_0, u)$ we recall the formula which we developed for the same purpose relative to $J(x_0, t_0, t_1, \tilde{u})$ in Theorem 2.5. For $[t_0, t_1] = [0, T]$ it reads

$$J(x_0, T, u_T) = J(x_0, T, \tilde{u}) - \int_0^T (\tilde{u}(t)+U^{-1}[B^* Q_T(t)+R^*]\tilde{x}(t))^* U(\tilde{u}(t)+U^{-1}[B^* Q_T(t)+R^*]\tilde{x}(t))dt \tag{3.34}$$

Let us now suppose $\tilde{u}$ to be a control for which $J(x_0, \tilde{u})$ is finite but such that

$$\tilde{u}(t) \neq -U^{-1}[B^* Q + R^*]\tilde{x}(t) \tag{3.35}$$

on some subset of $[0, \infty)$ of positive measure. Then there is a $T_0 > 0$ such that when $T > T_0$, $[0, T]$ includes a subset E of positive measure such that, for some $d > 0$

$$(\tilde{u}(t)+U^{-1}[B^*Q+R^*]\tilde{x}(t))^* U(\tilde{u}(t)+U^{-1}[B^*Q+R^*]\tilde{x}(t)) \geq d, \qquad t \in E. \tag{3.36}$$

Then, again using the uniform convergence argument on $[0, T_0]$,

$$\begin{aligned} J(x_0, \hat{u}) &= \lim_{T \to \infty} J(x_0, T, u_T) \leq \lim_{T \to \infty} J(x_0, T, \tilde{u}) \\ &\quad - \lim_{T \to \infty} \int_0^T (\tilde{u}(t)+U^{-1}[B^* Q_T(t)+R^*]\tilde{x}(t))^* U(\tilde{u}(t)+U^{-1}[B^* Q_T(t)+R^*]\tilde{x}(t))dt \\ &\leq J(x_0, \tilde{u}) - \lim_{T \to \infty} \int_0^{T_0} (\tilde{u}(t)+U^{-1}[B^* Q_T(t)+R^*]\tilde{x}(t))^* U(\tilde{u}(t)+U^{-1}[B^* Q_T(t)+R^*]\tilde{x}(t))dt \\ &= J(x_0, \tilde{u}) - \int_0^{T_0} (\tilde{u}(t)+U^{-1}[B^* Q+R^*]\tilde{x}(t))^* U(\tilde{u}(t)+U^{-1}[B^* Q+R^*]\tilde{x}(t))dt \\ &\leq J(x_0, \tilde{u}) - d\mu(E) < J(x_0, \tilde{u}). \end{aligned}$$

Since $\tilde{u}$ reduces to $\hat{u}$ if no set E can be found whereon (3. 36) is satisfied for some $d > 0$ and we conclude $\hat{u} = \tilde{u}$ in $L^2_m[0,\infty)$. Q. E. D.

To summarize the work of this section, we have seen that in order to find the control $\hat{u} \in L^2[0,\infty)$ minimizing

$$J(x_0, u) = \int_0^\infty [(Fx(t)+Gu(t))^* \tilde{W}(Fx(t)+Gu(t)) + u(t)^* U u(t)]\, dt$$

$$\equiv \int_0^\infty (x(t)^*, u(t)^*) \begin{pmatrix} W & R \\ R^* & U \end{pmatrix} \begin{pmatrix} x(t) \\ u(t) \end{pmatrix} dt$$

it is sufficient to find the solution Q described above of the matrix quadratic equation (3. 13), i. e.,

$$A^*Q + QA + W - (QB+R)U^{-1}(R^*+B^*Q) = 0 \qquad (3.\ 36)$$

then solve

$$\dot{\hat{x}} = (A - BU^{-1}[B^*Q+R^*])\hat{x}, \quad \hat{x}(0) = x_0,$$

the closed loop system arising from use of

$$\hat{u}(t) = -U^{-1}[B^*Q+R^*]\,\hat{x}(t)$$

in the original system $\dot{x} = Ax + Bu$.

As an example, let us consider the two dimensional system (1. 16), i. e.,

$$\begin{pmatrix} \dot{x}^1 \\ \dot{x}^2 \end{pmatrix} = \begin{pmatrix} 0 & 1 \\ -1 & 0 \end{pmatrix} \begin{pmatrix} x^1 \\ x^2 \end{pmatrix} + \begin{pmatrix} 0 \\ 1 \end{pmatrix} u$$

with the cost functional (1. 19), W, R, U being given by (1. 20), (1. 21), (1.22). Taking $T = \infty$, the relevant quadratic matrix equation is (cf. (3. 13))

$$\begin{pmatrix} 0 & -1 \\ 1 & 0 \end{pmatrix} \begin{pmatrix} q_1 & q_2 \\ q_2 & q_3 \end{pmatrix} + \begin{pmatrix} q_1 & q_2 \\ q_2 & q_3 \end{pmatrix} \begin{pmatrix} 0 & 1 \\ -1 & 0 \end{pmatrix} + \begin{pmatrix} \alpha+\beta & 0 \\ 0 & \alpha \end{pmatrix}$$

$$- \left[\begin{pmatrix} q_1 & q_2 \\ q_2 & q_3 \end{pmatrix} \begin{pmatrix} 0 \\ 1 \end{pmatrix} + \begin{pmatrix} -\beta \\ 0 \end{pmatrix} \right] \frac{1}{\delta+\beta} \left[(0,1) \begin{pmatrix} q_1 & q_2 \\ q_2 & q_3 \end{pmatrix} + (-\beta, 0) \right] = 0, \qquad (3.\ 37)$$

where we have used the known symmetry of Q to write it as

$$Q = \begin{pmatrix} q_1 & q_2 \\ q_2 & q_3 \end{pmatrix} .$$

From (3. 37) we have three scalar equations:

$$-2q_2 + \alpha+\beta - (q_2 - \beta)^2/(\delta+\beta) = 0$$

$$- q_3 + q_1 - (q_2 - \beta)q_3/(\delta+\beta) = 0 ,$$

$$2q_2 + \alpha - q_3^2/(\delta+\beta) = 0 .$$

The first, third and second of these equations are solved, in that order, to give

$$q_2 = - \delta \pm \sqrt{\delta^2 + \delta\alpha + \delta\beta + \alpha\beta}, \tag{3. 38}$$

$$q_3 = \pm \sqrt{(\delta + \beta)(2q_2 + \alpha)}, \tag{3. 39}$$

$$q_1 = q_3 + (q_2 - \beta)q_3/(\delta + \beta). \tag{3. 40}$$

It should be noted that, mathematically, there are four possible solutions. Which one should be used? The solution which we want may be identified from Proposition 3. 6, which says that the matrix Q which yields the optimal control is positive definite, together with a result which we will obtain in the next section to the effect that in many applications, including this one, there is only one positive definite solution Q of (3. 13). Assuming this result for now, we see that we must use the + sign in (3. 39). Now the determinant of Q is

$$\begin{aligned} q_1 q_3 - (q_2)^2 &= \text{(from (3. 40))} \\ &= q_3^2(1 + \frac{q_2 - \beta}{\delta+\beta}) - q_2^2 = \text{(from (3. 38))} \\ &= q_3^2(1 - \frac{\delta+\beta \mp \sqrt{\delta^2 + \delta\alpha + \delta\beta + \alpha\beta}}{\delta+\beta}) - q_2^2 \end{aligned} \tag{3. 41}$$

and since this must be positive if Q is positive definite, the minus sign must be used in (3. 41), corresponding to the + sign in (3. 38). Thus

$$q_2 = -\delta + \sqrt{\delta^2 + \delta\alpha + \delta\beta + \alpha\beta},$$

$$q_3 = \sqrt{(\delta+\beta)(2q_2 + \alpha)} = \sqrt{(\delta+\beta)(-2\delta + 2\sqrt{\delta^2 + \delta\alpha + \delta\beta + \alpha\beta} + \alpha)}$$

$$q_1 = q_3 + (q_2 - \beta)q_3/(\delta + \beta) =$$

$$= \sqrt{(\delta+\beta)(-2\delta + 2\sqrt{\delta^2 + \delta\alpha + \delta\beta + \alpha\beta} + \alpha)}\,\frac{\sqrt{\delta^2 + \delta\alpha + \delta\beta + \alpha\beta}}{\delta + \beta}.$$

The optimal control is generated by the feedback law

$$\hat{u}(t) = -U^{-1}(B^*Q + R^*)\hat{x}(t)$$

$$= -\frac{1}{\delta+\beta}\left((0,1)\begin{pmatrix} q_1 & q_2 \\ q_2 & q_3 \end{pmatrix} + (-\beta, 0)\right)\begin{pmatrix} x^1(t) \\ x^2(t) \end{pmatrix} = -\frac{(q_2-\beta)}{\delta+\beta}\hat{x}^1(t) - \frac{q_3}{\delta+\beta}\hat{x}^2(t)$$

$$= -\left(\frac{\sqrt{\delta^2 + \delta\alpha + \delta\beta + \alpha\beta}}{\delta + \beta} - 1\right)\hat{x}^1(t) - \sqrt{\frac{-2\delta + 2\sqrt{\delta^2 + \delta\alpha + \delta\beta + \alpha\beta} + \alpha}{\delta + \beta}}\,\hat{x}^2(t).$$

We see very quickly that different operational priorities, reflected in different weighting coefficients, α, β, δ, lead to different control laws. Let us take $\delta = 1$. If we emphasize system energy and neglect "ride quality", this is reflected in taking $\alpha > 0$, $\beta = 0$. For $\alpha = 1$, $\beta = 0$ the optimal feedback law is

$$\hat{u}(t) = -(\sqrt{2} - 1)\hat{x}^1(t) - (2\sqrt{2} - 1)\hat{x}^2(t). \tag{3.42}$$

If, on the other hand, we emphasize ride quality and neglect system energy, this is reflected in taking $\alpha = 0$, $\beta > 0$. For $\alpha = 0$, $\beta = 1$ the optimal feedback law is

$$\hat{u}(t) = -\left(\frac{1}{\sqrt{2}} - 1\right)\hat{x}^1(t) - \sqrt{(\sqrt{2} - 1)}\,\hat{x}^2(t). \tag{3.43}$$

It is natural to ask whether the control law (3. 42) really results in less system energy than (3. 43) and whether (3. 43) really gives a better ride quality than (3. 42). We will discuss a basis for such comparison in

Chapter VI and answers the question in Chapter VII.

4. Solution of Quadratic Matrix Equations

We have seen in the example at the end of the last section that the quadratic matrix equation (3.13) has more than one solution in general. The solution Q referred to in Theorem 3.5 is just one of these, singled out by the fact that it is the limit as $T \to \infty$ of $Q_T(0)$ (cf (3.14)). To cast further light on this situation we now examine a theory of quadratic matrix equations developed by Potter ([8]). We rewrite the equation (3.13) in the form

$$(A-BU^{-1}R^*)^*Q+Q(A-BU^{-1}R^*)+W-RU^{-1}R^* - QBU^{-1}B^*Q = 0.$$

Letting

$$B_1 = BU^{-1}B^*, \quad A_1 = A-BU^{-1}R^*, \quad W_1 = W - RU^{-1}R^* \tag{4.1}$$

the equation may be rewritten as

$$A_1^*Q + QA_1 + W_1 - QB_1Q = 0 . \tag{4.2}$$

Potter's theory relates the solutions of the quadratic matrix equation to certain properties of the $2n \times 2n$ matrix (cf. III-4)

$$M = \begin{pmatrix} A_1^* & W_1 \\ B_1 & -A_1 \end{pmatrix} . \tag{4.3}$$

In agreement with this block notation, we represent $2n$-dimensional vectors z in the form

$$z = \begin{pmatrix} x \\ y \end{pmatrix}, \quad x, y \in E^n .$$

To avoid algebraic complication, we assume for the greater part of this section that the matrix M has a diagonal Jordan form.

<u>Theorem 4.1.</u> <u>The solutions of the quadratic matrix equation</u> (4.2) <u>coincide</u>

with the set of matrices of the form

$$Q = XY^{-1} \tag{4.4}$$

where the $n \times n$ matrices

$$X = [x_1, x_2, \ldots, x_n],$$

and

$$Y = [y_1, y_2, \ldots, y_n],$$

are composed of the upper and lower halves, respectively, of n independent eigenvectors $z_1, z_2, \ldots, z_n$ of the matrix M and Y is nonsingular.

Proof. Let Q solve (4.2) and let

$$G = B_1 Q - A_1 \quad (= -(A - BU^{-1}[B^* Q + R^*])) \tag{4.5}$$

Then the equation (4.2) itself implies that

$$QG = QB_1 Q - QA_1 = W_1 + A_1^* Q. \tag{4.6}$$

Let Y be a nonsingular matrix which transforms G into its Jordan canonical form, J. Then

$$Y^{-1} GY = J, \quad \text{or} \quad GY = YJ. \tag{4.7}$$

Then let (cf. (4.4))

$$X = QY. \tag{4.8}$$

If we substitute (4.7) and (4.8) into (4.5) and (4.6) we can eliminate G and Q:

$$XJ = QYJ = QGY = W_1 Y + A_1^* QY = A_1^* X + W_1 Y, \tag{4.9}$$

$$YJ = GY = (B_1 Q - A_1)Y = B_1 X - A_1 Y. \tag{4.10}$$

Rewriting (4.9) and (4.10) in block matrix notation we have

$$MZ \equiv \begin{pmatrix} A_1^* & W_1 \\ B_1 & -A_1 \end{pmatrix} \begin{pmatrix} X \\ Y \end{pmatrix} = \begin{pmatrix} X \\ Y \end{pmatrix} J \equiv ZJ . \tag{4.11}$$

If J is diagonal, then the columns $\begin{pmatrix} x_i \\ y_i \end{pmatrix}$ of $\begin{pmatrix} X \\ Y \end{pmatrix}$ are clearly eigenvectors of M which must coincide with the diagonal entries of J.

To show that J is, indeed, diagonal, let $(z_1, z_2, \ldots, z_n)$ be the columns of $Z = \binom{X}{Y}$. If J were not diagonal it would have a block of the form

$$\begin{pmatrix} \lambda_1 & 1 & \ldots & 0 & 0 \\ 0 & \lambda_1 & \ldots & 0 & 0 \\ \vdots & \vdots & & \vdots & \vdots \\ 0 & 0 & \ldots & \lambda_1 & 1 \\ 0 & 0 & \ldots & 0 & \lambda_1 \end{pmatrix}.$$

If this block begins with the k-th diagonal element of J then the k-th column of ZJ is λz_k and the k-th column of MJ is Mz_k so we have, from (4.11)

$$(M - \lambda_1 I) z_k = 0. \tag{4.12}$$

On the other hand, for the (k+1)st column

$$(M - \lambda_1 I) z_{k+1} = z_k. \tag{4.13}$$

Since M is assumed to have diagonal canonical form we have, by the Cayley-Hamilton theorem,

$$(M - \lambda_r I)(M - \lambda_{r-1} I) \ldots (M - \lambda_2 I)(M - \lambda_1 I) = 0$$

where $\lambda_1, \lambda_2, \ldots, \lambda_r$ are the distinct eigenvalues of M. Multiplying on the left by z_{k+1} gives, using (4.12) and (4.13)

$$\prod_{j=2}^{r} (\lambda_1 - \lambda_j) z_k = 0$$

and then since $\lambda_1 \neq \lambda_j$, $j = 2, \ldots, r$, $z_k = 0$, which, since it implies $y_k = 0$, contradicts the nonsingularity of Y. We conclude therefore that J is diagonal and the columns of Z are eigenvectors of M as indicated earlier. It follows that $Q = XY^{-1}$ has the form indicated in the theorem.

On the other hand, if $z_1, z_2, \ldots, z_n$ are independent eigenvectors of M corresponding to eigenvalues $\lambda_1, \lambda_2, \ldots, \lambda_n$, not necessarily distinct, then with $J = \text{diag}(\lambda_1, \lambda_2, \ldots, \lambda_n)$ and $Z = (z_1, z_2, \ldots, z_n)$ we have

$$MZ = ZJ$$

which reduces to

$$A_1^* X + WY = XJ, \quad B_1 X - A_1 Y = YJ .$$

Since Y is assumed nonsingular we may multiply these equations on the right by Y^{-1} to get

$$A_1^* XY^{-1} + W = XJY^{-1},$$

or

$$A_1^* XY^{-1} + W = XY^{-1} YJY^{-1} \tag{4.14}$$

and

$$YJY^{-1} = B_1 XY^{-1} - A_1 . \tag{4.15}$$

Letting $Q = XY^{-1}$ and substituting (4.15) in (4.14)

$$XY^{-1}(B_1 XY^{-1} - A_1) = A_1^* XY^{-1} + W$$

or

$$A_1^* Q + QA_1 + W - QB_1 Q = 0 . \qquad \text{Q. E. D.}$$

We remark that the matrix

$$Q = XY^{-1}$$

depends only on the subspace spanned by the columns of Z. Any other selection of n eigenvalues spanning the same space (which can be done non-trivially, e.g., when M has multiple eigenvalues) corresponds to replacing Z by ZC, where C is some nonsingular $n \times n$ matrix. Then X and Y are replaced by XC and YC and

$$XC(YC)^{-1} = XCC^{-1}Y^{-1} = XY^{-1} .$$

Theorem 4.1 shows that the maximum (and generic, or typical) number of solutions of (4.2) is the number of combinations of the $2n$ eigenvalues of M taken n at a time, i.e.

$$\binom{2n}{n} = \frac{2n!}{n!\,n!} = \frac{2n(2n-1)\ldots(n+1)}{n(n-1)\ldots 1} .$$

When $n = 1$, so that (4.2) becomes the scalar equation

$$2a_1q + w - b_1q^2 = 0,$$

$\binom{2}{1} = 2$, the generic number of solutions of a standard scalar quadratic equation.

When the quadratic equation (4.2) is set in the optimal control context of Section 3 we are looking for a certain symmetric positive definite solution of that equation. Thus the question arises: among the (possibly) $\binom{2n}{n}$ solutions of (4.2) which one corresponds to the solution identified in Theorem 3.5? Potter's analysis enables us to answer this question as well.

<u>Theorem 4.2.</u> <u>If (as in the case</u> (4.1)) W_1 <u>and</u> B_1 <u>are hermitian matrices, then each matrix</u> $Q = XY^{-1}$ <u>constructed as in Theorem</u> 4.1, <u>with the additional property that</u>

$$\lambda_j \neq -\lambda_k, \quad j,k = 1,2,\ldots,n, \tag{4.16}$$

<u>for the eigenvalues</u> $\lambda_k, \lambda_2, \ldots, \lambda_n$ <u>of</u> M <u>selected, is also hermitian.</u>

<u>Proof.</u> Let $Q = XY^{-1}$ as in Theorem 4.1 and set

$$P = Y^*X.$$

Since

$$(Y^{-1})^*PY^{-1} = XY^{-1} = Q$$

and $(Y^{-1})^*PY^{-1}$ is Hermitian if and only if P is, we see that Q is Hermitian if and only if P is.

Let T be the $2n \times 2n$ matrix

$$T = \begin{pmatrix} 0 & -I \\ I & 0 \end{pmatrix}$$

and let the entries of P by p_k^j, $j,k = 1,2,\ldots,n$. Then clearly,

$$p_k^j = y_j^* x_k, \quad p_j^k = y_k^* x_j$$

and

$$p_k^j - \bar{p}_j^k = y_j^* x_k - \overline{y_k^* x_j} = y_j^* x_k - x_j^* y_k = (x_j^*, y_j^*)T\begin{pmatrix} x_k \\ y_k \end{pmatrix} = z_j^* T z_k .$$

Since we have assumed (4.16),

$$p^j_k - \bar{p}^k_j = (\bar{\lambda}_j + \lambda_k)^{-1}[\bar{\lambda}_j z_j^* T z_k + z_j^* T(\lambda_k z_k)] = (\bar{\lambda}_j + \lambda_k)^{-1}[z_j^*(MT+TM)z_k]. \tag{4.17}$$

But, using the assumed symmetry of W_1 and B_1,

$$M^*T+TM = \begin{pmatrix} A_1 & B_1 \\ W_1 & -A_1^* \end{pmatrix}\begin{pmatrix} 0 & -I \\ I & 0 \end{pmatrix} + \begin{pmatrix} 0 & -I \\ I & 0 \end{pmatrix}\begin{pmatrix} A_1^* & W_1 \\ B_1 & -A_1 \end{pmatrix}$$

$$= \begin{pmatrix} B_1 & -A_1 \\ -A_1^* & -W_1 \end{pmatrix} + \begin{pmatrix} -B_1 & A_1 \\ A_1^* & W_1 \end{pmatrix} = 0 . \tag{4.18}$$

Substituting (4.18) in (4.17) we have $p^j_k = \bar{p}^k_j$. Q. E. D.

<u>Corollary 4.3. If (as in (4.1)) B_1 and W_1 are hermitian, the eigenvalues of M occur in "pairs" $\pm\mu + i\nu$. If in addition A_1, B_1 and W_1 are real, the eigenvalues occur in "quadruples"</u>

$$\pm\mu \pm i\nu .$$

<u>Remark.</u> The quotation marks are used because the "pair" degenerates to one eigenvalue if $\mu = 0$. Similarly the "quadruple" $\pm\mu \pm i\nu$ degenerates to a pair or a single value if μ or $\nu = 0$ or μ and $\nu = 0$, respectively.

<u>Proof.</u> From (4.18), i.e.

$$\begin{pmatrix} A_1 & B_1 \\ W_1 & -A_1^* \end{pmatrix}\begin{pmatrix} 0 & -I \\ I & 0 \end{pmatrix} + \begin{pmatrix} 0 & -I \\ I & 0 \end{pmatrix}\begin{pmatrix} A_1^* & W_1 \\ B_1 & -A_1 \end{pmatrix} = 0 ,$$

there follows, since $\begin{pmatrix} 0 & -I \\ I & 0 \end{pmatrix}^{-1} = \begin{pmatrix} 0 & -I \\ I & 0 \end{pmatrix}$,

$$M = \begin{pmatrix} A_1^* & W_1 \\ B_1 & -A_1 \end{pmatrix} = \begin{pmatrix} 0 & -I \\ I & 0 \end{pmatrix}\begin{pmatrix} A_1 & B_1 \\ W_1 & -A_1^* \end{pmatrix}\begin{pmatrix} 0 & -I \\ I & 0 \end{pmatrix}$$

$$= \begin{pmatrix} 0 & -I \\ I & 0 \end{pmatrix}^{-1}\left[-\begin{pmatrix} A_1^* & W_1 \\ B_1 & -A_1 \end{pmatrix}^*\right]\begin{pmatrix} 0 & -I \\ I & 0 \end{pmatrix} = T^{-1}(-M^*)T. \tag{4.19}$$

Hence M and $-M^*$ are similar and consequently have the same eigenvalues. If $\mu + i\nu$ is an eigenvalue of M, $-\mu + i\nu$ is an eigenvalue of $-M^*$ and hence also of M. If A_1, B_1, W_1 are all real, M is real and the eigenvalues also occur in complex conjugate pairs $\mu \pm i\nu$. Q. E. D.

The importance of the fact that the eigenvalues of M occur in pairs $\pm \mu + i\nu$ when B_1 and W_1 are hermitian lies in the fact that, under such circumstances there is at most one subset $\{\lambda_1, \lambda_2, \ldots, \lambda_n\}$ of n eigenvalues of M such that $\operatorname{Re}(\lambda_k) > 0$, $k = 1, 2, \ldots, n$. We use this result in

Theorem 4. 4. *Let B_1 and W_1 be hermitian and let $Q = XY^{-1}$ be constructed, as in Theorem 4. 1, the columns of $Z = \binom{X}{Y}$ being eigenvectors of M corresponding to eigenvalues $\lambda_1, \lambda_2, \ldots, \lambda_n$. Suppose further that*

$$W_1 + QB_1Q \geq 0 \tag{4. 20}$$

and for every solution $x(t)$ of $\dot{x} = (A_1 - B_1Q)x$

$$\{x(t)^*(W_1 + QB_1Q)x(t) \equiv 0, \quad t \geq 0\} \Rightarrow x(t) \equiv 0, \quad t \geq 0. \tag{4. 21}$$

Then Q is hermitian and positive definite if and only if

$$\operatorname{Re}(\lambda_k) > 0, \qquad k = 1, 2, \ldots, n. \tag{4. 22}$$

Proof. From the proof of Theorem 4. 1, $\lambda_1, \lambda_2, \ldots, \lambda_n$ are the eigenvalues of

$$G = B_1Q - A_1.$$

Thus $-\lambda_1, -\lambda_2, \ldots, -\lambda_n$ are the eigenvalues of $A_1 - B_1Q$.

Suppose we have (4. 22). Then Theorem 4. 2 applies to show that Q is hermitian and, since Q satisfies (4. 2), rearranging gives

$$(A_1 - B_1Q)^*Q + Q(A_1 - B_1Q) + W_1 + QB_1Q = 0. \tag{4. 23}$$

Since the eigenvalues $-\lambda_1, -\lambda_2, \ldots, -\lambda_n$ of $A_1 - B_1Q$ have negative real parts, the procedure used in III- Theorem 1. 5 shows that

$$Q = \int_0^\infty e^{(A_1 - B_1Q)^*t}(W_1 + QB_1Q)e^{(A_1 - B_1Q)t}\,dt .$$

The positive definiteness of Q then follows immediately from (4. 20) and (4. 21).

On the other hand, if Q is hermitian and positive definite we have (4. 23) and every solution of $\dot{x} = (A_1 - B_1Q)x$ satisfies

$$\frac{d}{dt}(x(t)^* Qx(t)) = -x(t)^* (W_1 + QB_1Q)x(t) \leq 0$$

and the basic argument of III - Theorem 1. 5 again applies to show that the eigenvalues of $A_1 - B_1Q$ have negative real parts, hence that (4. 22) obtains. Q. E. D.

Now suppose (4. 2) coincides with the equation (3. 13) of Section 3 and (A, B) is stabilizable, (F, A) observable. From III - Proposition 3. 7 we know that

$$A_1 - B_1Q = A - BU^{-1}(B^* Q + R^*)$$

is a stability matrix so that its eigenvalues, which we will call $-\lambda_1, -\lambda_2, \ldots, -\lambda_n$ have negative real parts. Then $\lambda_1, \lambda_2, \ldots, \lambda_n$ is, from Corollary 4. 3, the unique subset of n eigenvalues of M with $\mathrm{Re}(\lambda_k) > 0$, $k = 1, 2, \ldots, n$. Since (4. 20) and (4. 21) follow in this case from the arguments used in Proposition 3. 7, Theorem 4. 4 applies to show that (4. 2) has exactly one positive definite hermitian solution - which must be the positive definite symmetric solution identified in Section 3 as $\lim_{T \to \infty} Q_T(0)$.

The foregoing results have been proved under the assumption that the Jordan form of M is diagonal. This is not at all necessary. We sketch in the following paragraphs the modifications necessary when this is not the case.

The proof of Theorem 4. 1 is word for word the same up to formula (4. 11). The proof that J is diagonal is then replaced by a result to the effect that the Jordan blocks of J have no greater dimension than the corresponding blocks of the Jordan form for M. The z_k are now permitted to be generalized eigenvectors of M but the selection of n vectors $z_1, z_2, \ldots, z_n$ is then subject to some limitations. Suppose M has eigenvalues $\lambda_1, \lambda_2, \ldots, \lambda_r$ and corresponding generalized eigenvectors $z_{1,1}, \ldots z_{1,\mu_1}, z_{2,1}, \ldots, z_{2,\mu_2}, \ldots, z_{r,1}, \ldots, z_{r,\mu_r}$, $\mu_1 + \mu_2 + \ldots + \mu_r \geq n$ and such that, for $\ell = 1, 2, \ldots, r$

$$Mz_{\ell,1} = \lambda_\ell z_{\ell,1}$$

$$Mz_{\ell,2} = \lambda_\ell z_{\ell,2} + z_{\ell,1}$$

$$\vdots$$

$$Mz_{\ell,\mu_\ell} = \lambda_\ell z_{\ell,\mu_\ell} + z_{\ell,\mu_\ell - 1} .$$

The vectors chosen as columns of Z must then be taken from the "front" of each sequence $z_{\ell,1}, \ldots, z_{\ell,\mu_\ell}$. That is

$$Z = (z_{1,1}, \ldots, z_{1,\nu_1}, \ldots, z_{r,1}, \ldots, z_{r,\nu_r}), \quad \nu_\ell \leq \mu_\ell, \quad \sum_{\ell=1}^{r} \nu_\ell = n .$$

This limits the selection process and the number of solutions is no longer $\binom{2n}{n}$. This is the higher dimensional counterpart of what happens when a scalar quadratic equation has a root of multiplicity 2.

The remaining theorems are essentially the same as above, except that in Corollary 4.3 the multiplicity of an eigenvalue $\lambda = \mu + i\nu$ is also preserved under the indicated reflections in the real and imaginary axes.

Actually, choosing the matrix J to be the Jordan form of G (cf. (4.5)) has little significance except that it identifies Q with eigenvectors or generalized eigenvectors - which are "standard" mathematical entities - of the matrix M. In fact, as the proof of Theorem 4.1 is examined it will become apparent that the significant result is really this: if Q satisfies (4.2) and G is given by (4.5), then a $2n \times n$ matrix

$$Z = \begin{pmatrix} X \\ Y \end{pmatrix} = \begin{pmatrix} QY \\ Y \end{pmatrix}, \tag{4.24}$$

with Y a nonsingular $n \times n$ matrix, satisfies

$$MZ = Z(Y^{-1}GY) . \tag{4.25}$$

Since the columns of the matrix on the right hand side of (4.25) are linear combinations of the columns of Z, we conclude that the columns of Z span an n-dimensional subspace $\mathfrak{Z}$ of E^{2n} which is invariant under M, i.e., $M : \mathfrak{Z} \to \mathfrak{Z}$. The fact that Y is nonsingular implies that the projection

$$P_2 : \mathscr{X} \to E^n, \quad P_2\binom{x}{y} = y ,$$

maps $\mathscr{X}$ onto E^n. It is easy to see that the converse is also true. If $\mathscr{X}$ is an n-dimensional invariant subsapce for M spanned by the columns of the $2n \times n$ matrix

$$Z = \binom{X}{Y}$$

with Y a nonsingular $n \times n$ matrix, so that

$$MZ = ZJ$$

for some $n \times n$ matrix J, then a solution Q of (4. 2) may be constructed from (4. 14) and (4. 15) as carried out above. These observations become significant when we consider the problem of numerical solution of (4. 2) by use of the algebraic method which we have outlined here. We will study numerical solutions in Chapter V.

The matrix Q can also be viewed as a "decoupling" matrix in the light of III-4. The 2n-dimensional system (cf. (2. 17), (2. 19))

$$\dot{\psi} = -A_1^*\psi - W_1 x, \qquad \dot{x} = -B_1\psi + A_1 x,$$

can be decoupled by setting

$$\begin{pmatrix} \psi \\ x \end{pmatrix} = \begin{pmatrix} I & Q \\ 0 & I \end{pmatrix}\begin{pmatrix} \theta \\ x \end{pmatrix} .$$

If Q satisfies (same as (3. 13))

$$A_1 Q + QA_1 + W_1 - QBQ = 0 \tag{4.26}$$

the resulting system is

$$\dot{\theta} = (-A_1^* + QB_1)\theta, \qquad \dot{x} = -B_1\theta + (A_1 - B_1 Q)\hat{x} .$$

Solutions tending to zero as $t \to \infty$ are obtained by taking $\theta(t) \equiv 0$, provided Q is taken as the unique symmetric positive definite solution of (4. 26).

With $\theta(t) \equiv 0$ we have

$$\psi = Q\hat{x}, \qquad \dot{\hat{x}} = (A_1 - B_1 Q)\hat{x}$$

and the $\hat{x}$ satisfy the optimal closed loop system equations. The subspace of E^{2n} specified by setting $\theta = 0$ is precisely the subspace spanned by the eigenvalues of

$$-M = \begin{pmatrix} -A_1^* & -W_1 \\ -B_1 & A_1 \end{pmatrix}$$

which have negative real parts, in agreement with the development presented above. The matrix $Q(t)$ of Section 2 can also be interpreted in a comparable light.

5. Related Topics: Optimization and Liapounov's Method, Dynamic Programming, Regulation of Nonlinear Systems

In this section we will be concerned, for the most part, with an autonomous (i. e., time independent) system of ordinary differential equations

$$\dot{x} = F(x, u), \qquad x \in E^n, \ u \in E^m, \tag{5.1}$$

where $F : E^{n+m} \to E^n$ is assumed continuously differentiable (at least) in the region of interest. If the control, u, is given as a function of the state,

$$u = K(x) \tag{5.2}$$

with $K : E^n \to E^m$ continuously differentiable, the closed loop system obtained by substituting (5. 2) into (5. 1) is,

$$\dot{x} = F(x, K(x))$$

and thus has the form

$$\dot{x} = g(x) \tag{5.3}$$

with $g : E^n \to E^n$ continuously differentiable.

We have noted earlier in I-2 that a point $x_0 \in E^n$ is an equilibrium point for (5. 3) just in case $g(x_0) = 0$. One of the problems of persistent interest in the theory of ordinary differential equations concerns the asymptotic stability of such an equilibrium point.

Definition 5.1. The equilibrium point $x_0 \in E^n$ is locally asymptotically stable if there is a neighborhood, N, of x_0 in E^n, such that for each initial state $x_1 \in N$, the solution $x(t)$ of (5.3) with $x(0) = x_1$ has the properties

(i) $x(t) \in N$ for all $t \geq 0$;

(ii) $\lim_{t \to \infty} x(t) = x_0$.

The equilibrium point x_0 is globally asymptotically stable if N can be taken to be all of E^n.

For linear systems

$$\dot{x} = Ax, \tag{5.4}$$

with A an $n \times n$ matrix, only global asymptotic stability need be considered; there are no systems (5.4) for which $x = 0$ is locally asymptotically stable but not globally asymptotically stable. For nonlinear systems local asymptotic stability may very well obtain without global asymptotic stability. As an example, consider the scalar equation

$$\dot{x} = -ax + bx^2, \quad x \in E^1, \quad a, b > 0. \tag{5.5}$$

This equation may be solved by the standard method of separation of variables to give, for $x(0) \neq 0$, $bx(0) - a \neq 0$,

$$x(t) = \frac{a}{b - e^{at+c}}, \quad c = \log\left(\frac{bx(0) - a}{x(0)}\right), \tag{5.6}$$

or

$$x(t) = \frac{a}{b + e^{at+c}}, \quad c = \log\left(-\frac{bx(0) - a}{x(0)}\right), \tag{5.7}$$

(5.6) valid when $(bx(0) - a)/x(0) > 0$, (5.7) valid when $(bx(0) - a)/x(0) < 0$. From these formulae it may be verified that

$$\lim_{t \to \infty} x(t) = 0$$

if $x(0) < 0$ or $0 < x(0) < \frac{a}{b}$. If $x(0) > \frac{a}{b}$, $x(t)$ escapes to $+\infty$ in finite time. The points $0, \frac{a}{b}$ are both equilibrium points, 0 being locally asymptotically stable while $\frac{a}{b}$ is not. Here we could take N to be $-\infty < x < \frac{a}{b}$.

The method most widely used to study asymptotic stability of an equilibrium point is due to A. A. Liapounov (III-[2], see III-[3] or [13] below for a good exposition) and involves the use of a Liapounov function.

Definition 5.2. The function $V : E^n \to E^1$ is a Liapounov function for (5.3) in the neighborhood M of x_0 if V is defined in M and

(i) V is continuously differentiable in M;

(ii) $V(x_0) = 0$;

(iii) $\{x \in M, x \neq x_0\} \Rightarrow V(x) > 0$;

(iv) $\left.\frac{d}{dt} V(x(t))\right|_{t=\tau} \leq 0$ whenever $x(t)$ is a solution of (5.3) and $x(\tau) \in M$.

A wide variety of theorems relating asymptotic stability to properties of the Liapounov function, $V(x)$, may be found in the differential equations literature. The reader is referred to III-[2], III-[3], and [13] below, in particular. For our purposes the following theorem will be adequate.

Theorem 5.3. Suppose V is a Liapounov function for (5.1) in the neighborhood M of x_0 and there is a function $U : E^n \to E^1$, also defined in M, such that

(i) U is continuous in M;

(ii) $U(x) \leq 0$, $x \in M$;

(iii) if $x(t)$ is a solution of (5.1) on an open, non-empty interval $t_0 < t < t_1$, $x(t) \in M$, $t_0 < t < t_1$, and

$$U(x(t)) \equiv 0, \quad t_0 < t < t_1,$$

then $x(t) \equiv x_0$ for all t.

Then, if condition (iv) of Definition 5.2 is strengthened to

$$\frac{d}{dt} V(x(t)) \left(\equiv \frac{\partial V}{\partial x}(x(t))\, g(x(t))\right) \leq U(x(t)), \tag{5.8}$$

$x(t) \in M$, (where $\frac{\partial V}{\partial x}$ denotes the gradient of V)

the equilibrium point x_0 is locally (at least) asymptotically stable and globally asymptotically stable if M can be taken to be E^n and

$$\lim_{\|x\| \to \infty} V(x) = \infty. \tag{5.9}$$

Proof. Since M is a neighborhoof of x_0 it includes the set

$\{x \in E^n \mid \|x - x_0\| \leq \epsilon\}$ for some $\epsilon > 0$. Let

$$V_\epsilon = \min_{\|x - x_0\| = \epsilon} \{V(x)\}$$

and let

$$N = \{x \in E^n \mid \|x - x_0\| \leq \epsilon, \quad V(x) < V_\epsilon\}. \tag{5.10}$$

Condition (iii) of Definition 5.2 shows that N is invariant under (5.3). $V(x(t))$ is non-increasing if $x(t)$ is a solution of (5.3) so such solutions cannot leave N if $x(0) \in N$; for $V(x(0)) < V_\epsilon \Rightarrow V(x(t)) < V_\epsilon$ for $t \geq 0$. Thus condition (i) of Definition 5.1 is fulfilled. Since $V(x(t))$ is non-increasing for $t \geq 0$, the monotone convergence theorem gives

$$\lim_{t \to \infty} V(x(t)) = v_0 \geq 0 . \tag{5.11}$$

Let N_0 be the closed subset of V defined by

$$N_0 = \{x \in V \mid V(x) \leq V(x(0))\}.$$

Then N_0 is closed and bounded and (see [7], e.g.) there must be a sequence t_k with $\lim_{k \to \infty} t_k = \infty$ such that

$$\lim_{k \to \infty} x(t_k) = \hat{x} \in N_0 . \tag{5.12}$$

Continuity of V together with (5.11) gives

$$V(\hat{x}) = v_0 .$$

Let $x(t)$ be the solution of (5.3) with

$$\hat{x}(0) = \hat{x} .$$

Since (5.3) is time independent,

$$\tilde{x}_k(t) = x(t + t_k) \tag{5.13}$$

is a solution of (5.3) for all k and, since $\tilde{x}_k(0) = x(t_k)$ for all k, (5.12) gives

$$\lim_{k \to \infty} \tilde{x}_k(0) = \hat{x} . \tag{5.14}$$

This result together with the standard regularity theory for ordinary differential equations shows that, given any $T > 0$

$$\lim_{k\to\infty} \|\tilde{x}_k(t) - \hat{x}(t)\| = 0$$

uniformly for $t_k \le t \le t_k + T$. The monotonicity of $V(x(t))$ shows that

$$V(\tilde{x}_k(0)) \ge V(\tilde{x}_k(t)) \ge v_0, \qquad 0 \le t < \infty.$$

Letting $k \to \infty$ and making use of (5.14) and the continuity of V there follows

$$V(\hat{x}(t)) \equiv v_0 . \tag{5.15}$$

Then

$$0 \equiv \frac{d}{dt} V(\hat{x}(t)) \le U(\hat{x}(t)), \qquad t \ge 0 .$$

Condition (ii) of our theorem gives $U(\hat{x}(t)) \equiv 0$ and then condition (iii) gives

$$\hat{x}(t) \equiv x_0, \qquad t \ge 0.$$

But then, from (ii) of Definition 5.2,

$$v_0 \equiv V(\hat{x}(t)) \equiv V(x_0) = 0 .$$

Then (5.11) becomes

$$\lim_{t\to\infty} V(x(t)) = 0$$

and (ii), (iii) of Definition 5.2 are easily seen to imply that

$$\lim_{t\to\infty} x(t) = x_0 ,$$

so that (ii) of Definition 5.1 is satisfied. We conclude that x_0 is at least locally asymptotically stable. If we can take $M = E^n$, (5.9) implies that the neighborhoods (5.10) expand to fill all of E^n as $\epsilon \to \infty$ and we have global asymptotic stability of x_0. Q. E. D.

The graph of $V(x)$ may usefully be envisioned as a "bowl" with x_0 at the center. Condition (iv) of Definition 5.2 shows that the point $(x(t), V(x(t)))$ on this graph moves in such a way that $V(x(t))$ decreases and, as we have seen, must eventually decrease to zero, forcing $x(t)$ to approach x_0. Figure IV.3 may be helpful. In it the horizontal plane represents the

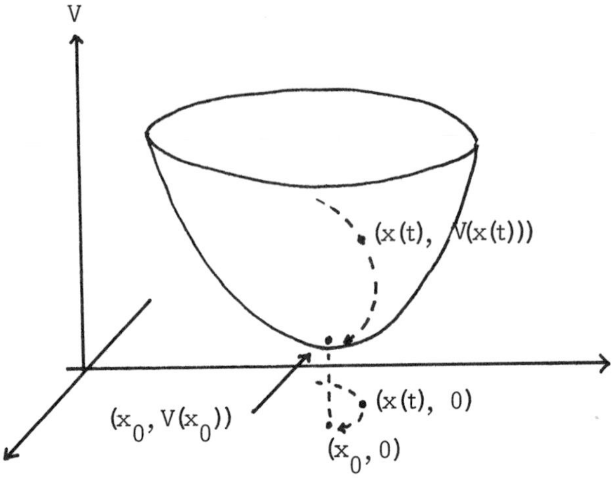

Fig. IV. 3. Liapounov function

space E^n, wherein the state $x(t)$, lies, while the vertical axis indicates values of V.

If (5. 3) is a scalar equation such that

$$(x - x_0)g(x) < 0, \qquad x \neq 0,$$

in some interval which includes $x = x_0$, a Liapounov function, $V(x)$, may be readily defined by

$$V(x) = -\int_{x_0}^{x} g(x)dx.$$

Then

$$\frac{d}{dt}V(x(t)) = \frac{dV}{dx}(x(t))\frac{dx}{dt} = -g(x(t))^2 \equiv U(x(t)).$$

Here $U(x) > 0$ unless $x = x_0$ and the local asymptotic stability of the point x_0 is obtained immediately from Theorem 5. 3. For the equation (5. 5) with $a = b = 1$, the graph of $V(x)$ is shown in Fig. IV. 4. Note that the failure of asymptotic stability to extend beyond $x = 1$ at the right is signalled by $V(x)$ ceasing to increase beyond that point. Here $M = (-\infty, 1)$.

A further example, this time in E^{2n}, is afforded by the second order system

$$\ddot{x} + g(x, \dot{x}) + h(x), \qquad x, \dot{x} \in E^n, \tag{5. 16}$$

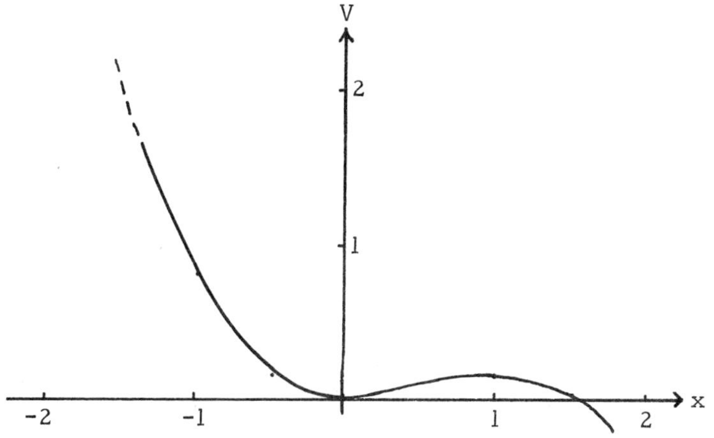

Fig. IV. 4. $V(x) = \frac{x^2}{2} - \frac{x^3}{3}$

where $h : E^n \to E^n$, $g : E^{2n} \to E^n$ are continuously differentiable and have the properties

$$\dot{x}^* g(x, \dot{x}) > 0, \qquad \dot{x} \neq 0, \tag{5.17}$$

$$\frac{\partial h^i}{\partial x^j} = \frac{\partial h^j}{\partial x^i}, \qquad i, j = 1, 2, \ldots, n, \tag{5.18}$$

$$x^* h(x) > 0, \qquad x \neq 0. \tag{5.19}$$

It should be noted that (5.17) and (5.19) imply, respectively,

$$g(x, 0) = 0, \quad x \in E^n \text{ (in particular } g(0, 0) = 0),$$

$$h(0) = 0,$$

so that $x = 0$, $\dot{x} = 0$ is an equilibrium point in E^{2n} for (5.16). "Exactness", (5.18), implies that the row vector $h(x)^*$ is the gradient of a function $H(x)$, $H : E^n \to E^1$, which may be specified by the curvilinear integral

$$H(x) = \int_{C_x} h(\xi)^* d\xi, \tag{5.20}$$

over any arc, C_x, joining 0 to x in E^n. A standard theorem given

$$\frac{\partial H}{\partial x} = h(x)^*$$

and, since the integral over the straight line joining 0 to x can be written as

$$\int_0^1 h(tx)^* x \, dt ,$$

(5.19) gives $H(x) > 0$. We define

$$V(x, \dot{x}) = \tfrac{1}{2}\|\dot{x}\|^2_{E^n} + H(x) \tag{5.21}$$

and compute

$$\frac{d}{dt} V(x(t), \dot{x}(t)) = \frac{\partial V}{\partial x}(x(t), \dot{x}(t))\dot{x}(t) + \frac{\partial V}{\partial \dot{x}}(x(t), \dot{x}(t))\ddot{x}(t) = h(x(t))^* \dot{x}(t)$$

$$+ \dot{x}(t)^* \ddot{x}(t) = (\text{using (5.16) to evaluate } \ddot{x}(t))$$

$$= \dot{x}(t)^* [h(x(t)) - g(x(t), \dot{x}(t)) - h(x(t))] = -\dot{x}(t)^* g(x(t), \dot{x}(t)) \equiv U(x(t), \dot{x}(t)) \leq 0 .$$

Suppose we have a solution of (5.16) such that $U(x(t), \dot{x}(t)) \equiv 0$ on an interval. Then (5.17) implies that $\dot{x}(t) \equiv 0$, $g(x(t), \dot{x}(t)) \equiv 0$ on that interval. Then $\ddot{x}(t) \equiv 0$ there and (5.16) becomes

$$h(x(t)) = 0$$

from which, via (5.19), we see that $x(t) \equiv 0$ there. Thus condition (iii) of Theorem 5.3 is satisfied and we have at least local asymptotic stability for the equilibrium point $x = 0$, $\dot{x} = 0$. If (5.19) is strengthened to

$$x^* h(x) \geq \delta \|x\|^p , \qquad x \in E^n ,$$

for some $\delta > 0$ and some $p > 0$, then (5.20), (5.21) imply

$$V(x, \dot{x}) \to \infty \quad \text{as} \quad \left\| \begin{pmatrix} x \\ \dot{x} \end{pmatrix} \right\|_{E^{2n}} \to \infty ,$$

and we have global asymptotic stability.

It will be clear to the reader that III-Theorem 1.5 is, at least as far as the sufficiency part is concerned, a special case of Theorem 5.3 above with

$$V(x) = x^* X x, \quad U(x) = -x^* H^* H x .$$

For the nonlinear case, condition (iii) of Theorem 5.3 replaces the condition of III-Theorem 1.5 that the pair (H, A) be observable.

There are a number of theorems in the differential equations literature demonstrating the existence of an appropriate Liapounov function for a given system (5.3) near an asymptotically stable equilibrium point. See [13] for example. However, except in the linear case and some relatively simple nonlinear cases such as those discussed above, finding $V(x)$ in practice tends to be difficult and as much of an art as a science. One of the great strengths of the optimization approach in control theory lies in the fact that, for a wide variety of control settings, the solution of the optimal control problem automatically yields a Liapounov function for the (presumed asymptotically stable) optimal closed loop plant. This is particularly true with regard to the approach to optimization theory which is called "dynamic programming". This method is associated with the name of Richard Bellman and is ably presented in references [10] and [12], e.g. There are a number of technical mathematical difficulties connected with this approach but it has the advantages of ease of application and transparency of motivation which make it highly useful in many applications.

Let us consider the general control system (5.1) and a cost functional

$$J(\xi_\tau, u, \tau, T) = \int_\tau^T G(x(t), u(t))dt + D(x(T)) \tag{5.22}$$

for a twice continuously differentiable cost integrand $G : E^{n+m} \to E^1$ and terminal cost $D : E^n \to E^1$. It is assumed here that $x(t)$, $u(t)$ together satisfy (5.1) with $x(\tau) = \xi_\tau$. Ordinarily the problem of interest at the outset is

$$\min_{u \in U_0} J(\xi_0, u, 0, T), \tag{5.23}$$

where U_0 is some class of admissible controls, here we admit piecewise continuous $u(t)$ such that $u(t) \in \Omega_u$ (a region in E^m) for all $t \in [0, T]$. We will suppose that this problem has a unique solution

$$\hat{u}(t) \ (= \hat{u}(\xi_0, 0, T, t)) \tag{5.24}$$

for all ξ_0 in some region $\Omega_x \subset E^n$ which contains, in addition to each initial state ξ_0, the resulting optimal trajectory, $\hat{x}(t)$.

The essential insight of the dynamic programming approach lies in "embedding" the problem (5.23) in the more general class of problems

$$\min_{u \in U_\tau} J(\xi_\tau, u, \tau, T), \qquad 0 \le \tau \le T, \tag{5.25}$$

where U_τ is the natural restriction of U_0 to $[\tau, T]$, and in the "principal of optimality": that the solution (5.24) of (5.23), restricted to $[\tau, T]$, must coincide with the solution of (5.25) if we take

$$\xi_\tau = \hat{x}(\tau),$$

where $\hat{x}(t)$ is the optimal trajectory for (5.1), (5.23) with $\hat{x}(0) = \xi_0$. The principal of optimality is a simple consequence of the additivity of the cost functional with respect to a decomposition $[0, T] = [0, \tau] \cup [\tau, T]$ of the time interval:

$$J(\xi, u, 0, T) = \int_0^\tau G(x(t), u(t))dt + J(\hat{x}(\tau), u, \tau, T), \tag{5.26}$$

together with the fact that the specification of u on τ, T does not affect the first term in (5.26).

Suppose the problem (5.23) has been solved for a given ξ_0 (and hence, suitably restricted, for all $\xi_\tau = \hat{x}(\tau)$ on the optimal trajectory $\hat{x}(t)$, $0 \le t \le T$). We define the <u>optimal value function</u>

$$V(\xi_\tau, \tau) = J(\xi_\tau, \hat{u}, \tau, T), \quad \xi_\tau = \hat{x}(\tau), \quad 0 \le \tau \le T, \tag{5.27}$$

where $\hat{u}$ is the optimal control for (5.23) (and hence, suitably restricted, for all $\xi_\tau = \hat{x}(\tau)$). This defines $V(\xi, \tau)$ for all (ξ, τ) on this optimal trajectory and if we now suppose, as we shall, that the set of all such optimal trajectories covers $\Omega_x \times [0, T]$, by examining all such trajectories we define $V(\xi, \tau)$ for $\xi \in \Omega_x$, $0 \le \tau \le T$.

Now we make a crucial assumption, which is at the heart of most of the technical difficulties associated with the method; we assume: $V(\xi, \tau)$ <u>is continuously differentiable for</u> $\xi \in \Omega_x$, $0 \le \tau \le T$. There are many important problems for which this assumption is not valid and which require suitable re-interpretation of some details of the method. But the assumption is valid for the class of problems which we will study here.

We are now in a position to develop what is known as "Bellman's equation".

<u>Theorem 5.4.</u> <u>Let</u> $V(\xi, \tau)$ <u>have the properties described above for</u> $\xi \in \Omega_X$, $0 \leq \tau \leq T$, <u>and suppose that the optimization problem</u>

$$\min_{u \in \Omega_u} \left(G(\xi, u) + \frac{\partial V}{\partial x}(\xi, \tau)\, F(\xi, u)\right) \tag{5.28}$$

<u>has the unique solution</u> $\hat{u}_{\xi,\tau}$ <u>for all</u> $\xi \in \Omega_X$, $0 \leq \tau \leq T$, <u>which varies continuously with</u> ξ, τ, $\xi \in \Omega_X$, $0 \leq \tau \leq T$. <u>Then</u> $V(\xi, \tau)$ <u>satisfies Bellman's equation</u>

$$0 = G(\xi, \hat{u}_{\xi,\tau}) + \frac{\partial V}{\partial x}(\xi, \tau) F(\xi, \hat{u}_{\xi,\tau}) + \frac{\partial V}{\partial t}(\xi, \tau), \tag{5.29}$$

$V(\xi, \tau)$ <u>satisfies the terminal condition</u>

$$V(\xi, T) = D(\xi), \quad \xi \in \Omega_X, \tag{5.30}$$

<u>the optimal control</u> $\hat{u}(\tau)$ <u>for the problem</u> (5.23) <u>is characterized by</u>

$$\hat{u}(\tau) = \hat{u}_{\hat{x}_1(\tau), \tau}, \tag{5.31}$$

<u>and is a continuous function of</u> τ, $0 \leq \tau \leq T$.

<u>Proof.</u> The terminal condition (5.30) is clear. For the rest, we consider a point $\xi \in \Omega_X$, an instant τ, $0 \leq \tau < T$ and another instant $\tau + \delta$ with $\tau < \tau + \delta \leq T$. Each admissible control $u(t)$, $\tau \leq t \leq T$, is divided in two parts, $u_1(t)$ is the restriction to $[\tau, \tau + \delta)$ and $u_2(t)$ is the restriction to $[\tau + \delta, T]$. Then, as in (5.26),

$$J(\xi, u, \tau, T) = \int_{\tau}^{\tau+\delta} G(x_1(t), u_1(t))dt + J(x_1(\tau+\delta), u_2, \tau + \delta, T), \tag{5.32}$$

where $x_1(t)$ is the solution of (5.1) corresponding to the initial state

$$x_1(\tau) = \xi$$

and the control $u_1(t)$ on $\tau \leq t \leq \tau + \delta$. Whatever $x_1(\tau + \delta)$ may be, we select $u_2(t) \equiv \hat{u}_2(t)$, the optimal control for $J(x_1(\tau + \delta), u_2, \tau + \delta, T)$, i.e.

$$J(x_1(\tau + \delta), \hat{u}_2, \tau+\delta, T) = \min_{u_2} J(x_1(\tau+\delta), u_2, \tau+\delta, T)$$

$$= (\text{by definition}) = V(x_1(\tau + \delta), \tau + \delta).$$

Then, with u_2 so specified, we have

$$J(\xi, u, \tau, T) = J(\xi, u_1, \hat{u}_2, \tau, T) = \int_{\tau}^{\tau+\delta} G(x_1(t), u_1(t))dt + V(x_1(\tau+\delta), \tau+\delta). \tag{5.33}$$

If we further specify

$$u_1(t) = \hat{u}_1(t),$$

where $\hat{u}_1(t)$ is the restriction of the optimal control $\hat{u}(t)$, defined on $\tau \leq t \leq T$, to $[\tau, \tau + \delta]$, then we have, of course

$$V(\xi, \tau) = \int_{\tau}^{\tau+\delta} G(\hat{x}_1(t), \hat{u}_1(t))dt + V(\hat{x}_1(\tau+\delta), \tau + \delta),$$

$\hat{x}_1(t)$ being the restriction of the optimal trajectory to $[\tau, \tau+\delta]$.

To establish (5. 31) we note, first of all, that if we assume the admissible controls $u(t)$ are piecewise continuous, we may as well also assume them continuous from the right at each point of discontinuity t_d, since the value of $u(t_d)$ itself plays no role in determining the solution of (5. 1) or the integral appearing in (5. 22). Thus we suppose, without loss of generality, that

$$u_1(\tau) = u(\tau) = \lim_{t \downarrow \tau} u(t) = \lim_{t \downarrow \tau} u_1(t)$$

for all controls considered in (5. 33) and, of course, also for the optimal control segment $\hat{u}_1(t)$. Thus, for any u_1,

$$\max_{t \in [\tau, \tau+\delta]} \|u_1(t) - u_1(\tau)\| = o(\delta), \qquad \delta \downarrow 0 . \tag{5.34}$$

Since (5. 1) gives, for $\tau \leq t \leq \tau + \delta$,

$$x_1(t) = \xi + \int_{\tau}^{t} F(x_1(t), u_1(t))dt,$$

the continuity of x_1 at τ and the continuity of F at (ξ, τ) implies that

$$x_1(t) = \xi + (t - \tau)F(\xi, \tau) + o(t - \tau), \qquad t \downarrow \tau . \tag{5.35}$$

(which we could also obtain from Taylor's formula). Using (5. 35) in (5. 33) and using the assumed continuous differentiability of V we see that

$$J(\xi, u_1, \hat{u}_2, \tau, T) = \delta G(\xi, u_1(\tau)) + V(\xi, \tau)$$

$$+ \delta \frac{\partial V}{\partial x}(\xi, \tau) F(\xi, u_1(\tau)) + \delta \frac{\partial V}{\partial t}(\xi, \tau) + o(\delta), \quad \delta \downarrow 0, \tag{5.36}$$

From this it follows that if we were to have

$$G(\xi, u_1(\tau)) + \frac{\partial V}{\partial x}(\xi, \tau) F(\xi, u_1(\tau)) > G(\xi, \hat{u}_{\xi,\tau}) + \frac{\partial V}{\partial x}(\xi, \tau) F(\xi, \hat{u}_{\xi,\tau}),$$

where $\hat{u}_{\xi,\tau}$ has been defined as the unique solution of (5. 28), then $J(\xi, u_1, \hat{u}_2, \tau, T)$ could be decreased by replacing $u_1(t)$ by the constant function $u(t) = \hat{u}_{\xi,\tau}$ on $[\tau, \tau+\delta]$, provided δ is sufficiently small. As this is impossible for the optimal control $\hat{u}_1(t)$, we conclude that

$$\hat{u}_1(\tau) = \hat{u}_{\xi,\tau},$$

from which (5. 31) follows when we take $\xi = \hat{x}(\tau)$, the point reached by the optimal trajectory, for (5. 1), (5. 23) at τ. The same argument, applied at instants t with $\tau < t < \tau + \delta$ in place of τ, gives $\hat{u}_1(t) \equiv \hat{u}_{\hat{x}_1(t), t}$. Now clearly,

$$J(\xi, \hat{u}_1, \hat{u}_2, \tau, T) = J(\xi, \hat{u}, \tau, T) = V(\xi, \tau) .$$

Using this in (5. 33) with $u_1(t) = \hat{u}_1(t)$, dividing the result by δ and letting $\delta \downarrow 0$ we obtain (5. 29) for $0 \le \tau < T$.

Since (5. 31) is satisfied for all τ with $0 \le \tau < T$ and $\hat{u}(\tau)$ is piecewise continuous, the continuity of $\hat{u}_{\xi,\tau}$ with respect to ξ, τ shows that $\hat{u}_1(\tau)$ must be a continuous function of τ for $0 \le \tau < T$. Since the value of $u(T)$ plays no role in itself, we may assume $\hat{u}(\tau)$ continuous for $0 \le \tau \le T$ and then (5. 31) holds for $0 \le \tau \le T$. Q. E. D.

From the continuous differentiability of $F(x, u)$ and $G(x, u)$ and our assumption that $\hat{u}_{\xi,\tau}$ solves (5. 31), it follows that

$$\frac{\partial G}{\partial u}(\xi, \hat{u}_{\xi,\tau}) + \frac{\partial V}{\partial x}(\xi, \tau) \frac{\partial F}{\partial u}(\xi, \hat{u}_{\xi,\tau}) = 0 . \tag{5.37}$$

It is this equation which is ordinarily used to determine $\hat{u}_{\xi,\tau}$, and hence $\hat{u}(\tau)$ via (5. 31).

The equations (5. 28), (5. 29) and (5. 30) together provide a complete characterization of the optimal controls $\hat{u}(\tau)$ and the optimal trajectories

$\hat{x}(\tau)$ for the problems which we study here.

For the linear-quadratic case, wherein (5. 1) becomes

$$\dot{x} = F(x, u) = Ax + Bu, \tag{5.38}$$

$$G(x, u) = (Hx + Ju)^* \tilde{W}(Hx + Ju) + u^* \tilde{U} u, \tag{5.39}$$

and

$$D(x) = x^* D x, \tag{5.40}$$

the solution of (5. 23) has already been obtained in Section 2. That result can be re-derived by trying to obtain a solution $V(\xi, \tau)$ of (5. 28), (5. 29), (5. 30) in the form

$$V(\xi, \tau) = \xi^* Q(\tau) \xi . \tag{5.41}$$

Substitution of (5. 41) into (5. 28), (5. 29), (5. 30), with $F(x, u)$, $G(x, u)$, $D(x)$ given by (5. 38), (5. 39), (5. 40), leads directly to the same result as before: the equations (5. 29), (5. 30) yield (see Section 2 for W, R, U)

$$\frac{dQ}{d\tau} + A^* Q + QA + W - (QB + R)U^{-1}(B^* Q + R^*) = 0, \tag{5.42}$$

$$Q(T) = D, \tag{5.43}$$

respectively, and (5. 28) (via (5. 37)) gives

$$\hat{u}(\tau) = - U^{-1}(B^* Q + R^*)\hat{x}(\tau). \tag{5.44}$$

Verification of these statements is left as Exercise 8 below.

All of what we have said here concerning the dynamic programming method applies equally well to time varying systems $\dot{x} = F(x, u, t)$ and time varying cost integrands $G(x, u, t)$. We have not discussed this explicitly because our real interest lies in the autonomous case and the problem (5. 23) modified so that the cost is computed over an infinite interval as in Section 3.

Let us suppose that the point (y_0, w_0) is an equilibrium point for $F(y, w)$ and that (cf. III-4)

$$F(y, w) = A(y - y_0) + B(w - w_0) + f(y, w), \quad y \in E^n, \quad w \in E^m, \tag{5.45}$$

where

$$A = \frac{\partial F}{\partial y}(y_0, w_0), \quad B = \frac{\partial F}{\partial w}(y_0, w_0) \tag{5.46}$$

and

$$f(y, w) = \mathcal{O}(\|y - y_0\| + \|w - w_0\|), \qquad y, w \to y_0, w_0. \tag{5.47}$$

We assume the pair (A, B) to be stabilizable. We take $D(y)$ (cf. (5.22)) $\equiv 0$ in this case, while we suppose that

$$G(y, w) = (H(y - y_0) + J(w - w_0))^* \tilde{W}(H(y - y_0) + J(w - w_0))$$

$$+ (w - w_0)^* \tilde{U}(w - w_0) + g(y, w), \tag{5.48}$$

$$g(y, w) = \mathcal{O}(\|y - y_0\|^2 + \|w - w_0\|^2), \qquad y, w \to y_0, w_0 .$$

We will assume that the pair (F, A) is observable.

We pose the problem

$$\min_w J(\eta, w), \quad J(\eta, w) = \int_0^\infty G(y(t), w(t))dt , \tag{5.49}$$

wherein it is assumed that $y(t)$ is the solution of

$$\dot{y} = F(y, w), \tag{5.50}$$

$$y(0) = \eta \in E^n , \tag{5.51}$$

corresponding to the control $w(t)$, $0 \le t < \infty$.

The first question, of course, is whether or not there exist controls w for which $J(\eta, w)$ is finite. In Exercise 9 below we ask the reader to extend the work done in III-4 slightly to prove

Proposition 5.5. _Assuming_ (A, B) _stabilizable, let the_ $n \times m$ _matrix_ K _be selected so that_ $A + BK$ _is a stability matrix._ _Let the linear control law_

$$w(t) - w_0 = K(y(t) - y_0)$$

be implemented in (5.50) _so that it becomes_ _(cf._ (5.45))

$$\dot{y} = (A + BK)(y - y_0) + f(y, w_0 + K(y - y_0)) \equiv (A + BK)(y - y_0) + \varphi(y - y_0), \tag{5.52}$$

$$\varphi(y - y_0) = \mathcal{O}(\|y - y_0\|), \qquad y \to y_0 . \tag{5.53}$$

Then there is a neighborhood N_K _of_ y_0 _in_ E^n _which is invariant for_ (5.52) _(i.e., solutions of_ (5.52) _initiating in_ N_K _remain in_ N_K),

$$\lim_{t\to\infty} y(t) = y_0, \quad y(t) \text{ solving (5.50), (5.51) for } \eta \in N_K,$$

(thus y_0 is an asymptotically stable critical point for the closed loop system (5.52)) and there exists $M_K > 0$, $\mu_K > 0$, such that

$$\|y(t)\| \leq M_K e^{-\mu_K t} \|\eta\|, \quad \eta \in N_K. \tag{5.54}$$

Using (5.54) together with the form (5.48) of $G(y, w)$, we see very easily that $J(\eta, u_k)$ ($\equiv w_0 + k(y - y_0)$)) is finite for (at least) all $\eta \in N_K$. In establishing the existence of N_K, as in Exercise 9 below, it is ordinarily necessary to assume N_K very small. In practice, asymptotic stability and finite cost may be expected for η in a substantially larger region - in particular for all $\eta \in E^n$ for which the solution of (5.51), (5.52) eventually enters the smaller region N_K rigorously established.

The dynamic programming technique is applied to the problem (5.49) in much the same way as to the earlier problem (5.23) - with one fundamental difference. If we define

$$J(y(\tau), w, \tau) = \int_\tau^\infty G(y(t), w(t))dt$$

for $y(t)$ a solution of (5.50) on $[\tau, \infty)$ with initial state $y(\tau)$, it will be observed that J really only depends upon $y(\tau)$ and u, not upon τ itself. Indeed, F and G are both independent of time, t, so $y(t - \tau)$ is a solution of (5.50) with $\eta = y(\tau)$ and

$$\int_\tau^\infty G(y(t), w(t))dt = \int_0^\infty G(y(t-\tau), w(t-\tau))dt.$$

Thus $J(y(\tau), w, \tau) = J(y(\tau), w)$. So when we write (5.26) now, it becomes, for $0 \leq \tau$,

$$J(y(\tau), w) = \int_\tau^{\tau+\delta} G(y(t), w_1(t))dt + J(y_1(\tau + \delta), w_2)$$

and when w_2 is chosen as the optimal control relative to the initial state $y_1(\delta)$, we have

$$J(y(\tau), w_1, \hat{w}_2) = \int_\tau^{\tau+\delta} G(y(t), w_1(t)dt + V(y_1(\tau + \delta)) = (\text{cf. (5.36)}\ \delta G(y(\tau), w_1(\tau))$$

$$+ \delta \frac{\partial V}{\partial y}(y(\tau))F(y(\tau), w_1(\tau)) + o(\delta), \qquad \delta \downarrow 0.$$

Here we have used the fact that V now depends only on the state vector:

$$V(\eta) = \min_{u} J(\eta, w) = \min_{u} \int_{0}^{\infty} G(y(t),\ w(t))dt\ .$$

Choosing

$$\hat{w}_1(\tau) = \hat{w}_{y_1(\tau)}$$

the presumed unique solution of

$$\min_{u \in \Omega_u} \{G(y(\tau),\ w) + \frac{\partial V}{\partial y}(y(\tau))F(y(\tau),\ w)\}, \tag{5.55}$$

which is seen, as before, to be a necessary condition for minimization, and letting $\delta \downarrow 0$, we obtain

$$0 = G(y(\tau), \hat{w}_{y(\tau)}) + \frac{\partial V}{\partial y}(y(\tau))F(y(\tau),\ \hat{w}_{y(\tau)}) \tag{5.56}$$

as the autonomous form of Bellman's equation. Again, (5. 55) is normally written and solved in the form

$$\frac{\partial G}{\partial w}(y(\tau),\ \hat{w}_{y(\tau)}) + \frac{\partial V}{\partial y}(y(\tau))\frac{\partial F}{\partial w}(y(\tau),\ \hat{w}_{y(\tau)}) = 0\ . \tag{5.57}$$

There remains the question of the existence of functions $V(y)$, $\hat{w}_y$, satisfying the equations (5. 55), (5. 56) in a suitable neighborhood of $y = y_0$ and the question as to whether or not (5. 55), (5. 56) are <u>sufficient</u> conditions rather than merely necessary conditions. We will take this up shortly. For the moment, let us content ourselves with two observations on the nature of the equations (5. 55), (5. 56), (5. 57).

First of all, (5. 57) indicates that the optimal control will be obtained in feedback form, as already anticipated by the notation. That is

$$\hat{w}(\tau) = \hat{w}_{y(\tau)} \equiv K(y(\tau))\,. \tag{5.58}$$

Feedback control is seen to be a natural feature for autonomous systems with a cost of the form (5. 49) - indeed, the control function, $\hat{w}$, depends only on the initial state $y(0)$, so, by extension, the value $\hat{w}(0)$ should depend only on $y(0)$. Then, since the problem starting at time τ is the same as the one starting at time 0, except that $y(0)$ is replaced by $y(\tau)$, we should expect $\hat{w}(\tau)$ to have the same functional dependence on $y(\tau)$ as $\hat{w}(0)$ does upon $\hat{y}(0)$.

Secondly, (5. 56) is already in a form which indicates that V is likely to play the role of a Liapounov function. With

$$\hat{w}(t) = K(y(t)),$$

$$g(y(t)) = F(y(t), \hat{w}(t)) = F(y(t), K(y(t)),$$

(5. 56) takes the form

$$\frac{\partial V}{\partial y}(y(t))g(y(t)) = - G(y(t), K(y(t))) \equiv U(y(t)),$$

which is the same as (iv) of Definition 5. 2. We ordinarily expect the cost integrand $G(y, w)$ to be non-negative; if so, U is non-positive. Thus the verification of asymptotic stability of the equilibrium point $y = y_0$ of the closed loop system

$$\dot{y} = F(y, K(y)) \equiv g(y) \tag{5. 59}$$

requires only the verification of $V(y) > 0$ for $y \neq y_0$ and of (5. 8) - which amounts to a generalization of the observability condition assumed in Theorem 5. 3. Indeed, for the linear constant coefficient case ((5. 38), (5. 39), (5. 40) with cost (5. 49)), the eqations (5. 56), (5. 57) yield precisely the equations

$$A^*Q + QA + W - (QB + R)U^{-1}(B^*Q + R^*) = 0,$$

$$\hat{u}(t) = - U^{-1}(B^*Q + R^*)\hat{x}(t),$$

when we try for a solution of the form ($y \equiv x$ here)

$$V(x) = x^*Qx.$$

Again, we leave the verification as Exercise 10. The condition (5. 8) for

$$U(y) = - G(y, K(y)) = y^*[(F - GU^{-1}(B^*Q + R^*))^*\tilde{W}\cdot(F - GU^{-1}(B^*Q + R^*)) + (QB + R)U^{-1}\tilde{U}U^{-1}(B^*Q + R^*)]\,y$$

is satisfied if the assumptions of Proposition 3. 7 are made.

The basic existence, uniqueness, necessity and sufficiency results for (5.55), (5.56), (5.57) were obtained by D. Lukes [11]. The development is not, however, carried out in the framework of dynamic programming and the proofs are beyond the scope of the present book. We merely state

Theorem 5.6. Let the functions $F(y, w)$, $G(y, w)$ in (5.45), (5.48) be at least twice continuously differentiable in some neighborhood of the equilibrium point $(y_0, w_0) \in E^{n+m}$ and let (A, B) be stabilizable, (H, A) observable. Then there is a neighborhood N of y_0 in E^n such that

(i) the equations (5.56), (5.57) have a unique solution $V(y)$, $\hat{w}_y = K(y)$ in N, $V(y)$ being at least twice continuously differentiable in N, $K(y)$ at least continuously differentiable there;

(ii) N is an invariant neighborhood for the closed loop system (5.59) (i.e., $\dot{y} = F(y, K(y))$);

(iii) for each $\eta \in N$ the solution $y(t)$ of (5.59) with $y(0) = \eta$ has the property

$$\lim_{t \to \infty} y(t) = 0;$$

indeed, when $K(y)$ is written as

$$K(y) = \hat{K}(y - y_0) + k(y), \quad \hat{K} = \frac{\partial K}{\partial y}(y_0), \quad k(y) = o(\|y - y_0\|), \quad y \to y_0,$$

$A + B\hat{K}$ is a stability matrix, $\hat{K}$ being the optimal feedback matrix for the truncated system (cf. (5.45), (5.48))

$$\dot{x} = Ax + Bu,$$

$$J_0(\xi, u) = \int_0^\infty ((Hx + Ju)^* \tilde{W}(Hx + Ju) + u^* \tilde{U} u)dt$$

formed from the linear and quadratic terms in (5.1), (5.49), respectively;

(iv) for each $\eta \in N$ the control

$$\hat{u}(t) = K(y(t)),$$

where $y(t)$ is the solution of (5.59) with $y(0) = \eta$ is the unique solution of the minimization problem (5.49) for which the trajectory $y(t)$ lies in N, $0 \le t < \infty$.

We remark that $V(y)$, $K(y)$ have continous partial derivatives of all orders, or are real analytic, if $F(y, w)$, $G(y, w)$ have continous partial derivatives of all orders, or are real analytic, respectively.

Though not explicitly stated here, the nonlinear control law $w = K(y)$ can actually be shown to extend (though sometimes not as a single-valued function of y) to the set of all points $y = \eta$ for which the problem (5.49) has a finite solution - that is, to all points η for which there is a control w steering a solution of (5.50) from η to the origin during $[0, \infty)$, or equivalently, to some small neighborhood of y_0 in finite time. To prove this one has to study (5.56), (5.57) in the context of hyperbolic partial differential equations, the optimal solutions $\hat{y}(t)$ playing the role of "characteristics". See [14], [15] and standard works on partial differential equations for details. Thus one can expect the nonlinear control law $w = K(y)$ to stabilize (5.50) in a larger region as compared with the linearized control law $w = w_0 + \hat{K}(y - y_0)$ used in the same nonlinear system.

We present now an elementary, scalar example in which the equations (5.56), (5.57) can be solved explicitly. One important aim in presenting this example is to show, as we have already indicated that we should expect, that this theory can lead to more globally effective stabilization for nonlinear systems than one obtains by using the optimal linear feedback control policy based on the linearized system.

Consider the nonlinear scalar control system

$$\dot{y} = F(y, w) = y + y^2 + w. \tag{5.60}$$

We define the cost integral by setting

$$G(y, u) = y^2 + w^2. \quad (W = 1, \ U = 1).$$

Thus the cost integrand remains quadratic, the departure from the theory of Section 3 lies only in the fact that (5.60) is nonlinear in the state variable y. The linearized equation is

$$\dot{x} = x + u \quad (A = 1, \ B = 1). \tag{5.61}$$

The optimal control based on the linearized system is obtained by solving the scalar quadratic equation (cf. (3.13))

$$2Q + 1 - Q^2 = 0 \, ,$$

which has the unique positive solution

$$Q = 1 + \sqrt{2} .$$

The optimal control for (5.61) is then

$$u = -(1 + \sqrt{2})x .$$

Now, if we use $w = -(1 + \sqrt{2})y$ in (5.60) the nonlinear closed loop system is

$$\dot{y} = -\sqrt{2}y + y^2 . \quad (5.62)$$

From the discussion following Definition 5.1 we know that the "region of asymptotic stability" for (5.62) is the interval

$$-\infty < y < \sqrt{2} .$$

In particular, use of the optimal linear feedback law for (5.61) in (5.60) results in a system (5.62) for which $y = 0$ is merely a locally asymptotically stable equilibrium point.

If we work with (5.60) directly and use the equations (5.56), (5.57), those equations become, respectively,

$$y^2 + (\hat{u}_y)^2 + \frac{\partial V}{\partial y}(y)(y + y^2 + \hat{u}_y) = 0 , \quad (5.63)$$

$$2\hat{u}_y + \frac{\partial V}{\partial y}(y) = 0 . \quad (5.64)$$

We use (5.64) to solve for $\hat{u}_y$ in terms of $\frac{\partial V}{\partial y}$ and substitute the result in (5.63) to obtain

$$-\tfrac{1}{4}\left(\frac{\partial V}{\partial y}(y)\right)^2 + (y + y^2)\frac{\partial V}{\partial y}(y) + y^2 = 0 ,$$

and the quadratic formula gives

$$\frac{\partial V}{\partial y}(y) = 2(y + y^2) \pm 2\sqrt{(y+y^2)^2 + y^2} . \quad (5.65)$$

We want $V(0) = 0$ and $V(y) > 0$ for $y \neq 0$. The latter dictates use of the + sign in (5.65) for $y \geq 0$, the − sign for $y < 0$. Then (5.64) gives

$$u_y \equiv H(y) = -\tfrac{1}{2}\frac{\partial V}{\partial y}(y) = \begin{cases} -(y+y^2) - \sqrt{(y+y^2)^2 + y^2} , & y \geq 0 . \\ -(y+y^2) + \sqrt{(y+y^2)^2 + y^2} , & y < 0 . \end{cases}$$

The closed loop system is

$$\dot{y} = \pm\sqrt{(y+y^2)^2 + y^2} \quad \begin{pmatrix} + & \text{for } y < 0 \\ - & \text{for } y \geq 0 \end{pmatrix}$$

for which $y = 0$ is a globally asymptotically stable critical point.

Apart from existence and uniqueness of solutions of (5.55), (5.56), (5.57) and questions of the necessity and sufficiency of these conditions for optimal control, there is the matter of approximate representation of the solution with a view to actual implementation of the optimal feedback law (5.58) determined by (5.57). In [11] Lukes has not neglected to treat this practical question along with the theoretical ones already considered in Theorem 5.6. Suppose that in (5.45) and (5.48), $F(y, w)$ and $G(y, w)$ can be developed in multivariable Taylor series, convergent in some neighborhood N of the equilibrium point (y_0, w_0) in E^{n+m}. For convenience of notation we let

$$x = y - y_0, \tag{5.66}$$

$$u = w - w_0 . \tag{5.67}$$

Then (5.45) becomes (cf. (5.46) for A, B)

$$\begin{aligned} F(y_0 + x, w_0 + u) &= Ax + Bu + f(y_0 + x, w_0 + u) \\ &= Ax + Bu + \sum_{k=2}^{\infty}\left(\sum_{\|(p,q)\|=k} f_{(p,q)}(x^1)^{p_1} \ldots (x^n)^{p_n} (u^1)^{q_1} \ldots (u^m)^{q_m}\right) \end{aligned} \tag{5.68}$$

Here

$$(p, q) = (p^1 \ldots p^n, q^1, \ldots, q^m)$$

is an "index vector" with non-negative integer components and

$$\|(p, q)\| = p^1 + \ldots + p^n + q^1 + \ldots + q^m .$$

For each such index vector p, $f(p, q)$ is a vector in E^n. Similarly, from (5.48),

$$\begin{aligned} G(y_0 + x, w_0 + u) &= (Hx + Ju)^* \tilde{W} (Hx + Ju) + u^* \tilde{U} u + g(y_0 + x, w_0 + u) \\ &\equiv x^* W x + 2x^* R u + u^* U u + g(y_0 + x, w_0 + u) = x^* W x + 2x^* R u + u^* U u \\ &+ \sum_{k=3}^{\infty}\left(\sum_{\|(p,q)\|=k} g_{(p,q)}(x^1)^{p_1} \ldots (x^n)^{p_n} (u^1)^{q_1} \ldots (u^m)^{q_m}\right) \end{aligned} \tag{5.69}$$

with the same conventions as before, but now $g_{(p,q)}$ is a scalar for all index vectors (p, q).

Following [11], we attempt to obtain a solution of (5.56), (5.57) in the form (here $p = (p_1, p_2, \dots, p_n)$)

$$V(y_0 + x) = x^* Q x + \sum_{k=3}^{\infty} \left(\sum_{\|p\| = k} v_p (x^1)^{p_1} \dots (x^n)^{p_n} \right). \tag{5.70}$$

$$\hat{w}(y) = \hat{K}(y_0 + x) = w_0 + \hat{u}(y_0 + x) = w_0 + \hat{K}x + \sum_{k=2}^{\infty} \left(\sum_{\|p\| = k} \ell_p (x^1)^{p_1} \dots (x^n)^{p_n} \right). \tag{5.71}$$

This is done by substituting the formulae (5.68), (5.69), (5.70), (5.71) into (5.56), (5.57) and collecting the coefficients of like terms $(x^1)^{p_1} \dots (x^n)^{p_n} (u^1)^{q_1} \dots (u^m)^{q_m}$, or $(x^1)^{p_1} \dots (x^n)^{p_n}$, as the case may be. Identification of coefficients is carried out in an alternating fashion, using (5.56) to determine the coefficients in (5.70) for $\|p\| = k + 1$ and then using that result in (5.57) to get the coefficients in (5.71) for $\|p\| = k$.

We note that with $y = y_0 + x$, $w = w_0 + u$,

$$\frac{\partial F}{\partial y}(y_0 + x, w_0 + u) = \frac{\partial F}{\partial x}(y_0 + x, w_0 + u),$$

$$\frac{\partial F}{\partial u}(y_0 + x, w_0 + u) = \frac{\partial F}{\partial u}(y_0 + x, w_0 + u),$$

and similarly for $G(y_0 + x, w_0 + u)$, $V(y_0 + x)$, $K(y_0 + x)$, etc. Thus, forming gradients (note that no negative powers of the x^i actually occur)

$$\frac{\partial V}{\partial y}(y_0 + x) = \frac{\partial V(y_0 + x)}{\partial x} = 2x^* Q + \sum_{k=3}^{\infty} \left(\sum_{\|p\| = k} v_p \left(\frac{p_1}{x^1}, \dots, \frac{p_n}{x^n} \right) (x^1)^{p_1} \dots (x^n)^{p_n} \right), \tag{5.72}$$

$$\frac{\partial G}{\partial w}(y_0 + x, w_0 + u) = \frac{\partial G(y_0 + x, w_0 + u)}{\partial u} = 2x^* R + 2u^* U$$

$$+ \sum_{k=3} \left(\sum_{\|(p,q)\| = k} g_{(p,q)} \left(\frac{q_1}{u^1}, \dots, \frac{q_m}{u^m} \right) (x^1)^{p_1} \dots (x^n)^{p_n} (u^1)^{q_1} \dots (u^m)^{q_m} \right), \tag{5.73}$$

$$\frac{\partial F}{\partial w}(y_0 + x, w_0 + u) = \frac{\partial F(y_0 + x, w_0 + u)}{\partial u} = Bu$$

$$+ \sum_{k=2} \left(\sum_{\|(p,q)\| = k} f_{(p,q)} \left(\frac{q_1}{u^1}, \dots, \frac{q_m}{u^m} \right) (x^1)^{p_1} \dots (x^n)^{p_n} (u^1)^{q_1} \dots (u^m)^{q_m} \right). \tag{5.74}$$

$$2x^*R+2u^*U+\sum_{k=3}^{\infty}\left(\sum_{\|(p,q)\|=k} g_{(p,q)}\left(\frac{q_1}{u^1},\ldots,\frac{q_m}{u^m}\right)(x^1)^{p_1}\ldots(x^n)^{p_n}(u^1)^{q_1}\ldots(u^m)^{q_m}\right.$$
$$+\left[2x^*Q+\sum_{k=3}^{\infty}\sum_{\|p\|=k} v_p\left(\frac{p_1}{x^1},\ldots,\frac{p_n}{x^n}\right)(x^1)^{p_1}\ldots(x^n)^{p_n}\right]\times$$
$$\times\left[B+\sum_{k=2}^{\infty}\left(\sum_{\|(p,q)\|=k} f_{(p,q)}\left(\frac{q_1}{u^1},\ldots,\frac{q_m}{u^m}\right)(x^1)^{p_1}\ldots(x^n)^{p_n}(u^1)^{q_1}\ldots(u^m)^{q_m}\right)\right]=0\,. \tag{5.75}$$

It should be noted here that $g_{(p,q)}\left(\frac{q_1}{u^1},\ldots,\frac{q_m}{u^m}\right)$ and $v_p\left(\frac{p_1}{x^1},\ldots,\frac{p_n}{x^n}\right)$ are row vectors of dimension m, n, respectively, while $f_{(p,q)}\left(\frac{q_1}{u^1},\ldots,\frac{q_m}{u^m}\right)$, being the product of the n dimensional column vector $f_{(p,q)}$ with an m dimensional row vector, is an $n \times m$ matrix. Formula (5.75) can, in principle, be solved to give u as a series in x with coefficients ℓ_p which depend on the (as yet unknown) v_p:

$$u=\hat{u}(y_0+x)=\hat{K}x+\sum_{k=2}^{\infty}\left(\sum_{\|p\|=k}\ell_p(x^1)^{p_1}\ldots(x^n)^{p_n}\right). \tag{5.76}$$

The equation (5.56) becomes, after use of (5.68), (5.69), (5.70),

$$x^*Wx+x^*Ru+u^*Rx+u^*Uu+\sum_{k=3}^{\infty}\left(\sum_{\|(p,q)\|=k} g_{(p,q)}(x^1)^{p_1}\ldots(x^n)^{p_n}(u^1)^{q_1}\ldots(u^m)^{q_m}\right)$$
$$+\left[2x^*Q+\sum_{k=3}^{\infty}\left(\sum_{\|p\|=k} v_p\left(\frac{p_1}{x^1},\ldots,\frac{p_n}{x^n}\right)(x^1)^{p_1}\ldots(x^n)^{p_n}\right)\right]\times$$
$$\times\left[Ax+Bu+\sum_{k=2}^{\infty}\left(\sum_{\|(p,q)\|=k} f_{(p,q)}(x^1)^{p_1}\ldots(x^n)^{p_n}(u^1)^{q_1}\ldots(u^m)^{q_m}\right)\right]=0. \tag{5.77}$$

The series (5.76) is substituted into (5.77), coefficients of resulting terms $(x^1)^{p_1}\ldots(x^n)^{p_n}$ are formed, as algebraic expressions in the v_p, after which, in principle, the v_p can be computed by setting these algebraic expressions equal to zero. Then, returning to (5.76) with the (now numerical valued) v_p, the ℓ_p can be given numerical values and the nonlinear feedback law given as a multi-dimensional power series in $x^1, x^2, \ldots, x^n$. The whole process is computationally very complex and it is clear that one cannot expect to obtain general closed form recursion formulae for the v_p, ℓ_p in any compact or usable form.

For computational purposes the following procedure is recommended. Assume that at the r-th stage $\hat{K}$, ℓ_p, $\|p\| \leq r$, Q, v_p, $\|p\| \leq r+1$ have already been determined. Set

$$u = \hat{K}x + \sum_{k=2}^{r}\left(\sum_{\|p\|=k} \ell_p (x^1)^{p_1} \dots (x^n)^{p_n}\right) + \sum_{\|p\|=r+1} \ell_p (x^1)^{p_1} \dots (x^n)^{p_n} + \dots \quad (5.78)$$

Substitute (5.78) into (5.75), truncated at $k = r+1$ in the following sense: include terms involving $g_{(p,q)}$ only for $\|(p,q)\| \leq r+2$, include terms involving v_p only for $\|p\| \leq r+2$. This is a finite, though tedious process. When complete, discard all terms $(x^1)^{p_1} \dots (x^n)^{p_n}$ in the result for which $\|p\| > r+1$. Terms with $\|p\| < r+1$ will already be zero. As a result, (5.75) will take the form, as far as coefficients of $(x^1)^{p_1} \dots (x^n)^{p_n}$, $\|p\| = r+1$, are concerned

$$\sum_{\|p\|=r+1} (2\ell_p^* U + \varphi_p^*)(x^1)^{p_1} \dots (x^n)^{p_n} = 0 , \quad (5.79)$$

where φ_p depends on $f_{(p,q)}$, $g_{(p,q)}$, $\|(p,q)\| < r+1$, ℓ_p, $\|p\| \leq r$, and v_p, $\|p\| \leq r+2$. The v_p, $\|p\| = r+2$, are as yet undetermined.

Now substitute (5.78), the ℓ_p, $\|p\| = r+1$ being given as functions of the v_p, $\|p\| = r+2$, into the equation (5.77) with only terms involving $f(p,q)$, $\|(p,q)\| \leq r+1$, $g_{(p,q)}$, $\|(p,q)\| \leq r+2$, v_p, $\|p\| \leq r+2$, retained in that equation. This again is a finite but tedious procedure, after which terms with $\|p\| > r+2$ are discarded and those with $\|p\| < r+2$ will already be zero if earlier identification has been carried out correctly. Setting terms with $\|p\| = r+2$ equal to zero results in a system of equations for the v_p, $\|p\| = r+2$, which can be shown to have a unique solution.

This all sounds pretty bad. The actual work is greatly simplified by a very fortunate observation made in [11]. It turns out that when (5.78) is substituted for u in the truncated form of (5.77), the terms involving the ℓ_p, $\|p\| = r+1$, all vanish for $r \geq 1$. The terms involving the ℓ_p, $\|p\| = r+1$ are just

$$(2x^*R + 2x^*\hat{K}^*U + 2x^*QB)\left(\sum_{\|p\|=r+1} \ell_p (x^1)^{p_1} \dots (x^n)^{p_n}\right),$$

and this is zero because

$$\hat{K} = -U^{-1}(B^* Q + R^*).$$

Consequently, though not obvious at first, the last term in (5.78) can in fact be omitted when (5.78) is substituted into (5.77). Since the ℓ_p, $\|p\| = r + 1$, are the terms which depend on the v_p, $\|p\| = r + 2$, this results in considerable saving of effort in calculating the equations satisfied by the v_p, $\|p\| = r + 2$. Once the v_p, $\|p\| = r + 2$, are computed however, the expressions for the ℓ_p, $\|p\| = r + 1$, in terms of the v_p, $\|p\| = r + 2$, are used to provide numerical values for the ℓ_p, $\|p\| = r + 1$.

Let us see how this goes for the first step, $r = 0$. We begin by setting (cf. (5.78))

$$u = \hat{K}x + \ldots \tag{5.80}$$

Substituting (5.80) into (5.75) with $g_{(p,q)}$ retained only for $\|(p,q)\| \le 2$, $f(p,q)$ retained only for $\|(p,q)\| \le 1$, v_p retained only for $\|p\| \le 2$ (in other words, none of these are in fact retained in (5.75) in view of the summation limits) we have

$$2x^* R + 2x^* \hat{K}^* U + 2x^* QB = 0 ,$$

giving

$$\hat{K} = -U^{-1}(B^* Q + R^*). \tag{5.81}$$

Note that Q is as yet undetermined - its entries play the role of the v_p for $\|p\| = r + 2 = 2$ in this case. Substituting (5.81) in (5.80) and the result of that into (5.77), again appropriately truncated, we have (after transposing certain terms in (5.77))

$$x^* Wx + x^* R(-U^{-1}(B^* Q + R^*))x + x^*(-(QB + R)U^{-1})R^* x + x^*(QB+R)U^{-1}(B^* Q + R^*)x$$
$$+ x^* Q(A - BU^{-1}(B^* Q + R^*))x + x^*(A - BU^{-1}(B^* Q + R^*))^* Qx = 0 ,$$

from which we have

$$A^* Q + QA + W - (QB + R)U^{-1}(B^* Q + R^*) = 0 ,$$

equation (3.13) again. Not surprizingly, the first order approximation (5.80), (5.81) to the nonlinear optimal control law is the linear optimal control law for the linearized system with the cost truncated to quadratic terms only.

This fact is important in showing the solvability of the truncated equation (5.77) (as described above) for the v_p, $\|p\| = r + 2$, in general. After terms with $\|p\| > r + 2$ have been discarded, assuming earlier terms have been correctly identified, the resulting equation for the v_p, $\|p\| = r + 2$, has the form

$$\frac{\partial}{\partial x}\left(\sum_{\|p\|=r+2} v_p (x^1)^{p_1} \dots (x^n)^{p_n}\right)(A - BU^{-1}(B^*Q + R^*))x$$
$$= \sum_{\|p\|=r+2} h_p (x^1)^{p_1} \dots (x^n)^{p_n} \equiv H_{r+2}(x) \tag{5.82}$$

for appropriate scalar coefficients h_p, $\|p\| = r+2$, which are independent of the v_p, $\|p\| = r + 2$. Admittedly, careful inspection of (5.75), (5.76), (5.77) is required to see that this is the case. From the theory of Section 3 we know that

$$A - BU^{-1}(B^*Q + R^*) \equiv \hat{A}$$

is a stability matrix. Consider then the function

$$V_{r+2}(x) \equiv - \int_0^\infty H_{r+2}(e^{\hat{A}t}x)dt \ . \tag{5.83}$$

Forming the gradient of both sides and multiplying on the right by $\hat{A}x$ we have

$$\frac{\partial V_{r+2}}{\partial x}\hat{A}x = - \int_0^\infty \frac{\partial H_{r+2}}{\partial x}(e^{\hat{A}t}x)e^{\hat{A}t}dt\,\hat{A}x = - \int_0^\infty \frac{\partial H_{r+2}}{\partial x}(e^{\hat{A}t}x)\hat{A}e^{\hat{A}t}x\,dt \ . \tag{5.84}$$

Since $\hat{A}$ is a stability matrix, $e^{\hat{A}t}x$ traces a curve C_x in E^n from the origin to x as t ranges from ∞ to 0. Letting $u = e^{\hat{A}t}x$, (5.84) can be expressed in terms of a curvilinear integral over C_x:

$$\frac{\partial V_{r+2}}{\partial x}\hat{A}x = \int_{C_x} \frac{\partial H_{r+2}}{\partial x}(u)du = (\text{since } H_{r+2}(0) = 0) = H_{r+2}(x) \ .$$

Thus $V_{r+2}(x)$ solves (5.82) and can be seen from (5.83) and the form of

$H_{r+2}(x)$ shown in (5. 82) to have the form

$$V_{r+2}(x) = \sum_{\|p\|=r+2} \hat{v}_p (x^1)^{p_1} \ldots (x^n)^{p_n} .$$

It follows that the system of linear algebraic equations in the v_p which results from identification of coefficients in (5. 82), and which is the same as the truncated form of (5. 77), has the solution $\hat{v}_p$, $\|p\| = r + 2$, and this solution can be shown to be unique. Thus the v_p, $\|p\| = r + 2$, can be obtained by identification of coefficients in the truncated form of (5. 77), which we have already described, and the result will be that the v_p equal the $\hat{v}_p$ shown above.

The end result of the above is, in fact, a system of linear algebraic equations for the v_p, $\|p\| = r + 2$. The order of the system increases quite rapidly with n, the system dimension, and the index r + 2, corresponding to the order of the coefficients of V(y) being determined. The complexity of the operations is such that in most cases only - relatively small values of r can be tackled - even if more could be computed the problem of storage of the ℓ_p would eventually be a limiting factor in applications.

Let us consider a simple example: a pendulum with horizontal control force w, as shown in Fig. IV. 5. The (massless) rod is stiff and the pendulum may be rotated through an arbitrary angle θ, $-\infty < \theta < \infty$. One may see that, if the length, mass and gravitational constant are all taken equal to unity, then

$$\ddot{\theta} + \sin\theta = \cos\theta\, w ,$$

or, in first order form, with $y^1 = \theta$, $y^2 = \dot{\theta}$,

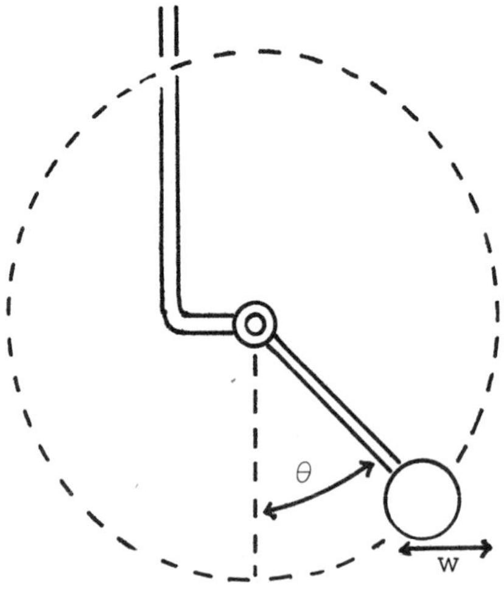

Fig. IV. 5. Pendulum with Horizontal Control Force

$$\dot{y} = \begin{pmatrix} \dot{y}^1 \\ \dot{y}^2 \end{pmatrix} = \begin{pmatrix} y^2 \\ -\sin y^1 + (\cos y^1) w \end{pmatrix} = F(y, w). \qquad (5.85)$$

The point $y^1 = y^2 = w = 0$ is clearly an equilibrium point. As cost integrand we will use

$$G(y, w) = 2(1 - \cos y^1) + (y^2)^2 + w^2 , \qquad (5.86)$$

the first two terms being twice the energy for the system. The linearized equations of motion are

$$\begin{pmatrix} \dot{x}^1 \\ \dot{x}^2 \end{pmatrix} = \begin{pmatrix} 0 & 1 \\ -1 & 0 \end{pmatrix} \begin{pmatrix} x^1 \\ x^2 \end{pmatrix} + \begin{pmatrix} 0 \\ 1 \end{pmatrix} u \equiv Ax + Bu , \qquad (5.87)$$

and the quadratic approximation to the cost integrand (5.86), expressed in terms of x^1, x^2, u, is

$$(x^1)^2 + (x^2)^2 + u^2 = (x^1, x^2) \begin{pmatrix} 1 & 0 \\ 0 & 1 \end{pmatrix} \begin{pmatrix} x^1 \\ x^2 \end{pmatrix} + 1 \cdot u^2, \quad W = \begin{pmatrix} \frac{1}{2} & 0 \\ 0 & 0 \end{pmatrix}, \quad U = 1 .$$

Here the solution of the quadratic matrix equation (3.13) is (using (3.37)ff. with $\alpha = 1$, $\beta = 0$, $\delta = 1$)

$$Q = \begin{pmatrix} 1.91228 & .41421 \\ .41421 & 1.35219 \end{pmatrix}$$

and the optimal feedback matrix is

$$\hat{K} = -1 \cdot (0, 1)\, Q = (-.41421, \ -1.95663) , \qquad (5.88)$$

so that the optimal closed loop matrix is

$$\hat{A} = A + B\hat{K} = \begin{pmatrix} 0 & 1 \\ -1.41421 & -1.35219 \end{pmatrix} .$$

In this example G is an even function of y, w and F is an odd function of y, w, from which one may verify by inspection of (5.56), (5.57) that V will be an even function of y and $K(y)$ will be an odd function of y. Also, since $y_0 = 0$, $w_0 = 0$ in this example, the distinction between y, w and

x, u made in the general example is not necessary. Thus the fourth order and third order approximations to $V(y)$, $K(y)$, respectively, are

$$V(y) = y^{*}Qy + \sum_{\|p\|=4} v_p (y^1)^{p_1}(y^2)^{p_2} + \ldots = 1.91228(y^1)^2 + .82842 y^1 y^2 + 1.24219(y^2)^2$$

$$+ v_{4,0}(y^1)^4 + v_{3,1}(y^1)^3 y^2 + v_{2,2}(y^1)^2(y^2)^2 + v_{1,3} y^1 (y^2)^3 v_{0,4}(y^2)^4 + \ldots \quad (5.89)$$

$$K(y) = \hat{K}y + \sum_{\|p\|=3} \ell_p (y^1)^{p_1}(y^2)^{p_2} = -.41421 y^1 - 1.35219 y^2$$

$$+ \ell_{3,0}(y^1)^3 + \ell_{2,1}(y^1)^2 y^2 + \ell_{1,2} y^1 (y^2)^2 + \ell_{0,3}(y^2)^3 + \ldots . \quad (5.90)$$

Using (5.90) in (5.75) with appropriate truncation and

$$\frac{\partial V}{\partial y} = (3.82456 y^1 + .82842 y^2 + 4v_{4,0}(y^1)^3 + 3v_{3,1}(y^1)^2 y^2$$

$$+ 2v_{22} y^1 (y^2)^2 + v_{1,3}(y^2)^3, \ .82842 y^1 + 2.70438 y^2$$

$$+ v_{3,1}(y^1)^3 + 2v_{2,2}(y^1)^2 y^2 + 3v_{1,3} y^1 (y^2)^2 + 4v_{0,4}(y^2)^3 + \ldots , \quad (5.91)$$

$$G(y, w) = (y^1)^2 + (y^2)^2 + w^2 - \frac{1}{12}(y^1)^4 + \ldots \quad (5.92)$$

$$\frac{\partial G}{\partial w} = 2w + \ldots \quad (5.93)$$

$$F(y, w) = \begin{pmatrix} y^2 \\ -y^1 + \frac{1}{6}(y^1)^3 + w - \frac{1}{2}(y^1)^2 w + \ldots \end{pmatrix}, \quad (5.94)$$

$$\frac{\partial F(y, w)}{\partial w} = \begin{pmatrix} 0 \\ 1 - \frac{1}{2}(y^1)^2 \end{pmatrix}, \quad (5.95)$$

we obtain

$$2(-.41421 y^1 - 1.35219 y^2 + \ell_{3,0}(y^1)^3 + \ell_{2,1}(y^1)^2 y^2 + \ell_{1,2} y^1 (y^2)^2 + \ell_{0,3}(y^2)^3 + \ldots)$$

$$+ (.82842 y^1 + 2.70438 y^2 + v_{3,1}(y^1)^3 + 2v_{2,2}(y^1)^2 y^2 + 3v_{1,3} y^1 (y^2)^2 + 4v_{0,4}(y^2)^3 + \ldots)$$

$$\times (1 - \tfrac{1}{2}(y^1)^2) = 0.$$

The first order terms vanish, as we would expect, and the equation obtained by retaining only third order terms is

$$(2\ell_{3,0}+v_{3,1}-.41421)(y^1)^3+(2\ell_{2,1}+2v_{2,2}-1.35219)(y^1)^2y^2$$
$$+(2\ell_{1,2}+3v_{1,3})y^1(y^2)^2+(2\ell_{0,3}+4v_{0,4})(y^2)^3=0 .$$

Solving for the ℓ_p, we have

$$\ell_{3,0}=.20710-.5v_{3,1},\quad \ell_{2,1}=.67609-v_{2,2},$$
$$\ell_{1,2}=1,5v_{1,3},\quad \ell_{0,3}=-2v_{0,4} . \tag{5.96}$$

Now, using (5.91), (5.92), (5.94) in (5.77), we obtain

$$(y^1)^2+(y^2)^2+w^2-\frac{1}{12}(y^1)^4+\ldots$$
$$+(3.82456y^1+.82842y^2+4v_{4,0}(y^1)^3+3v_{3,1}(y^1)^2y^2+2v_{22}y^1(y^2)^2$$
$$+v_{1,3}(y^2)^3+\ldots)y^2+(.82842y^1+2.70438y^2+v_{3,1}(y^1)^3$$
$$+2v_{2,2}(y^1)^2y^2+3v_{1,3}y^1(y^2)^2+4v_{0,4}(y^2)^3+\ldots)\times$$
$$\times(-y^1+\frac{1}{6}(y^1)^3+w-\tfrac{1}{2}(y^1)^2w+\ldots)=0 . \tag{5.97}$$

In (5.97) we now set $w = K(y)$, $K(y)$ being given by (5.90) and the ℓ_p, $\|p\| = 3$, given in terms of the v_p, $\|p\| = 4$, by (5.96). In the result it is easy to check that all quadratic terms vanish and we are left with terms of fourth, or higher, order. The vanishing of the ℓ_p, $\|p\| = 3$, from the expression can also be varified here, so that, in fact, the ℓ_p, $\|p\| = 3$, could have been omitted when (5.90) was substituted in (5.97). Retaining only fourth order terms from our expression, (5.97) becomes

$$(.30963-1.41421v_{3,1})(y^1)^4+(1.57090+4v_{4,0}-2.82842v_{2,2}-1.35219v_{3,1})(y^1)^3y^2$$
$$+(1.82840-4.24263v_{1,3}-2.70438v_{22})(y^1)^2(y^2)^2$$
$$+(2v_{22}-5.65684v_{0,4}-4.04657v_{1,3})y^1(y^2)^3+(v_{1,3}-5.40876v_{0,4})(y^2)^4=0 .$$

Setting the coefficients of the individual terms equal to zero we obtain

$$\begin{pmatrix} 0 & 1.41421 & 0 & 0 & 0 \\ -4 & 1.35219 & 2.82842 & 0 & 0 \\ 0 & 0 & 2.70438 & 4.24263 & 0 \\ 0 & 0 & 0 & -4.05657 & -5.65684 \\ 0 & 0 & 0 & 1 & -5.40876 \end{pmatrix} \begin{pmatrix} v_{4,0} \\ v_{3,1} \\ v_{2,2} \\ v_{1,3} \\ v_{0,4} \end{pmatrix} = \begin{pmatrix} .30963 \\ 1.57090 \\ 1.82840 \\ 0 \\ 0 \end{pmatrix} .$$

We immediately have

$$v_{3,1} = .21894 .$$

The last three, $v_{2,2}$, $v_{1,3}$, $v_{0,4}$, are obtained by solving three equations in three unknowns, with the result

$$v_{2,2} = .34380 ,$$

$$v_{1,3} = .21180 ,$$

$$v_{0,4} = .03033 .$$

Then we can obtain

$$v_{4,0} = -.07561 ,$$

and we have, as our fourth order representation of $V(y)$

$$V(y) = 1.91228(y^1)^2 + .82842 y^1 y^2 + 1.35219(y^2)^2 - .07561(y^1)^4 + .21894(y^1)^3 y^2 + .34380(y^1)^2(y^2)^2 + .21180 y^1 (y^2)^3 + .03033(y^2)^4 + \dots . \tag{5.98}$$

The values for $v_{4,0}$, $v_{3,1}$, $v_{2,2}$, $v_{1,3}$ and $v_{0,4}$ are now substituted into (5.96) to give

$$\ell_{3,0} = .09763, \quad \ell_{2,1} = .33229, \quad \ell_{1,2} = -.31770 ,$$

$$\ell_{0,3} = -.06066$$

and we have as the third order representation of the optimal nonlinear control law

$$w = K(y) = -.41421 y^1 - 1.35219 y^2 + .09763(y^1)^3 + .33229(y^1)^2 y^2 - .31770 y^1 (y^2)^3 - .06066(y^2)^3 + \dots . \tag{5.99}$$

Having done all of this work, we should hope that the nonlinear control law, even though a truncated one, would do a better job than the linear one. We can do this by analyzing equilibria other than $y^1 = y^2 = 0$. With use of the linear control law

$$w = -.41421y^1 - 1.35219y^2 \tag{5.100}$$

in the nonlinear system (5.85) there are equilibria (y^1, y^2) which are given by $y^2 = 0$, y^1 satisfying

$$(-41421y^1)\cos y^1 = \sin y^1. \tag{5.101}$$

The solutions closest to $y^1 = 0$ (but not equal to zero) are

$$y^1 = \pm 2.34073 .$$

The equilibria $(\pm 2.34073, 0)$ are unstable. Initial states $(y_0^1, 0)$ with $y_0^1 > 2.34073$ or $y_0^1 < -2.34073$ yield solutions which tend away from y_0^1, $|y^1(t)|$ increasing, while states $(y_0^1, 0)$ with $-2.34073 < y_0^1 < 2.34073$ yield solutions which tend to $(0, 0)$ - our objective of course. In any case, the points $(\pm 2.34073, 0)$ lie on the boundary of the region of asymptotic stability of the origin. When the nonlinear control law (5.99) is used, (5.101) is replaced by

$$(-.41421y^1 + .09763(y^1)^3)\cos y^1 = \sin y^1$$

with solutions

$$y^1 = \pm 4.57948 .$$

It may be verified that for states $(y_0^1, 0)$, $-4.57948 < y_0^1 < 4.57948$, the resulting solution of the nonlinear closed loop system obtained by substituting the nonlinear control law (5.99) into the nonlinear system (5.85) tends to $(0, 0)$ as $t \to \infty$. Thus the boundary points $(\pm 2.34073, 0)$ for the linear control law in the nonlinear system are "pushed out" to $(\pm 4.57948, 0)$ when the nonlinear control law is used in the nonlinear system. In Exercise 12 we suggest a numerical procedure for "mapping out" the regions of asymptotic stability of the origin in each case.

The power series method discussed here has a number of disadvantages. The first, the complexity of the operations which must be carried out to calculate the coefficients, is already abundantly evident. Another disadvantage is that there is no reason to believe that this power series converges in the whole of the region where $V(y)$ is actually defined - i. e., the complete region of asymptotic stability of the point (y_0, w_0). This type of situation is already familiar from parallel phenomena in the case of functions of a complex variable. Some sort of "analytic continuation" might be attempted but seems unreasonably complex. For large regions (5. 56), (5. 57) are probably best treated by numerical methods for partial differential equations (see [14], [15], e. g.) using the power series to provide starting values near the equilibrium point. But such methods will not be discussed here.

Bibliographical Notes, Chapter IV

[1] R. E. Kalman: "Contributions to the theory of optimal control", Bol. Soc. Mat. Mexicana 5 (1960), pp. 102-119.

[2] R. E. Kalman and R. S. Bucy: "New results in linear prediction and filtering theory", J. Basic Eng., Trans. ASME, Ser. D, 83 (1961), pp. 95-100.

The foregoing are generally acknowledged as the papers which gave birth to the now widely used "linear-quadratic" techniques discussed in this chapter and in Chapters V, VI and VII. Reference III- [4], while not an "early" paper, constitutes a very fine mathematical treatment, particularly in regard to the relationship between the optimization problem on the infinite interval and the problem of stabilization via linear feedback.

[3] R. E. Kalman, P. L. Falb and M. A. Arbib: "Topics in Mathematical System Theory", McGraw-Hill Book Co., New York, 1969.

[4] W. M. Wonham: "Linear Multivariable Control: A Geometric Approach", Lecture Notes in Economics and Mathematical Systems, Vol. 101, Springer-Verlag, New York, 1974.

The optimization problem studied in this chapter is treated via the dynamic programming approach in Chapter 12 of Wonham's book.

[5] H. K. Wilson: "Ordinary Differential Equations", Addison Wesley Pub. Co., Reading, Mass., 1971.

Chapter 8 of Wilson's book contains a very readable account of the theory of "maximal solutions" of general ordinary differential equations.

[6] Fred Brauer and John A. Nohel: "Ordinary Differential Equations" A First Course", W. A. Benjamin, Inc., New York, 1967.

[7] R. C. Buck: "Advanced Calculus", McGraw-Hill Book Co., New York, 1956.

[8] J. E. Potter: "Matrix quadratic solutions", J. SIAM Appl. Math., 14(1966), pp. 496-501.

[9] H. Rund: "The Hamilton-Jacobi Theory in the Calculus of Variations", D. Van Nostrand Co., Toronto, New York, Princeton, 1966.

[10] G. Hadley: "Nonlinear and Dynamic Programming", Addison-Wesley Pub. Co., Reading, Mass., 1964.

[11] D. L. Lukes: "Optimal regulation of nonlinear dynamical systems", SIAM J. Control 7 (1969), pp. 75-100.

[12] L. D. Berkovitz: "Optimal Control Theory", Applied Mathematical Sciences, Vol. 12, Springer-Verlag, New York, Heidelberg, Berlin 1974.

A very readable account of Liapounov's method, including a proof of the existence of a Liapounov function near an asymptotically stable equilibrium point, is contained in

[13] J. Hale: "Ordinary Differential Equations", Vol. XXI of "Pure and Applied Mathematics", Wiley-Interscience, New York 1969.

[14] G. Forsythe and W. Wasow: "Finite Difference Methods for Partial Differential Equations", John Wiley and Sons, Inc., New York, 1960.

[15] L. Collatz: "The Numerical Treatment of Differential Eqations", Springer-Verlag, Berlin, New York, 1966.

Exercises, Chapter IV

1. Re-prove Theorem 2.1 in the case where all quantities involved are allowed to be complex. (Take ϵ real and note that for $\hat{w}, w_1 \in L_n^2[t_0, t_1]$

$$\|\hat{w} + \epsilon w_1\|^2 = \|\hat{w}\|^2 + 2\epsilon \operatorname{Re}(\hat{w}, w_1) + \epsilon^2 \|w_1\|^2 .)$$

2. Referring to I-Exercise 2, and the parameters in part (iii) of that problem, we have the system

$$\dot{w} = \left[\frac{2(\frac{r}{w})^2}{1 + (\frac{r}{w})^2} - 1 - u^1 \right] w ,$$

$$\dot{r} = \left[\frac{2(\frac{v}{r})^2}{1 + (\frac{v}{r})^2} - 1 - u^2 \right] r - \frac{1}{4} w ,$$

$$\dot{v} = 2 - v - r .$$

The coefficients -1 in the first and second equations include normal wolf shooting and rabbit trapping levels. The controls u^1, u^2 correspond to a change $u^1 w$ in the number of wolves shot and a change $u^2 r$ in the number of rabbits allowed to be trapped. The equilibrium point of interest, corresponding to $u^1 = u^2 = 0$ is

$w_0 = .8730$ (hundred wolves)

$r_0 = .8730$ (thousand rabbits)

$v_0 = 1.1270$ ($\times 10^4$ acres vegetation)

Let $x^1 = w - w_0$, $x^2 = r - r_0$, $x^3 = v - v_0$. The linearized equations about the above equilibrium point, including the controls u^1, u^2, are

$$\begin{pmatrix} \dot{x}^1 \\ \dot{x}^2 \\ \dot{x}^3 \end{pmatrix} = \begin{pmatrix} -2 & 2 & 0 \\ -.25 & -2.1706 & 1.8752 \\ 0 & -1 & -1 \end{pmatrix} \begin{pmatrix} x^1 \\ x^2 \\ x^3 \end{pmatrix} + \begin{pmatrix} -.8730 & 0 \\ 0 & -.8730 \\ 0 & 0 \end{pmatrix} \begin{pmatrix} u^1 \\ u^2 \end{pmatrix}$$

$$\equiv Ax + Bu .$$

The responses of interest are $\frac{r}{w}, \frac{v}{r}$. The equilibrium values are $1, \sqrt{5/3}$. For

$$y^1 = \frac{r}{w} - 1, \quad y^2 = \frac{v}{r} - \sqrt{5/3}$$

we have the linearizations

$$\left.\begin{aligned} y^1 &= -1.1454x^1 + 1.1545x^2 \\ y^2 &= -1.4788x^2 + 1.1454x^3 \end{aligned}\right\} y = Fx .$$

Suppose at time $t = 0$ a census is taken, giving

$$w = .5, \quad r = 2, \quad v = .5 .$$

Set up in detail the optimization problem which one must solve in order to compute the controls $u^1(t)$, $u^2(t)$, $0 \leq t \leq 1$ ($t = 1$ corresponds, let us say, to 10 years) minimizing

$$\int_0^1 [4(u^1(t))^2 + (u^2(t))^2]\, dt + 4(y^1(1))^2 + (y^2(1))^2 .$$

Save your formulation for use in Chapter V.

3. (a) Evaluate the integral in (2.43) by the method indicated there and obtain explicit formulae for $\hat{u}(t)$, $\hat{x}(t)$.

(b) Analyze the behavior of $\hat{u}(t)$, $\hat{x}(t)$, $\hat{x}(t) - \hat{u}(t)$ under the following circumstances, using the formulae obtained in (a):

(i) $\alpha = \beta = 1$, $\delta \to \infty$,

(ii) $\alpha = \delta = \gamma = 1$, $\beta \to \infty$,

(iii) $\alpha = \delta = \beta = 1$, $\gamma \to \infty$,

(iv) $\delta = \beta = \gamma = 1$, $\alpha \to \infty$.

4. Consider the system $\dot{x} = A(t)x + B(t)u$, $t_0 \leq t \leq t_1$. Let $\xi(t)$, $t_0 \leq t \leq t_1$, be a given function in $L_n^2[t_0, t_1]$. Consider the problem

$$\min_{u \in L_m^2[t_0, t_1]} \{ \int_{t_0}^{t_1} [(x(t)-\xi(t))^* W(x(t)-\xi(t)) + u(t)^* U u(t)]\, dt$$

$$+ (x(t_1)-\xi(t_1))^* V(x(t_1) - \xi(t_1))\}, \quad x(t_0) = x_0,$$

where W and U are symmetric positive definite $n \times n$ and $m \times m$ matrices, respectively and V is a non-negative symmetric $n \times n$ matrix.

(a) Develop a set of necessary conditions, in the form of a two point boundary value problem, which must be satisfied if the control $\hat{u} \in L_m^2[t_0, t_1]$ is to solve this problem.

(b) Show that the optimal cost is given by an expression $x_0^* Q(0) x_0 + p(0)^* x_0 + r(0)$, where $Q(t)$, $p(t)$, $r(t)$ are $n \times n$ matrix, n-vector, scalar functions for $t \in [t_0, t_1]$. Obtain differential equations for $Q(t)$, $p(t)$, $r(t)$ and express the optimal control $u(t)$ in terms of these and the optimal trajectory $\hat{x}(t)$.

5. Show that if Q is a symmetric nonsingular solution of

$$A_1^* Q + QA_1 + W_1 - QB_1 Q = 0$$

then the matrix

$$M = \begin{pmatrix} A_1^* & W_1 \\ B_1 & -A_1^* \end{pmatrix}$$

is similar to the matrix $\begin{pmatrix} A_1^* - QB & 0 \\ B_1 & B_1Q - A_1 \end{pmatrix}$. If the Jordan form of $G \equiv B_1 Q - A_1$ is known, what can be inferred concerning the Jordan form of M?

6. Suppose A is an $n \times n$ matrix with eigenvalues $\lambda_1, \lambda_2, \ldots, \lambda_r$ with negative real part and eigenvalues $\lambda_{r+1}, \ldots, \lambda_n$ with positive real part. Suppose also

$$B_1 = BU^{-1}B^*, \quad (A, B) \text{ controllable.}$$

(a) Show that the quadratic matrix equation

$$A^*Q + QA - QB_1Q = 0$$

has a non-negative symmetric solution Q such that the eigenvalues of $A - BU^{-1}B^*Q$ are precisely

$$\lambda_1, \lambda_2, \ldots, \lambda_r, \ -\lambda_{r+1}, \ldots, -\lambda_n \, .$$

(b) Show that Q can be computed quite directly once the eigenvectors of A have been computed. You may assume the Jordan form of A is diagonal. (Hint: let

$$P^{-1}AP = \Lambda = \text{diag}\,(\lambda_1, \lambda_2, \ldots, \lambda_n).$$

Then study the eigenvectors of

$$\begin{pmatrix} P^* & 0 \\ 0 & P^{-1} \end{pmatrix} \begin{pmatrix} A^* & 0 \\ B_1 & -A \end{pmatrix} \begin{pmatrix} (P^{-1})^* & 0 \\ 0 & P \end{pmatrix}. \text{)}$$

(c) Use this method to find a two dimensional row vector k^* such that

$$A + bk^*$$

is a stability matrix, where

$$A = \begin{pmatrix} 0 & 1 \\ 1 & 0 \end{pmatrix}, \quad b = \begin{pmatrix} 0 \\ 1 \end{pmatrix}.$$

7.* Let A, B_1, B, Q be as in Problem 6. Show that if $\hat{x}(t)$, $\hat{u}(t)$ together satisfy

$$\dot{\hat{x}} = (A - BU^{-1}B^*Q)\hat{x}, \qquad \hat{x}(0) = x_0$$

$$\hat{u} = -U^{-1}B^*Q\hat{x}(t),$$

then

$$\int_0^\infty \|u(t)\|^2_{E^m}\, dt \le \int_0^\infty \|u(t)\|^2_{E^m}\, dt$$

whenever x, u together satisfy

$$\dot{x} = Ax + Bu, \qquad x(0) = 0$$

$$u \in L_m^2[0,\infty), \qquad \lim_{t\to\infty} \|x(t)\|_{E^n} = 0 .$$

8. Show that the equations (5. 29), (5. 30), (5. 37) yield the Riccati matrix differential equation (5. 42), the terminal condition (5. 43) and the optimal control law (5. 44) when one tries a solution of the form (5. 41). Hint: Since V in (5. 41) is a quadratic form in ξ, Q can be assumed symmetric and asymmetric terms such as $2\xi^* QA\xi$ can be written in the symmetric form $\xi^*(A^*Q + QA)\xi$.

9.* Prove Proposition 5. 5. Hint: adapt part of III - Theorem 1. 5 in the following way. Let X solve

$$(A+BK)^* X + X(A + BK) + I = 0 \quad (I = n \times n \text{ identity}),$$

and then compute $\frac{d}{dt}((y(t) - y_0)^* X(y(t) - y_0))$, where $y(t)$ is a solution of (5. 52). Show that this quantity is $\le -\mu(y(t)-y_0)^* X(y(t) - y_0)$ if $y(t)$ lies in a neighborhood of y_0 of the form

$$(y - y_0)^* X(y - y_0) < \epsilon$$

with $\epsilon > 0$ chosen sufficiently small.

10. Show that the equations

$$A^* Q + QA + W - (QB + R)U^{-1}(B^* Q + R^*) = 0,$$

$$u(t) = - U^{-1}(B^* Q + R^*)\hat{x}(t),$$

result from attempting a solution of (5. 56), (5. 57) in the linear constant coefficient case (5.38), (5.39) when we assume a solution of the form $(y \equiv x$ here) $x^* Vx$. Here $W = H^*\tilde{W}H$, $R = H^*\tilde{W}J$, $U = \tilde{U} + J^*\tilde{W}J$.

11. Show that if the power series method is applied to the example (5. 60) ff. the series for the nonlinear optimal control law is convergent for $|y| < \sqrt{2}$, divergent for $|y| > \sqrt{2}$, even though that control law is defined as an analytic function of real values of y for $-\infty < y < \infty$.

12. The region of asymptotic stability of an equilibrium point y_0 for a system $\dot{y} = g(y)$ can be "mapped out" approximately by the following numerical procedure. Select a small sphere about y_0, say $\|y - y_0\| = \epsilon$, $\epsilon > 0$ and small. Let y_k, $k = 1, 2, \ldots, K$ be a collection of "representative points" on this sphere. Solve $\dot{y} = g(y)$ "backwards" from these points using a good numerical integration routine (one of the Runge-Kutta methods ([15]) for example). That is, solve

$$\dot{y} = -g(y), \quad y(0) = y_k, \quad k = 1, 2, \ldots, K,$$

over a long time interval in each case. The resulting solutions will more or less (depending on the density and distribution of the y_k) fill the region of asymptotic stability of y_0. Use this method to compare the control laws (5.99), (5.100) in the system (5.85). In this case the region can be sketched as a two dimensional region inside a closed curve in the y^1, y^2 plane.

13. Compute the third order approximation to the optimal nonlinear control law for the nonlinear system of Exercise 2 above near the equilibrium point w_0, r_0, v_0 shown there. Use as the cost to be minimized the integral

$$\int_0^\infty [(\tfrac{r}{w} - 1)^2 + (\tfrac{v}{r} - \sqrt{5/3})^2 + (u^1)^2 + (u^2)^2]\, dt .$$

Decide upon, and carry out, a procedure for comparing this nonlinear control law with its linearized counterpart when both are used in the nonlinear system.

14. Re-develop the theory of Section 3 under weaker hypotheses. Specifically, let it be assumed that the cost functional takes the form

$$J(x_0, u) = \int_0^\infty [(Fx + Gu)^* \tilde{W} (Fx + Gu) + u^* \tilde{U} u]\, dt$$

with $\tilde{W} > 0$, $\tilde{U} > 0$ (both symmetric) and (F, A) detectable (rather than observable). Show that (3.13) continues to have a solution $Q \geq 0$ such that the control law (3.14) stabilizes the system and yields a minimum for the cost functional. Show by example that Q need not be > 0 in this case. Hint: reduce the system to the from (4.27) shown in III-4 and consider control inputs Bu which do not affect the unobservable part of the system.

CHAPTER V

COMPUTATIONAL METHODS

1. Solution of Quadratic Matrix Equations Using Potter's Method

The optimal linear feedback control law

$$u = \hat{K}x,$$

$$\hat{K} = -U^{-1}(B^*Q + R^*),$$

developed in IV-3 is widely used in a variety of control applications. Since computation of $\hat{K}$ requires that we know the unique symmetric positive definite solution Q of the quadratic matrix equation IV-(3.13), numerical techniques for solving this equation are of considerable importance.

In this section we study a method based on the theory of IV-4. When the pair (A, B) is stabilizable and (F, A) is observable and A_1, B_1, W_1 are formed as in IV-(4.1), we have seen that the $2n \times 2n$ real matrix

$$M = \begin{pmatrix} A_1^* & W_1 \\ B_1 & -A_1 \end{pmatrix}$$

has exactly n eigenvalues $\lambda_1, \lambda_2, \ldots, \lambda_n$, with positive real parts. The unique positive definite solution Q of IV-(3.13) is then

$$Q = XY^{-1},$$

where the columns of

$$Z = \begin{pmatrix} X \\ Y \end{pmatrix}$$

are eigenvectors of M corresponding to these eigenvalues. Generally speaking, however, it is not efficient to calculate these eigenvectors. We have discussed in Chapter III a method for triangularization of a matrix. Employing that method for n steps (rather than $2n-1$, as would be required to completely triangularize M) results in a matrix P such that

$$P^{-1}MP = \begin{pmatrix} M_1 & M_3 \\ 0 & M_2 \end{pmatrix} ,$$

where we may arrange that M_1 is upper triangular with the above eigenvalues $\lambda_1, \lambda_2, \ldots, \lambda_n$ appearing on its diagonal. We then have

$$MP = P \begin{pmatrix} M_1 & M_3 \\ 0 & M_2 \end{pmatrix}$$

and, letting $\hat{Z}$ be the $2n \times n$ matrix consisting of the first n columns of P,

$$M\hat{Z} = \hat{Z}M_1 .$$

Let J be the Jordan form of M_1. Then there exists a nonsingular $n \times n$ matrix C such that $C^{-1}M_1C = J$ and

$$M\hat{Z}C = \hat{Z}C(C^{-1}M_1C) = \hat{Z}CJ ,$$

and setting $Z = \hat{Z}C$ we have IV-(4.11). From the remarks at the end of Section IV we know, therefore, that the desired solution Q of IV-(3.13) has the form

$$Q = XY^{-1} = (\hat{X}C)(\hat{Y}C)^{-1} = \hat{X}\hat{Y}^{-1},$$

where

$$\hat{Z} = \begin{pmatrix} \hat{X} \\ \hat{Y} \end{pmatrix} .$$

We may compute Q, therefore, by partial triangularization of M, as outlined in III-1; it is not necessary to compute the Jordan form J.

When some of $\lambda_1, \lambda_2, \ldots, \lambda_n$ are complex, as will frequently be the case, the triangularization procedure may be modified slightly to avoid complex arithmetic, if desired. Suppose

$$\lambda_k = \mu_k + i\nu_k , \tag{1.1}$$

μ_k and ν_k being the real and imaginary parts of λ_k, respectively. Then M_1 is similar to a real matrix (see II-[1], e.g.)

$$\tilde{M}_1 = \begin{pmatrix} \mu_1 & -\nu_1 & * & * & \cdots \\ \nu_1 & \mu_1 & * & * & \cdots \\ 0 & 0 & \mu_2 & -\nu_2 & \cdots \\ 0 & 0 & \nu_2 & \mu_2 & \cdots \\ \vdots & \vdots & \vdots & \vdots & \end{pmatrix} .$$

The matrix M_1 is upper block diagonal with diagonal blocks of dimension at most 2, as shown. Any eigenvalue λ_k which is real appears as a 1×1 block, of course. The columns of the matrix $\tilde{Z}$ such that

$$M\tilde{Z} = \tilde{Z}\tilde{M}_1$$

are linear combinations of the columns of M_1 and we still have

$$Q = \tilde{X}\tilde{Y}^{-1}, \qquad \tilde{Z} = \begin{pmatrix} \tilde{X} \\ \tilde{Y} \end{pmatrix} .$$

The columns of $\tilde{Z}$ can be calculated without recourse to complex arithmetic. When λ is a real eigenvalue of M, this is the usual procedure for finding a real eigenvector (see [1] e.g.). When $\lambda = \mu + i\nu$, $\bar{\lambda} = \mu - i\nu$ are complex conjugate eigenvalues of M, M has complex conjugate eigenvectors

$$x = \xi + i\eta ,$$

$$\bar{x} = \mu - i\eta ,$$

and the equations

$$Mx = \lambda x, \qquad M\bar{x} = \overline{\lambda x}$$

are equivalent to the real equations

$$M\xi = \mu\xi - \nu\eta, \quad \text{or} \quad (M - \mu I)\xi = -\nu\eta \tag{1.2}$$

$$M\eta = \nu\xi + \mu\eta, \quad \text{or} \quad (M-\mu I)\eta = \nu\xi \ . \tag{1.3}$$

Substituting either of (1.2), (1.3) into the other we find that ξ, η both satisfy the real system of n equations in n unknowns

$$(M^2 - 2\mu M + (\mu^2 + \nu^2)I)y = 0 \ . \tag{1.4}$$

Assuming $\lambda, \bar{\lambda}$ to be simple eigenvalues of M, the matrix in (1.4) has rank $n-2$. The vectors ξ, η may be taken to be any two linearly independent solutions of (1.4) which may be found by the usual method of Gaussian elimination, for example.

With this theoretical background, we proceed to an example. Let the control system in question be

$$\begin{pmatrix} \dot{x}^1 \\ x^2 \end{pmatrix} = \begin{pmatrix} 0 & 1 \\ 0 & 0 \end{pmatrix} \begin{pmatrix} x^1 \\ x^2 \end{pmatrix} + \begin{pmatrix} 0 \\ 1 \end{pmatrix} u$$

and the quadratic cost

$$\int_0^\infty [\,(x^1(t))^2 + (x^2(t))^2 + (u(t))^2]\, dt \ ,$$

so that

$$W_1 = \begin{pmatrix} 1 & 0 \\ 0 & 1 \end{pmatrix}, \qquad U = 1, \qquad B_1 = BU^{-1}B^* = \begin{pmatrix} 0 & 0 \\ 0 & 1 \end{pmatrix}.$$

Here $R = 0$, so $A_1 = A = \begin{pmatrix} 0 & 1 \\ 0 & 0 \end{pmatrix}$, and

$$M = \begin{pmatrix} 0 & 0 & 1 & 0 \\ 1 & 0 & 0 & 1 \\ 0 & 0 & 0 & -1 \\ 0 & 1 & 0 & 0 \end{pmatrix} .$$

The characteristic polynomial of M is

$$\det \begin{pmatrix} \lambda & 0 & -1 & 0 \\ -1 & \lambda & 0 & -1 \\ 0 & 0 & \lambda & 1 \\ 0 & -1 & 0 & \lambda \end{pmatrix} = \lambda^4 - \lambda^2 + 1 .$$

The quadratic formula immediately gives

$$\lambda^2 = \frac{1}{2} \pm i\frac{\sqrt{3}}{2} = e^{\pm i\frac{\pi}{3}}$$

so that the possible values of λ with positive real part are

$$\lambda = e^{\pm i\frac{\pi}{6}} = \frac{\sqrt{3}}{2} \pm i\frac{1}{2} .$$

This gives

$$\mu = \frac{\sqrt{3}}{2}, \quad \nu = \frac{1}{2} .$$

Then

$$M^2 - 2\mu M + \mu^2 + \nu^2 = M^2 - \sqrt{3}M + I$$

$$= \begin{pmatrix} 0 & 0 & 0 & -1 \\ 0 & 1 & 1 & 0 \\ 0 & -1 & 0 & 0 \\ 1 & 0 & 0 & 1 \end{pmatrix} - \sqrt{3}\begin{pmatrix} 0 & 0 & 1 & 0 \\ 1 & 0 & 0 & 1 \\ 0 & 0 & 0 & -1 \\ 0 & 1 & 0 & 0 \end{pmatrix} + \begin{pmatrix} 1 & 0 & 0 & 0 \\ 0 & 1 & 0 & 0 \\ 0 & 0 & 1 & 0 \\ 0 & 0 & 0 & 1 \end{pmatrix} = \begin{pmatrix} 1 & 0 & -\sqrt{3} & -1 \\ -\sqrt{3} & 2 & 1 & -\sqrt{3} \\ 0 & -1 & 1 & \sqrt{3} \\ 1 & -\sqrt{3} & 0 & 2 \end{pmatrix} .$$

Multiplying the equation $(M^2 - \sqrt{3}M + I)y = 0$ on the right by

$$\begin{pmatrix} 1 & 0 & 0 & 0 \\ \sqrt{3} & 1 & 0 & 0 \\ 0 & 0 & 1 & 0 \\ -1 & 0 & 0 & 1 \end{pmatrix}$$

(this is equivalent to Gaussian elimination) we obtain

$$\begin{pmatrix} 1 & 0 & -\sqrt{3} & -1 & y^1 \\ 0 & 2 & -2 & -2\sqrt{3} & y^2 \\ 0 & -1 & 1 & \sqrt{3} & y^3 \\ 0 & -\sqrt{3} & \sqrt{3} & 3 & y^4 \end{pmatrix} = 0 \quad .$$

The last two equations are the same as the second. We are left with

$$\begin{aligned} y^1 \quad -\sqrt{3}y^3 - \quad y^4 &= 0 \\ 2y^2 - 2y^3 - 2\sqrt{3}y^4 &= 0 \quad . \end{aligned}$$

Letting $y^3 = 1$, $y^4 = 1$ gives

$$y^1 = \sqrt{3} + 1, \quad y^2 = \sqrt{3} + 1.$$

Letting $y^3 = 1$, $y^4 = -1$, gives

$$y^1 = \sqrt{3} - 1, \quad y^2 = 1 - \sqrt{3} \; .$$

Consequently we let

$$\tilde{Z} = (\xi, \eta) = \begin{pmatrix} \sqrt{3}+1 & \sqrt{3}-1 \\ \sqrt{3}+1 & 1-\sqrt{3} \\ 1 & 1 \\ 1 & -1 \end{pmatrix} = \begin{pmatrix} \tilde{X} \\ \tilde{Y} \end{pmatrix}$$

and

$$Q = \tilde{X}\tilde{Y}^{-1} = \begin{pmatrix} \sqrt{3}+1 & \sqrt{3}-1 \\ \sqrt{3}+1 & 1-\sqrt{3} \end{pmatrix} \begin{pmatrix} \frac{1}{2} & \frac{1}{2} \\ \frac{1}{2} & -\frac{1}{2} \end{pmatrix} = \begin{pmatrix} \sqrt{3} & 1 \\ 1 & \sqrt{3} \end{pmatrix}.$$

This is clearly positive definite and symmetric. To verify we note that IV-(3.13) in this case is (with $Q = \begin{pmatrix} q_1 & q_2 \\ q_2 & q_3 \end{pmatrix}$)

$$\begin{pmatrix} 0 & 0 \\ 1 & 0 \end{pmatrix} \begin{pmatrix} q_1 & q_2 \\ q_2 & q_3 \end{pmatrix} + \begin{pmatrix} q_1 & q_2 \\ q_2 & q_3 \end{pmatrix} \begin{pmatrix} 0 & 1 \\ 0 & 0 \end{pmatrix} + \begin{pmatrix} 1 & 0 \\ 0 & 1 \end{pmatrix} - \begin{pmatrix} q_2^2 & q_2 q_3 \\ q_2 q_3 & q_3^2 \end{pmatrix} = 0$$

or

$$1 - q_2^2 = 0$$

$$q_1 - q_2 q_3 = 0$$

$$2q_2 + 1 - q_3^2 = 0$$

which does, indeed, have the solution

$$q_2 = 1, \quad q_3 = \sqrt{3}, \quad q_1 = \sqrt{3} .$$

Along with other methods this technique runs into some difficulty when there are eigenvalues of M with small real part. For then the eigenvalues $\lambda = \mu + i\nu$ and $-\mu + i\nu$ are nearly equal and the problem of determining them and the associated partial triangularization of M becomes ill-conditioned. In such cases the calculations must be carried out with a high degree of accuracy - often entailing double precision work when an electronic computer is used. Since this can frequently be a nuisance, other methods, based on iteration, are often favored in practice, particularly if they are of the "self-correcting" variety. A further disadvantage associated with the method presented here is the introduction of a matrix of dimension $2n \times 2n$ rather than $n \times n$. The number of computational operations required increases proportionately with the cube of the matrix dimension, so treatment of a $2n \times 2n$ matrix can be an order of magnitude more time consuming than comparable treatment of an $n \times n$ matrix.

2. The Kleinman-Newton Method

The most widely used method for finding the unique positive definite symmetric solution Q of IV-(3.13) involves a modification, due to Kleinman, of the familiar Newton method for solving algebraic equations III-[1]. It appears in [4] and is detailed in this section. In the course of treating the method we obtain several identities and inequalities which are of frequent use and will, in fact, be employed in Chapter VI. We begin with a proposition which just restates a result already discussed in IV-1.

Proposition 2.1. *Let* (A, B) *be stabilizable and let* K *be an* $m \times n$ *feedback matrix such that* $A + BK$ *is a stability matrix. For given* $x_0 \in E^n$, *let* $x_K(t)$ *be the solution of*

$$\dot{x}_K(t) = (A + BK)x_K(t), \quad x_K(0) = x_0,$$

obtained by the use of the feedback law

$$u_K(t) = Kx_K(t)$$

in the control system

$$\dot{x} = Ax + Bu. \tag{2.1}$$

Then (cf. IV-(3.19))

$$J(x_0, u_K) = x_0^* Q_K x_0 \tag{2.2}$$

where Q_K *is the unique solution of the Liapounov equation*

$$(A + BK)^* Q_K + Q_K(A + BK) + (F + GK)^* \tilde{W}(F + GK) + K^* \tilde{U} K = 0. \tag{2.3}$$

Proof. The cost, from IV-(3.19), is

$$J(x_0, u_K) = \int_0^\infty [(Fx_K(t) + Gu_K(t))^* \tilde{W}(Fx_K(t) + Gu_K(t)) + u_K(t)^* \tilde{U} u_K(t)]\, dt.$$

Since

$$x_K(t) = e^{(A+BK)t}x_0$$

and

$$U_K(t) = Kx_K(t) = Ke^{(A+BK)t}x_0\,,$$

this becomes

$$J(x_0,u_K) = x_0^*\int_0^\infty e^{(A+BK)^*t}[(F+GK)^*\tilde{W}(F+GK)+K^*\tilde{U}K]\,e^{(A+BK)t}dtx_0 \equiv x_0^*Q_Kx_0\,. \tag{2.4}$$

That Q_K is the unique solution of (2. 3) follows from the arguments already presented in Theorem 1. 5 of Chapter III. Q. E. D.

Before proving the main theorem of this section we present the following lemma which will provide a tool repeatedly used in the sequel.

Proposition 2. 2. Let $\tilde{Q},\hat{Q}$ be solutions of (2. 3) corresponding to $\tilde{K},\hat{K}$ respectively. Then

$$(A+B\hat{K})^*(\tilde{Q}-\hat{Q})+(\tilde{Q}-\hat{Q})(A+B\hat{K})+(U\tilde{K}+B^*\tilde{Q}+R^*))^*U^{-1}(U\tilde{K}+(B^*\tilde{Q}+R^*))$$
$$-(U\hat{K}+(B^*\tilde{Q}+R^*))^*U^{-1}(U\hat{K}+(B^*\tilde{Q}+R^*)) = 0\,. \tag{2.5}$$

Proof. We let $Q = \tilde{Q}$, $K = \tilde{K}$ in (2. 3) and then rearrange to obtain

$$(A+B\hat{K})^*\tilde{Q}+\tilde{Q}(A+B\hat{K})+(F+G\tilde{K})^*\tilde{W}(F+G\tilde{K})+\tilde{K}^*U\tilde{K}$$
$$+\tilde{K}^*B^*\tilde{Q}+\tilde{Q}B\tilde{K}-\hat{K}^*B^*\tilde{Q}-\tilde{Q}B\hat{K} = 0\,. \tag{2.6}$$

Now taking $Q = \hat{Q}$, $K = \hat{K}$ in (2. 3) and subtracting the resulting equation from (2. 6) we obtain

$$(A+BK)^*(\tilde{Q}-\hat{Q})+(\tilde{Q}-\hat{Q})(A+B\hat{K})+\tilde{K}^*(B^*\tilde{Q}+R^*)+(\tilde{Q}B+R)\tilde{K}$$
$$+\tilde{K}^*U\tilde{K}-\hat{K}^*(B^*\tilde{Q}+R^*)-(\tilde{Q}B+R)\hat{K}-\hat{K}^*U\hat{K} = 0\,. \tag{2.7}$$

In obtaining (2. 7) we have used the identities

$$W = F^*\tilde{W}F,\quad R = F^*\tilde{W}G,\quad U = \tilde{U}+G^*\tilde{W}G\,.$$

Now adding $(\tilde{Q}B + R)U^{-1}(B^*\tilde{Q} + R^*)$ to the third, fourth and fifth terms in (2.7) and subtracting the same quantity from the sixth, seventh and eighth terms, (2.5) follows immediately. Q. E. D.

Theorem 2.3. Let (A, B) be stabilizable and (F, A) observable. Let K_0 be an $m \times n$ matrix such that $A + BK_0$ is a stability matrix and let sequences $\{Q_i\}$, $\{K_i\}$ of $n \times n$, $n \times m$ matrices, respectively, be generated by letting Q_i be the unique solution of the Liapounov equation

$$(A+BK_i)^* Q_i + Q_i(A+BK_i) + (F+GK_i)^* \tilde{W}(F+GK_i) + K_i^* UK_i = 0, \quad i = 0, 1, 2, 3, \ldots, \tag{2.8}$$

and setting

$$K_{i+1} = -U^{-1}[B^* Q_i + R^*], \quad i = 0, 1, 2, 3, \ldots . \tag{2.9}$$

Then

(i) Each matrix $A + BK_i$ is a stability matrix;

(ii) Each matrix Q_i is symmetric and positive definite with

$$Q_{i+1} \le Q_i, \quad i = 0, 1, 2, 3, \ldots ; \tag{2.10}$$

(iii) we have

$$\lim_{i \to \infty} Q_i = Q \tag{2.11}$$

where Q is the unique positive definite solution of the quadratic matrix equation IV-(3.13).

(iv) The convergence (2.11) takes place quadratically, i.e., there is an $M > 0$ such that

$$\|Q - Q_{i+1}\| \le M \|Q - Q_i\|^2, \quad i = 0, 1, 2, 3, \ldots . \tag{2.12}$$

Proof. We will proceed by induction, assuming K_i to be such that $A + BK_i$ is a stability matrix and Q_i, the unique solution of (2.8), to be positive definite. Since for $i = 0$ we have assumed $A + BK_0$ stable and since Q_0 satisfies (2.8) and (H, A) is observable, the arguments employed earlier in III-Theorem 1.5 show that Q_0 is positive definite. Thus the induction assumption holds for $i = 0$.

Assuming for a given value of i that Q_i, K_i satisfy the above hypothesis, we define K_{i+1} by (2.9). Rearranging (2.8) much as in Proposition 2.2 we have

$$(A+BK_{i+1})^* Q_i + Q_i(A+BK_{i+1}) + (F+GK_i)^* \tilde{W}(F+GK_i) + K_i^* \tilde{U} K_i$$
$$+ K_i^* B^* Q_i + Q_i BK_i - K_{i+1}^* B^* Q_i - Q_i BK_{i+1} = 0.$$

Further rearrangement together with (2.9) and the identities $W = F^* \tilde{W} F$, $R = F^* \tilde{W} G$, $U = \tilde{U} + G^* \tilde{W} G$ gives

$$(A+BK_{i+1})^* Q_i + Q_i(A+BK_{i+1})^* + (UK_i + B^* Q_i + R^*)U^{-1}(UK_i + B^* Q_i + R^*)$$
$$+ (I, -(Q_i B+R)U^{-1}) \begin{pmatrix} W & R \\ R & U \end{pmatrix} \begin{pmatrix} I \\ -U^{-1}(B^* Q_i + R^*) \end{pmatrix} = 0.$$

Using $(UK_i+B^*Q_i+R^*)U^{-1}(UK_i+B^*Q_i+R^*) \geq 0$, $Q_i < 0$ and the argument of III-Theorem 1.5 we conclude that $A + BK_{i+1}$ is a stability matrix.

To show that $Q_i \geq Q_{i+1}$ we note that

$$(A+BK_i)^* Q_i + Q_i(A+BK_i) + (F+GK_i)^* \tilde{W}(F+GK_i) + K_i^* \tilde{U} K_i = 0$$

and

$$(A+BK_{i+1})^* Q_{i+1} + Q_{i+1}(A+BK_{i+1}) + (F+GK_{i+1})^* \tilde{W}(F+GK_{i+1}) + K_{i+1}^* \tilde{U} K_{i+1} = 0. \qquad (2.13)$$

These agree with the equations satisfied by $\tilde{Q}, \tilde{K}, \hat{Q}, \hat{K}$ in Proposition 2.2 if we let $\tilde{Q} = Q_i$, $\tilde{K} = K_i$, $\hat{Q} = Q_{i+1}$, $\hat{K} = K_{i+1}$. Then (2.5) becomes

$$(A+BK_{i+1})^*(Q_i - Q_{i+1}) + (Q_i - Q_{i+1})(A+BK_{i+1})$$
$$+ (UK_i + (B^* Q_i + R^*))^* U^{-1}(UK_i + (B^* Q_i + R^*))$$
$$- (UK_{i+1} + (B^* Q_i + R^*))^* U^{-1}(UK_{i+1} + (B^* Q_i + R^*)) = 0. \qquad (2.14)$$

Using (2.9) the last line of (2.14) vanishes and, since the second line is non-negative and $A + BK_{i+1}$ is a stability matrix, $Q_i - Q_{i+1}$ is non-negative and (2.10) follows. The positive definiteness of Q_{i+1} follows from (2.13).

Since the Q_i are symmetric, have the monotonicity property (2.10) and are bounded below by the unique positive definite solution Q of IV-(3.13) (each Q_i, by Proposition 2.1, yields the cost associated with the feedback matrix K_i and this cost must be greater than or equal to the optimal cost) we conclude, adapting IV-Lemma 3.4, that

$$\lim_{i \to \infty} Q_i = \hat{Q} , \tag{2.15}$$

where $\hat{Q}$ is a symmetric positive definite $n \times n$ matrix. Since K_{i+1} and Q_i are related by (2.9) we have

$$\lim_{i \to \infty} K_i = \hat{K} . \tag{2.16}$$

Using (2.15), (2.16) in (2.9) and (2.13) we conclude that

$$\hat{K} = - U^{-1}[B^*\hat{Q} + R^*] \tag{2.17}$$

and

$$(A + B\hat{K})^*\hat{Q} + \hat{Q}(A + B\hat{K}) + F^*\tilde{W}F + QBU^{-1}B^*Q - RU^{-1}R^* = 0 . \tag{2.18}$$

Substituting (2.17) into (2.18) and using the definitions of W, R, U (IV-(1.13) ff.) we have

$$A^*\hat{Q} + \hat{Q}A + W - (\hat{Q}B + R)U^{-1}(B^*\hat{Q} + R^*) = 0 \tag{2.19}$$

and since (see remarks following statement of IV-Theorem 4.4), Q is the unique symmetric positive definite solution of this equation, we finally have

$$\lim_{i \to \infty} Q_i = Q$$

as calimed. Further $\hat{K} = K$ given by (2.17) is the optimal feedback matrix.

Finally we establish that the convergence is quadratic. We use the equation (2.13) satisfied by Q_{i+1} and the following equation, obtained by letting $i \to \infty$ in (2.13):

$$(A + BK)^*Q + Q(A + BK) + (F + GK)^*\tilde{W}(F + GK) + K^*\tilde{U}K = 0 . \tag{2.20}$$

Letting $\hat{K} = K_{i+1}$, $\hat{Q} = Q_{i+1}$, $\tilde{K} = K$, $\tilde{Q} = Q$ in Proposition (2.2) we arrive at

$$(A+BK_{i+1})^*(Q-Q_{i+1})+(Q-Q_{i+1})(A+BK_{i+1})$$

$$+ (UK + (B^*Q+R^*))^*U^{-1}(UK + (B^*Q + R^*))$$

$$- (UK_{i+1} + (B^*Q + R^*))^*U^{-1}(UK_{i+1} + (B^*Q + R^*)) = 0. \tag{2.21}$$

The second line in (3. 44) is zero because $k = -U^{-1}(B^*Q + R^*)$. Then using (2. 9) in the third line we have, finally

$$(A+BK_{i+1})^*(Q-Q_{i+1})+(Q-Q_{i+1})(A+BK_{i+1}) = (Q-Q_i)BU^{-1}B^*(Q-Q_i) . \tag{2.22}$$

Since $\lim_{i\to\infty} K_{i+1} = K$ and $A + BK$ is stable, one can see (using formulae of the form (2. 4)) that there is an $M_0 > 0$ s. t. $\|X_i\| \leq M_0 \|Y_i\|$, uniformly in i, if

$$(A+BK_{i+1})^*X_i + X_i(A+BK_{i+1}) = Y_i .$$

Applying this with $Y_i = (Q-Q_i)BU^{-1}B^*(Q-Q_i)$ and

$$\|(Q-Q_i)BU^{-1}B^*(Q-Q_i)\| \leq M_1 \|Q-Q_i\|^2$$

we finally have (2. 12) with $M = M_0M_1$. Q. E. D.

It is evident from (2. 8) that the single most important tool for application of the method outlined here must be an efficient routine for solving the successive Liapounov equations which arise. In Section 3 we present theory and methods relevant to this requirement. Of course we already have one such method in III-Corollary 1. 4. The requirement that $A + BK_0$ should be a stability matrix lends new significance to the stabilization methods discussed in III-2 as well. They may now be regarded as providing "starting values" for the method outlined in this section.

To present a simple example we again take the system IV-(1. 16) with the cost functional IV-(1. 19), $\alpha = \delta = 1$, $\beta = 0$. Then

$$A = \begin{pmatrix} 0 & 1 \\ -1 & 0 \end{pmatrix}, \quad B = \begin{pmatrix} 0 \\ 1 \end{pmatrix}, \quad \tilde{U} = 1,$$

$$(F+GK_i)^* \tilde{W}(F+GK_i) = F^*\tilde{W}F = \begin{pmatrix} 1 & 0 \\ 0 & 1 \end{pmatrix}, \quad R = 0$$

and the equations (2. 8), (2. 9) become, with

$$K_i = (k_{1i}, k_{2i}),$$

$$\begin{pmatrix} 0 & -1+k_{1i} \\ 1 & k_{2i} \end{pmatrix}\begin{pmatrix} q_{1i} & q_{2i} \\ q_{2i} & q_{3i} \end{pmatrix} + \begin{pmatrix} q_{1i} & q_{2i} \\ q_{2i} & q_{3i} \end{pmatrix}\begin{pmatrix} 0 & 1 \\ -1+k_{1i} & k_{2i} \end{pmatrix} + \begin{pmatrix} 1 & 0 \\ 0 & 1 \end{pmatrix}$$

$$+ \begin{pmatrix} k_{1i}^2 & k_{1i}k_{2i} \\ k_{1i}k_{2i} & k_{2i}^2 \end{pmatrix} = 0 , \tag{2. 23}$$

$$k_{1i} = -q_{2i}, \quad k_{2i} = -q_{3i} . \tag{2. 24}$$

In this very simple case the successive Liapounov equations just amount to three linear equations in three unknowns q_{1i}, q_{2i}, q_{3i}, and are readily solved. An initial stabilizing feedback matrix is

$$(k_{10}, k_{20}) = (0, -1).$$

The results of the iteration method of Theorem 2. 3 applied to this case are summarized in the following table.

i	k_{1i}	k_{2i}	q_{1i}	q_{2i}	q_{3i}
0	0	-1	2. 000	. 5000	1. 5000
1	-. 5000	-1. 5000	1. 9166	. 4166	1. 3611
2	-. 4166	-1. 3611	1. 9123	. 4142	1. 3522
3	-. 4142	-1. 3522	1. 9122	. 4142	1. 3521
4	-. 4142	-1. 3521	1. 9122	. 4142	1. 3521

(2. 25)

Here the method is seen to converge in just four steps. The final values of k_1, k_2 yield the feedback law

$$\hat{u}(t) = -.4142\hat{x}^1(t) - 1.3521\hat{x}^2(t)$$

which, to four decimal places, agrees with our earlier computation, IV-(3.42).

3. Iterative Solution of Linear Matrix Equations

Let A, B, C be given $n \times n$, $m \times m$ and $n \times m$ matrices and consider the problem of determining an $n \times m$ matrix X such that

$$AX + XB = C . \tag{3.1}$$

The Liapounov equation (2.8), or, more generally, the equation III-(1.23), is obtained as a special case with $n = m$, A, B, C being replaced by A^*, A, $-W$, respectively. We already have one method for solving such equations, outlined in III-Corollary 1.4 and III-Exercise 4. It is one of a class of what are called "transformational" methods, which rely on the determination of nonsingular matrices P, Q such that

$$P^{-1}AP = \tilde{A}, \quad Q^{-1}BQ = \tilde{B}$$

are in such a form that the equation

$$\tilde{A}\tilde{X} + \tilde{X}\tilde{B} = \tilde{C}, \quad \tilde{X} = P^{-1}\tilde{X}Q, \quad \tilde{C} = P^{-1}CQ ,$$

can be solved more or less directly. These methods have a number of desirable features, one of which we will point out in Chapter VII. Perhaps the most sophisticated method of this type is due to Bartels and Stewart [5]. It involves more in the way of numerical linear algebra than we wish to discuss here but the point of that method is to alleviate the ill-conditioning which arises when A and B have eigenvalues λ, μ, respectively, which nearly sum to zero, by a very sophisticated choice of P and Q.

The method which we wish to discuss here is due to Smith [6] and is iterative in character. It is quite widely applicable; in particular it will work whenever A and B have eigenvalues restricted to some common half plane $\mathrm{Re}(\alpha z) < 0$, where α is a non-zero complex number. However, multiplying (µ.1) by α, we may as well assume the eigenvalues λ_k, μ_j of

A, B, respectively, satisfy

$$\mathrm{Re}(\lambda_k) < 0, \quad k = 1, 2, \ldots, n, \quad \mathrm{Re}(\mu_j) < 0, \quad j = 1, 2, \ldots, m, \tag{3.2}$$

and this is the assumption under which we will proceed here.

Let r be a negative number and observe that (3.1) is the same as

$$(rI_n - A)X(rI_m - B) - (rI_n + A)X(rI_m + B) = -2rC . \tag{3.3}$$

Since $r < 0$ and we have (3.2) for the eigenvalues of A and B, $rI_n + A$ and $rI_m + B$ are both invertible. Multiplying (3.3) on the right and left by $(rI_n + A)^{-1}$, $(rI_m + B)^{-1}$, respectively, we have

$$X = (rI_n + A)^{-1}(rI_n - A)X(rI_m - B)(rI_m + B)^{-1} + 2r(rI_n + A)^{-1}C(rI_m + B)^{-1} . \tag{3.4}$$

Letting

$$U = (rI_n + A)^{-1}(rI_n - A), \tag{3.5}$$

$$V = (rI_m - B)(rI_m + B)^{-1}, \tag{3.6}$$

$$Y = 2r(rI_n + A)^{-1}C(rI_m + B)^{-1}, \tag{3.7}$$

(3.4) takes the form

$$X = UXV + Y . \tag{3.8}$$

Lemma 3.1. The matrices U and V have the property

$$\lim_{k \to \infty} \|U^k\| = 0, \tag{3.9}$$

$$\lim_{k \to \infty} \|V^k\| = 0 . \tag{3.10}$$

Proof. Indeed, the result is true for any matrix M whose eigenvalues have moduli strictly less than one. Let us see that this is true for U, V and then proceed with the demonstration. It is clearly enough to consider U. This matrix has the form

$$U = f(A)$$

for the analytic function

$$f(\lambda) = \frac{r - \lambda}{r + \lambda} .$$

From III-Exercise 12 and accompanying remarks we conclude that the eigenvalues γ_k of U have the form

$$\gamma_k = \frac{r - \lambda_k}{r + \lambda_k} \qquad (3.11)$$

where λ_k is an eigenvalue of A. Writing

$$\lambda_k = \alpha_k + \beta_k i$$

we have

$$|\gamma_k|^2 = \frac{(r-\alpha_k+\beta_k i))(r-\alpha_k-\beta_k i))}{(r+(\alpha_k+\beta_k i))(r+(\alpha_k-\beta_k i))} = \frac{(r-\alpha_k)^2 + \beta_k^2}{(r+\alpha_k)^2 + \beta_k^2} < 1 \qquad (3.12)$$

because r and $\alpha_k = \mathrm{Re}(\lambda_k)$ are both negative. Since this is true for every k, there is a real number γ, $0 \leq \gamma < 1$, such that

$$|\gamma_k| \leq \gamma, \quad k = 1, 2, \ldots, n.$$

If, as in III-Exercise 12 and Section III-1,

$$P^{-1}AP = \Lambda$$

is a block-diagonalized matrix similar to A, then (III-Exercise 12 again)

$$\Omega \equiv P^{-1}UP = \begin{pmatrix} \gamma_1 I_1 + M_1 & 0 & \cdots & 0 \\ 0 & \gamma_2 I_2 + M_2 & \cdots & 0 \\ \vdots & \vdots & & \vdots \\ 0 & 0 & \cdots & \gamma_\nu I_\nu + M_\nu \end{pmatrix} \qquad (3.13)$$

where $\gamma_1, \gamma_2, \ldots, \gamma_\nu$ are the distinct eigenvalues of U and $M_1, M_2, \ldots, M_\nu$ are strictly superdiagonal, nilpotent matrices. Since

$$U^k = P\Omega^k P^{-1} \tag{3.14}$$

it is enough to look at the matrices

$$(\gamma_j I_j + M_j)^k = \sum_{\ell=0}^{p_j-1} \binom{k}{\ell} (\gamma_j)^{k-\ell} M_j^\ell .$$

Since

$$\left| \binom{k}{\ell} (\gamma_j)^{k-\ell} \right| < \frac{k^\ell}{\ell!} \gamma^{k-\ell} \tag{3.15}$$

we conclude readily that

$$\lim_{k \to \infty} \| (\gamma_j I_j + M_j)^k \| = 0$$

and thus

$$\lim_{k \to \infty} \| \Omega^k \| = 0 .$$

Then (3.9) follows from (3.14). Q. E. D.

From (3.14) and (3.15) one may readily conclude that for any ρ with

$$0 \leq \gamma < \rho < 1$$

there is a positive number $M(\rho)$ such that

$$\| U^k \| \leq M(\rho)\rho^k , \quad k = 0, 1, 2, \ldots . \tag{3.16}$$

A comparable inequality applies to V:

$$\| V^k \| \leq N(\sigma)\sigma^k , \quad 0 \leq \left| \frac{r - \mu_j}{r + \mu_j} \right| \leq \nu < \sigma < 1, \quad j = 1,2,\ldots, m . \tag{3.17}$$

Now we can prove the theorem upon which the proposed iterative method depends.

__Theorem 3.2.__ __Let__ X_0 __be an arbitrary__ $n \times m$ __matrix and let a sequence__ $\{X_k\}$ __of__ $n \times m$ __matrices be generated by__

$$X_{k+1} = U X_k V + Y, \quad k = 0, 1, 2, 3, \ldots . \tag{3.18}$$

Then

$$\lim_{k \to \infty} X_k = X \, , \tag{3.19}$$

where X *is the unique solution of* (3.8) *(and hence of* (3.1)). *Moreover, the convergence is linear: there are real numbers* a, b, a < 0, *such that*

$$\log \|X_k - X\| \le ak + b \, , \quad k = 0,1,2,\ldots \, . \tag{3.20}$$

Proof. The proof is essentially the contraction fixed point theorem (III-[1], or [2] below). Nevertheless, we give details. We have, for k = 1,2,3,...

$$X_{k+1} - X_k = U(X_k - X_{k-1})V \, .$$

Applying this for k = 1, 2, . . . , K we have

$$X_{K+1} - X_K = U^K(X_1 - X_0)V^K \, .$$

Then using (3.16) and (3.17) we have

$$\|X_{K+1} - X_K\| \le \|U^K\| \, \|X_1 - X_0\| \, \|V^k\| \le M(\rho)N(\sigma)(\rho\sigma)^k \|X_1 - X_0\|. \tag{3.21}$$

Then if k, ℓ are non-negative integers, $k \ge \ell$, we have

$$\begin{aligned} \|X_k - X_\ell\| &\le \sum_{j=\ell}^{k-1} \|X_{j+1} - X_j\| \\ &\le M(\rho)N(\sigma)\Big(\sum_{j=\ell}^{k-1} (\rho\sigma)^j\Big) \|X_1 - X_0\| \\ &\le M(\rho)N(\sigma)\Big(\sum_{j=\ell}^{\infty} (\rho\sigma)^j\Big) \|X_1 - X_0\| = \frac{M(\rho)N(\sigma)(\rho\sigma)^\ell}{1-\rho\sigma} \|X_1 - X_0\| \end{aligned} \tag{3.22}$$

and, since $|\rho\sigma| < 1$, we conclude that $\{X_k\}$ is a Cauchy sequence in the space E^{nm} of $n \times m$ matrices. There is then an $n \times m$ matrix X such that (3.19) is true. Using (3.19) in (3.18) we see that X is a solution of (3.8), and hence of (3.1). That (3.8) has only one solution follows from

$$\|X - \tilde{X}\| \le M(\rho)N(\sigma)(\rho\sigma)^k \|X - \tilde{X}\| \, , \tag{3.23}$$

which is valid for any solutions $X, \tilde{X}$ of (3. 8) and any positive integer k - following essentially the same argument as led to (3. 22), but using (3. 8) instead of (3. 18). For k sufficiently large $M(\rho)N(\sigma)(\rho\sigma)^k < 1$ and (3. 23) implies $\|X - \tilde{X}\| = 0$, i. e., $X = \tilde{X}$.

Finally, using (3. 8) and (3. 9) we have, for $k = 0, 1, 2, \ldots$

$$X_{k+1} - X = U(X_k - X)V$$

whence

$$\|X_k - X\| \leq M(\rho)N(\sigma)(\rho\sigma)^k \|X_0 - X\|$$

so that

$$\log \|X_k - X\| \leq k \log(\rho\sigma) + b \equiv ak + b, \quad b = \log(M(\rho)N(\sigma)\|X_0 - X\|),$$

which is (3. 20). Q. E. D.

The rate of convergence in this theorem depends on the product $\rho\sigma$ and that product, in turn, depends on the numbers $|\gamma_k|$ computed in (3. 11) and similar quantities associated with B and V. If the α_k are small relative to the β_k, $|\gamma_k|$ is near 1 for $r = 0$. On the other hand, $|\gamma_k| \to 1$ also as $r \to \infty$. Thus r needs to be selected in some intermediate range. Ordinarily when applying this method the eigenvalues of A, B are not actually computed so one would not attempt to select r according to the criterion that it minimize the largest of the expressions (3. 12). A reasonable rule of thumb is to compute

$$\alpha = \max_{k=1,2,\ldots,n} \sum_{\ell=1}^{n} |a_\ell^k|$$

$$\beta = \max_{k=1,2,\ldots,m} \sum_{\ell=1}^{m} |b_\ell^k|$$

and take $r = -\hat{r} \max\{\alpha, \beta\}$ for some small positive integer $\hat{r} - 1, 2, 3$, e. g.. In repeated applications $\hat{r}$ can be adjusted to improve performance.

In some cases it may be true that numbers α, β are known _a priori_ such that

$$\mathrm{Re}(\lambda_k) \le -\alpha, \quad |\mathrm{Im}(\lambda_k)| \le \beta, \quad k = 1, 2, \ldots, n, \tag{3.24}$$

$$\mathrm{Re}(\mu_j) \le -\alpha, \quad |\mathrm{Im}(\mu_j)| \le \beta, \quad j = 1, 2, \ldots, n. \tag{3.25}$$

For β we can always take $\max\{\|A\|, \|B\|\}$, though this is rather crude. Estimation of α is somewhat more subtle (see VII-6). If we assume that r is to be taken less than $\mathrm{Re}(\lambda_k)$, $\mathrm{Re}(\mu_j)$ for all k, j, we can verify from (3.12) that for the eigenvalues γ_k of U

$$|\gamma_k|^2 \le |\gamma|^2 = \frac{(r+\alpha)^2 + \beta^2}{(r-\alpha)^2 + \beta^2}, \tag{3.26}$$

and the same is true of the moduli of the eigenvalues of V. The right hand side of (3.24) is minimized by taking

$$r = -\sqrt{\alpha^2 + \beta^2} \tag{3.27}$$

as may be easily verified. We conclude that if (3.24), (3.25) are ture and, in addition, the eigenvalues of A and B lie to the right of r, then (3.27) is the best available choice for r. In this case $\rho\sigma$ is any number satisfying

$$1 > \rho\sigma > |\gamma|^2 = \frac{\sqrt{1 - (\beta/\alpha)^2} - 1}{\sqrt{1 + (\beta/\alpha)^2} + 1}.$$

We see conclusively from (3.28) that when the matrices A and B have eigenvalues close to the imaginary axis in the left half plane, the number $\rho\sigma$ will be close to 1. When this is the case the convergence (3.19) may be quite slow. Another result in [6] can be used to accelerate convergence in such cases - but with other risks, as we will note.

<u>Theorem 3.3.</u> <u>Let</u> $Y_0 = Y$ <u>and let sequences</u> $\{Y_j\}$, $\{U_j\}$, $\{V_j\}$ <u>of</u> $n \times m$, $n \times n$, $m \times m$ <u>matrices, respectively, be generated by</u>

$$Y_{j+1} = U_j Y_j V_j + Y_j, \tag{3.29}$$

$$U_0 = U,\ V_0 = V,\ U_{j+1} = (U_j)^2,\quad V_{j+1} = (V_j)^2,\quad j = 0,1,2,3,\dots\ . \tag{3.30}$$

<u>Then</u>

$$\lim_{j\to\infty} Y_j = X \tag{3.31}$$

<u>where</u> X <u>is the unique solution of</u> (3.8) <u>and the convergence is quadratic</u>: <u>for</u> a <u>as in</u> (3.20) <u>and</u> $\tilde{b}$ <u>real we have</u>

$$\log \|Y_j - X\| \le a 2^j + \tilde{b},\quad j = 0,1,2,\dots\ . \tag{3.32}$$

<u>Proof.</u> Repeatedly substituting (3.18) into itself, we obtain, for $k = 0,1,2,\dots,$

$$X = U^k X V^k + \sum_{\ell=0}^{k-1} U^\ell Y V^\ell .$$

From (3.9), (3.10)

$$\lim_{k\to\infty} U^k X V^k = 0$$

and hence

$$\lim_{k\to\infty} \left(\sum_{\ell=0}^{k-1} U^\ell Y V^\ell\right) = X .$$

We define

$$Y_j = \sum_{\ell=0}^{2^j-1} U^\ell Y V^\ell .$$

Then (3.31) is certainly true and

$$Y_{j+1} = \sum_{\ell=0}^{2^{j+1}-1} U^\ell Y V^\ell = \sum_{\ell=0}^{2^j-1} U^\ell Y V^\ell + U^{2^j}\left(\sum_{\ell=0}^{2^j-1} U^\ell Y V^\ell\right)V^{2^j} = Y_j + U_j Y_j V_j , \tag{3.33}$$

where

$$U_j = U^{2^j},\quad V_j = V^{2^j}$$

clearly satisfy (3.30). Equation (3.33) is the same as (3.29). It is easy to verify that if we take $X_0 = Y$ in Theorem 3.2 the resulting sequence $\{X_k\}$

is given by

$$X_k = \sum_{\ell=0}^{k} U^\ell Y V^\ell$$

so we conclude that for this special starting value $X_0 = Y$ we have

$$X_{2^j - 1} = Y_j . \tag{3.34}$$

Substituting (3. 34) into (3. 20) we have

$$\log \|Y_j - X\| = \log \|X_{2^j-1} - X\| \leq a(2^j - 1) + b = a\,2^j + b - a$$

which is (3. 32) with $\tilde{b} = b - a$. Q. E. D.

The quadratic rate of convergence for the $\{Y_j\}$ is, of course, much more rapid than the linear rate of convergence for the X_k. Nevertheless, the use of (3. 29), (3. 30) cannot be recommended without reservation. The method (3. 18) is self-correcting : if at some stage we make a mistake in calculating X_k, due to round-off error or some other factor, (3. 19) will still be true. In effect the erroneous X_k just replaces X_0, which, we recall, is arbitrary, and the procedure starts over "from scratch". Repeated small errors correspond to use of

$$X_{k+1} = U X_k V + Y + E_k$$

where E_k is a small "error" matrix. One can show that if $\|E_k\| < \epsilon$ for all k, then there is a $\delta > 0$, which tends to zero with ϵ, such that for all sufficiently large k

$$\|X_k - X\| < \delta .$$

The method (3. 29), (3. 30) does not have the self-correcting property. For we see from repeated use of (3. 29) that

$$Y_{j+1} = U_j Y_j V_j + U_{j-1} Y_{j-1} V_{j-1} + \ldots + U_1 Y_1 V_1 + U_0 Y_0 U_0 + Y_0 .$$

Thus errors in Y_ℓ, U_ℓ, V_ℓ, $0 \leq \ell \leq j$, can accumulate, being never corrected in the subsequent calculations. The lack of the self correcting property is signalled here, as in other situations where that is the case, by the fact that the problem data (i. e., the matrices U, V and Y) do not appear directly in the formulae defining the successive iterations. If the method of Theorem 3. 3 is used, care must be taken to perform each step with adequate precision. This can be particularly troublesome when it comes to keeping U_j close to U^{2^j} when the matrix U has eigenvalues whose moduli are almost equal to 1, as a result of A having eigenvalues $\alpha_k + i\beta_k$ for which the quantity $|\beta_k/\alpha_k|$ becomes large. This is frequently the case when $\dot{x} = Ax$ models a vibratory physical system with a wide range of natural frequencies. It goes without saying that these considerations apply to the matrix V as well.

We illustrate the material of this section by going back to (2. 23) and taking the values

$$k_{1i} = 0, \quad k_{2i} = -1$$

which we used as starting values to generate the table (2. 25). Here (2.23) becomes

$$\begin{pmatrix} 0 & -1 \\ 1 & -1 \end{pmatrix}\begin{pmatrix} q_1 & q_3 \\ q_2 & q_3 \end{pmatrix} + \begin{pmatrix} q_1 & q_2 \\ q_2 & q_3 \end{pmatrix}\begin{pmatrix} 0 & 1 \\ -1 & -1 \end{pmatrix} + \begin{pmatrix} 1 & 0 \\ 0 & 2 \end{pmatrix} = 0,$$

corresponding to

$$B = \begin{pmatrix} 0 & 1 \\ -1 & -1 \end{pmatrix}, \quad A = B^* = \begin{pmatrix} 0 & -1 \\ 1 & -1 \end{pmatrix}, \quad C = \begin{pmatrix} -1 & 0 \\ 0 & -2 \end{pmatrix},$$

in (3. 1). The eigenvalues of A and B are, in both cases

$$-\frac{1}{2} \pm i\frac{\sqrt{3}}{2}$$

so our formula (3. 27) can be applied and we take

$$r = -\sqrt{\left(\frac{1}{2}\right)^2 + \left(\frac{\sqrt{3}}{2}\right)^2} = -1 .$$

We then have

$$rI_2 + A = \begin{pmatrix} -1 & -1 \\ 1 & -2 \end{pmatrix}, \quad rI_2 - A = \begin{pmatrix} -1 & 1 \\ -1 & 0 \end{pmatrix}.$$

So (cf. (3. 5))

$$U = \begin{pmatrix} -1 & -1 \\ 1 & -2 \end{pmatrix}^{-1} \begin{pmatrix} -1 & 1 \\ -1 & 0 \end{pmatrix} = \begin{pmatrix} -\frac{2}{3} & \frac{1}{3} \\ -\frac{1}{3} & -\frac{1}{3} \end{pmatrix} \begin{pmatrix} -1 & 1 \\ -1 & 0 \end{pmatrix} \begin{pmatrix} \frac{1}{3} & -\frac{2}{3} \\ \frac{2}{3} & -\frac{1}{3} \end{pmatrix}$$

and (cf. (3. 6)), since $B = A^*$ here,

$$V = U^* = \begin{pmatrix} \frac{1}{3} & \frac{2}{3} \\ -\frac{2}{3} & -\frac{1}{3} \end{pmatrix}.$$

From (3. 7)

$$Y = -2 \begin{pmatrix} -\frac{2}{3} & \frac{1}{3} \\ -\frac{1}{3} & -\frac{1}{3} \end{pmatrix} \begin{pmatrix} -1 & 0 \\ 0 & -2 \end{pmatrix} \begin{pmatrix} -\frac{2}{3} & -\frac{1}{3} \\ \frac{1}{3} & -\frac{1}{3} \end{pmatrix} = \begin{pmatrix} \frac{4}{3} & 0 \\ 0 & \frac{2}{3} \end{pmatrix}$$

and (3. 8) becomes

$$X = \begin{pmatrix} \frac{1}{3} & -\frac{2}{3} \\ \frac{2}{3} & -\frac{1}{3} \end{pmatrix} X \begin{pmatrix} \frac{1}{3} & \frac{2}{3} \\ -\frac{2}{3} & -\frac{1}{3} \end{pmatrix} + \begin{pmatrix} \frac{4}{3} & 0 \\ 0 & \frac{2}{3} \end{pmatrix}.$$

Letting

$$X = \begin{pmatrix} x & y \\ y & z \end{pmatrix}$$

the method of Theorem 3. 2 amounts to

$$x_{k+1} = \frac{1}{9}x_k - \frac{4}{9}y_k + \frac{4}{9}z_k + \frac{4}{3},$$

$$y_{k+1} = \frac{2}{9}x_k - \frac{5}{9}y_k + \frac{2}{9}z_k,$$

$$z_{k+1} = \frac{4}{9}x_k - \frac{4}{9}y_k + \frac{1}{9}z_k + \frac{2}{3} .$$

Let us start with

$$x_0 = 1, \quad y_0 = 0, \quad z_0 = 1,$$

i. e., $X_0 = I$. Carrying out the indicated iterations, we obtain the following table.

k	x_k	y_k	z_k
0	1	0	1
1	1.8888	.4444	1.2222
2	1.8888	.4444	1.4444
3	1.9876	.4938	1.4691
4	1.9876	.4938	1.4938
5	1.9986	.4993	1.4965
6	1.9986	.4993	1.4993
7	1.9998	.4999	1.4996
8	1.9998	.4999	1.4999
9	1.9999	.4999	1.4999

(3.35)

This agrees, of course, with the values q_{10}, q_{20}, q_{30} in the first row of (2.25).

In order to compare the accelerated method of Theorem 3.3 it is not necessary, in this example, to carry out (3.29), (3.30) explicitly, because we know the Y_j are the same as X_{2^j-1}. If the table (3.35) is extended it will be found that the last row is repeated exactly in all further iterations. Consequently

$$Y_0 = \begin{pmatrix} 1 & 0 \\ 0 & 1 \end{pmatrix} = X_{2^0-1} = X_0 ,$$

$$Y_1 = \begin{pmatrix} 1.8888 & .4444 \\ .4444 & 1.8888 \end{pmatrix} = X_{2^1-1} = X_1 ,$$

$$Y_2 = \begin{pmatrix} 1.9876 & .4938 \\ .4938 & 1.4691 \end{pmatrix} = X_{2^2-1} = X_3 ,$$

$$Y_3 = \begin{pmatrix} 1.9998 & .4999 \\ .4999 & 1.4996 \end{pmatrix} = X_{2^3-1} = X_7 ,$$

$$Y_4 = \begin{pmatrix} 1.9999 & .4999 \\ .4999 & 1.4999 \end{pmatrix} = X_{2^4-1} = X_{15} .$$

Thus in this case (3. 29), (3. 30) requires four iterations as compared with nine for (3. 18). However, in Exercise 4 we invite the reader to repeat the calculation of Y_1, Y_2, Y_3, Y_4, using four decimal places in all intermediate calculations, to see what the effect of the accumulated errors would be in this case and to compare that with a table (3. 35) generated in the same way.

4. Integration of the Matrix Riccati Differential Equation in the Constant Coefficient Case

We consider the constant coefficient linear control system

$$\dot{x} = Ax + Bu \tag{4. 1}$$

and the quadratic cost functional, also defined with constant coefficient matrices

$$\begin{aligned} J(x_0, u, T) &= \int_0^T [(Fx(t)+Gu(t))^* \tilde{W} (Fx(t)+Gu(t)) + u(t)^* \tilde{U} u(t)]\, dt \\ &= \int_0^T (x(t)^*, u(t)^*) \begin{pmatrix} W & R \\ R^* & U \end{pmatrix} \begin{pmatrix} x(t) \\ u(t) \end{pmatrix} dt . \end{aligned} \tag{4. 2}$$

We have seen in IV-2 that minimization of $J(x_0, u, T)$ for finite $T > 0$ is accomplished via solution of the Riccati differential equation IV-(2. 28), thereby obtaining a positive definite symmetric matrix valued function $Q_T(t)$. In IV-3 we have seen that for $T = \infty$ the solution is given in terms of

$$Q = \lim_{T \to \infty} Q_T(0).$$

For the purposes of the present section, and for numerical treatment generally, it is more convenient to define

$$P(t) = Q(T - t). \tag{4.3}$$

Then, rearranging as in IV-4,

$$\dot{P} = A_1^* P + PA_1 + W_1 - PB_1 P . \tag{4.4}$$

The condition IV-(3. 8) is replaced by

$$P(0) = 0 . \tag{4.5}$$

In this section we will consider two questions relative to (4. 4), (4. 5). The first concerns the replacement of (4. 4), which involves, even after symmetry has been taken into account, $n(n+1)/2$ coupled scalar differential equations, by another system which often in practice involves significantly fewer differential equations. This technique has been developed by a number of authors, of whom we particularly note J. Casti ([8]- [10]) and T. Kailath ([7]). Our exposition is taken from Kailath's paper. The second question concerns the matter of numerical approximation of

$$Q = \lim_{t \to \infty} P(t)$$

by a sequence of matrices $\{P_k\}$ which arise from numerical integration of (4. 4).

It is frequently true that the matrices W_1 and B_1 in (4. 4) have ranks which are small compared to n, the dimension of the state vector, x. Indeed, IV-(4. 1) shows that the rank of B_1 is less than or equal to m, the dimension of the control vector u. From IV-(1. 13)ff, IV-(4. 1), we see that the rank of W_1 is less than or equal to q, the dimension of the output vector $y = Fx + Gu$. Perhaps it is not surprising, then, that the n^2 equations in the system (4. 4) can be replaced by a set of equations reflecting the fact that W_1 and B_1 have these ranks m and q - which are frequently much smaller than n.

Before stating the theorem which makes the above precise we note that W_1 can be written in terms of two $q \times n$ matrices:

$$W_1 = F_1^* F_2 . \tag{4.6}$$

In general we can take (cf. IV-(1.13)ff., IV-(4.1))

$$F_1 = F, \quad F_2 = (\tilde{W} - \tilde{W} G U^{-1} G^* \tilde{W}) F. \tag{4.7}$$

When W_1 is positive definite it can be written in terms of a single $q \times n$ matrix

$$W_1 = F_1^* F_1 . \tag{4.8}$$

For details of this decomposition, the reader is referred to [1].

In order to simplify some of the computations in Theorem 4.1 below, we make the change of variable

$$u = U^{-\frac{1}{2}} \tilde{u} .$$

This allows us, without loss of generality, to take $U = I$ in (4.2), provided we replace B, G and R by

$$\tilde{B} = BU^{-\frac{1}{2}} ,$$

$$\tilde{G} = GU^{-\frac{1}{2}} ,$$

$$\tilde{R} = RU^{-\frac{1}{2}} .$$

In terms of these matrices we have (cf. IV-(4.1))

$$A_1 = A - \tilde{B}\tilde{R}^* ,$$

$$W_1 = W - \tilde{R}\tilde{R}^* ,$$

$$B_1 = \tilde{B}\tilde{B}^* .$$

The feedback law IV-(2.31) becomes, in terms of $\tilde{u}$,

$$\tilde{u}(t) = -[\tilde{B}^* Q(t) + \tilde{R}^*] x(t) = [\tilde{K}(T - t) - \tilde{R}^*] x(t)$$

if we let

$$\tilde{K}(t) = -\tilde{B}^* P(t) . \tag{4.9}$$

<u>Theorem 4.1.</u> <u>Let</u> $\tilde{K}(t)$, $L_1(t)$, $L_2(t)$ <u>satisfy the differential equations</u>

$$\dot{\tilde{K}}(t) = -\tilde{B}^* L_1(t)^* L_2(t), \tag{4.10}$$

$$\dot{L}_i(t) = L_i(t)(A_1 + \tilde{B}\tilde{K}(t)), \qquad i = 1, 2 \tag{4.11}$$

<u>with initial conditions</u> (cf. (4.6))

$$\tilde{K}(0) = 0, \quad L_i(t) = F_i, \quad i = 1, 2 . \tag{4.12}$$

<u>Then</u>

$$P(t) = \int_0^t L_1^*(s) L_2(s) ds, \tag{4.13}$$

$$\tilde{K}(t) = -\tilde{B}^* P(t) , \tag{4.14}$$

<u>where</u> $P(t)$ <u>is the solution of</u> (4.10), (4.11), (4.12).

<u>Remarks.</u> In all (4.10), (4.11) comprise $n(m + 2q)$ coupled scalar differential equations. When

$$m + 2q < \frac{n+1}{2} ,$$

which is frequently the case for large dimensional systems (we will give an example at the end of the section) a significant reduction in computational effort may be realized. When W_1 is positive definite, so that (4.8) applies, we have $L_1(t) \equiv L_2(t)$ and the number of scalar equations involved is reduced to $n(m + q)$.

<u>Proof of Theorem 4.1.</u> Let $P(t)$ be the solution of (4.4), (4.5) on $0 \leq t < \infty$ and let $\tilde{K}(t)$ be defined by (4.9). We differentiate (4.4) and (4.9) to obtain

$$\dot{\tilde{K}}(t) = -\tilde{B}^* \dot{P}(t) \tag{4.15}$$

$$\ddot{P}(t) = A_1^*\dot{P} + \dot{P}A_1 - PB_1\dot{P} - \dot{P}B_1P = (A_1 + \tilde{B}\tilde{K}(t))^*\dot{P} + \dot{P}(A_1 + \tilde{B}\tilde{K}(t)). \tag{4.16}$$

This is a homogeneous linear matrix differential equation for $\dot{P}(t)$. Letting $\Phi(t)$ denote the fundamental matrix solution of

$$\dot{\Phi} = \Phi(A_1 + \tilde{B}\tilde{K}(t)) \tag{4.17}$$

which satisfies $\Phi(0) = I$, it may be verified just by differentiation and uniqueness of solutions for (4.16) that

$$\dot{P}(t) = \Phi(t)^*\dot{P}(0)\Phi(t) . \tag{4.18}$$

Now $P(0) = 0$, so (4.4) gives

$$\dot{P}(0) = W_1$$

and (4.18) becomes

$$\dot{P}(t) = \Phi(t)^*W_1\Phi(t) = \Phi(t)^*F_1^*F_2\Phi(t) . \tag{4.19}$$

Defining

$$L_i(t) = F_i\Phi(t), \quad i = 1, 2, \tag{4.20}$$

we have, from (4.17),

$$\dot{L}_i(t) = F_i\Phi(t)(A_1 + \tilde{B}\tilde{K}(t)) = L_i(t)(A_1 + \tilde{B}\tilde{K}(t)),$$

which is (4.11). Multiplying (4.19) on the left by $-\tilde{B}^*$ and using (4.15), (4.20) we have

$$\dot{\tilde{K}}(t) = -\tilde{B}^*K_1^*(t)L_2(t) \quad (= -\tilde{B}^*\dot{P}(t)) \tag{4.21}$$

and this is (4.10). The formula (4.14) comes from integration of (4.21), using $\tilde{K}(0) \doteq 0$, and (4.13) follows from (4.20) and integration of (4.19) with $P(0) = 0$. Q. E. D.

In terms of the original control variable u, the optimal feedback law is

$$u(t) = U^{-\frac{1}{2}}\tilde{u}(t) = [U^{\frac{1}{2}}\tilde{K}(T-t)-R^*]\,x(t) \equiv [K(T-t)-R^*]\,x(t), \tag{4.22}$$

where

$$K(t) = U^{\frac{1}{2}}\tilde{K}(t) = -U^{\frac{1}{2}}\tilde{B}^* P(t) = -B^* P(t)\,.$$

Multiplying (4.10) by $U^{\frac{1}{2}}$ we have

$$\dot{K}(t) = -B^* L_1(t)L_2(t). \tag{4.23}$$

Transposing and using $K(t)$, B in (4.11), (4.12) we have

$$\dot{L}_i(t)^* = (A_1 + BU^{-1}K(t))^* L_i(t)^*\,, \tag{4.24}$$

$$L_i(0)^* = F_i^*\,, \quad i = 1, 2\,. \tag{4.25}$$

The equations are most conveniently solved in this form. The optimal feedback matrix K (cf. IV-(3.17)) is then

$$K = U^{-1}(K_\infty - R^*), \quad K_\infty = \lim_{t\to\infty} K(t)\,. \tag{4.26}$$

We have indicated that this procedure is most advantageous when m and q are small as compared with n. This is very often the case when a "distributed parameter" system is modelled by a finite dimensional "lumped parameter" system $\dot{x} = Ax + Bu$. See [11] for a general discussion.

For example, a stretched string fixed at its left hand endpoint, $x = 0$, and free to move vertically at its right hand endpoint, $x = 1$, has linearized dynamics expressed by the partial differential equation

$$\frac{\partial^2 y}{\partial t^2} = \frac{k}{\rho}\frac{\partial^2 y}{\partial x^2} \qquad 0 \le x \le 1, \quad t \ge 0 \tag{4.27}$$

and boundary conditions

$$y(0,t) \equiv 0\,, \quad k\frac{\partial y}{\partial x}(1,t) = u(t)\,, \tag{4.28}$$

$y(x,t)$ denoting the vertical displacement of the string, k the elastic modulus, ρ the linear mass density, and $u(t)$ a force acting in the vertical direction at $x = 1$. The system is shown in Figure V-1. This is one of the

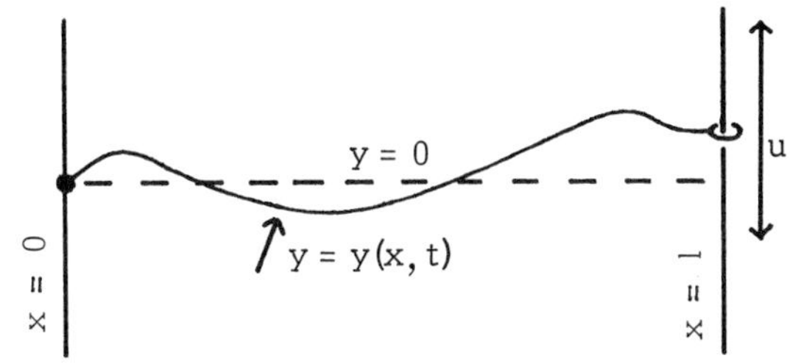

Fig. V-1. Stretched String

most familiar examples of what are described in control literature as "distributed parameter" systems. The state $y(x,t)$, $\frac{\partial y}{\partial t}(x,t)$ is indexed by the parameter x which has infinitely many values "distributed" between 0 and 1. A finite dimensional approximation is obtained by considering n "stations" along the string

$$x_j = \frac{j}{n}, \quad j = 1, 2, \ldots, n.$$

Then the system (4.27), (4.28) can be approximated by

$$\frac{d^2 y^j}{dt^2} = \frac{k}{\rho} n^2 (y_{k+1} - 2y_k + y_{k-1}), \quad j = 2, \ldots, n-1$$

$$\frac{d^2 y^1}{dt^2} = \frac{k}{\rho} n^2 (y_2 - 2y_1),$$

$$\frac{d^2 y^n}{dt^2} = -\frac{k}{\rho} n^2 (y_n - y_{n-1}) + \frac{n}{\rho} u(t),$$

leading to a control system of the form

$$\dot{y} = Ay + bu, \quad y, b \in E^{2n}.$$

Here y is the vector whose components are y^j, $j = 1, 2, \ldots, n$, $\dot{y}^j$, $j = 1, 2, \ldots, n$. The components y^j are approximations to the displacements $y(x_j, t)$, $j = 1, 2, \ldots, n$. The matrices A and B are given by

$$A = \begin{pmatrix} 0 & I \\ A_1 & 0 \end{pmatrix}, \quad 2n \times 2n \ ,$$

where

$$A_1 = n^2 \begin{pmatrix} -2k/\rho & k/\rho & 0 & \cdots & 0 & 0 \\ k/\rho & -2k/\rho & k/\rho & \cdots & 0 & 0 \\ 0 & k/\rho & -2k/\rho & \cdots & 0 & 0 \\ \vdots & \vdots & \vdots & & \vdots & \vdots \\ 0 & 0 & 0 & \cdots & -2k/\rho & k/\rho \\ 0 & 0 & 0 & \cdots & k/\rho & -2k/\rho \end{pmatrix}, \tag{4.29}$$

$$b^* = (0, 0, \ldots, 0, 0, 0, 0, \ldots, 0, \frac{n}{\rho}) \ . \tag{4.30}$$

Let us consider, for each n, a cost based on the string velocity at $x = 1$. For this we construct a quadratic cost function of the form

$$\int_0^T [(\dot{y}^n)^2 + (u(t))^2] \, dt \ ,$$

corresponding to

$$\tilde{W} = \tilde{U} = 1 \ , \tag{4.31}$$

$$F = (0, 0, \ldots \ 0, 0, 0, 0, \ldots, 0, 1) \ , \tag{4.32}$$

$$G = 0. \tag{4.33}$$

The number of differential equations which must be solved in (4. 4) is $n(n+1)/2$ while $2n$ equations suffice for the method of Theorem 4. 1. This comparison favors the latter method quite strongly in applications of this sort because fairly large values of n (say, 10 or 20) are required for reasonable approximation to the original distributed system. For $n = 10$ ($n = 20$) solution of (4. 4) involves 55 (210) scalar equations while the method (4. 23), (4. 24) involves 20 (40) scalar equations.

We do not wish to give here any detailed treatment of the numerical solution of the differential equations (4. 4) or (4. 23), (4. 24). Numerical methods for differential equations in general are discussed in [12] and [13] as well as many other texts. We will point out, however, that rather different considerations predominate when IV-(2. 28) integrated over a comparatively short interval $[t_0, t_1]$ to obtain the solution of the finite interval problem discussed in IV-2, as compared with those which prevail when (4. 4) is integrated over a long interval $[0, T]$ in order to approximate the solution Q of IV-(3. 13). In the first instance the formula for the optimal control, IV-(2. 38), involves the matrix $Q_T(t)$ throughout the interval $[t_0, t_1]$.

In this situation it may be worthwhile to expend considerable effort to ensure that the numerical method employed is quite accurate over the entire range of integration, $[t_0, t_1]$. Some of the very accurate predictor-corrector methods ([12], [13]) may be appropriate under these circumstances. The fourth order Runge-Kutta method ([13]) is often adequate. This type of effort is rarely profitable in the second instance where (4. 4) is integrated over a long interval $[0, T]$ in order to approximate the desired solution Q of IV-(3. 13). In this latter case all that matters is how well the <u>final</u> value approximates Q. This often dictates use of a different numerical integration scheme.

For most applications solution of IV-(3. 13) is best carried out by the methods outlined in Sections 1 - 3 of the present chapter. Nevertheless there are circumstances where numerical solution of (4. 4), or (4. 23), (4. 24), over a long interval may be appropriate. Increasingly we see the use of small programmable electronic calculators by those working in the field. Commonly these calculators have rather small amounts of random access memory. The method of Section 1 requires, ordinarily, storage of the full $2n \times 2n$ matrix M. In addition, the process of finding eigenvalues of such a matrix and carrying out the required triangulation may be stretching the programming capabilities of such machines. It is fairly safe to say that this is a method for the large "number crunchers" if n is even modestly large. The Newton-Kleinman method of Section 2 is less restricted in this respect but it does require solution of successive Liapounov equations, usually by the method of Section 3 or by the Bartels-Stewart method ([5]). The latter is definitely out of the question for small computers or programmable calculators. (See the code in [5]). The method of Section 3 is easily programmed but it does require inversion of the matrix $(rI_n + A)$. This and the requirement for considerable matrix multiplication makes it hard to use unless substantial random access memory is available.

Numerical integration of (4.4), or (4.23), (4.24) is usually less efficient timewise than the methods of Sections 1, 2 but it has the advantages of being simple to program and of requiring relatively small amounts of random access memory. This is particularly true of the method of Theorem 4.1 involving the equations (4.23), (4.24). In both cases, and particularly in the case of (4.23), (4.24), it is relatively easy to take advantage of "sparseness" properties of A, B, F (the situation arising when most of the entries of these matrices are zero, as in (4.29), (4.30), (4.23) when n is large.

To give some idea as to why a highly accurate numerical integration routine may not be particularly advantageous in solving (4.4) and to show what considerations are important, we consider a class of numerical integration schemes for the vector first order differential equation

$$\frac{dy}{dt} = f(y) . \tag{4.34}$$

Letting $t_{k+1} - t_k = h$ be the step length and letting $y(t_k)$ be approximated by y_k, we consider the methods

$$y_{k+1} = y_k + h[\mu f(y_{k+1}) + (1 - \mu)f(y_k)] . \tag{4.35}$$

For $\mu = 0$ we have the simple Euler method

$$y_{k+1} = y_k + hf(y_k), \tag{4.36}$$

for $\mu = 1$ we have the "implicit" Euler method

$$y_{k+1} = y_k + hf(y_{k+1}) \tag{4.37}$$

and for $\mu = 1$ we have the "implicit" trapezoidal rule

$$y_{k+1} = y_k + \frac{h}{2}[f(y_{k+1}) + f(y_k)] \, . \tag{4.38}$$

The first two have discretization error $\mathcal{O}(h)$ while the last has error $\mathcal{O}(h^2)$.

Let $\frac{\partial f}{\partial y}$ denote the Jacobian matrix of f. It is well known (I - [3]) that if

$$f(\eta) = 0$$

and $\frac{\partial f}{\partial y}(\eta)$ has only eigenvalues with negative real parts, then solutions of (4. 34) initiating near η have the property

$$\lim_{t \to \infty} y(t) = \eta \, .$$

Any of the numerical methods (4. 35) may be viewed as an iterative procedure for approximation of η. Assuming (4. 35) can be solved for y_{k+1}, so that it becomes equivalent to

$$y_{k+1} = F(h, \mu, y_k) \equiv F(y_k) \, , \tag{4.39}$$

the fact that $f(\eta) = 0$ is seen to imply that η is a fixed point for $F(y)$. The general contraction fixed point theorem ([2], III-[1]), guarantees that

$$\lim_{k \to \infty} y_k = \eta$$

if (i) some y_{k_0} is "sufficiently close" to η and (ii) $F_y(\eta)$ has eigenvalues of modulus less than one. Moreover, the rate of convergence is like $|\gamma_1|^k$, where γ_1 is the eigenvalue of $F_y(\eta)$ of largest modulus. Viewed in this light the accuracy of the numerical integration routine is important only to the extent that it guarantees (i) - it has very little to do with (ii).

Write (4. 35) as

$$g_1(y_{k+1}) = g_0(y_k)$$

$$g_1(y) = y - h\mu f(y), \quad g_0(y) = y + h(1-\mu)f(y) \, .$$

Then (cf. (4. 39))

$$F(y) = g_1^{-1}(g_0(y))$$

and

$$F_y(\eta) = g_{1y}(\eta)^{-1} g_{0y}(\eta)$$
$$= [I - h\mu f_y(\eta)]^{-1} [I + h(1-\mu) fy(\eta)] .$$

It follows that the eigenvalues of $F_y(\eta)$ have the form (III-Exercise 12)

$$\gamma = \frac{1 + h(1-\mu)\lambda}{1 - h\mu\lambda} .$$

Assuming

$$\mathrm{Re}(\lambda) < 0 \tag{4.40}$$

for all eigenvalues λ of $fy(\eta)$, we consider the methods (4. 36), (4. 37), (4. 38) in turn. For the simple Euler method, $\mu = 0$,

$$\gamma = 1 + h\lambda .$$

The condition $|\gamma| < 1$ follows from (4. 40) only for small values of h. This means that h may have to be taken very small, even if λ is a complex number with large negative real part. Making h small usually means that the y_k converge rather slowly to η. (See [2]). Thus (4. 36) cannot be recommended in such circumstances.

The implicit Euler method, $\mu = 1$, gives

$$\gamma = \frac{1}{1 - h\lambda} ,$$

$$|\gamma|^2 = \frac{1}{(1-h\ \mathrm{Re}(\lambda))^2 + (h\ \mathrm{Im}(\lambda))^2} < 1 \tag{4.41}$$

for any $h > 0$ if (4. 40) is true. For this method we only need to have y_{k_0} sufficiently close to η for some value of k_0. However, the fact that the discretization error is $\mathcal{O}(h)$ means that this method is not highly accurate and may have trouble reaching this latter objective in some cases.

The implicit trapezoidal rule, $\mu = \frac{1}{2}$, gives

$$\gamma = \frac{1 + \frac{h\lambda}{2}}{1 - \frac{h\lambda}{2}} ,$$

$$|\gamma|^2 = \frac{(1+\frac{h}{2}\operatorname{Re}(\lambda))^2 + (\frac{h}{2}\operatorname{Im}(\lambda))^2}{(1-\frac{h}{2}\operatorname{Re}(\lambda))^2 + (\frac{h}{2}\operatorname{Im}(\lambda))^2} < 1 \tag{4.42}$$

when (4. 40) is true. Moreover, the discretization error being $\mathcal{O}(h^2)$, it may be easier to guarantee that some y_{k_0} is sufficiently close to η. Observe, however, that the improved accuracy of (4. 38) over (4. 37) is not reflected in an improved rate of convergence. In fact, for any h and any λ satisfying (4. 40), (4. 41) is less than (4. 42), i. e., as far as rate of convergence is concerned, the less accurate implicit Euler method is superior to the more accurate implicit trapezoidal rule.

In the case of the differential equation (4. 4), f(y) corresponds to the right hand side of (4. 4) and η corresponds to

$$Q = \lim_{t\to\infty} P(t) .$$

It can be shown (see Chapter VI) that $\frac{\partial f}{\partial y}(\eta)$ corresponds to the linear operator on matrices P defined by

$$T(P) = (A_1 - BU^{-1}B^*Q)^*P + P(A_1 - BU^{-1}B^*Q).$$

The eigenvalues of T, as an operator on E^{n^2}, are of the form $\lambda_i + \overline{\lambda}_j$, where λ_i, λ_j are eigenvalues of $A_1 - BU^{-1}B^*Q$. Since the latter is a stability matrix, $\operatorname{Re}(\lambda_i + \overline{\lambda}_j) < 0$ and the above analysis is applicable to the differential equation (4. 4). We have noted that it is rarely advantageous to solve IV-(3. 13) by numerical integration of (4. 4). If this is done, however, the above discussion indicates why a highly accurate method, i. e., one with small discretization error, need not be the right approach.

We have indicated that in some cases the method (4. 23), (4. 24) may be useful for calculating the optimal feedback matrix for the problem on the infinite interval. But the differential equations (4. 10), (4. 11) are not of the type discussed above. If we represent K, L_1^*, L_2^* in vector notation as y_K, y_1, y_2, so that (4. 34) becomes

$$\dot{y} \equiv \begin{pmatrix} \dot{y}_K \\ y_1 \\ y_2 \end{pmatrix} = \begin{pmatrix} f^K(y_K, y_1, y_2) \\ f^1(y_K, y_1, y_2) \\ f^2(y_K, y_1, y_2) \end{pmatrix} = f(y) ,$$

it follows that the Jacobian of f has the form

$$\frac{\partial f}{\partial y} = \begin{pmatrix} f^K_{y_K} & f^K_{y_1} & f^K_{y_2} \\ f^1_{y_K} & f^1_{y_1} & f^1_{y_2} \\ f^2_{y_K} & f^2_{y_1} & f^2_{y_2} \end{pmatrix} .$$

The form of (4.23) shows that

$$f^K_{y_K} \equiv 0 ,$$

because K does not occur in (4.23). When (4.24) is solved on $[0, \infty)$, the fact that $A_1 + BU^{-1}K(t)$ is, in the limit, a stability matrix, shows that

$$\lim_{t \to \infty} L_1 = \lim_{t \to \infty} L_2 = 0 .$$

Then the form of (4.23) shows that

$$\lim_{t \to \infty} f^K_{y_1} = \lim_{t \to \infty} f^K_{y_2} = 0 .$$

It follows that, in the limit, $\frac{\partial f}{\partial y}$ has the form

$$\begin{pmatrix} 0 & 0 & 0 \\ * & * & * \\ * & * & * \end{pmatrix}$$

and hence has zero as an eigenvalue - the multiplicity being equal to the number of components in K, i.e., mn. Thus (4.40) is never true for the system (4.23), (4.24). Indeed, every triple: K arbitrary, $L_1 = L_2 = 0$, satisfies $f(y) = 0$ in this case. There is no unique fixed point for F as defined by (4. 39) and the contraction fixed point theory is not applicable here. This is reflected in the fact that the method is not self-correcting, which in turn is signalled by the fact that W_1 does not appear in (4. 23), (4. 24) - only in the initial conditions (4. 25) via (4. 8).

In the strict sense it cannot be proved that numerical solutions of (4.23), (4. 24) tend to the same limit as the exact solution when $t \to \infty$. If we suppose that K_i, L_{1i}, L_{2i} are numerical approximations to $K(t_i)$, $L_1(t_i)$, $L_2(t_i)$, $t_i = ih$, $i = 0, 1, 2, 3, \dots$, obtained by some valid numerical integration scheme, it turns out that

$$\lim_{i \to \infty} L_{1i} = \lim_{i \to \infty} L_{2i} = 0 ,$$

$$\lim_{i \to \infty} K_i = K_\infty + \Delta K ,$$

where (cf. (4. 26))

$$K_\infty = \lim_{t \to \infty} K(t) .$$

In general $\Delta K \neq 0$. Concerning its size one can prove the following: the error ΔK can be bounded by a constant times

$$\frac{1}{\gamma} h^n, \quad h \text{ small}, \ \gamma \text{ small}, \tag{4. 43}$$

if the numerical method used has discretization error $\mathcal{O}(h^n)$ and the eigenvalues of $A + BK = A_1 + BU^{-1}K_\infty$ all lie to the left of the line

$$\text{Re}(\lambda) = -\gamma, \quad \gamma > 0 ,$$

in the complex plane. A rigorous proof of this assertion would carry us too far afield at this point. A short heuristic discussion proceeds as follows. The truncation error (the error made at each step) for an n-th order numerical integration scheme can be bounded by

$$h^{n+1} y^{(n+1)} \tag{4.44}$$

where $y^{(n+1)}$ denotes the (n+1)-st derivative of the actual solution $y(t)$.

If $y(t)$ stands for $L_1(t)$, $L_2(t)$ in (4. 24), $y^{n+1}(t)$ will tend to zero exponentially like $e^{-\gamma t}$ for large values of t. Now as $t \to \infty$ the accumulated discretization errors $E_{1i} = L_{1i} - L_1(t)$, $E_{2i} = L_{2i} - L_2(t)$ behave much like the solution of an equation

$$\dot{E} = (A_1 + BU^{-1}K_\infty)E + \text{truncation error} ,$$

from which it can be shown that E_{1i}, E_{2i} will behave like $e^{-\gamma t}h^n$ for large $t = t_i$. Now $K(t)$ is obtained by integration of $-B^*L_1(t)^*L_2(t)$ and K_i is correspondingly obtained by numerical integration of $-B^*L_1^*(t)L_2(t)$, but with that quantity replaced by $-B^*L_{1i}^*L_{2i}$. At each step of the integration there is the error attributable to the numerical scheme itself, which will be of the order (4. 44) plus the error in approximation of $-B^*L_1^*(t)L_2(t)$ by $-B^*L_{1i}^*L_{2i}$. Since this latter error behaves like $e^{-\gamma t}h^n$ and each step has length h, it introduces an error like $e^{-\gamma t}h^{n+1}$ at each step. Combined with the fact that $y^{(n+1)}$ decays like $e^{-\gamma t}$ we see that the numerical integration error at each step is comparable to $e^{-\gamma t}h^{n+1}$. The integrated error will be like

$$\sum_{k=1}^{\infty} (e^{-\gamma kh}h^{n+1}) = h^{n+1}\sum_{k=1}^{\infty}\left(\frac{1}{e^{\gamma h}}\right)^k = h^{n+1}\frac{1}{1-e^{\gamma h}} = h^{n+1}\,\mathcal{O}\left(\frac{1}{\gamma h}\right) ,$$

which leads to the extimate (4. 43). The foregoing is no substitute for a proof but it does essentially reflect the way in which the error comes about.

The implications are clear. When (4. 23), (4. 24) is used to generate an approximation to K_∞, the accuracy of the numerical integration scheme does affect the error ΔK. Thus use of a highly accurate scheme is important here, whereas we saw that need not be the case in numerical integration of (4. 4) for the same purpose. The factor $\frac{1}{\gamma}$ also indicates trouble when the eigenvalues of $A_1 + BU^{-1}K_\infty$ lie close to the imaginary axis. This causes difficulties with all of the other methods examined in this chapter also, but for integration of (4. 4) or use of the method of Sections II, III it manifests itself in terms of rate of convergence while for the method (4. 23), (4. 24) it actually aggravates the final error.

We will complete this section by examining the implicit trapezoidal rule as it applies to the system (4. 23), (4. 24). For brevity we consider only the symmetric case where $L_1(t) \equiv L_2(t) \equiv L(t)$. Applying the method (4. 38) to (4. 23), (4. 24) we obtain the following equations for the approximations K_i, L_i to $K(t_i)$, $L(t_i)$, $t_i = ih$, $i = 0, 1, 2, 3, \ldots$

$$K_0 = 0,$$

$$L_0 = F_1^* \quad (W_1 = F_1^* F_1 \text{ (cf. (48))}$$

and thereafter

$$K_{i+1} = K_i - \frac{h}{2} B^* [L_{i+1}^* L_{i+1} + L_i^* L_i] \tag{4.45}$$

$$L_{i+1}^* = L_i^* + \frac{h}{2} [(A_1 + BU^{-1} K_{i+1})^* L_{i+1}^* + (A_1 + BU^{-1} K_i)^* L_i^*] . \tag{4.46}$$

The first equation can be substituted in the second to give

$$L_{i+1}^* = L_i^* + \frac{h}{2}[(A_1 + BU^{-1}K_i)^* L_{i+1}^* + (A_1 + BU^{-1}K_i)^* L_i] - \frac{h^2}{4}[L_{i+1}^* L_{i+1} + L_i^* L_i] BU^{-1} B^* L_{i+1}^*$$

or

$$L_{i+1}^* = [I - \frac{h}{2}(A_1 + BU^{-1}K_i)^*]^{-1}([I + \frac{h}{2}(A_1 + BU^{-1}K_i)^*] L_i^* - \frac{h^2}{4}[L_{i+1}^* L_{i+1} + L_i^* L_i] BU^{-1} B^* L_{i+1}^*). \tag{4.47}$$

The presence of the rather small term $\frac{h^2}{4}$ allows (4. 47) to be solved for L_{i+1}^* by ordinary fixed point iteration (III- [1], or [2] below) in most cases. The eigenvalues of the matrix

$$[I - \frac{h}{2}(A_1 + BK_i)^*]^{-1} [I + \frac{h}{2}(A_1 + BK_i)^*]$$

will be of the form (see Section 3 above)

$$\frac{1 + \frac{h}{2}\bar{\lambda}}{1 - \frac{h}{2}\bar{\lambda}} \tag{4.48}$$

where λ is an eigenvalue of $A + BK_i$. As K_i converges to $K_\infty + \Delta K$, these eigenvalues will assume negative real parts if ΔK is small. The modulus of (4.48) is then < 1 for all λ and stability of the numerical procedure results. One would ordinarily not invert the matrix in (4.47). Instead one would write

$$[I-\frac{h}{2}(A_1+BU^{-1}K_i)^*]L_{i+1}^* = [I+\frac{h}{2}(A_1+BU^{-1}K_i)^*]L_i^* - \frac{h^2}{4}[L_{i+1}^*L_{i+1}+L_i^*L_i]BU^{-1}B^*L_{i+1}^* \tag{4.49}$$

and solve the successive linear equations of the fixed point iteration by Gaussian elimination. For small h one iteration will frequently be sufficient for the accuracy required.

Let us examine the behavior of this method when applied to the scalar system

$$\dot{x} = ax + u, \quad a > 0,$$

with the cost functional

$$\int_0^\infty [a^2(x(t))^2 + (u(t))^2]\, dt .$$

Here $A = A_1 = a$, $B = 1$, $W = W_1 = 1$, $U = 1$. The equation satisfied by $Q \equiv q$ is

$$2aq + a^2 - q^2 = 0 ,$$

which has the positive solution

$$q = (1 + \sqrt{2})a$$

and the resulting optimal feedback constant

$$\hat{k} = -(1 + \sqrt{2})a \tag{4.50}$$

leading to the closed loop system

$$\dot{x} = -\sqrt{2}\, ax .$$

In this case, unlike the high dimensional example discussed earlier, there is no dimensional advantage in using (4.23), (4.24) in place of (4.4) In fact (4.4) consists of one equation in this case while (4.23), (4.24) becomes two equations

$$\dot{k} = -\ell^2 ,$$

$$\dot{\ell} = (a + k)\ell .$$

We integrate these to illustrate the use of (4.23), (4.24) and to verify the error behavior. The initial conditions are

$$k(0) = k_0 = 0 ,$$

$$\ell(0) = \ell_0 = a .$$

The equations (4.45), (4.47) are

$$k_{i+1} = k_i - \frac{h}{2}(\ell_{i+1}^2 + \ell_i^2) \tag{4.51}$$

$$\ell_{i+1} = [1 - \frac{h}{2}(a + k_i)]^{-1}([1 + \frac{h}{2}(a + k_i)]\ell_i - \frac{h^2}{4}(\ell_{i+1}^2 + \ell_i^2)\ell_{i+1}) . \tag{4.52}$$

The computations were carried out with five decimal place accuracy on a programmable calculator with the results summarized in the following tables. Some intermediate steps are omitted for brevity.

Table 1a

a = 1, h = .2

i	ℓ_i	k_i
0	1.00000	.00000
1	1.19026	-.24167
2	1.33908	-.56265
3	1.40612	-.93969
4	1.37006	-1.32512
5	1.24260	-1.66723
⋮	⋮	⋮
13	.17434	-2.41031
14	.13119	-2.41507
15	.09864	-2.41777
16	.07413	-2.41929
17	.05570	-2.42015
⋮	⋮	⋮
26	.00424	-2.42126
27	.00318	-2.42126
28	.00239	-2.42126
29	.00179	-2.42126
30	.00134	-2.42126

Table 1b

a = 1, h = .1

i	ℓ_i	k_i
0	1.00000	.00000
2	1.19297	-.24191
4	1.34464	-.56651
6	1.41158	-.95000
8	1.37091	-1.34092
10	1.31332	-1.52113
⋮	⋮	⋮
26	.17267	-2.40535
28	.13023	-2.40994
30	.09816	-2.41255
32	.07395	-2.41403
34	.05570	-2.41487
⋮	⋮	⋮
52	.00433	-2.41612
54	.00326	-2.41613
56	.00245	-2.41613
58	.00185	-2.41613
60	.00139	-2.41613

Table 2a

a = 2, h = .2

i	ℓ_i	k_i
0	2.00000	.00000
1	2.63847	-1.09615
2	2.73015	-2.53768
3	2.19576	-3.76520
4	1.45141	-4.45800
5	.85906	-4.74245
⋮	⋮	⋮
13	.00768	-4.88443
14	.00424	-4.88444
15	.00234	-4.88444

Table 2b

a = 2, h = .1

i	ℓ_i	k_i
0	2.00000	.00000
2	2.67817	-1.12531
4	2.74013	-2.65024
6	2.12124	-3.86827
8	1.36556	-4.48550
10	.80753	-4.72300
⋮	⋮	⋮
26	.00848	-4.84262
28	.00478	-4.84264
30	.00269	-4.84264

The correct values are, from (4. 50), $\hat{k} = -(1 + \sqrt{2}) = -2.41421\ldots$ for $a = 1$ and $-2(1 + \sqrt{2}) = 4.82842\ldots$ for $a = 2$. Now it is known (see [14], e. g.) that for a second order method like we are using here the error, as a function of the step size, h, has the form

$$\text{error} = ch^2 + \mathcal{O}(h^3).$$

This enables us to improve our results by the method of "extrapolation to the limit". If for $t = ih = 2i(\frac{h}{2})$ our numerical results have the form

$$(\text{Step length } h)\quad k_i = k + ch^2 + \ldots$$

$$(\text{Step length } \tfrac{h}{2})\quad \hat{k}_i = k + c(\tfrac{h}{2})^2 + \ldots$$

we can solve for k, c (approximately) in terms of the numerical values k_i, $\hat{k}_i$

$$k \approx \frac{4}{3}\hat{k}_i - \frac{1}{3}k_i,$$

$$c \approx \frac{4}{3}\left(\frac{k_i - \hat{k}_i}{h^2}\right).$$

Using this procedure we obtain

$$a = 1 : \ k \approx \frac{4}{3}(-2.41613) - \frac{1}{3}(-2.42126) = -2.41442$$

$$c \approx -.17100$$

$$a = 2 : \ k \approx \frac{4}{3}(-4.84264) - \frac{1}{3}(-4.88444) = -4.82871$$

$$(= 2(-2.41435))$$

$$c \approx -5.57333 .$$

The improved values are clearly quite reasonable. Often this process is much to be preferred to a further integration of the differential equations with a smaller step-size. It should be noted that c bears little demonstrable relationship to $\frac{1}{\gamma}$ (= 1/1. 41442 in the first case and 1/3. 82871 in the second). The expression (4. 43) is given by way of a bound, valid for small γ, not an asymptotic expression for the error. The reader will also note the more rapid convergence for the more highly damped case, $a = 2$, as compared with the case $a = 1$.

Bibliographical Notes, Chapter V

An easily accessible account of various numerical procedures for the solution of problems involving matrices and linear algebra generally is

[1] L. Fox: "An Introduction to Numerical Linear Algebra", Oxford University Press, New York, 1965.

In this chapter and also in Chapter VII some familiarity with optimization methods for functions of several variables will be found useful. We list two (out of many possible) references. Reference [3] is much more complete while reference [2], in this author's less than disinterested opinion, is easier to read.

[2] D. L. Russell: "Optimization Theory", W. A. Benjamin, Inc., New York, 1970.

[3] D. J. Wilde and C. S. Beightler: "Foundations of Optimization", Prentice-Hall, Inc., Englewood Cliffs, N. J., 1967.

[4] D. L. Kleinman: "On an iterative technique for Riccati equation computations", IEEE Trans. Auto. Control, AC-13 (1968), pp. 114-115.

[5] R. H. Bartels and G. W. Stewart: Algorithm 432, "Solution of the matrix equation AX + XB = C", Commun. ACM, 15(1972), pp. 820-826.

[6] R. A. Smith: "Matrix equation XA + BX = C", SIAM J. Appl. Math., 16(1968), pp. 198-201.

[7] T. Kailath: "Some Chandrasekhar-type algorithms for quadratic regulator problems", Proc. 1972 Decision and Control Conf., (New Orleans), pp. 219-223, Dec. 1972.

[8] ______: "Some new algorithms for recursive linear estimation in constant linear systems", IEEE Trans. Inf. Thy. IT-19 (1973).

[9] J. Casti: "Matrix Riccati equations, dimensionality reduction and generalized X-Y functions", Utilitas Mathematica (1974).

[10] ______: "Reduction of dimensionality for systems of linear two-point boundary value problems with constant coefficients", J. Math. Anal. Appl., 45 (1974).

[11] J. Casti and L. Ljung: "Some analytic and computational results for operator Riccati equations", AISM J. Control, 13 (1975), pp. 817-826.

From the wealth of books available which treat the numerical solution of ordinary differential equations we select only two here. Reference [12] develops error estimates and convergence and stability theories in a very thorough and yet accessible manner. It is very clearly written but perhpas a little austere. Reference [13] is a little more conversational in tone. Both include discussion of two point boundary value problems as well as initial value problems.

[12] P. Henrici: "Discrete Variable Methods in Ordinary Differential Equations", John Wiley and Sons, Inc., New York, 1962.

[13] F. B. Hildebrand: "Finite-Difference Equations and Simulations", Prentice-Hall, Inc., Englewood Cliffs, N. J., 1968.

[14] L. F. Richardson and J. A. Gaunt: "The deferred approach to the limit", Philos. Trans. Roy. Soc. London. Ser. A., Vol. 226 [1927], pp. 229-361.

Exercises, Chapter V

1. Use Potter's method to compute the optimal feedback matrix $\hat{K}$ for the system III-(3.12) relative to the cost

$$\int_0^\infty [(\dot{y})^2 + (\ddot{y})^2 + (u)^2 dt] = \int_0^\infty [(x^1)^2 + (-\tfrac{3}{5}x^2 + \tfrac{4}{5}u)^2 + (u)^2] \, dt .$$

This problem should be viewed as a computer programming assignment since the eigenvalues of a 6×6 matrix will be required. We suggest five decimal place accuracy.

2. Repeat 1. but use the Newton-Kleinman method of Section 2, solving the Liapounov equations by the iterative procedures described in Section 3. Again, this should be viewed as a programming assignment. Compare the methods described in Theorems 3.2, 3.3 with respect to computer time and accuracy. Again we suggest that five decimal place accuracy be the goal. Additionally, one might experiment with various values of r as described in (3.3)ff.

3. Use the method of Theorem 4.1 with the implicit trapezoidal rule, to recalculate the optimal feedback matrix for the example given in Section 1 following (1.4). Take $h = .1, .05$, and use the process of extrapolation to the limit described at the end of Section 4 to improve your result. If a reasonably good programmable calculator is available, we suggest its use for this problem.

4. Carry out the problem described at the end of Section 3. Here the Y_k are to be calculated, using four decimal places in all intermediate steps, directly from the formulae (3.29), (3.30) and not extracted from a table of the matrices X_k - the latter table, used for comparison, should be computed separately. A programmable calculator should be quite adequate for this problem.

5. (Very ambitious computational assignment!) Use the method of Theorem 4.1 to obtain the optimal feedback matrix for the system

$$\begin{pmatrix} \dot{y} \\ \dot{y} \end{pmatrix} = \begin{pmatrix} 0 & I \\ A_1 & 0 \end{pmatrix} \begin{pmatrix} y \\ \dot{y} \end{pmatrix} + bu$$

where A_1, b are described in (4. 29), (4. 30). Take $k = \rho = 1$ and use the cost

$$\int_0^\infty [(\dot{y}^n)^2 + (u)^2]\, dt .$$

Experiment with various values of the step length h; generally speaking h should be somewhat smaller than $\frac{1}{n}$. Try values $n = 5, 10, 15, \ldots$, as time and computational facilities permit. Can you draw any conclusions about the limiting optimal feedback law as $n \to \infty$?

6. Complete the work begun in IV- Exercise 2 by calculating the optimal controls $\hat{u}^1$, $\hat{u}^2$ on the time interval $[0,1]$. In this case the matrix Riccati differential equation (2. 28) should be integrated numerically. We suggest the fourth order Runge-Kutta method ([12], [13]) with $h = .05$.

CHAPTER VI

OPTIMAL CONTROL AND ESTIMATION FOR LINEAR STOCHASTIC SYSTEMS

1. Preliminaries

Many of the factors which affect an operating system cannot be specified precisely. When we originally presented linear control systems we wrote them as

$$\dot{x} = A(t)x + B(t)u + C(t)v,$$

where x represents the state vector, u the control vector and v a disturbance vector. So far we have said nothing about the disturbance v except to assume in most of our discussions that it is equal to zero. We propose to repair this defect now by taking v into account explicitly but- since by its very nature it is not precisely known - as a statistical entity which we call "white noise".

As a by-product of our work we obtain another derivation of the optimal control law $\hat{u}(t) = -U^{-1}(B^*Q + R^*)\hat{x}(t)$ studied in Chapter IV and, by also treating the initial state as a random vector, we are able to assign scalar values to the performance characteristics of controlled systems, thereby establishing criteria for comparison of various control schemes.

We begin this work by reviewing some of the elements of probability theory which we will use. Let x represent a vector random variable, which may take on many values, possibly all values, in the space E^n, but which assumes these values with certain relative frequencies or probabilities. These relative frequencies we make precise by the use of a probability density function $\sigma : E^n \to [0, \infty)$ which is Lebesgue measurable ([1], [2]), at least, and

$$\sigma : E^n \to [0, \infty)$$

satisfies

$$\int_{E^n} \sigma(x)\, dx = 1 .$$

There results from this a probability distribution, P_σ, which assigns to every measurable set $\Sigma \subseteq E^n$ the number

$$P_\sigma(\Sigma) = \int_\Sigma \sigma(x) dx \equiv \text{Prob}\,\{x \in \Sigma\},$$

expressing the probability that the vector x lies in the subset Σ.

Given a continuous function (random variable)

$$f : E^n \to E^1$$

we may define the average, or expected value of $f(x)$ as

$$E(f) = \int_{E^n} f(x)\, \sigma(x)\, dx .$$

A particularly important example occurs for $f(x) = x^i$, the i-th component of x:

$$E(x^i) \equiv e^i = \int_{E^n} x^i \sigma(x)\, dx .$$

These definitions can be extended componentwise to vector functions. In particular

$$E(x) \equiv \int_{E^n} x\sigma(x) dx \equiv \begin{pmatrix} \int_{E^n} x^1 \sigma(x) dx \\ \vdots \\ \int_{E^n} x^n \sigma(x) dx \end{pmatrix} .$$

If f is square integrable over E^n with respect to σ, i.e.,

$$\int_{E^n} |f(x)|^2 \sigma(x)dx < \infty$$

and g(x) is another function with the same property, we can define the covariance of f and g:

$$\begin{aligned}
\text{cov}(f,g) &= E((f - E(f))(g - E(g))) = \int_{E^n} (f(x) - E(f))(g(x) - E(g))\sigma(x)dx \\
&= \int_{E^n} f(x)g(x)\sigma(x)dx - E(f)\int_{E^n} g(x)\sigma(x)dx \\
&\qquad - E(g)\int_{E^n} f(x)\sigma(x)dx + E(f)E(g)\int_{E^n} \sigma(x)dx \\
&\qquad - E(fg) - E(f)E(g) - E(g)E(f) + E(f)E(g) \\
&= E(fg) - E(f)E(g) .
\end{aligned} \tag{1. 1}$$

The standard deviation of f is the square root of the covariance of f with itself:

$$\text{s. d. } f = \sqrt{E(f^2) - (E(f))^2} .$$

For vector valued functions

$$f(x) = \begin{pmatrix} f^1(x) \\ \vdots \\ f^p(x) \end{pmatrix}, \qquad g(x) = \begin{pmatrix} g^1(x) \\ \vdots \\ g^q(x) \end{pmatrix},$$

with square integrable components, we introduce the covariance matrix of dimensions $p \times q$ =

$$\text{cov}(f,g) = \begin{pmatrix} \text{cov}(f^1, g^1) & \dots & \text{cov}(f^1, g^q) \\ & & \vdots \\ \text{cov}(f^p, g^1) & \therefore & \text{cov}(f^p, g^q) \end{pmatrix}$$

$$= E(fg^*) - E(f)E(g)^* . \tag{1. 2}$$

It should be noted that

$$\text{cov}(g, f) = (\text{cov}(f, g))^*, \tag{1.3}$$

* in this case ordinarily indicating transpose. It is rarely necessary to consider complex functions f, g. When we do, (1.1) is replaced by

$$\text{cov}(f, g) = E((f - E(f))(\overline{g} - E(\overline{g}))) = E(f\overline{g}) - E(f)E(\overline{g})$$

and, when that modification is carried over into vector functions in (1.2), * then becomes "adjoint", i. e., conjugate transpose. When f and g are equal, the covariance matrix (1.2), now $\text{cov}(f, f)$, becomes a symmetric (or hermitian) non-negative $p \times p$ matrix.

A number of important special cases arise. When either f or g has "zero mean"

$$E(f) = 0 \quad \text{or} \quad E(g) = 0$$

(1.1) becomes

$$\text{cov}(f, g) = E(fg).$$

In the vector case if $E(f) = 0$ or $E(g) = 0$ we have in place of (1.2)

$$\text{cov}(f, g) = E(fg^*) \tag{1.4}$$

where fg^* is the $p \times q$ matrix with entries

$$(fg^*)^i_j = f^i g^j$$

and the expected value in (1.4) is defined entrywise.

In some instances the function f may depend on only a subset of the components of x, say $\hat{x} = (x^1, \ldots, x^r)$ and g may depend only on $\tilde{x} = (x^{r+1} \ldots x^n)$ while $\sigma(x)$ factors into

$$\sigma(x) = \hat{\sigma}(x)\tilde{\sigma}(\tilde{x}), \qquad \hat{\sigma} \geq 0, \quad \tilde{\sigma} \geq 0,$$

with

$$\int_{E^r} \hat{\sigma}(\hat{x})d\hat{x} = \int_{E^{n-r}} \tilde{\sigma}(\tilde{x})\, d\tilde{x} = 1.$$

This is equivalent to the statement that the random variables f and g are independent (see [3], [4]). We then have

$$E(f) = \int_{E^n} f(x)\sigma(x)dx = \int_{E^r} f(\hat{x})\hat{\sigma}(\hat{x})d\hat{x} \int_{E^{n-r}} \tilde{\sigma}(\tilde{x})d\tilde{x} = \int_{E^r} f(\hat{x})\hat{\sigma}(\hat{x})d\hat{x}$$

and similarly

$$E(g) = \int_{E^{n-r}} g(\tilde{x})\tilde{\sigma}(\tilde{x})d\tilde{x}$$

while

$$E(fg) = \int_{E^r}\int_{E^{n-r}} f(\hat{x})g(\tilde{x})\hat{\sigma}(\hat{x})\tilde{\sigma}(\tilde{x})d\tilde{x}\,d\hat{x}$$

$$= \left(\int_{E^r} f(\hat{x})\hat{\sigma}(\hat{x})d\hat{x}\right)\left(\int_{E^{n-r}} g(\tilde{x})\tilde{\sigma}(\tilde{x})d\tilde{x}\right) = E(f)\,E(g) .$$

Under these circumstances it is clear from (1.1) that

$$\operatorname{cov}(f, g) = 0. \tag{1.5}$$

The covariance matrix has many uses, as will become apparent as this chapter progresses. Suppose we take $f(x) = x$, assume that $\int_{E^n} (x^k)^2 \sigma(x)dx < \infty$, $k = 1, 2, \ldots, n$, and define the $n \times n$ covariance matrix

$$X = \operatorname{cov}(f, f) = \operatorname{cov}(x, x).$$

Assuming that $E(x) = 0$, which we shall henceforth unless explicitly stated to the contrary,

$$X = E(xx^*). \tag{1.6}$$

Then

$$E(\|x\|^2 = \int_{E^n} [(x^1)^2 + \ldots + (x^n)^2]\,\sigma(x)dx$$

$$= \sum_{k=1}^{n} \int_{E^n} (x^k)^2\sigma(x)dx = \sum_{k=1}^{n} \operatorname{cov}(x^k, x^k) = \operatorname{Tr}\operatorname{cov}(x, x) = \operatorname{Tr} X, \tag{1.7}$$

where Tr denotes the trace of an $n \times n$ matrix - the sum of the diagonal entries.

We conclude this introductory section by reviewing some of the important properties of the normal distribution, certainly the one most commonly used in practice. On E^1 it is defined by the density function (see [3], [4] for this as well as for proofs of properties cited below)

$$\sigma_{\mu,\nu^2}(x) = \frac{1}{\nu\sqrt{2\pi}} \exp\left(-\frac{1}{2}\left(\frac{x-\mu}{\nu}\right)^2\right), \quad \mu \text{ real}, \quad \nu > 0 . \tag{1.8}$$

For this density function we have

$$E(x) = \mu, \quad \operatorname{cov}(x,x) = \nu^2, \quad \text{s.d. } x = \nu . \tag{1.9}$$

It should be noted that once the probability distribution is known (or assumed) to be normal, it is completely determined by $E(x) = \mu$ and $\operatorname{cov}(x,x) = \nu^2$. If $E(x) = \mu = 0$, as is often the case, $\sigma_{0,\nu} = \sigma_\nu$ is completely determined by ν. This will play an important role in our later analysis.

For n independent normally distributed random variables $x^1, x^2, \ldots, x^n$ the <u>joint normal distribution</u> is

$$\sigma_{\mu,N}(x) = \prod_{k=1}^{n} \left[\frac{1}{\nu^k\sqrt{2\pi}} \exp\left(-\frac{1}{2}\left(\frac{x^k-\mu^k}{\nu^k}\right)^2\right)\right] . \tag{1.10}$$

Here

$$\mu = \begin{pmatrix} \mu^1 \\ \mu^2 \\ \vdots \\ \mu^n \end{pmatrix}, \quad x = \begin{pmatrix} x^1 \\ x^2 \\ \vdots \\ x^n \end{pmatrix} .$$

The mean or expected value of the vector x is

$$E(x) = \mu$$

and

$$N = \operatorname{cov}(x,x) = \operatorname{diag}\left((\nu^1)^2, (\nu^2)^2, \ldots, (\nu^n)^2\right)$$

the $n \times n$ diagonal matrix with the indicated diagonal entries.

If x is a random vector in E^n and the continuous relationship

$$y = f(x), \quad f = E^n \to E^m$$

holds, y becomes a random vector in E^m. For any measurable subset $\Sigma \subseteq E^m$

$$\text{Prob}\,\{y \in \Sigma\} = \text{Prob}\,\{x \,|\, f(x) \in \Sigma\} = \text{Prob}\,\{x \in f^{-1}(\Sigma)\},$$

where $f^{-1}(\Sigma)$ is precisely the set of x in E^n such that $f(x) \in \Sigma$. We are particularly interested in the case where

$$y = Fx,$$

F being a nonsingular $n \times n$ matrix. In this case we observe that

$$E(y) = F E(x), \quad \text{i.e.,} \quad E(y) = F\mu, \tag{1.11}$$

$$\begin{aligned} Y \equiv \text{cov}(y, y) &= E(yy^*) - E(y)E(y)^* = E(Fxx^*F^*) - F\mu\mu^*F^* \\ &= F(E(xx^*) - \mu\mu^*)F^* = F\,\text{cov}\,(x, x)F^* = FXF^*. \end{aligned} \tag{1.12}$$

If $\Sigma \subseteq E^m$ we have

$$\text{Prob}\,\{y \in \Sigma\} = \text{Prob}\,\{x \in F^{-1}\Sigma\} = \int_{F^{-1}(\Sigma)} \sigma_{\mu,\nu}(x)dx\,.$$

Using the standard rule for change of variable in multidimensional integrals we have

$$\int_{F^{-1}\Sigma} \sigma_{\mu,\nu}(x)dx = \int_{\Sigma} \sigma_{\mu,\nu}(F^{-1}y)d(F^{-1}y) = |\det F^{-1}| \int_{\Sigma} \sigma_{\mu,\nu}(F^{-1}y)dy. \tag{1.13}$$

These observations allow us to discuss the joint normal distribution for (not necessarily independent) random variables $y^1, y^2, \ldots, y^n$. Let us take $\nu^1 = \nu^2 = \ldots = \nu^n = 1$ in (1.10), $\mu = F^{-1}\mu_0$. Then the integrand in (1.13) becomes using (1.10), the <u>joint normal probability density</u>

$$\sigma_{\mu_0, FF^*} = \frac{1}{(\det 2\pi FF^*)^{1/2}} \exp(-\tfrac{1}{2}(F^{-1}y - F^{-1}\mu_0)^*(F^{-1}y - F^{-1}\mu_0))$$

$$= \frac{1}{(\det 2\pi FF^*)^{1/2}} \exp(-\tfrac{1}{2}(y-\mu_0)^*(FF^*)^{-1}(y-\mu_0)) . \tag{1.14}$$

The expected value and covariance of y have already been identified in (1.11) and (1.12). Since $\mu = F^{-1}\mu_0$, (1.11) gives

$$E(y) = FF^{-1}\mu_0 = \mu_0 . \tag{1.15}$$

The choice $\nu^1 = \nu^2 = \ldots = \nu^n = 1$ gives $X = \operatorname{cov}(x,x) = I$ and (1.12) then shows that

$$\operatorname{cov}(y,y) = FF^* . \tag{1.16}$$

We summarize in

Proposition 1.1. *If y is a vector random variable with joint normal probability density (1.14) then*

$$E(y) = \mu_0$$

$$\operatorname{cov}(y,y) = FF^* .$$

A converse to the above proposition, which may also serve as our definition of "joint normal distribution", is

Proposition 1.2. *Let it be assumed that the random variable y has a joint normal distribution and*

$$E(y) = \mu , \quad \operatorname{cov}(y,y) = Y , \tag{1.17}$$

for some symmetric positive definite matrix Y. Then the density function associated with y is

$$\sigma_{\mu,Y}(y) = \frac{1}{(\det 2\pi Y)^{1/2}} \exp(-\tfrac{1}{2}((y-\mu)^* Y^{-1}(y-\mu)) . \tag{1.18}$$

Next, suppose x is a random vector with vector mean μ and covariance matrix X.

Let a^* be the non-zero rwo vector

$$a^* = (a^1, a^2, \dots, a^n).$$

In standard texts on probability theory (see [4] particularly) the reader will find that

$$\xi = a^* x$$

is normally distributed in E^1. Now

$$E(\xi) = E(a^1x^1 + \dots + a^nx^n) = a^1E(x^1) + \dots + a^nE(x^n) = a^*\mu \tag{1.19}$$

$$\operatorname{cov}(\xi, \xi) = \sum_{k=1}^{n} \sum_{\ell=1}^{n} a^k \bar{a}^\ell \operatorname{cov}(x^k, x^\ell) = a^* X a, \tag{1.20}$$

where $a = (a^*)^*$ is the associated column vector and X, as indicated, is the covariance matrix of x. Since ξ is normally distributed, comparing (1.19), (1.20) with (1.18) we have

__Proposition 1.3.__ _If x is normally distributed with vector mean μ and covariance matrix X, then for each non-zero row vector a^*, $\xi = a^*x$ is normally distributed with density function_

$$\sigma_{a^*\mu,\, (a^*Xa)^{1/2}}(\xi) = \frac{1}{(a^*Xa)^{1/2}\sqrt{2\pi}} \exp\left(-\frac{1}{2}\,\frac{(\xi - a^*\mu)^2}{a^*Xa}\right). \tag{1.21}$$

If we consider an m vector, $m \le n$,

$$\xi = Ax \equiv \begin{pmatrix} a_1^* \\ \vdots \\ a_m^* \end{pmatrix} x, \tag{1.22}$$

with rank $A = m$, it is shown similarly that

$$E(\xi) = A\mu \equiv \mu_0$$

$$\mathrm{cov}(\xi, \xi) = AX A^* .$$

The distribution is normal and must have the form (1.18). Hence, comparing with (1.17), we have

Proposition 1.4. Let x be as in Proposition 1.3 and let ξ be given by (1.22). Then ξ is normally distributed with density function

$$\sigma_{\mu_0, (AXA^*)^{-1}}(\xi) = \frac{1}{(\det 2\pi AXA^*)^{1/2}} \exp(-\tfrac{1}{2}(\xi-\mu_0)(AXA^*)^{-1}(\xi-\mu_0)) . \quad (1.23)$$

In the sequel we consider only random variables having some sort of joint normal probability distribution. While many different distributions arise in various contexts, the central limit theorem and its accompanying results ([3], [4]) indicate that a very wide class of phenomena may be treated in this way.

2. Linear Systems with White Noise Disturbance

Let us again consider the linear system

$$\dot{x} = A(t)x + B(t)u + C(t)v, \quad x \in E^n, \quad u \in E^m, \quad v \in E^p, \quad (2.1)$$

where $u \in L_m^2[t_0, t_1]$ is a control function and $v \in L_p^2[t_0, t_1]$ is a disturbance affecting the system. We will suppose that u is generated by a linear feedback law

$$u(t) = K(t)x(t), \quad K(t) \ m \times n . \quad (2.2)$$

Then, with

$$S(t) = A(t) + B(t)K(t), \quad (2.3)$$

(2.1) becomes

$$\dot{x} = S(t)x + C(t)v . \quad (2.4)$$

We have noted that disturbances are, by their very nature, unpredictable, i.e., of a random nature. For this reason we want to consider v as a random element of $L_p^2[t_0, t_1]$. To do so in complete rigor requires the definition of a probability density on the infinite dimensional (Hilbert) space $L_p^2[t_0, t_1]$, a problem which cannot be treated in detail within the scope of this book. We will proceed by an alternate route which makes use of probability densities in finite dimensional spaces, albeit of increasing dimension.

Let $L_{p,N}^2[t_0, \tau]$, $t_0 < \tau \leq t_1$, denote the space of p-vector valued functions v(t) defined on $[t_0, \tau]$ assuming constant values $v_1, v_2, \ldots, v_N$ on the intervals

$$[t_0, t_0 + \frac{L}{N}], \ (t_0 + \frac{L}{N}, t_0 + \frac{2L}{N}], \ldots \ (t_0 + (N-1)\frac{L}{N}, \ \tau], \tag{2.5}$$

respectively, where $L = \tau - t_0$. We denote these intervals by I_k, k = 1, 2, . . . , N. Thus

$$v(t) \equiv v_k, \quad t \in I_k.$$

We assume that each v_k is a random vector in E^p with associated probability density $\sigma_k(v_k)$ and having zero mean:

$$E(v_k) = \int_{E^p} v_k \sigma_k(v_k)\, dv_k = 0.$$

Then $L_{p,N}^2[t_0, \tau]$ (really just E^{pN}) can be assigned a probability density

$$\bar{\sigma}(v) = \bar{\sigma}(v_1, v_2, \ldots, v_N) = \prod_{k=1}^{N} \sigma_k(v_k). \tag{2.6}$$

Functions v in $L_{p,N}^2[t_0, \tau]$ then become random variables with this probability density. If Σ_k, k = 1, 2, . . . , N, denotes a measurable subset of E^p, then $\prod_{k=1}^{N} \Sigma_k \equiv \Sigma$ denotes the Cartesian product subset of $L_{p,N}^2[t_0, \tau]$ (or E^{pN}) and

$$\text{Prob}\{v \in \Sigma\} = \prod_{k=1}^{N} \text{Prob}\{v_k \in \Sigma_k\}.$$

This construction corresponds to the assumption that for $k \neq \ell$, v_k and v_ℓ are independent random vectors and, as we saw in (1.5),

$$\text{cov}(v_k, v_\ell) = 0.$$

Let us define

$$\mathrm{cov}\,(v_k, v_k) = V_k, \quad k = 1, 2, \ldots, N. \tag{2.7}$$

Each v_k is then a symmetric, non-negative $p \times p$ matrix.

Let us consider the solution, $x(t)$, of (2.4) corresponding to $v \in L^2_{p,N}[t_0, \tau]$ with

$$x(t_0) = x_0 \in E^n. \tag{2.8}$$

Consistent with our statistical approach, we assume that x_0 is also a random variable with probability density function $\sigma_0(x_0)$ defined on E^n. We suppose x_0 has zero mean and

$$\mathrm{cov}\,(x_0, x_0) \equiv X_0 . \tag{2.9}$$

Then (x_0, v) is a random element of $E^n \times L^2_{p,N}[t_0, \tau]$ with (assuming x_0, v mutually independent) probability density (cf. (2.6))

$$\sigma(x_0, v) = \sigma_0(x_0)\bar{\sigma}(v).$$

Letting $\Phi(t, s)$ denote the fundamental matrix solution of

$$\dot{\Phi} = S(t)\,\Phi$$

with $\Phi(s, s) = I$, the solution $x(t)$ referred to above has, at τ, the value

$$x(\tau) = \Phi(\tau, t_0)x_0 + \sum_{k=1}^{N} \int_{t_{k-1}}^{t_k} \Phi(\tau, s)C(s)ds\, v_k \equiv \Psi_0 x_0 + \sum_{k=1}^{N} \Psi_k v_k, \tag{2.10}$$

where $t_k = t_0 + \frac{kL}{N}$ (cf. (2.5)) and the definitions of Ψ_k, $k = 0, 1, 2, \ldots, N$, are clear. We see, then, that $x(\tau)$ becomes an n-dimensional random variable defined on $E^n \times L^2_{p,N}[t_0, \tau]$, the functional form, indicating the dependence on (x_0, v), or $x_0, v_1, v_2, \ldots, v_N)$, being given by the right hand side of (2.10). Since we have assumed that $x_0, v_1, v_2, \ldots, v_N$ all have zero mean, it is easy to see that $x(\tau)$ likewise has zero mean, i.e., $E(x(\tau)) = 0$. What we are primarily interested in is the symmetric non-negative matrix

$$X_N(\tau) \equiv \operatorname{cov}(x(\tau),\ x(\tau)) = E(x(\tau)x(\tau)^*) \ . \tag{2.11}$$

It turns out that $X_N(\tau)$ can be computed quite readily.

Proposition 2.1. Let $x(t)$ be the solution of (2.4), (2.8) for $(x_0, v) \in E^n \times L^2_{p,N}[t_0, \tau]$. Then

$$X_N(\tau) = \Psi_0 X_0 \psi_0^* + \sum_{k=1}^{N} \Psi_k V_k \Psi_k^* \ . \tag{2.12}$$

Further, if Y is any symmetric $n \times n$ matrix

$$E(x(\tau)^* Y x(\tau)) = \operatorname{Tr}(YX_N(\tau)) \ . \tag{2.13}$$

Proof. The linearity of $\operatorname{cov}(f, g)$ with respect to f and g (conjugate linearity with respect to g if complex values are admitted) allows us to see that

$$\operatorname{cov}(x(\tau), x(\tau)) = \operatorname{cov}(\Psi_0 x_0 + \sum_{k=1}^{N} \psi_k v_k,\ \Psi_0 x_0 + \sum_{k=1}^{N} \psi_k v_k)$$

$$= \operatorname{cov}(\Psi_0 x_0, \Psi_0 x_0) + \sum_{k=1}^{N} [\operatorname{cov}(\Psi_0 x_0, \Psi_k v_k) + \operatorname{cov}(\Psi_k v_k, \Psi_0 x_0)] + \sum_{k=1}^{N} \sum_{j=1}^{N} \operatorname{cov}(\Psi_k v_k, \Psi_j v_j). \tag{2.14}$$

Now

$$E(\Psi_k v_k) = \int_{E^p} \Psi_k v_k \sigma_k(v_k) dv_k$$

$$= \Psi_k \int_{E^p} v_k \sigma_k(v_k) dv_k = \Psi_k E(v_k) = 0, \qquad k = 1, 2, \ldots, n$$

so

$$\operatorname{cov}(\Psi_k v_k, \Psi_j v_j) = E(\Psi_k v_k v_j^* \Psi_j^*)$$

$$= \int_{E^{2p}} \Psi_k v_k v_j^* \Psi_j^* \sigma_k(v_k)\sigma_j(v_j) dv_k dv_j$$

$$= \Psi_k \int_{E^{2p}} v_k v_j^* \sigma(v_k)\sigma(v_j) dv_k dv_j \Psi_j^*$$

$$= \Psi_k \operatorname{cov}(v_k, v_j)\Psi_j^* = \begin{cases} 0, & k \neq j \\ \Psi_k V_k \Psi_k^*, & k = j \end{cases} \quad (2.15)$$

Similar reasoning applies to show that

$$\operatorname{cov}(\Psi_0 x_0, \Psi_k v_k) = 0, \qquad k = 1, 2, \ldots, N, \quad (2.16)$$

$$\operatorname{cov}(\Psi_0 x_0, \Psi_0 x_0) = \Psi_0 X_0 \Psi_0^* . \quad (2.17)$$

Then (2.12) follows immediately from (2.14) - (2.17).

For (2.13) we use the result (see Section 3 below) to the effect that

$$\operatorname{Tr} AB = \operatorname{Tr} BA$$

whenever A, B are matrices of dimension $r \times s$, $s \times r$ (so that AB and BA are both defined). Then

$$E(x(\tau)^* Y x(\tau)) = E(\operatorname{Tr}(x(\tau)^* Y x(\tau))$$

(since $x(\tau)^* Y x(\tau)$ is a 1×1 matrix)

$$= E(\operatorname{Tr} Y x(\tau) x(\tau)^*$$

$$= \operatorname{Tr}(Y E(x(\tau) \times (\tau)^*)) = \operatorname{Tr}(Y X_N(\tau))$$

and we have (2.13). Q. E. D.

Our objective now is to see what happens as N, the number of subintervals of $[t_0, \tau]$, increases toward infinity. In order to obtain a meaningful result we must make some further hypotheses. We assume that there is a continuous $n \times n$ symmetric non-negative matrix valued function V(t) defined on some interval which includes $[t_0, \tau]$. For each $N = 1, 2, 3, \ldots$ we consider elements $v = (v_{N,1}, v_{N,2}, \ldots, v_{N,N}) \in L^2_{p,N}[t_0, \tau]$ and suppose that

$$\operatorname{cov}(v_{N,k}, v_{N,j}) = \begin{cases} 0, & k \neq j \\ \frac{N}{L} V(\tau_{N,k}), & k = j \end{cases}, \tag{2.18}$$

where

$$\tau_{N,k} \in [t_{N,k-1}, t_{N,k}] \equiv [t_0 + (k-1)\frac{L}{N}, t_0 + k\frac{L}{N}] . \tag{2.19}$$

It is not easy to motivate the term $\frac{N}{L}V(\tau_{N,k})$ at this point. We will make an attempt in this direction at the end of this section. The important point now is to see what happens to $\operatorname{cov}(x(\tau), x(\tau)) = E(x(\tau)x(\tau)^*) = X_N(\tau)$ as we permit N to grow without bound.

<u>Theorem 2.2.</u> <u>For any choices of</u> $\tau_{N,k} \in [t_{N,k-1}, t_{N,k}]$ <u>the limit</u>

$$\lim_{N \to \infty} X_N(\tau) \equiv X(\tau)$$

<u>exists and</u>

$$X(\tau) = \Phi(\tau, t_0) X_0 \Phi(\tau, t_0)^* + \int_{t_0}^{\tau} \Phi(\tau, s) C(s) V(s) C(s)^* \Phi(\tau, s)^* ds . \tag{2.20}$$

<u>Further,</u>

$$E(x(\tau)^* Y x(\tau)) = \operatorname{Tr} Y X(\tau) \tag{2.21}$$

<u>for each</u> $n \times n$ <u>symmetric matrix</u> Y. <u>Regarded as a function of</u> τ, $X(\tau)$ <u>satisfies the matrix ordinary differential equation</u>

$$\frac{dX(\tau)}{d\tau} \equiv \dot{X}(\tau) = S(\tau)X(\tau) + X(\tau)S(\tau)^* + C(\tau)V(\tau)C(\tau)^* \tag{2.22}$$

<u>and the initial condition</u>

$$X(0) = X_0 = \operatorname{cov}(x_0, x_0).$$

Proof. Let the $X_N(\tau)$ be defined as in Proposition 1.1 with the $V_k = V_{N,k} = \operatorname{cov}(v_{N,k}, v_{N,k})$ being given by (2.18). Then

$$X_N(\tau) = \Psi_0 X_0 \Psi_0^* + \sum_{k=1}^{N} \Psi_k V_k \Psi_k^* = \quad \text{(cf. (2.10))}$$

$$= \Phi(\tau, t_0) X_0 \Phi(\tau, t_0)^* + \frac{N}{L} \sum_{k=1}^{N} \int_{t_{N,k-1}}^{t_{N,k}} \int_{t_{N,k-1}}^{t_{N,k}} \Phi(\tau, s) C(s) V(\tau_{N,k}) C(\sigma)^* \Phi(\tau, \sigma)^* d\sigma ds.$$

From the uniform continuity of $\Phi(\tau, t)$, $C(t)$ on $t_0 \le t \le \tau$ $(= t_0 + L)$ it follows that

$$\frac{N}{L} \int_{t_{N,k-1}}^{t_{N,k}} \int_{t_{N,k-1}}^{t_{N,k}} \Phi(\tau, s) C(s) V(\tau_{N,k}) C(\sigma)^* \Phi(\tau, \sigma)^* d\sigma ds$$

$$= \frac{L}{N} [\Phi(\tau, \tau_{N,k}) C(\tau_{N,k}) V(\tau_{N,k}) C(\tau_{N,k})^* \Phi(\tau, \tau_{N,k})^* + \delta_{N,k}] \qquad (2.24)$$

where

$$\lim_{N \to \infty} \| \delta_{N,k} \| = 0 \quad \text{uniformly}, \quad k = 1, 2, \ldots, N. \qquad (2.25)$$

But the uniform continuity of $\Phi(\tau, t)$, $V(t)$, $C(t)$ for $t_0 < t < \tau$ shows that

$$\int_{t_{N,k-1}}^{t_{N,k}} \Phi(\tau, s) C(s) V(s) C(s)^* \Phi(\tau, s)^* ds$$

$$= \frac{L}{N} [\Phi(\tau, \tau_{N,k}) C(\tau_{N,k}) V(\tau_{N,k}) C(\tau_{N,k})^* \Phi(\tau, \tau_{N,k})^* + d_{N,k}] \qquad (2.26)$$

where also

$$\lim_{N \to \infty} \| d_{N,k} \| = 0 \quad \text{uniformly}, \quad k = 1, 2, \ldots, N. \qquad (2.27)$$

Combining (2.23), (2.24), (2.26) we see that

$$X_N(\tau) = \Phi(\tau, t_0) X_0 \Phi(\tau, t_0) + \int_{t_0}^{\tau} \Phi(\tau, s) C(s) V(s) C(s)^* \Phi(\tau, s) ds + \sum_{k=1}^{N} \frac{L}{N} (\delta_{N,k} - d_{N,k}).$$

Now

$$\left\|\sum_{k=1}^{N} \frac{L}{N}(\delta_{N,k} - d_{N,k})\right\| \leq L \max_{k=1,\dots,N} \{ \|\delta_{N,k}\| + \|d_{N,k}\| \} .$$

Thus (2.25) and (2.27) imply

$$\lim_{N\to\infty} \left\|\sum_{k=1}^{N} \frac{L}{N}(\delta_{N,k} - d_{N,k})\right\| = 0$$

and we have

$$X(\tau) \equiv \lim_{N\to\infty} X_N(\tau)$$

$$= \Phi(\tau, t_0)X_0\Phi(\tau, t_0) + \int_{t_0}^{\tau} \Phi(\tau, s)C(s)V(s)C(s)^*\Phi(\tau, s)^* ds . \qquad (2.28)$$

This establishes (2.20). The result (2.21) follows immediately from (2.13) and the first part of (2.28). Differentiating (2.20) with respect to τ and using

$$\frac{\partial \Phi(\tau, s)}{\partial \tau} = S(\tau)\Phi(\tau, s), \qquad \Phi(\tau, \tau) = I,$$

together with (2.20) again we have

$$\dot{X}(\tau) = \frac{\partial \Phi(\tau, t_0)}{\partial \tau} X_0 \Phi(\tau, t_0) + \int_{t_0}^{\tau} \frac{\partial \Phi(\tau, s)}{\partial \tau} C(s)V(s)C(s)^*\Phi(\tau, s)^* ds$$

$$+ \Phi(\tau, t_0)X_0 \frac{\partial \Phi(\tau, t_0)^*}{\partial \tau} + \int_{t_0}^{\tau} \Phi(\tau, s)C(s)V(s)C(s)^* \frac{\partial \Phi(\tau, s)^*}{\partial \tau} ds$$

$$+ \Phi(\tau, \tau)C(\tau)V(\tau)C(\tau)^*\Phi(\tau, \tau)^*$$

$$= S(\tau)X(\tau) + X(\tau)S(\tau)^* + C(\tau)V(\tau)C(\tau)^*,$$

which is (2.22). Q. E. D.

Now, the question remaining is this. Can we meaningfully write

$$X(\tau) = \operatorname{cov}(x(\tau), x(\tau)) ? \qquad (2.29)$$

For each finite value of N we have seen that

$$X_N(\tau) = \text{cov}\,(x(\tau), x(\tau)). \tag{2.30}$$

In the case of (2. 30) $x(\tau)$ is a random vector on $E^n \times L^2_{p,N}[t_0, \tau]$, depending, as it does, on $(x_0, v) = (x_0, v_1, v_2, \ldots, v_N)$. The probability distribution is $\sigma(x_0, v) = \sigma_0(x_0)\bar{\sigma}(v) = \sigma_0(x_0)\sigma_1(v_1)\sigma_2(v_2)\ldots\sigma_N(v_N)$. Thus $\text{cov}(x(\tau), x(\tau))$ can be rigorously defined and, as we have seen, it turns out to be the matrix $X_N(\tau)$ given by (2. 12). The problem in connection with (2. 29) is that $x(\tau)$ there depends on $x_0 \in E^n$, $v \in L^2_p[t_0, \tau]$. The latter is an infinite dimensional Hilbert space and we have avoided the question of defining a probability distribution on this space. Having done this, we have no probability distribution with respect to which $\text{cov}\,(x(\tau), x(\tau))$ can be defined in this case. It turns out, however, that such a probability distribution exists quite generally and, with respect to that probability distribution $X(\tau)$ is, indeed, equal to $\text{cov}(x(\tau), x(\tau))$. Below we discuss only the case wherein the $v_{N,k}$ are independent and normally distributed.

We will refer to the space on which the probability density function is defined as the "sample" space. (This will suffice here: more precisely the term sample space is often used to refer to the collection of "events", in this case the set of measurable subsets of, say, E^n, or $E^n \times L^2_{p,N}[t_0, \tau]$, etc.) Now $x(\tau)$ is a random variable defined on $E^n \times L^2_{p,N}[t_0, \tau]$ for each $N = 1, 2, 3, \ldots$, and we have seen that the covariance matrix is $X_N(\tau)$, given by (2. 12). Now $x_n(\tau)$ is related to the independent random vectors $x_0, v_{N,1}, \ldots, v_{N,N}$ by

$$x(\tau) = \Phi(\tau, t_0)x_0 + \sum_{k=1}^{N} \int_{t_{N,k-1}}^{t_{N,k}} \Phi(\tau, s)C(s)ds\, v_{N,k} \equiv F_N \begin{pmatrix} x_0 \\ v_N \end{pmatrix}. \tag{2.31}$$

If we assume that the x_0, $v_{N,k}$ are independent and normally distributed then, following the procedure outlined in Section 1, a normal probability density function associated with $x(\tau)$ can be defined in E^n. We know that the mean is zero and the covariance matrix is $X_N(\tau)$. Thus for $x_0, v_N \in E^n \times L^2_{p,N}[t_0, \tau]$, $x(\tau)$ is a random vector and, if we "lift" the probability density, via F_N, to E^n, as in Section 1, E^n becomes the new sample space and the probability density on E^n for $x(\tau)$ must, by virtue of Proposition 1. 2, be

$$\sigma_{0,X_N(\tau)}(x) \equiv \frac{1}{(\det 2\pi X_N(\tau))^{1/2}} \exp(-\frac{1}{2}x^* X_N(\tau)^{-1} x) . \tag{2. 32}$$

The statement

$$\operatorname{cov}(x(\tau), x(\tau)) = X_N(\tau)$$

is true if we use the density $\sigma(x_0, v)$ on $E^n \times L^2_{p,N}[t_0, \tau]$ or if we use the density (2. 32) on E^n. In the first instance we are computing the covariance of the random variable (2. 31) defined on $E^n \times L^2_{p,N}[t_0, \tau]$ with respect to the probability density $\sigma(x_0, v_N)$ in the second instance the random variable is the identity on E^n and the probability density is given by (2. 32). It is a complicated matter to study the behavior of $\sigma(x_0, v_N)$ as N tends to infinity but it is easy to see what happens to (2. 32) as $N \to \infty$. We have

$$\lim_{N \to \infty} X_N(\tau) = X(\tau),$$

with $X(\tau)$ given by (2. 20). The probability density (2. 32) then has the behavior

$$\lim_{N \to \infty} \sigma_{0,X_{N(\tau)}}(x) \equiv \sigma_{0,X(\tau)}(x) = \frac{1}{(\det 2\pi X(\tau))^{1/2}} \exp(-\frac{1}{2}x^* X(\tau)^{-1} x) .$$

With respect to this probability density it is true that

$$\operatorname{cov}(x(\tau), x(\tau)) = X(\tau).$$

This still (deliberately) skirts the question of a probability distribution on $E^n \times L^2_p[t_0, \tau]$. In order to satisfy the reader that this question is not wholly intractible, we ask him to consider the following argument.

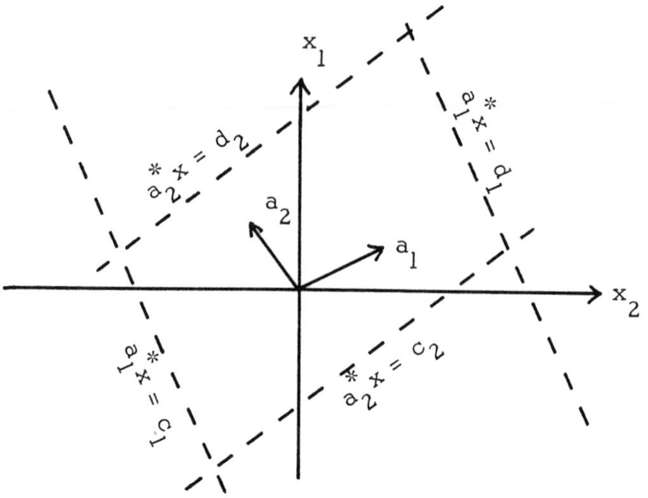

Fig. VI. 1. A "box" in E^2

Let $a_1, a_2, \ldots, a_r$, $r \leqslant n$, be independent vectors in E^n. An "r-box" in E^n is a set

$$B = \{x \in E^n | c_k \leq a_k^* \leq d_k, \; c_k < d_k, \; c_k, d_k, \text{ real}, \; k = 1, 2, \ldots, r. \}.$$

Note that the box is infinite in $n - r$ directions if $r < n$.

A probability density on E^n assigns a probability "measure" to B. If the probability density on E^n is

$$\sigma_{0,X}(x) = \frac{1}{\det 2\pi X)^{1/2}} \exp(-\frac{1}{2} x^* X^{-1} x)$$

then

$$\text{Prob}\{x \in B\} = \int_B \sigma(x)\, dx \; .$$

Equally well, if we let

$$\xi = Ax = \begin{pmatrix} a_1^* \\ a_2^* \\ \vdots \\ a_r^* \end{pmatrix} x$$

then, with (cf. Proposition 1.4 and Exercise 1 below)

$$\text{Prob}\,\{x \in B\} = \text{Prob}\,\{c_k \le \xi^k \le d_k,\ k = 1, 2, \dots, r\}$$
$$= \int_{c_r}^{d_r} \cdots \int_{c_1}^{d_1} \sigma_{0, AXA^*}(\xi)\, d\xi^1 \dots d\xi^r \tag{2.33}$$

where

$$\sigma_{0, AXA^*}(\xi) = \frac{1}{(\det 2\pi AXA^*)^{1/2}} \exp\left(-\frac{1}{2}\xi^*(AXA^*)^{-1}\xi\right).$$

Now it is well known ([5], [6]) that if $a = a(s) \in L_p^2[t_0, \tau]$ then

$$\ell(u) = \int_{t_0}^{\tau} a(s)^* v(s) ds = (v, a)_{L_p^2[t_0, \tau]}$$

defines a continuous linear functional on $L_p^2[t_0, \tau]$. An "r-box" in $L_p^2[t_0, \tau]$ is a set

$$B = \{v \in L_p^2[t_0, \tau] \mid c_k \le (v, a_k)_{L_p^2[t_0, \tau]} \le d_k,\ c_k < d_k,\ c_k, d_k \text{ real},\ k = 1, 2, \dots, r\}. \tag{2.34}$$

Now we pose the question: can we assign a probability measure to B? We will argue that we can do this for any continuous, positive, symmetric $p \times p$ matrix function $V(s)$ defined on $[t_0, \tau]$. The process is essentially the same as what we have used already in Theorem 2.2.

Each element $v \in L_p^2[t_0, \tau]$ can be approximated arbitrarily closely in that space by an element $(v_{N,1}, \dots, v_{N,N}) = v_N \in L_{p,N}^2[t_0, \tau]$. Moreover

$$\lim_{N \to \infty} \sum_{k=1}^{N} \frac{L}{N} v_{N,k}^* v_{N,k} = \int_{t_0}^{\tau} v(s)^* v(s) ds = \|v\|^2_{L_p^2[t_0, \tau]},$$

$$\lim_{N \to \infty} \sum_{k=1}^{N} \frac{L}{N} a_{N,k}^* v_{N,k} = \int_{t_0}^{\tau} a(s)^* v(s) ds = (v, a)_{L_p^2[t_0, \tau]}.$$

Let

$$\xi_j = \int_{t_0}^{\tau} a_j(s)^* v(s) ds, \qquad j = 1, 2, \ldots, r.$$

$$\xi_j = \lim_{N \to \infty} \sum_{k=1}^{N} \frac{L}{N} (a_j)^*_{N,k} v_{N,k} \equiv \lim_{N \to \infty} \xi_j^N .$$

Suppose we assume ($\tau_{N,k}$ as in Theorem 2.2)

$$\operatorname{cov}(v_k, v_\ell) = \begin{cases} 0, & k \neq \ell \\ \frac{N}{L} V(\tau_{N,k}) \equiv \frac{N}{L} V_{N,k}, & k = \ell \end{cases}$$

Then it is easy to see that ($A_{N,k} = \begin{pmatrix} (a_1)^*_{N,k} \\ \vdots \\ (a_r)^*_{N,k} \end{pmatrix}$)

$$\operatorname{cov}(\xi^N, \xi^{N'}) = \sum_{k=1}^{N} \frac{L}{N} A_{N,k} V_{N,k} A^*_{N,k} .$$

If we assume the $v_{N,k}$ normally distributed with zero mean, then the ξ^N are also normally distributed with zero mean. The probability density must be

$$\sigma^N(\xi) = \frac{1}{\left(\det 2\pi \sum_{k=1}^{N} \frac{L}{N} A_{N,k} V_{N,k} A^*_{N,k}\right)^{1/2}} \times \exp\left(-\frac{1}{2} \xi^* \left(\sum_{k=1}^{N} \frac{L}{N} A_{N,k} V_{N,k} A^*_{N,k}\right)^{-1} \xi\right) .$$

As $N \to \infty$ we have (cf. Theorem 2.2)

$$\lim_{N \to \infty} \sum_{k=1}^{N} \frac{L}{N} A_{N,k} V_{N,k} A^*_{N,k} = A$$

where $A = (a^i_j)$,

$$a^i_j = \int_{t_0}^{\tau} a_i(s)^* V(s) a_j(s) \, ds .$$

(hence

$$A = \int_{t_0}^{\tau} A(s)V(s)A(s)^* ds, \quad A(s) = \begin{pmatrix} a_1(s) \\ \vdots \\ a_r(s) \end{pmatrix}).$$

The limiting probability density is

$$\sigma_{0,A}(\xi) = \frac{1}{(\det 2\pi A)^{1/2}} \exp\left(-\frac{1}{2}\xi^* A^{-1}\xi\right). \tag{2.35}$$

The probability measure of the box (2. 34) in $L_p^2[t_0, \tau]$ is defined to be (cf. (2. 33))

$$\text{Prob}\,\{v \in B\} = \int_{c_r}^{d_r} \cdots \int_{c_1}^{d_1} \sigma_{0,A}(\xi) d\xi^1 \ldots d\xi^r .$$

Definition 2. 3. Disturbances $v \in L_p^2[t_0, \tau]$ constitute (normally distributed) white noise with covariance matrix $V(s)$ if these disturbances occur in such a manner that for any fixed elements $a_1, a_2, \ldots, a_r \in L_p^2[t_0, \tau]$ we have

$$\text{Prob}\,\{c_k \leq (v, a_k) \leq d_k, \quad k = 1, 2, \ldots, r\} = \int_{c_r}^{d_r} \cdots \int_{c_1}^{d_1} \sigma_{0,A}(\xi) d\xi^1 \ldots d\xi^r$$

where $\sigma_{0,A}(\xi)$ is the probability density on E^r defined by (2. 35).

The meaning of the term "white noise" is (hopefully) elucidated in Exercise 2.

Definition 2. 4. Disturbances v defined on $[0, \infty)$ and lying in $L_p^2[0, T]$ for each $T > 0$ constitute (normally distributed) steady state white noise with covariance matrix V (constant) if the restrictions of v to $[0, T]$ constitute white noise with covariance V, in the sense of Definition 2. 3, for each $T > 0$.

We now consider the constant coefficient system

$$\dot{x} = Sx + Cv, \quad x \in E^n, \; v \in E^p \tag{2.36}$$

and we suppose that the matrix S is a stability matrix. Supposing v to be steady state white noise with covariance matrix V we have

Theorem 2.5. Let $X(\tau) = \mathrm{cov}(x(\tau), x(\tau))$, where $x(\tau)$ is the solution of (2.36) with

$$x(0) = x_0, \quad \mathrm{cov}(x_0, x_0) = X_0. \tag{2.37}$$

Then

$$\lim_{\tau \to \infty} X(\tau) = X$$

where X is the solution of the Liapounov equation

$$SX + XS^* + CVC^* = 0 . \tag{2.39}$$

Proof. Using (2.20) we see that for each $\tau \geq 0$

$$SX(\tau) + X(\tau)S^* = S\Phi(\tau, 0)X_0\Phi(\tau, 0)^*$$
$$+ \Phi(\tau, 0)X_0\Phi(\tau,0)^*S^* + \int_0^\tau S\Phi(\tau,s)CVC^*\Phi(\tau,s)^*ds + \int_0^\tau \Phi(\tau,s)CVC^*\Phi(\tau,s)^*S^*ds .$$

Since

$$\frac{\partial \Phi(\tau,s)}{\partial s} = \frac{\partial}{\partial s}(e^{S(\tau-s)}) = -Se^{S(\tau-s)} = -S\Phi(\tau, s),$$

we have

$$SX(\tau) + X(\tau)S^* = Se^{S\tau}X_0e^{S^*\tau}$$
$$+ e^{S\tau}X_0e^{S^*\tau}S^* - \int_0^\tau \frac{\partial}{\partial s}(\Phi(\tau, s)CVC^*\Phi(\tau, s)^*)ds$$
$$= Se^{S\tau}X_0e^{S^*\tau} + e^{S\tau}X_0e^{S^*\tau}S^* + e^{S\tau}CVC^*e^{S^*\tau} - CVC^* .$$

Since S is a stability matrix

$$\lim_{\tau \to \infty} e^{S\tau} = 0 . \tag{2.40}$$

Letting X be the unique solution of (2.39) (known to exist from Chapter III), we have

$$S(X - X(\tau)) + (X - X(\tau))S^* + Se^{S\tau}X_0e^{S^*\tau} + e^{S\tau}X_0e^{S^*\tau}S^* + e^{S\tau}CVC^*e^{S^*\tau} = 0$$

whence (cf. III - 1)

$$X-X(\tau)=\int_0^\infty e^{S\sigma}[Se^{S\tau}X_0e^{S^*\tau}+e^{S\tau}X_0e^{S^*\tau}S^*+e^{S\tau}CVC^*e^{S^*\tau}]e^{S\sigma}d\sigma$$

$$= e^{S\tau}\{\int_0^\infty e^{S\sigma}[SX_0+X_0S^*+CVC^*]e^{S^*\sigma}d\sigma\}e^{S^*\tau} = e^{S\tau}Ze^{S^*\tau} \tag{2.41}$$

where Z is the fixed (with respect to τ) matrix displayed in brackets. With (2.40), (2.41) clearly gives

$$\lim_{\tau\to\infty}(X-X(\tau)) = \lim_{\tau\to\infty} e^{S\tau}Ze^{S^*\tau} = 0. \qquad \text{Q. E. D.}$$

It should be noted that X is independent of the initial state covariance X_0.

For each τ, the probability density associated with $x(\tau)$ is

$$\sigma_{0,X(\tau)}(x) = \frac{1}{(\det 2\pi X(\tau))^{1/2}}\exp(-\frac{1}{2}x^*X(\tau)^{-1}x).$$

Since $\lim_{\tau\to\infty} X(\tau) = X$, the limiting probability density for $x(\tau)$ is

$$\sigma_{0,X}(x) = \frac{1}{(\det 2\pi X)^{1/2}}\exp(-\frac{1}{2}x^*X^{-1}x). \tag{2.42}$$

In other words, for Σ a measurable subset of E^n,

$$\lim_{\tau\to\infty}(\text{Prob}\,\{x(\tau)\in\Sigma\}) = \int_\Sigma \sigma_{0,X}(x)\,dx\ .$$

There is another way in which we can interpret the result of Theorem 2.5. Let χ_Σ be the "characteristic function" of Σ, i.e.,

$$\chi_\Sigma(x) = \begin{cases} 1, & x\in\Sigma \\ 0, & x\notin\Sigma. \end{cases}$$

Then for any τ_1, τ_2,

$$T_\Sigma = \int_{\tau_1}^{\tau_2} \chi_\Sigma(x(\tau))d\tau$$

represents the amount of time which $x(\tau)$ spends in Σ and

$$F_\Sigma = T_\Sigma/(\tau_2 - \tau_1)$$

represents the fraction of the time spend in Σ. The quantities T_Σ, F_Σ can be regarded as random variables and one can justify the following formal computation: the expected value of T_Σ is

$$E(T_\Sigma) = \int_{\tau_1}^{\tau_2} E(\chi_\Sigma(x(\tau)))d\tau = \int_{\tau_1}^{\tau_2}\int_{E^n} \chi_\Sigma(x)\sigma_{0,X(\tau)}(x)dxd\tau = \int_{\tau_1}^{\tau_2}\int_\Sigma \sigma_{0,X(\tau)}(x)\,dx\,d\tau \;.$$

If we let τ_1, τ_2 both tend to ∞, we see that

$$\lim_{\tau_1,\tau_2\to\infty} E(T_\Sigma) = \int_{\tau_1}^{\tau_2}\int_\Sigma \sigma_{0,X}(x)dx\,d\tau = (\tau_2-\tau_1)\int_\Sigma \sigma_{0,X}(x)\,dx \;.$$

$$\lim_{\tau_1,\tau_2\to\infty} E(F_\Sigma) = \int_\Sigma \sigma_{0,X}(x)\,dx \;. \tag{2.43}$$

(In fact it is easy to see that (2. 43) can be replaced by

$$\lim_{\tau_2\to\infty} E(F_\Sigma) = \int_\Sigma \sigma_{0,X}(x)\,dx \;.)$$

Thus the matrix X and the probability density (2. 42) associated with X have the significance that they permit computation of the average time which $x(\tau)$ spends in Σ, either for intervals $[\tau_1, \tau_2]$ with τ_1, τ_2 both large or just τ_2 large, assuming $x(\tau)$ obeys (2. 36) with v normally distributed steady state white noise with covariance matrix V. Thus if Σ denotes a "danger area" for plant operation, representing operating states which, perhaps, cause damage of some sort, one can expect to estimate the average time spent in the danger area, expected cumulative damage, etc. For these and other reasons the probabilistic interpretation of the equation

(2.39) is extremely important.

3. The Calculus of Matrix Functions

In many applications it is useful and convenient to consider $n \times m$ matrices, X, as elements of the vector space E^{nm}. The entries of X can be taken in any pre-determined order and made into the components of a vector of dimension nm. Whenever this is necessary, let us agree to do it column by column. The fact is, however, that it is normally undesirable to actually express X in this form. It is preferable to keep X in matrix form and re-express the various important properties of E^{nm} in matrix terms.

Definition 3.1. For matrices $X, Y \in E^{nm}$ the inner product is defined by

$$(X, Y)_{E^{nm}} = \operatorname{Tr} Y^* X \tag{3.1}$$

where "Tr" denotes the trace of the $m \times m$ matrix in question, i.e., the sum of the diagonal entries.

To show that this is consistent with the usual inner product in E^{nm}, note that for $X = (x^i_j)$, $Y = (y^i_j)$,

$$(Y^*X)^j_j = \sum_{i=1}^{n} (Y^*)^j_i x^i_j = \sum_{i=1}^{n} x^i_j \bar{y}^i_j$$

and hence

$$\operatorname{Tr} Y^* X = \sum_{j=1}^{m} \sum_{i=1}^{n} x^i_j \bar{y}^i_j , \tag{3.2}$$

and the right hand side of (3.2) is the inner product which we would obtain if X and Y were written out, column by column, as nm-dimensional vectors.

The norm is then

$$\|X\| = (X, X)^{1/2} = \operatorname{Tr} X^* X$$

Various properties of the inner product are reflected in properties of the trace operation. Thus

$$\operatorname{Tr} M^* = \overline{\operatorname{Tr} M} \tag{3.3}$$

gives

$$(Y, X)_{E^{nm}} = \operatorname{Tr} X^* Y = \overline{\operatorname{Tr} Y^* X} = \overline{(X, Y)}_{E^{nm}} .$$

If M_1 and M_2 are $m \times n$, $n \times m$ matrices, then

$$\operatorname{Tr} M_1 M_2 = \sum_{i=1}^{m} \left(\sum_{j=1}^{n} (M_1)^i_j (M_2)^j_i \right) = \sum_{j=1}^{n} \left(\sum_{i=1}^{m} (M_2)^j_i (M_1)^i_j \right) = \operatorname{Tr} M_2 M_1 . \tag{3.4}$$

Thus

$$(X, Y)_{E^{nm}} = \operatorname{Tr} Y^* X = \operatorname{Tr} XY^* = \operatorname{Tr} (X^*)^* Y^* = (Y^*, X^*)_{E^{nm}} . \tag{3.5}$$

A wide variety of linear operators may be defined on E^{nm}, of course. Those of greatest interest to us, in view of the importance of equations of the form $Ax + XB = C$ in our subject, are linear operators, T, defined by the formula

$$T(x) = AX + XB, \tag{3.6}$$

where A is an $n \times n$ matrix and B is an $m \times m$ matrix. We then have

<u>Proposition 3.2.</u> <u>The linear operator</u> T <u>on</u> E^{nm} <u>has the adjoint</u>

$$T^* Y = A^* Y + YB^*, \tag{3.7}$$

<u>i. e., with this definition of</u> T^*,

$$(T(x), Y)_{E^{nm}} = (X, T^*(Y))_{E^{nm}} .$$

<u>The operator</u> T <u>has an inverse</u> T^{-1} <u>just in case</u> A <u>and</u> $(-B)$ <u>have no common eigenvalues.</u>

<u>Proof.</u> We have

$$\begin{aligned}(T(x), Y)_{E^{nm}} &= \operatorname{Tr} Y^* T(x) = \operatorname{Tr} Y^* (AX + XB) = \operatorname{Tr} Y^* AX + \operatorname{Tr} Y^* XB \\ &= \overline{\operatorname{Tr} X^* A^* Y} + \overline{\operatorname{Tr} B^* X^* Y} = \overline{\operatorname{Tr} X^* A^* Y} + \overline{\operatorname{Tr} X^* YB^*} \\ &= \overline{\operatorname{Tr} X^* (A^* Y + YB^*)} = \overline{(A^* Y + YB^*, X)}_{E^{nm}} = (X, A^* Y + YB^*)_{E^{nm}} = (X, T^*(Y))_{E^{nm}} .\end{aligned}$$

The condition for the existence of T^{-1} is just III- Corollary 1. 4. The proof is III-Exercise 4. Q. E. D.

Remark. In general the formula for T^{-1} involves block diagonalization of A and B and formulae of the type III-(1. 20). However, if A and B are $n \times n$ and $m \times m$ stability matrices, respectively, the formula for T^{-1} is simpler. Following the lines of the argument in III-(1. 26), ff., we let

$$X = -\int_0^\infty e^{At} Ce^{Bt} dt \tag{3. 8}$$

and compute

$$Ax + XB = -\int_0^\infty [Ae^{At} C e^{Bt} + e^{At} C e^{Bt} B]\, dt$$

$$= -\int_0^\infty \frac{d}{dt}(e^{At} C e^{Bt})dt = C - \lim_{T \to \infty} e^{AT} C e^{BT} = C$$

and we conclude that when A and B are stability matrices

$$T^{-1}(C) = -\int_0^\infty e^{At} Ce^{Bt} dt \, . \tag{3. 9}$$

The following result has numerous applications.

Proposition 3. 3. *Let* $T(X) = AX + XB$ *be nonsingular, so that the equation*

$$T(X) = AX + XB = C \tag{3. 10}$$

has a unique $n \times m$ *solution* X *for each given* $n \times m$ *matrix* C. *Let* Y *be any* $m \times n$ *matrix. Then*

$$\text{Tr } YX = \text{Tr } PC$$

where P *is the unique solution of*

$$BP + PA = Y \, . \tag{3. 11}$$

Proof. We have, from (3. 10) and the nonsingularity of T,

$$X = T^{-1}(C).$$

Then

$$\mathrm{Tr}\ YX = (Y^*, T^{-1}(C))_{E^{nm}} = ((T^*)^{-1}Y^*, C)_{E^{nm}} = \mathrm{Tr}\ PC$$

where

$$P^* = (T^*)^{-1}Y^* . \tag{3.12}$$

The equation (3.12) means

$$T^*P^* = Y^* ,$$

that is, from Proposition 3.2,

$$A^*P^* + P^*B^* = Y^* . \tag{3.13}$$

Transposing (3.13) we have

$$BP + PA = Y$$

which is (3.11). Q. E. D.

An alternate proof, valid when (3.9) holds, may also be of interest. We have

$$\mathrm{Tr}\ YX = \mathrm{Tr}\ Y(T^{-1}(C)) = -\mathrm{Tr}\ Y\int_0^\infty e^{At}Ce^{Bt}dt = -\int_0^\infty \mathrm{Tr}\ Ye^{At}Ce^{Bt}dt$$

$$= -\int_0^\infty \mathrm{Tr}\ e^{Bt}Ye^{At}C\,dt = -\mathrm{Tr}\int_0^\infty e^{Bt}Ye^{At}dt\,C = \mathrm{Tr}\ PC$$

where

$$P = -\int_0^\infty e^{Bt}Ye^{At}\,dt ,$$

and hence satisfies (3.11).

In the sections to follow in this chapter and in Chapter VII we will have frequent occasion to consider functions

$$F : E^{nm} \to E^1 ,$$

i. e., scalar valued functions of an $n \times m$ matrix. Such functions obviously present nothing really new, they are just functions of an nm dimensional vector variable. Again, however, it is desirable to continue to treat that variable in matrix form. The most important aspect of the theory for us is the re-interpretation of the gradient operation.

Definition 3.4. Let F $(= F(x)) : E^{nm} \to E^1$. If the domain of F includes a neighborhood in E^{nm} of X_0, we say that F is differentiable at X_0 if and only if there is an $m \times n$ matrix, which we denote by $F_X(X_0)$, or $\frac{\partial F}{\partial X}(X_0)$, such that

$$|F(X_0 + \Delta X) - F(X_0) - \operatorname{Tr} F_X(X_0)\Delta X| = o(\|\Delta X\|) \tag{3.14}$$

as the norm, $\|\Delta X\|$, of the $n \times m$ matrix ΔX tends to 0.

We will say that $F_X(X_0)$ is the gradient of F with respect to X at X_0.

The simplest way to calculate $F_X(X_0)$ in practice is summarized as follows:

(i) Form $F(X_0 + \epsilon\Delta X)$, ϵ real.

(ii) Compute the ordinary derivative $\frac{dF}{d\epsilon}(X_0 + \epsilon\Delta X)|_{\epsilon = 0}$.

(iii) Using the properties of the inner product (3.1), express $\frac{dF}{d\epsilon}(X_0 + \epsilon\Delta X)|_{\epsilon = 0}$ in the form $\operatorname{Tr} D\Delta X$ for some $m \times n$ matrix D. Then $D = F_X(X_0)$.

Differentiation may be carried out either explicitly or implicitly. We give two examples here.

As an example of explicit differentiation, consider the quadratic function of $X \in E^{nm}$ (A, B, X real matrices here)

$$F(x) = \operatorname{Tr} X^* AX + \operatorname{Tr} BX + c,$$

where A is an $n \times n$ matrix, B is an $m \times n$ matrix, and c is a scalar. We compute

$$F(X_0 + \epsilon\Delta X) = \operatorname{Tr} X_0^* AX_0 + \epsilon \operatorname{Tr} \Delta X^* AX_0$$
$$+ \epsilon \operatorname{Tr} X_0^* A\Delta X + \epsilon^2 \operatorname{Tr} \Delta X^* A\Delta X + \operatorname{Tr} BX_0 + \epsilon \operatorname{Tr} B\Delta X + c.$$

Then

$$\frac{dF}{d\epsilon}(X_0 + \epsilon \Delta X)\Big|_{\epsilon=0} = \text{Tr } \Delta X^* A X_0 + \text{Tr } X_0^* A \Delta X + \text{Tr } B \Delta X$$

$$= \text{Tr } X_0^* A^* \Delta X + \text{Tr } X_0^* A \Delta X + \text{Tr } B\Delta X = \text{Tr}([X_0^*(A^* + A) + B]\,\Delta X)$$

and hence

$$F_X(X_0) = X_0^*(A^* + A) + B.$$

To give an example of implicit differentiation, suppose the pair of real matrices (A, B) is stabilizable (A $n \times m$, B $n \times m$) and K_0 is a real $m \times n$ matrix such that $A + BK_0$ is a stability matrix. Let W be a real $n \times n$ matrix and let $Q(K)$ be the unique solution of

$$(A + BK)^* Q(K) + Q(K)(A + BK) + W + RK + K^* R^* + K^* UK = 0 \tag{3.15}$$

for $K = K_0 + \epsilon \Delta K$ near K_0. Let M be an arbitrary $n \times n$ matrix and let

$$F(K) = \text{Tr } M\, Q(K). \tag{3.16}$$

What is $F_K(K_0)$?

We rewrite (3.15) in the form

$$(A + B(K_0 + \epsilon\Delta K))^* Q(K_0 + \epsilon\Delta K) + Q(K_0 + \epsilon\Delta K)(A + B(K_0 + \epsilon\Delta K))$$
$$+ W + R(K_0 + \epsilon\Delta K) + (K_0 + \epsilon\Delta K)^* R^* + (K_0 + \epsilon\Delta K)^* U(K_0 + \epsilon\Delta K) = 0$$

Differentiating this matrix valued function of the scalar variable ϵ, we find that, with

$$Q_\epsilon \equiv \frac{d}{d\epsilon} Q(K_0 + \epsilon\Delta K)\Big|_{\epsilon=0}$$

we have

$$(A + BK_0)^* Q_\epsilon + Q_\epsilon(A + BK_0) + \Delta K^* B^* Q(K_0)$$
$$+ Q(K_0) B \Delta K + R\Delta K + \Delta K^* R + K_0^* U \Delta K + \Delta K^* U K_0 = 0. \tag{3.17}$$

Let

$$T_0(Q) = (A + BK_0)^* Q + Q(A + BK_0).$$

Now

$$\frac{d}{d\epsilon} F(K_0 + \epsilon \Delta K)\Big|_{\epsilon = 0} = \frac{d}{d\epsilon} \operatorname{Tr} MQ(K_0 + \epsilon\Delta K) = \operatorname{Tr} M Q_\epsilon .$$

From (3.17) we have

$$Q_\epsilon = -(T_0)^{-1}(\Delta K^* B^* Q(K_0) + Q(K_0)B\Delta K$$
$$+ R\Delta K + \Delta K^* R^* + K_0^* U\Delta K + \Delta K^* U K_0) \equiv -(T_0)^{-1} Z(\Delta K). \qquad (3.18)$$

Therefore, using Proposition 3.3 (with trivial sign changes)

$$\operatorname{Tr} M Q_\epsilon = \operatorname{Tr} PZ(\Delta K), \qquad (3.19)$$

where

$$(A + BK)P + P(A + BK)^* + M = 0. \qquad (3.20)$$

Now

$$\operatorname{Tr} PZ(\Delta K) = \operatorname{Tr} P[\Delta K^* B^* Q(K_0)$$
$$+ Q(K_0)B\Delta K + R\Delta K + \Delta K^* R^* + K_0^* U\Delta K + \Delta K^* U K_0]$$
$$= \operatorname{Tr} [\Delta K^* B^* Q(K_0)P + PQ(K_0)B\Delta K + \Delta K^* R^* P$$
$$+ PR\Delta K + \Delta K^* U K_0 P + PK_0^* U\Delta K] = \text{(since all quantities are real)}$$
$$= \operatorname{Tr}[P(Q(K_0)B + R + K_0^* U) + P^*(Q(K_0) B + R + K_0^* U)]\Delta K . \qquad (3.21)$$

We conclude therefore that

$$F_K(K_0) = P(Q(K_0)B + R + K_0^* U) + P^*(Q(K_0)B + R + K_0^* U). \qquad (3.22)$$

When M, and hence P, is symmetric, (3.22) becomes

$$F_K(K_0) = 2P(Q(K_0)B + R + K_0^* U) . \qquad (3.23)$$

4. Optimal Feedback Control for Linear Stochastic Systems on a Finite Interval

We again consider the (possibly time varying) linear control system

$$\dot{x} = A(t)x + B(t)u + C(t)v, \quad x \in E^n, \quad u \in E^m, \quad v \in E^p \tag{4.1}$$

and assume that v is white noise on $[t_0, t_1]$ with continuous covariance matrix $V(t)$. We consider only controls generated by continuous linear feedback relations

$$u(t) = K(t)x(t), \quad K(t) \ m \times n, \quad t \in [t_0, t_1] \tag{4.2}$$

and find, as in Section 2, that $\text{cov}\,(x(t), x(t)) \equiv X(t)$ satisfies

$$\dot{X} = S(t)X(t) + X(t)X(t)^* + C(t)V(t)C(t)^*, \quad t \in [t_0, t_1], \tag{4.3}$$

$$X(t_0) = X_0, \tag{4.4}$$

where

$$X_0 = \text{cov}\,(x_0, x_0),$$

$$S(t) = (A(t) + B(t)K(t)).$$

We return now to the quadratic cost functional of Chapter IV, with the same assumptions on the coefficient matrices, adopting notation reflecting the fact that it is now the matrix $K(t)$ which is the determining factor in system performance,

$$J(x_0, t_0, t_1, v, K) = \int_{t_0}^{t_1} [(F(t)x(t) + G(t)u(t))^* \tilde{W}(t)(F(t)x(t) + G(t)u(t))$$

$$+ u(t)^* \tilde{U}(t)u(t)]\,dt + x(t_1)^* Z x(t_1)$$

$$= \int_{t_0}^{t_1} x(t)^* [(F(t) + G(t)K(t))^* \tilde{W}(t)(F(t) + G(t)K(t)) + K(t)^* \tilde{U}(t)K(t)]\,x(t)dt$$

$$+ x(t_1)^* Z x(t_1) = \int_{t_0}^{t_1} x(t)^* Y(t)x(t)dt + x(t_1)^* Z x(t_1), \tag{4.5}$$

where Z is an $n \times n$ symmetric non-negative matrix and

$$Y(t) = (F(t) + G(t)K(t))^* \tilde{W}(t)(F(t) + G(t)K(t)) + K(t)^* \tilde{U} K(t)$$

$$= \text{(cf. IV-(2. 6))}\ W(t)+R(t)K(t)+K(t)^* R(t)^* +K(t)^* U(t)K(t).$$

We define $J(X_0, t_0, t_1, V, K)$ (= $J(K)$ when the other arguments are understood) as the expected value of the cost (4. 5). Thus, from (2. 21) of Theorem 2. 2,

$$J(X_0, t_0, t_1, V, K) = E\left(\int_{t_0}^{t_1} x(t)^* Y(t)x(t)dt + x(t_1)^* Z x(t_1)\right)$$

$$= \int_{t_0}^{t_1} E(x(t)^* Y(t)x(t))dt + E(x(t_1)^* Z x(t_1))$$

$$= \int_{t_0}^{t_1} (\mathrm{Tr}\ Y(t)X(t))dt + \mathrm{Tr}\ ZX(T) \tag{4. 6}$$

where $X(t)$ is given by (4. 3), (4. 4). The <u>stochastic optimal control problem</u> is

$$\min_{\substack{K \\ (= K(t),\ t \in [t_0, t_1]}} J(X_0, t_0, t_1, V, K) . \tag{4. 7}$$

This problem can be viewed in an entirely deterministic setting if we regard the covariance matrix, $X(t)$, as the system state and the feedback matrix, $K(t)$, as the control. The differential equation is (4. 3), the initial state is (4. 4) and the cost is (4. 6). Observe, however, that (4. 3) is not a linear control system. Since

$$S(t)X(t) = (A(t) + B(t)K(t))X(t)$$

the right hand side of (4. 3) contains (matrix) products of the control and state, $K(t)$ and $X(t)$.

We now have the following theorem.

<u>Theorem 4. 1.</u> <u>Let</u> $Q(t)$ <u>satisfy the matrix Riccati differential equation</u> IV-(2. 28) <u>on</u> $[t_0, t_1]$, <u>i. e.</u>,

$$-\dot{Q}(t) = A(t)^* Q(t) + Q(t)A(t) + W(t) - (Q(t)B(t)+R(t))U(t)^{-1}(B(t)^* Q(t)+R(t)) \tag{4. 8}$$

with

$$Q(t_1) = Z . \tag{4. 9}$$

Then

$$\hat{K}(t) = - U(t)^{-1}[B(t)^* Q(t) + R(t)^*] \tag{4. 10}$$

is the unique solution of (4. 7) for every choice of $X_0 > 0$, V (= $V(t)$, $t \in [t_0, t_1]$) and the minimal cost is given by

$$J(\hat{K}) = \operatorname{Tr} Q(t_0)X_0 + \int_{t_0}^{t_1} \operatorname{Tr} (Q(t)C(t)V(t)C(t)^*)\, dt . \tag{4. 11}$$

Proof. For each continuous $K(t)$, let $S(t) = A(t) + B(t)K(t)$ and let $Q_K(t)$ satisfy

$$-\dot{Q}_K(t) = S(t)^* Q_K(t) + Q_K(t)S(t) + Y(t), \tag{4. 12}$$

$$Q_K(t_1) = Z. \tag{4. 13}$$

For $X(t)$ satisfying (4. 3), (4. 4) we have

$$\operatorname{Tr} Q_K(t_0)X_0 - \operatorname{Tr} ZX(t_1) = - \int_{t_0}^{t_1} \frac{d}{dt}(\operatorname{Tr} Q_K(t)X(t))dt . \tag{4. 14}$$

Now

$$\frac{d}{dt}\operatorname{Tr} Q_K(t)X(t) = \operatorname{Tr}\left(\frac{d}{dt}(Q_K(t)X(t))\right) = \operatorname{Tr}(\dot{Q}_K(t)X(t) + Q_K(t)\dot{X}(t))$$

$$= \operatorname{Tr}([-S(t)^* Q_K(t) - Q_K(t)S(t) - Y(t)]X(t)$$

$$+ Q_K(t)[S(t)X(t) + X(t)S(t)^* + C(t)V(t)C(t)^*])$$

= (using the property (3. 4) of the trace)

$$= \operatorname{Tr} Q_K(t)C(t)V(t)C(t)^* - \operatorname{Tr} Y(t)X(t). \tag{4. 15}$$

Using (4.15) in (4.14) we have

$$J(K) = \int_{t_0}^{t_1} (\mathrm{Tr}\, Y(t)X(t))dt + \mathrm{Tr}\, ZX(t_1)$$

$$= \mathrm{Tr}\, Q_K(t_0)X_0 + \int_{t_0}^{t_1} (\mathrm{Tr}\, Q_K(t)C(t)V(t)C(t)^*)dt \; . \tag{4.16}$$

We note that if P and Q are two symmetric matrices, both non-negative, we have

$$\mathrm{Tr}\, PQ = \mathrm{Tr}\; P^{1/2}QP^{1/2} = \mathrm{Tr}\, Q^{1/2}PQ^{1/2} \geq 0 \; . \tag{4.17}$$

If either P or Q is positive definite, equality holds in (4.17) if and only if the other is zero.

We now make use of a result similar to V- Proposition 2.2. If $Q_K(t)$, $Q_{\hat{K}}(t)$ are solutions of (4.12), (4.13) corresponding to feedback matrices $K(t)$, $\hat{K}(t)$, $S(t) = A(t) + B(t)K(t)$, $\hat{S}(t) = A(t) + B(t)\hat{K}(t)$, then

$$-(\dot{Q}_{\hat{K}}(t) - \dot{Q}_K(t)) = S(t)^*(Q_{\hat{K}}(t) - Q_K(t)) + (Q_{\hat{K}}(t) - Q_K(t))S(t)$$
$$+ (U(t)\hat{K}(t) + B(t)^*Q_{\hat{K}}(t) + R(t)^*)^*U(t)^{-1}(U(t)\hat{K}(t)+B(t)^*Q_{\hat{K}}(t)+R(t)^*)$$
$$- (U(t)K(t)+B(t)^*Q_{\hat{K}}(t)+R(t)^*)^*U(t)^{-1}(U(t)K(t)+B(t)^*Q_{\hat{K}}(t)+R(t)^*) \; . \tag{4.18}$$

The proof of this result goes exactly as in V - Proposition 2.2. If we suppose that

$$\hat{K}(t) = -\,U(t)^{-1}[B(t)^*Q_K(t) + R(t)^*] \tag{4.19}$$

then (4.18) gives

$$(\dot{Q}_K(t) - \dot{Q}_{\hat{K}}(t)) = -\,S(t)^*(Q_K(t)-Q_{\hat{K}}(t)) - (Q_K(t)-Q_{\hat{K}}(t))S(t)$$
$$- (U(t)K(t)+B(t)^*Q_{\hat{K}}(t)+R(t)^*)^*U(t)^{-1}(U(t)K(t)+B(t)^*Q_{\hat{K}}(t)+R(t)^*). \tag{4.20}$$

Letting $\Psi(t,\tau)$ denote the fundamental matrix solution of

$$\frac{d\Psi}{dt} = - S(t)^* \Psi$$

with

$$\Psi(\tau, \tau) = I$$

and noting that $Q_K(t_1) = Q_{\hat{K}}(t_1) = Z$, (4.20) implies

$$O = Q_K(t_1) - Q_{\hat{K}}(t_1) = \Psi(t_1, \tau)(Q_K(\tau) - Q_{\hat{K}}(\tau))\Psi(t, \tau)^*$$

$$- \int_{\tau}^{t_1} \Psi(t_1,t)(U(t)K(t)+B(t)^* Q_{\hat{K}}(t)+R(t)^*)^* U(t)^{-1}(U(t)K(t)+B(t)^* Q_{\hat{K}}(t)+R(t)^*)\Psi(t_1,t)^* dt,$$

and, using $\Psi(t_1, \tau)^{-1}\Psi(t_1, t) = \Psi(t_1, \tau)^{-1}\Psi(t_1, \tau)\Psi(\tau, t) = \Psi(\tau, t) = \Phi(t, \tau)^*$, where

$$\frac{d\Phi}{dt} = - S(t)\Phi, \quad \Phi(\tau, \tau) = I,$$

we have

$$Q_K(\tau) - Q_{\hat{K}}(\tau)$$
$$= \int_{\tau}^{t_1} \Phi(t,\tau)^*(U(t)K(t)+B(t)^* Q_{\hat{K}}(t)+R(t)^*)^* U(t)^{-1}(U(t)K(t)+B(t)^* Q_{\hat{K}}(t)+R(t)^*)\Phi(t, \tau)dt \; . \tag{4.21}$$

It follows that

$$Q_K(\tau) - Q_{\hat{K}}(\tau) \geq 0, \quad \tau \in [t_0, t_1] \tag{4.22}$$

and essentially the same argument as was used in IV - Theorem 2.5 to establish the uniqueness of the optimal control $u(t)$ shows that equality holds in (4.22) for $t_0 \leq \tau \leq t_1$ if and only if $K(t) \equiv \hat{K}(t)$, $\tau \leq t \leq t_1$.

Going back to (4.16) now, we have

$$J(K) - J(\hat{K}) = \text{Tr}\,(Q_K(t_0) - Q_{\hat{K}}(t_0))X_0$$
$$+ \int_{t_0}^{t_1} (\text{Tr}\,(Q_K(t) - Q_{\hat{K}}(t))C(t)V(t)C(t)^*)dt \geq 0, \tag{4.23}$$

the inequality following from an argument like (4.17). Moreover, since we have assumed $X_0 > 0$, equality holds in (4.23) if and only if $Q_K(t_0) = Q_{\hat{K}}(t_0)$, i.e. $K(t) \equiv \hat{K}(t)$, $t_0 \leq t \leq t_1$. Thus, if $\hat{K}$ and $Q_{\hat{K}}(t)$ together satisfy (4.12), (4.13), (4.19),

$$J(\hat{K}) \leq J(K)$$

for all continuous $K = K(t)$, $t \in [t_0, t_1]$, and equality holds if and only if $K(t) \equiv \hat{K}(t)$. Now $\hat{K}(t)$, $Q_{\hat{K}}(t)$ satisfy (4.12), (4.13), (4.19) if and only if $Q_{\hat{K}}(t) \equiv Q(t)$, where $Q(t)$ satisfies (4.8), (4.9) and $\hat{K}(t)$ satisfies (4.10). Equation (4.11) follows from (4.16) with $K(t) = \hat{K}(t)$. Q. E. D.

Remark. The feedback relation (4.10) is still optimal if X_0 is only ≥ 0 but uniqueness is then a more complicated matter. See Exercise 3 at the end of this chapter.)

We have, then, the remarkable result that the same optimal control which solved the quadratic optimal control problem of Chapter IV solves the stochastic optimal control problem no matter what X_0 and $V(t)$ may be. This is a very important result in view of the fact that X_0 and $V(t)$ are often difficult or impossible to determine in practice. It should be noted, however, that though $\hat{K}(t)$ as given by (4.10) is optimal for every choice of X_0 and $V(t)$, the optimal value of the cost does depend on these matrices as we see from (4.11).

5. Optimal Feedback Control for Constant Linear Stochastic Systems on an Infinite Interval

We now pass to consideration of the constant coefficient system

$$\dot{x} = Ax + Bu + Cv, \quad x \in E^n, \; u \in E^m, \; v \in E^p. \tag{5.1}$$

We suppose the pair (A, B) to be stabilizable and restrict attention to controls generated by constant linear feedback relations

$$u(t) = Kx(t), \quad K \;\; m \times n, \tag{5.2}$$

such that

$$S = A + BK$$

is a stability matrix. We will suppose that v is steady state white noise with constant covariance matrix V. The $p \times p$ matrix V is symmetric and non-negative. Then $\mathrm{cov}(x(t), x(t)) = X(t)$ satisfies

$$\dot{X} = SX + XS^* + CVC^*,$$

on $[0,\infty)$ with

$$X(0) = X_0 = \mathrm{cov}\,(x_0, x_0)\ .$$

We have seen in Section 3 that

$$\lim_{t\to\infty} X(t) = X \tag{5.3}$$

where

$$SX + XS^* + CVC^* = 0. \tag{5.4}$$

We introduce the cost

$$J(x_0, v, K) = \int_0^\infty [\,(Fx(t)+Gu(t))^*\tilde{W}(Fx(t)+Gu(t))+u(t)^*\tilde{U}u(t)\,]\,dt$$

$$= \int_0^\infty x(t)^*[\,(F+GK)^*\tilde{W}(F+GK)+K^*\tilde{U}K]\,x(t)dt\ . \tag{5.5}$$

We are tempted immediately to define

$$J(X_0, V, K) = E(J(x_0, v, K)).$$

But this causes some difficulty. Proceeding formally we compute, as in the preceding section

$$J(X_0, V, K) = \int_0^\infty \mathrm{Tr}\ YX(t)\,dt\ , \tag{5.6}$$

where (cf. (4.5) ff)

$$Y = (F+GK)^*\tilde{W}(F+GK)+K^*\tilde{U}K = W+RK+K^*R^*+K^*UK. \tag{5.7}$$

But, since Y is constant and X(t) satisfies (5.3),

$$\lim_{t \to \infty} \operatorname{Tr} YX(t) = \operatorname{Tr} YX$$

and the integral (5. 6) is infinite in general. There are several ways in which this difficulty can be overcome. We discuss two of them.

We may consider the instantaneous cost

$$(Fx(t) + Gu(t))^* \tilde{W}(Fx(t)+Gu(t)) + u(t)^* \tilde{U} u(t)$$

$$= x(t)^* [\,(F+GK)^* \tilde{W}(F+GK) + K^* UK\,]\, x(t) = x(t)^* Y x(t). \tag{5. 8}$$

Its expected value is $\operatorname{Tr} YX(t)$ and, in the limit, this converges to $\operatorname{Tr} YX$. Thus $\operatorname{Tr} YX$ may be considered to be the steady state instantaneous cost covariance, representing the covariance of the instantaneous cost after all effects due to initial conditions have disappeared. Version 1 of the steady state stochastic optimal control problem is then

$$\min_{K} \operatorname{Tr} YX(K), \tag{5. 9}$$

$$S(K) = A + BK \quad \text{a stability matrix}, \tag{5. 10}$$

$$S(K)X(K) + X(K)X(K)^* + CVC^* = 0 . \tag{5. 11}$$

Since $S(K)$ is a stability matrix, the Liapounov equation

$$S(K)^* Q(K) + Q(K)S(K) + Y = 0 \tag{5. 12}$$

has a unique symmetric non-negative solution $Q(K)$, positive definite with further assumptions on Y as in Chapter IV. Then, applying Proposition 3. 3, we see, since $X(K)$ satisfies (5. 11), that

$$\operatorname{Tr} YX(K) = \operatorname{Tr} Q(K)\, CVC^* . \tag{5. 13}$$

It follows that (5. 9), (5. 11) can be restated as

$$\min_{K} \operatorname{Tr} Q(K)\, CVC^*, \tag{5. 14}$$

with Q_K satisfying (5. 12), $S(K)$ as in (5. 10).

We might call the above a "far-sighted" approach, since it concerns only the limiting behavior of $X(t)$ as $t \to \infty$.

Curiously enough, the second approach is "near-sighted", in a sense. We suppose that disturbances are worse now than they will be in the future. Instead of taking $v(t)$ to be steady state white noise with constant covariance V we assume it has covariance

$$V(t) = e^{-\mu t} V, \quad t \in [0, \infty).$$

Then we have

$$\dot{X}(t) = S(K)X(t) + X(t)S(K)^{*} + e^{-\mu t} CVC^{*}$$

$$X(0) = 0.$$

On a finite interval, $[0, T]$, we proceed as in the preceeding section to obtain

$$J(X_0, 0, T, e^{-\mu t}V, K) = \operatorname{Tr} Q_{K,T}(0)X_0 + \int_0^T e^{-\mu t} \operatorname{Tr}(Q_{K,T}(t)\, CVC^{*})\, dt \tag{5.15}$$

where $Q_{K,T}(t)$ satisfies (cf. (4.12), (4.13))

$$-\dot{Q}_{K,T} = S(K)^{*} Q_{K,T} + Q_{K,T} S(K) + Y,$$

$$Q_K(T) = 0.$$

Since $S(K)$ is a stability matrix, an argument similar to that used in Theorem 2.5 shows that (for any fixed τ, T_1)

$$\lim_{T \to \infty} Q_{K,T}(\tau) = \lim_{t \to -\infty} Q_{K,T_1}(t) = Q(K)$$

where $Q(K)$ is the unique solution of

$$S(K)^{*} Q(K) + Q(K)S(K) + Y = 0 . \tag{5.16}$$

Then

$$J(X_0, 0, T, e^{-\mu t}V, K) = \mathrm{Tr}\, Q_{K,T}(0)X_0$$

$$+ \int_0^T e^{-\mu t} dt\ \mathrm{Tr}\, Q(K)CVC^* + \int_0^T e^{-\mu t}\,\mathrm{Tr}((Q_{K,T}(t) - Q(K))CVC^*)dt.$$

Because $\lim_{t \to \infty} e^{-\mu t} = 0$ and $\lim_{T \to \infty} (Q_{K,T}(t) - Q_K) = 0$ it is not hard to see that the last integral converges to zero as $T \to \infty$. Hence

$$J(X_0, e^{-\mu t}V, K) \equiv \lim_{T \to \infty} J(X_0, 0, T, e^{-\mu t}V, K)$$

$$= \mathrm{Tr}\, Q(K)X_0 + \frac{1}{\mu}\,\mathrm{Tr}\, Q(K)CVC^* = \mathrm{Tr}\, Q(K)(X_0 + \frac{1}{\mu} CVC^*).$$

The problem

$$\min_K J(X_0, e^{-\mu t}V, K)$$

or

$$\min_K \mathrm{Tr}\, Q(K)(X_0 + \frac{1}{\mu} CVC^*) \tag{5.17}$$

thus has precisely the same form as (5.14) but with CVC^* replaced by $X_0 + \frac{1}{\mu} CVC^*$. Both problems are of the form

$$\min \mathrm{Tr}\, Q(K)M (= \mathrm{Tr}\, MQ(K)) \tag{5.18}$$

where $Q(K)$ satisfies (5.12) and M is an $n \times n$ symmetric non-negative matrix.

In (5.15), $e^{-\mu t}$ can be replaced by any non-negative function $g(t)$ such that

$$\int_0^\infty g(t)dt < \infty$$

with much the same results, the factor $\frac{1}{\mu}$ in (5.17) then being replaced by this integral. In particular we can take $g(t) \equiv 0$ and then the problem involves only X_0 and we have $M = X_0$ in (5.17).

Theorem 5.1. For every non-negative symmetric $n \times n$ matrix M the problem (5.18) has the solution

$$K = -U^{-1}(B^*Q + R^*), \tag{5.19}$$

where Q is the unique symmetric positive definite solution of (3.13), i.e.

$$A^*Q + QA + W - (QB + R)U^{-1}(B^*Q + R^*) = 0. \tag{5.20}$$

Proof. It is noteworthy that (5.19) can be shown to be necessary for $\hat{K}$ to minimize Tr MQ(K), M positive definite symmetric, by very elementary reasoning. Regarding (5.18) as a function F(K) defined on E^{mn}, the familiar theory of extrema of differentiable functions of several variables (V - [2], [3]) applies to show that a necessary condition for $\hat{K}$ to solve (5.18) is

$$F_K(\hat{K}) = 0. \tag{5.21}$$

Since M is symmetric, the example at the end of Section 3 applies to show that

$$F_K(\hat{K}) = P(\hat{K})(Q(\hat{K})B + R + \hat{K}^*U). \tag{5.22}$$

Here $P(\hat{K})$ satisfies

$$S(\hat{K})P(\hat{K}) + P(\hat{K})S(\hat{K})^* + M = 0 \tag{5.23}$$

(Hence if $M = CVC^*$, $P(\hat{K}) = X(\hat{K}) = \lim_{t \to \infty} \operatorname{cov}(x(t), x(t))$.) Clearly $P(\hat{K})$ is symmetric if M is symmetric, non-negative or positive definite as M is non-negative or positive definite, respectively. If M is positive definite, P is positive definite and (5.21), (5.22) together imply

$$Q(\hat{K})B + R + \hat{K}^*U = 0 \tag{5.24}$$

which immediately gives

$$\hat{K} = -U^{-1}(B^*Q(\hat{K}) + R^*). \tag{5.25}$$

Substituting (5.25) into (5.16) and rearranging we see immediately that $Q = Q(\hat{K})$ must be the unique (for (F, A) observable, (A, B) stabilizable) positive definite solution of (5.20).

That (5.19) is sufficient for $\hat{K}$ to minimize (5.18) follows from an argument similar to that used in the proof of Theorem 4.1. From (4.17),

$$\operatorname{Tr} MQ(K) \geq \operatorname{Tr} MQ(\hat{K}) \tag{5.26}$$

if

$$Q(K) \geq Q(\hat{K}). \tag{5.27}$$

As in Theorem 4.1, we go back to V - Proposition 2.2 to see (replacing $\tilde{K}$ there by $\hat{K}$, and $\hat{K}$ there by K) that

$$S(K)^*(Q(K) - Q(\hat{K})) + (Q(K) - Q(\hat{K}))S(K)$$

$$+ (UK + B^*Q(\hat{K}) + R^*)^* U^{-1}(UK + B^*Q(\hat{K}) + R^*) = 0,$$

provided $\hat{K}$, $Q(\hat{K})$ together satisfy (5.24). Since the last term is non-negative and $S(K)$ is a stability matrix, (5.26) and (5.27) follow. Q. E. D.

If $X_0 = \operatorname{cov}(x_0, x_0) > 0$, and $M = X_0 + \frac{1}{\mu} CVC^*$ as in (5.17), then $P > 0$ and it is easy to see that (5.19) is the unique solution of (5.18). If $M = CVC^*$, as in (5.14), and V is $p \times p$ with $p < n$, or if CVC^* is only non-negative, rather than positive definite, for some other reason, it is not easy to establish uniqueness <u>a priori</u>. But we do have

<u>Proposition 5.2.</u> <u>If</u> V <u>is a positive definite</u> $p \times p$ <u>matrix and</u>

$$(F, A) \text{ is observable}, \tag{5.28}$$

$$(A, B) \text{ is stabilizable}, \tag{5.29}$$

$$(S(\hat{K})(= A + B\hat{K}),\ C) \text{ is controllable}, \tag{5.30}$$

<u>then</u> (5.19) <u>is the unique solution of</u> (5.18).

<u>Proof.</u> The conditions (5.28), (5.29) are known from Chapter IV to imply that (5.20) has a unique positive definite solution Q. The condition (5.30) implies that $P(\hat{K})$ is positive definite for $M = CVC^*$ (III - Theorem 1.5 with $H = C^*$, $W = \alpha M = \alpha CVC^*$, $\alpha V > I$). Then (5.21) is true if and only if (5.24), and hence (5.25), is true. Then $Q(K)$ is a positive definite symmetric solution of (5.20) and there is only one such solution, Q. Thus (5.25) is the same as (5.19) and K is uniquely determined by (5.19).
Q. E. D.

The reader is invited in Exercise 4 to show that uniqueness may fail if (5. 30) is not true.

The fact that the formula (5. 19) for the optimal feedback matrix $\hat{K}$ can be obtained just by differentiation of the function Tr MQ(K) suggests that a number of other problems might be investigated by the same method. Let us return to a question considered in Chapter III. We have a system

$$\dot{x} = Ax + Bu, \quad x \in E^n, \ u \in E^m \tag{5.31}$$

and an observation

$$\omega = Hx, \quad \omega \in E^r. \tag{5.32}$$

We suppose that $r < n$ and, without loss of generality, that H has maximal rank. The "output feedback" problem concerns the properties of (5. 31) with u a linear function of the observation ω :

$$u = K\omega = KHx, \quad K \quad m \times r.$$

The closed loop system is then

$$\dot{x} = (A + BKH)x \equiv S(K)x \tag{5.33}$$

We have noted in Chapter III that it is not always possible to find K such that S(K) is a stability matrix. But in those cases where S(K) can be made a stability matrix by some choice of K it does make sense to pose the problem (5. 9) - (5. 11) again. That is

$$\min_K \operatorname{Tr} YX(K) \tag{5.34}$$

$$S(K) = A + BKH \text{ a stability matrix }, \tag{5.35}$$

$$S(K)X(K) + X(K)S(K)^* + CVC^* = 0 , \tag{5.36}$$

the $n \times n$ non-negative matrix Y being defined as before (cf. (5. 8)) but with K replaced by KH. Again (5. 34) can be replaced by

$$\min_K \operatorname{Tr} MQ(K) \ (\equiv F(K))$$

with

$$M = CVC^* \tag{5.37}$$

or the alternate problem involving the cost functional (5.15) can be posed, for which M takes the form

$$M = X_0 + \frac{1}{\mu} CVC^* .$$

In this case, of course, (5.12) becomes

$$(A + BKH)^* Q(K) + Q(K)(A + BKH) + Y = 0.$$

The necessary condition for a feedback matrix $\hat{K}$ to be optimal is again

$$F_K(\hat{K}) = 0.$$

The determination of the formula for F_K proceeds much as in Section III but with a slight difference. Instead of the formula for $Z(\Delta K)$ implied by (3.18) we have (replacing K_0 by $\hat{K} H$)

$$Z(\Delta K) = H^* \Delta K^* B^* Q(K) + Q(\hat{K}) B \Delta K H$$
$$+ H^* \Delta K^* R^* + R\Delta K H + H^* \hat{K}^* U \Delta K H + H^* \Delta K^* U \hat{K} H .$$

The formula (3.21) is replaced by- note that (cf. (3.20), (5.36)) $P = X$ -

$$\text{Tr } PZ(\Delta K) = \text{Tr}[H^* \Delta K^* B^* Q(\hat{K}) P$$
$$+ PQ(\hat{K}) B\Delta K H + H^* \Delta K^* R^* P + PR\Delta K H + PH^* \hat{K}^* U \Delta K H + H^* \Delta K^* U \hat{K} HP]$$
$$= \text{(assuming now that } M, \text{ and hence } P, \text{ is symmetric)}$$
$$= 2 \text{ Tr } [(HPQ(\hat{K})B + HPR + HPH^* \hat{K}^* U)\Delta K]$$

so that in this case

$$F_K(\hat{K}) = 2(HPQ(\hat{K})B + HPR + HPH^* \hat{K}^* U).$$

The condition $F_K(\hat{K}) = 0$ is then

$$\hat{K} = -U^{-1}(B^*Q(\hat{K})PH^* + R^*PH^*)(HPH^*)^{-1} . \tag{5.38}$$

It will be noted that (5.38) reduces to (5.19) when $H = I$. When H is $r \times n$ with $r < n$, however, the formula (5.38) is quite different from (5.19). Perhaps the most important distinction is that (5.38) explicitly involves the matrix P. Since P $(= P(K))$ satisfies (cf. (3.20))

$$(A + B\hat{K}H)P(\hat{K}) + P(\hat{K})(A+B\hat{K}H)^* + M = 0$$

and $M = CVC^*$ or $X_0 + \frac{1}{\mu}CVC^*$, the formula (5.38) for $\hat{K}$ involves the covariance matrices X_0 for the initial state x_0 and/or the covariance matrix V characterizing the white noise disturbance $v(t)$. This is not the case in Theorem 5.1 where (5.19) gives the optimal feedback matrix for <u>every</u> choice of M (hence for every choice of X_0, V).

We see, then, that if the problem (5.34) - (5.36) has a solution, $\hat{K}$, the $m \times r$ matrix $\hat{K}$ is characterized by the equations

$$\hat{K} = -U^{-1}(B^*Q(\hat{K})P(\hat{K})H^* + R^*P(\hat{K})H^*)(HP(\hat{K})H^*)^{-1}, \tag{5.39}$$

$$(A+B\hat{K}H)^*Q(\hat{K})+Q(\hat{K})(A+B\hat{K}H)+W+R\hat{K}H+H^*\hat{K}^*R^*+H^*\hat{K}^*U\hat{K}H = 0, \tag{5.40}$$

$$(A+B\hat{K}H)P(\hat{K})+P(\hat{K})(A+B\hat{K}H)^* + M = 0. \tag{5.41}$$

For the comparable problem in Theorem 5.1 the third equation does not play any role in characterizing $\hat{K}$.

A modification of the Newton-Kleinman method to systems of the form (5.39) - (5.41) has been developed by Axsäter [7]. With K_0 an initial stabilizing feedback matrix, successive $m \times r$ matrices $K_0, K_1, K_2, \ldots$, $n \times n$ matrices $Q_0, Q_1, Q_2, P_0, P_1, P_2$ are generated by

$$(A+BK_iH)^*Q_i+Q_i(A+BK_iH)+W+RK_iH+H^*K_i^*R^*+H^*K_i^*UK_iH = 0, \tag{5.42}$$

$$(A+BK_iH)P_i + P_i(A+BK_iH)^* + M = 0, \tag{5.43}$$

$$K_{i+1} = -U^{-1}(B^*Q_iP_iH^* + R^*P_iH^*)(HP_iH^*)^{-1} . \tag{5.44}$$

It is difficult to give any reasonably general result concerning convergence of (5.42) - (5.44) to a solution $\hat{K}$ of (5.39) - (5.40). (The reader is invited to examine some of the properties of the K_i, P_i, Q_i in Exercise 5 at the end of this chapter.) In general (5.42) - (5.44) is not Newton's method here and, in particular, there is no reason to expect quadratic convergence. Some analysis of this method and some alternate methods are suggested in [7], [17].

6. Optimal State Estimation (Filtering) and the Separation Theorem

In III-4 we have seen that the problem of stabilizing a linear control system

$$\dot{x} = Ax + Bu, \quad x \in E^n, \; u \in E^m \tag{6.1}$$

becomes difficult, or even impossible, if all we have available is a linear observation

$$\omega = Hx, \quad \omega \in E^r, \tag{6.2}$$

and we are restricted to feedback laws of the form

$$u = K_1\omega = K_1Hx, \quad K_1 \; m \times r. \tag{6.3}$$

We have seen also that this difficulty can be circumvented by the introduction of a state estimator system. Assuming that the use of a feedback relationship

$$u = Kx, \quad K \; m \times n,$$

is desirable in (6.1), we introduce the state estimator system

$$\dot{y} = (A + BK)y + L(\omega - Hy) = (A+BK-LH)y + LHx, \tag{6.5}$$

and replace (6.4) by

$$u = Ky. \tag{6.6}$$

The stability of the coupled system (6.1), (6.5), (6.6) was analyzed in III-Theorem 4.3 by a transformation which is equivalent to the introduction of a new variable

$$e = x - y. \tag{6.7}$$

The vector $e(t)$ is the error incurred when we estimate the state $x(t)$ by $y(t)$. Following the computations of III - Theorem 4.3 and substituting (6.6) into (6.1), x and e together satisfy

$$\begin{aligned} \dot{x} &= (A + BK)x - BKe \\ \dot{e} &= (A - LH)e \end{aligned} \tag{6.8}$$

and we have asymptotic stability of the combined original plant - state estimator system just in case $A + BK$ and $A - LH$ are both stability matrices.

If the pair (A, B) is stabilizable we can find K, $m \times n$, so that $A + BK$ is a stability matrix. Indeed, we can do this in a number of ways, in particular, we can choose K as in IV-3 so that a quadratic cost functional is minimized. If the pair (H, A) is observable or, more generally, detectable (III - Definition 4.1), then an $n \times r$ feed-forward matrix L can be found so that $A - LH$ is a stability matrix. However, as yet we have not developed optimality criteria for the selection of L. We proceed to rectify this defect in the present section, with the gratifying result that a theory completely dual to that of Sections 4 and 5 is obtained. It is possible, in fact, to develop the theory in a framework dual to that of Chapter IV. This was carried out originally in [9]. We elect, however, to follow a route dual, as we will see, to that of Sections 4 and 5 because that route introduces fewer complexities in the arguments involved.

Consider the (possibly) time varying system

$$\dot{x} = \tilde{A}(t)x + \tilde{B}(t)K(t)y + D(t)\nu \tag{6.9}$$

with observation

$$\omega = H(t)x + \tilde{J}(t)\nu + \mu \tag{6.10}$$

and associated state estimator system

$$\dot{y} = (\tilde{A}(t) + \tilde{B}(t)K(t))y + L(t)(\omega - H(t)y). \tag{6.11}$$

Here $\nu \in E^{\rho}$ is a disturbance affecting the original plant (6.1), which we assume to be white noise with non-negative symmetric $\rho \times \rho$ covariance matrix $\tilde{N}(t)$, while $\mu \in E^{\sigma}$ is a disturbance affecting the observation, or measurement, ω, which we assume also to be white noise with positive definite symmetric $\sigma \times \sigma$ covariance matrix $\tilde{M}(t)$. As indicated in (6.11), we allow ω to be affected by the "plant noise" ν as well as the "measurement noise" μ. We assume that μ and ν are independent.

Defining the estimation error as in (6.7), equations (6.9) - (6.11) give

$$\dot{e}(t) = (\tilde{A}(t) - L(t)H(t))e(t) + D(t)\nu - L(t)(\tilde{J}(t)\nu + \mu). \tag{6.12}$$

Proceeding as in Section 2, we find that if we let

$$E(t) = \mathrm{cov}(e(t), e(t))$$

then $E(t)$ satisfies the matrix differential equation

$$\begin{aligned}\dot{E} = {} & (\tilde{A}(t)-L(t)H(t))E + E(\tilde{A}(t)-L(t)H(t))^* + D(t)\tilde{N}(t)D(t)^* \\ & - D(t)\tilde{N}(t)\tilde{J}(t)^*L(t)^* - L(t)\tilde{J}(t)\tilde{N}(t)D(t)^* \\ & + L(t)\tilde{J}(t)\tilde{N}(t)\tilde{J}(t)^*L(t)^* + L(t)\tilde{M}(t)L(t)^* \\ \equiv {} & (\tilde{A}(t) - L(t)H(t))E + E(\tilde{A}(t) - L(t)H(t))^* \\ & + N(t) - J(t)^*L(t)^* - L(t)J(t) + L(t)M(t)L(t)^* .\end{aligned} \tag{6.13}$$

If we suppose $x(t_0) = x_0$ to be a random vector with

$$\mathrm{cov}(x_0, x_0) = X_0$$

and, as in III - 4, take $y(t_0) = 0$, then $e(0) \equiv e_0 = x_0 - 0 = x_0$ and

$$\mathrm{cov}(e_0, e_0) = X_0 \equiv E_0 .$$

We now pose the

Stochastic Optimal Estimation (Filtering) Problem

Let $\tilde{Y}(t)$ be a continuous $n \times n$ symmetric non-negative matrix defined on $[t_0, t_1]$ and let $\tilde{Z}$ be an $n \times n$ symmetric non-negative matrix.

We pose the problem of determining an $n \times r$ matrix function $\hat{L}(t)$ minimizing the expected value of the cost functional

$$\left[\int_{t_0}^{t_1} e(t)^* \tilde{Y}(t) e(t)dt + e(t_1)^* \tilde{Z} e(t_1)\right] = \operatorname{Tr}\left[\int_{t_0}^{t_1} \tilde{Y}(t)\, E(t)\, dt + \tilde{Z} E(t_1)\right] . \tag{6.14}$$

<u>Theorem 6.1.</u> <u>The expected value of</u> (6.14) <u>is minimized by setting</u>

$$\hat{L}(t) = [\,P(t)H(t)^* + J(t)^*\,]\, M(t)^{-1} \tag{6.15}$$

where $P(t)$ satisfies the Riccati matrix differential equation

$$\dot{P} = \tilde{A}(t)P + P\tilde{A}(t)^* + N(t) - (PH(t)^* + J(t)^*)M(t)^{-1}(H(t)P + J(t)) \qquad (6.15)$$

<u>on</u> $[t_0, t_1]$, $M(t)$, $N(t)$ <u>and</u> $J(t)$ <u>are defined in</u> (6.13), <u>and</u>

$$P(t_0) = E_0 . \tag{6.17}$$

<u>Proof.</u> The proof is almost word for word the same as that of Theorem 4.1 from (4.18) on. We let $E_L(t)$, $E_{\hat{L}}(t)$ satisfy (6.13) with feed-forward matrices $L(t)$, $\hat{L}(t)$ respectively. We let $\tilde{S}(t) = \tilde{A}(t) - L(t)H(t)$. Then

$$\begin{aligned}\dot{E}_{\hat{L}}(t) - \dot{E}_L(t) &= \tilde{S}(t)(E_{\hat{L}}(t) - E_L(t)) + (E_{\hat{L}}(t) - E_L(t))\tilde{S}(t)^* \\ &+ L(t)H(t)E_{\hat{L}}(t) - \hat{L}(t)H(t)E_{\hat{L}}(t) + E_{\hat{L}}(t)H(t)^*L(t)^* - E_{\hat{L}}(t)H(t)^*\hat{L}(t)^* \\ &- J(t)^*\hat{L}(t)^* - \hat{L}(t)J(t) + \hat{L}(t)M(t)\hat{L}(t)^* + J(t)^*L(t)^* + L(t)J(t) - L(t)M(t)L(t)^* .\end{aligned}$$

Rearranging, this gives

$$\begin{aligned}\dot{E}_{\hat{L}}(t) - \dot{E}_L(t) &= \tilde{S}(t)(E_{\hat{L}}(t) - E_L(t)) + (E_{\hat{L}}(t) - E_L(t))\tilde{S}(t)^* \\ &+ (\hat{L}(t)M(t) - E_{\hat{L}}(t)H(t)^* - J(t)^*)M(t)^{-1}(M(t)\hat{L}(t)^* - H(t)E_{\hat{L}}(t) - J(t)) \\ &- (L(t)M(t) - E_{\hat{L}}(t)H(t)^* - J(t)^*)M(t)^{-1}(M(t)L(t)^* - H(t)E_{\hat{L}}(t) - J(t)).\end{aligned} \tag{6.18}$$

If we require

$$\hat{L}(t) = [\,E_{\hat{L}}(t)H(t)^* + J(t)^*\,]\, M(t)^{-1} \tag{6.19}$$

and let $\Phi(t,\tau)$ be the fundamental solution of

$$\dot{\Phi} = (\tilde{A}(t)-L(t)H(t))\Phi, \quad \Phi(\tau,\tau) = I$$

then (6.17), (6.18) together with $E_L(t_0) = E_{\hat{L}}(t_0) = E_0$ gives

$$E_{\hat{L}}(t)-E_L(t)=$$

$$-\int_{t_0}^{t} \Phi(t,s)[(L(s)M(s)-E_{\hat{L}}(s)H(s)^*-J(s)^*)M(s)^{-1}(M(s)L(s)^*-H(s)E_{\hat{L}}(s)-J(s))]\Phi(t,s)^* ds$$

and, since $M(t) = \tilde{M}(t) + \tilde{J}(t)\tilde{N}(t)\tilde{J}(t)^*$ is positive definite, there follows

$$E_{\hat{L}}(t) \le E_L(t), \quad t \in [t_0, t_1] .$$

Then since $\tilde{Y}(t) \ge 0$, $t \in [t_0, t_1]$, $\tilde{Z} \ge 0$,

$$\mathrm{Tr}\,[\int_{t_0}^{t_1} \tilde{Y}(t)E_{\hat{L}}(t)dt + \tilde{Z}E_{\hat{L}}(t_1)] \le \mathrm{Tr}\,[\int_{t_0}^{t_1} \tilde{Y}E_L(t)dt + \tilde{Z}E_L(t_1)]$$

so that $\hat{L}(t)$ is optimal. Substituting (6.19) into (6.13) with $\hat{L}(t)$ replaced by $L(t)$ the equation (6.13) becomes (6.16) (with $P(t) = E_{\hat{L}}(t)$) and the formula (6.18) is the same as (6.15). Q. E. D.

A uniqueness result comparable to that of Theorem 4.1 can be obtained if $\tilde{Z} > 0$. (See Exercise 6).

In the constant coefficient case, where $\tilde{A}(t) \equiv \tilde{A}$, $\tilde{B}(t) \equiv \tilde{B}$, $H(t) \equiv H$, $\tilde{J}(t) \equiv \tilde{J}$, $\tilde{N}(t) \equiv \tilde{N}$, $\tilde{M}(t) \equiv \tilde{M}$, for each L such that $\tilde{A} - LH$ is a stability matrix we may consider

$$E = \lim_{t\to\infty} E(t)$$

as representing the steady-state estimation error covariance due to continuing disturbances ν, μ after all effects due to the initial error e_0 have disappeared. We find just as in Section 5 that

$$(\tilde{A}-LH)E + E(\tilde{A}-LH)^* + N - J^*L^* - LJ + LML^* = 0 \tag{6.20}$$

and we may the problem of choosing $\hat{L}$ so as to minimize the cost

$$\mathrm{Tr}\,\tilde{Z}E \tag{6.21}$$

for some non-negative $n \times n$ symmetric matrix $\tilde{Z}$.

Theorem 6.2. The problem of minimizing (6.21), where E satisfies (6.20), is solved by

$$\hat{L} = [PH^* + J^*]M^{-1} \tag{6.22}$$

where P is the unique positive definite symmetric solution of the quadratic matrix equation

$$\tilde{A}P + P\tilde{A}^* + N - (PH^* + J^*)M^{-1}(HP + J) = 0, \tag{6.23}$$

provided that $(H, \tilde{A}$ is detectable and $(\tilde{A}, D)$ is controllable.

The proof bears the same relationship to Theorem 6.1 as Theorem 5.1 bears to Theorem 4.1 and is left as Exercise 7.

For completeness in the proof of Theorem 6.1 one should, of course, prove the existence of a solution of (6.16), (6.17) on $[t_0, t_1]$ just as we did earlier in IV - Proposition 2.4. But there is no need to repeat the argument; IV - Proposition 2.4 is all that is needed. In fact, there is a one-to-one correspondence between optimal control problems as studied in Chapter IV, or in Sections 4 and 5 of the present chapter, and optimal estimation problems as studied in the present section. The matrices used to define the problems themselves and the matrices which constitute the solutions of those problems map into each other according to the following table.

Duality Table for Optimal Control and Optimal Estimation Problems

Optimal Control m (dimension u)	=	Optimal Estimation r (dimension ω)
t	$s = t_0 + (t_1 - t))$	s
$A(t)$	=	$\tilde{A}(s)^*$
$B(t)$	=	$H(s)^*$
$K(t)$	=	$-L(s)^*$
$C(t)V(t)C(t)^*$	=	$\tilde{Y}(s)$
$W(t)$	=	$N(s)$
$R(t)$	=	$J(s)$
$U(t)$	=	$M(s)$
Z	=	E_0
$Q(t)$	=	$P(s)$
X_0	=	$\tilde{Z}$
$X(t)$	=	$Z(s)$

We have not made explicit use of $Z(s)$ in Theorem 6.1. It is the solution of

$$-\dot{Z} = (\tilde{A}(t)-L(t)H(t))^*Z + Z(\tilde{A}(t) - L(t)H(t)) + \tilde{Y}(t)$$

with

$$Z(t_1) = \tilde{Z} .$$

It is possible to set up a different pairing between the elements in the columns of the above tables. See Exercise 8.

The table above modifies in the obvious way to relate Theorem 6.2 to the material of Section 5 (or IV - 3).

Combining Sections 4 and 5 with the foregoing material of the present section it might appear that everything is settled. We know how to find an optimal $\hat{K}$ for the control problem and we know how to find an optimal $\hat{L}$ for the estimation problem. But a complete design involves simultaneous choice of both K and L in the system (6.8), (6.9). Let us consider this problem for awhile. For simplicity we consider only the constant coefficient case, the time varying case being left as Exercise 9.

We consider, then, the constant coefficient control system (cf. (6.9))

$$\dot{x} = Ax + BKy + D\nu , \tag{6.24}$$

wherein ν is steady state white noise with covariance matrix (cf. Section 5) $\tilde{N} \geq 0$, and the control, u, is determined by linear feedback $u = Ky$ on the solution, y, of the state estimator system

$$\dot{y} = (A + BK)y + L(\omega - Hy) . \tag{6.25}$$

In (6.25) ω is a noise-disturbed observation

$$\omega = Hx + \tilde{J}\nu + \mu , \tag{6.26}$$

ν as described above and μ steady state white noise with covariance matrix $\tilde{M} > 0$. With W, R, U having the same properties as in Section 5, we will concern ourselves with the limit of the expected instantaneous cost:

$$\lim_{t \to \infty} E(x(t)^* W x(t) + x(t)^* R u(t) + u(t)^* R^* x(t) + u(t)^* U u(t))$$
$$= \lim_{t \to \infty} E(x(t)^* W x(t) + x(t)^* RKy(t) + y(t)^* K^* R^* x(t) + y(t)^* K^* UKy(t)) . \tag{6.27}$$

This expression for the cost is appropriate to (6.24), (6.25) since we are now assuming that $u = Ky$ rather than $u = Kx$. We again find it convenient to introduce the estimation error

$$e(t) = x(t) - y(t) . \tag{6.28}$$

If we then let

$$\left.\begin{aligned} X &= \lim_{t \to \infty} \operatorname{cov}(x(t), x(t)), \quad & Y &= \lim_{t \to \infty} \operatorname{cov}(y(t), y(t)), \\ Z &= \lim_{t \to \infty} \operatorname{cov}(x(t), y(t)), \quad & E &= \lim_{t \to \infty} \operatorname{cov}(e(t), e(t)), \end{aligned}\right\} \tag{6.29}$$

the cost (6.27) becomes

$$J(K, L) \equiv \operatorname{Tr}(XW + Z^* RK + K^* R^* Z + YK^* UK) \tag{6.30}$$

Since the equations (6.24), (6.25) satisfied by x and y depend upon both K and L, both of these matrices must be considered together in the optimal design problem which is

$$\min_{K,L} J(K, L), \tag{6.31}$$

with the constraint that $A + BK$ and $A - LH$ should both be stability matrices. The question, now, is this: How is the optimal solution, $\tilde{K}$, $\tilde{L}$, of (6.31) related to the individual optimal solutions $\hat{K}$, $\hat{L}$ of the stochastic control problem and the optimal filtering problem, respectively? The answer to this question is not at all obvious *a priori*; its resolution is embodied in what is called the *separation theorem*. (See [14] - [16].) This theorem has had the misfortune of being supplied with many easy (but wrong) pseudo-proofs and many correct, but impenetrable, actual proofs. Not infrequently it is stated in an ambiguous manner. Here, in the constant coefficient case, we provide an algebraic proof and, we hope, an unequivocal statement of what the theorem says in our restricted context.

Theorem 6.3. The individual solutions

$$\hat{K} = -U^{-1}(B^*Q + R^*) \tag{6.32}$$

(Q *the unique positive definite solution of* IV-(3.13)) *and*

$$\hat{L} = (PH^* + J)M^{-1}, \tag{6.33}$$

(P *the unique positive definite solution of* (6.23) *with* $\tilde{A} = A$) *provide, jointly, a global minimum for* $J(K, L)$ *as defined by* (6.30).

Remarks. The importance of this theorem lies in the fact that it says that optimal linear feedback control design and optimal filter design can be carried out separately without considering the possible effect that the one might have in interraction with the other - provided that the minimization of the cost $J(K, L)$ is, indeed, the actual goal of the control analysis. (Many a mathematically bent control theorist comes to grief on this point - there are usually all sorts of other considerations in the backs of the minds of the "real world" control designers with whom he works. Some of these are discussed in Chapter VII.)

<u>Proof of Theorem 6.3.</u> We introduce $\hat{y}$, the solution of the optimal state estimator for (6.24), (see Exercise [14])

$$\dot{\hat{y}} = A\hat{y} + BKy + \hat{L}(\omega - H\hat{y}), \tag{6.34}$$

with $\hat{L}$ defined by (6.33), and the optimal estimation error

$$\hat{e}(t) = x(t) - \hat{y}(t). \tag{6.35}$$

The triple $\hat{y}$, y, $\hat{e}$ (y given by (6.25)) is a solution of the stochastic linear system

$$\begin{pmatrix} \dot{\hat{y}} \\ y \\ \hat{e} \end{pmatrix} = \begin{pmatrix} A & BK & \hat{L}H \\ LH & A+BK-LH & LH \\ 0 & 0 & A-\hat{L}H \end{pmatrix} \begin{pmatrix} \hat{y} \\ y \\ \hat{e} \end{pmatrix} + \begin{pmatrix} \hat{L}(\mu + \tilde{J}\nu) \\ L(\mu + \tilde{J}\nu) \\ D\nu - \hat{L}(\mu + \tilde{J}\nu) \end{pmatrix}, \tag{6.36}$$

the second equation following from

$$\dot{y} = (A + BK)y + LH(x - y) + L(\mu + \tilde{J}\nu) = (A+BK-LH)y + LH(\hat{y} + \hat{e}) + L(\mu + \tilde{J}\nu).$$

Let Y be given by (6.29) and define

$$\hat{Y} = \lim_{t \to \infty} \text{cov}(\hat{y}(t), \hat{y}(t)), \quad P = \lim_{t \to \infty} \text{cov}(\hat{e}(t), \hat{e}(t)) \tag{6.37}$$

(note that P is independent of the choice of K and is the matrix appearing in (6.23), (6.33)). Further, let

$$Z_1 = \lim_{t \to \infty} \text{cov}(\hat{y}(t), y(t)), \quad Z_2 = \lim_{t \to \infty} \text{cov}(y(t), \hat{e}(t)), \quad Z_3 = \lim_{t \to \infty} \text{cov}(\hat{y}(t), \hat{e}(t)).$$

Following the developments of Section 5 and noting that the constraints imposed upon K and L in (6.31) ensure that the matrix in (6.36) is a stability matrix, we have

$$\begin{pmatrix} A & BK & \hat{L}H \\ LH & A+BK-LH & LH \\ 0 & 0 & A-\hat{L}H \end{pmatrix}\begin{pmatrix} \hat{Y} & Z_1 & Z_2 \\ Z_1^* & Y & Z_3 \\ Z_2^* & Z_3^* & P \end{pmatrix} + \begin{pmatrix} \hat{Y} & Z_1 & Z_2 \\ Z_1^* & Y & Z_3 \\ Z_2^* & Z_3^* & P \end{pmatrix}\begin{pmatrix} A^* & H^*L & 0 \\ K^*B^* & (A+BK-LH)^* & 0 \\ H^*\hat{L}^* & H^*L & (A-\hat{L}H)^* \end{pmatrix}$$

$$+\begin{pmatrix} \hat{L}M\hat{L}^* & \hat{L}ML^* & \hat{L}J-\hat{L}M\hat{L}^* \\ LM\hat{L} & LML^* & LJ-LM\hat{L}^* \\ J^*\hat{L}^*-\hat{L}M\hat{L}^* & J^*L^*-\hat{L}M\hat{L}^* & N-\hat{L}J-J^*\hat{L}+\hat{L}M\hat{L}^* \end{pmatrix} = 0\,, \tag{6.38}$$

where, as in (6.13),

$$M = \tilde{M} + \tilde{J}\tilde{N}\tilde{J}^*, \quad J = \tilde{J}\tilde{N}D^*, \quad N = D\tilde{N}D^*. \tag{6.39}$$

For Z_2, Z_3 we have

$$\begin{pmatrix} A & BK \\ LH & A+BK-LH \end{pmatrix}\begin{pmatrix} Z_2 \\ Z_3 \end{pmatrix} + \begin{pmatrix} Z_2 \\ Z_3 \end{pmatrix}(A-\hat{L}H)^* + \begin{pmatrix} \hat{L} \\ L \end{pmatrix}(HP + J - M\hat{L}^*) = 0\,.$$

But $\hat{L}$ is defined by (6.33) so $HP + J - M\hat{L}^* = 0$ and

$$\begin{pmatrix} A & BK \\ LH & A+BK-LH \end{pmatrix}\begin{pmatrix} Z_2 \\ Z_3 \end{pmatrix} + \begin{pmatrix} Z_2 \\ Z_3 \end{pmatrix}(A-\hat{L}H)^* = 0\,. \tag{6.40}$$

We know that $A - \hat{L}H$ is a stability matrix and, since K and L are restricted to values for which $A + BK$ and $A - LH$ are stability matrices, the matrix

$$\begin{pmatrix} A & BK \\ LH & A + BK - LH \end{pmatrix}$$

in (6.40) is a stability matrix - following the reasoning originally presented in III-4. Thus (6.40) has the unique solution

$$Z_2 = 0, \quad Z_3 = 0\,. \tag{6.41}$$

Then clearly

$$\begin{pmatrix} A & BK \\ LH & A+BK-LH \end{pmatrix}\begin{pmatrix} \hat{Y} & Z_1 \\ Z_1^* & Y \end{pmatrix} + \begin{pmatrix} \hat{Y} & Z_1 \\ Z_1^* & Y \end{pmatrix}\begin{pmatrix} A^* & H^*L^* \\ K^*B^* & (A+BK-LH) \end{pmatrix} + \begin{pmatrix} \hat{L}M\hat{L}^* & \hat{L}ML^* \\ LM\hat{L}^* & LML^* \end{pmatrix} = 0. \tag{6.42}$$

Now let us return to the formula for $J(K, L)$, i. e., (6. 30). Since $x = \hat{y} + \hat{e}$,

$$X = \lim_{t\to\infty} \operatorname{cov}(x(t), x(t)) = \lim_{t\to\infty} \operatorname{cov}(\hat{y}(t) + \hat{e}(t), \hat{y}(t) + \hat{e}(t))$$

$$= \hat{Y} + Z_2 + Z_2^* + P = (\text{cf. } (6.41)) = \hat{Y} + P$$

$$Z = \lim_{t\to\infty} \operatorname{cov}(x(t), y(t)) = \lim_{t\to\infty} \operatorname{cov}(\hat{y}(t) + \hat{e}(t), y(t))$$

$$= Z_1 + Z_3^* = (\text{cf. } (6.41)) = Z_1$$

so that

$$J(K, L) = \operatorname{Tr}(\hat{Y}W + Z_1^*RK + K^*R^*Z_1 + PW + YK^*UK)$$

$$= \operatorname{Tr}\begin{pmatrix} W & RK \\ K^*R^* & K^*UK \end{pmatrix}\begin{pmatrix} \hat{Y} & Z_1 \\ Z_1^* & Y \end{pmatrix} + \operatorname{Tr} PW . \tag{6.43}$$

The last term here is independent of both L and K and need not be treated further.

Now we introduce a transformation entirely analogous to the one used in III - 4. Since

$$\begin{pmatrix} I & 0 \\ I & -I \end{pmatrix}\begin{pmatrix} I & 0 \\ I & -I \end{pmatrix} = \begin{pmatrix} I & 0 \\ 0 & I \end{pmatrix}, \tag{6.44}$$

$$\operatorname{Tr}\begin{pmatrix} W & RK \\ K^*R^* & K^*UK \end{pmatrix}\begin{pmatrix} \hat{Y} & Z_1 \\ Z_1^* & Y \end{pmatrix}$$

$$= \operatorname{Tr}\begin{pmatrix} I & I \\ 0 & -I \end{pmatrix}\begin{pmatrix} W & RK \\ K^*R^* & K^*UK \end{pmatrix}\begin{pmatrix} I & 0 \\ I & -I \end{pmatrix}\begin{pmatrix} I & 0 \\ I & -I \end{pmatrix}\begin{pmatrix} \hat{Y} & Z_1 \\ Z_1^* & Y \end{pmatrix}\begin{pmatrix} I & I \\ 0 & -I \end{pmatrix}$$

$$\equiv \operatorname{Tr}\begin{pmatrix} W+RK+K^*R^*+K^*UK & -RK-K^*UK \\ -K^*R^*-K^*UK & K^*UK \end{pmatrix}\begin{pmatrix} \hat{Y} & Y_1 \\ Y_1^* & Y_2 \end{pmatrix}. \tag{6.45}$$

The equation for

$$\begin{pmatrix} \hat{Y} & Y_1 \\ Y_1^* & Y_2 \end{pmatrix} = \begin{pmatrix} I & 0 \\ I & -I \end{pmatrix}\begin{pmatrix} \hat{Y} & Z_1 \\ Z_1^* & Y \end{pmatrix}\begin{pmatrix} I & I \\ 0 & -I \end{pmatrix}$$

can be found by multiplying (6. 42) on the left by $\begin{pmatrix} I & 0 \\ I & -I \end{pmatrix}$ and on the right by $\begin{pmatrix} I & I \\ 0 & -I \end{pmatrix}$, introducing products like (6. 44) between matrices as appropriate, with the result

$$\begin{pmatrix} A+BK & -BK \\ 0 & A-LH \end{pmatrix}\begin{pmatrix} \hat{Y} & Y_1 \\ Y_1^* & Y_2 \end{pmatrix} + \begin{pmatrix} \hat{Y} & Y_1 \\ Y_1^* & Y_2 \end{pmatrix}\begin{pmatrix} (A+BK)^* & 0 \\ -K^*B^* & (A-LH)^* \end{pmatrix}$$
$$+ \begin{pmatrix} \hat{L}M\hat{L}^* & \hat{L}M(\hat{L}-L)^* \\ (\hat{L}-L)M\hat{L}^* & (\hat{L}-L)M(\hat{L}-L)^* \end{pmatrix} = 0 . \qquad (6.46)$$

Using Proposition 3. 3 together with (6. 45) we see that

$$\mathrm{Tr}\begin{pmatrix} W & RK \\ K^*R & K^*UK \end{pmatrix}\begin{pmatrix} \hat{Y} & Z_1 \\ Z_1^* & Y \end{pmatrix} = \mathrm{Tr}\begin{pmatrix} Q_1 & Q_2 \\ Q_2^* & Q_3 \end{pmatrix}\begin{pmatrix} \hat{L}M\hat{L}^* & \hat{L}M(\hat{L}-L)^* \\ (\hat{L}-L)M\hat{L}^* & (\hat{L}-L)M(\hat{L}-L)^* \end{pmatrix}$$

$$= \mathrm{Tr}\ (M^{\frac{1}{2}}L^*,\ M^{\frac{1}{2}}(\hat{L}-L)^*)\begin{pmatrix} Q_1 & Q_2 \\ Q_2^* & Q_3 \end{pmatrix}\begin{pmatrix} \hat{L}M^{\frac{1}{2}} \\ (\hat{L}-L)M^{\frac{1}{2}} \end{pmatrix}$$

$$= \mathrm{Tr}\ Q\hat{L}M\hat{L}^* + \mathrm{Tr}\left[(M^{\frac{1}{2}}\hat{L}^*,\ M^{\frac{1}{2}}(\hat{L}-L)^*)\begin{pmatrix} Q_1-Q & Q_2 \\ Q_2^* & Q_3 \end{pmatrix}\begin{pmatrix} \hat{L}M^{\frac{1}{2}} \\ (\hat{L}-L)M^{\frac{1}{2}} \end{pmatrix}\right] \qquad (6.47)$$

where $M^{\frac{1}{2}}$ is the positive square root of M (cf. (6. 39)), Q is the unique positive definite solution of IV-(3. 13) and

$$\begin{pmatrix} (A+BK)^* & 0 \\ -K^*B^* & (A-LH)^* \end{pmatrix}\begin{pmatrix} Q_1 & Q_2 \\ Q_2^* & Q_3 \end{pmatrix} + \begin{pmatrix} Q_1 & Q_2 \\ Q_2^* & Q_3 \end{pmatrix}\begin{pmatrix} A+BK & -BK \\ 0 & A-LH \end{pmatrix}$$
$$+ \begin{pmatrix} W+RK+K^*R+K^*UK & -RK-K^*UK \\ -K^*R-K^*UK & K^*UK \end{pmatrix} = 0 . \qquad (6.48)$$

Combining (6. 43) and (6. 47) we have

$$J(K, L) = \mathrm{Tr}(Q\hat{L}M\hat{L}^* + PW)$$
$$+ \mathrm{Tr}\left[(M^{\frac{1}{2}}\hat{L}^*, \; M^{\frac{1}{2}}(\hat{L}-L)^*) \begin{pmatrix} Q_1 - Q & Q_2 \\ Q_2 & Q_3 \end{pmatrix} \begin{pmatrix} \hat{L}M^{\frac{1}{2}} \\ (\hat{L}-L)M^{\frac{1}{2}} \end{pmatrix} \right] . \tag{6.49}$$

The first term is independent of either K or L - only $\hat{K}$, $\hat{L}$ are involved. Since (6.48) gives

$$(A+BK)^* Q_1 + Q_1(A+BK) + W + RK + K^*R + K^*UK = 0 , \tag{6.50}$$

and Q satisfies the same equation with K replaced by $\hat{K}$, the second term in (6.49) vanishes when $L = \hat{L}$ and $K = \hat{K}$. The inequality and identity

$$J(K, L) \geq J(\hat{K}, \hat{L}) = \mathrm{Tr}(Q\hat{L}M\hat{L}^* + PW) \tag{6.51}$$

consequently follow if we can show that

$$\begin{pmatrix} Q_1 - Q & Q_2 \\ Q_2^* & Q_3 \end{pmatrix} \geq 0 . \tag{6.52}$$

To do this we combine the equation satisfied by Q, i.e.,

$$(A + B\hat{K})^* Q + Q(A + B\hat{K}) + W + R\hat{K} + \hat{K}^*R + \hat{K}^*U\hat{K} \tag{6.53}$$

with (6.50) to obtain, as in (5.27) ff.,

$$(A+BK)^*(Q_1 - Q) + (Q_1 - Q)(A+BK) + (UK+B^*Q+R^*)^*U^{-1}(UK+B^*Q+R^*) = 0. \tag{6.54}$$

For Q_2 (6.48) gives

$$(A + BK)^* Q_2 + Q_2(A - LH) - Q_1BK - RK - K^*UK = 0$$

or, rearranging,

$$(A+BK)^* Q_2 + Q_2(A - LH) - (Q_1 - Q)BK - (QB+R+K^*U)K = 0 . \tag{6.55}$$

Combining (6.54) and (6.55) with the equation for Q_3 implied by (6.48) we have, recalling (6.32),

$$\begin{pmatrix} (A+BK)^* & 0 \\ -K^*B^* & (A-LH)^* \end{pmatrix}\begin{pmatrix} Q_1-Q & Q_2 \\ Q_2^* & Q_3 \end{pmatrix}+\begin{pmatrix} Q_1-Q & Q_2 \\ Q_2^* & Q_3 \end{pmatrix}\begin{pmatrix} A+BK & -BK \\ 0 & A-LH \end{pmatrix}$$

$$+\begin{pmatrix} (\hat{K}-K)^*U(\hat{K}-K) & (\hat{K}-K)^*UK \\ K^*U(\hat{K}-K) & K^*UK \end{pmatrix} = 0 . \tag{6.56}$$

Since the coefficient matrices are assumed to be stability matrices and

$$\begin{pmatrix} (\hat{K}-K)^*U(\hat{K}-K) & (\hat{K}-K)^*UK \\ K^*U(\hat{K}-K) & K^*UK \end{pmatrix} = \begin{pmatrix} (\hat{K}-K)^*U^{\frac{1}{2}} \\ K^*U^{\frac{1}{2}} \end{pmatrix}(U^{\frac{1}{2}}(\hat{K}-K),\ U^{\frac{1}{2}}K) \geq 0 ,$$

Liapounov's theorem (III- Theorem 1. 5) applies to give the inequality in (6. 52). We conclude that the pair $\hat{K}$, $\hat{L}$ minimizes the cost $J(K, L)$. Q. E. D.

Uniqueness of $\hat{K}, \hat{L}$ as the optimal solution of (6. 31) is a little delicate. We define

$$\mathcal{A}(K, L) = \begin{pmatrix} A+BK & -BK \\ 0 & A-LH \end{pmatrix}, \quad \mathcal{L}(L) = \begin{pmatrix} \hat{L}M^{\frac{1}{2}} \\ (\hat{L}-L)M^{\frac{1}{2}} \end{pmatrix}, \quad \mathcal{K}(K) = (U^{\frac{1}{2}}(\hat{K}-K),\ U^{\frac{1}{2}}K).$$

Then (6. 49), (6. 56) give

$$J(K, L) = \mathrm{Tr}(Q\hat{L}M\hat{L}^* + PW) + \mathrm{Tr}\int_0^\infty \mathcal{L}(L)^* e^{\mathcal{A}(K, L)^* t}\mathcal{K}(K)^*\mathcal{K}(K)e^{\mathcal{A}(K, L)t}\mathcal{L}(L)dt$$

and the inequality in (6. 51) is strict if and only if

$$\mathcal{K}(K)e^{\mathcal{A}(K, L)t}\mathcal{L}(L) \not\equiv 0, \quad t \in [0, \infty),$$

which is equivalent to (differentiating repeatedly and using the Cayley-Hamilton theorem as usual) the statement that at least one of the matrices

$$\mathcal{K}(K)(\mathcal{A}(K, L))^j\mathcal{L}(L), \quad j = 0, 1, \ldots, 2n-1, \tag{6.57}$$

is different from zero. Now, for $j = 0$, (6. 57) reduces to the matrix

$$U^{\frac{1}{2}}(\hat{K}-K)\hat{L}M^{\frac{1}{2}} + U^{\frac{1}{2}}K(\hat{L}-L)M^{\frac{1}{2}} = U^{\frac{1}{2}}(\hat{K}\hat{L} - KL)M^{\frac{1}{2}} \tag{6.58}$$

from which we see very easily that no pairs $(\hat{K}, L)$, $L \neq \hat{L}$, $(K, \hat{L})$, $K \neq \hat{K}$, can minimize (6. 31) if $\dim u \leq \dim x$, $\dim \omega \leq \dim x$ and $\hat{K}, \hat{L}$ have maximal rank. However, a general result appears difficult to obtain and one can show by example that there are cases where all of the matrices (6. 57) are equal to zero for $K \neq \hat{K}$, $L \neq \hat{L}$. (See Exercise 10.)

7. Formulation and Interpretation: An Example

In the foregoing sections we have introduced the reader to the elementary theory of linear stochastic control systems. There remains the matter of actually putting this theory to use. Rather than presenting a collection of general principles here, we will undertake a fairly concrete (though idealiz--ed) case study.

We begin with one of our favorite examples, that of a pendulum. A rod, r, is imbedded in a firm supporting device. At the end of the rod is a pivot, p, from which a pendulum is suspended. Assuming there is some friction at the pivot, and using appropriate physical units, the linearized equations of motion can be taken to be

$$\ddot{\theta} + \dot{\theta} + \theta = w, \qquad (7.1)$$

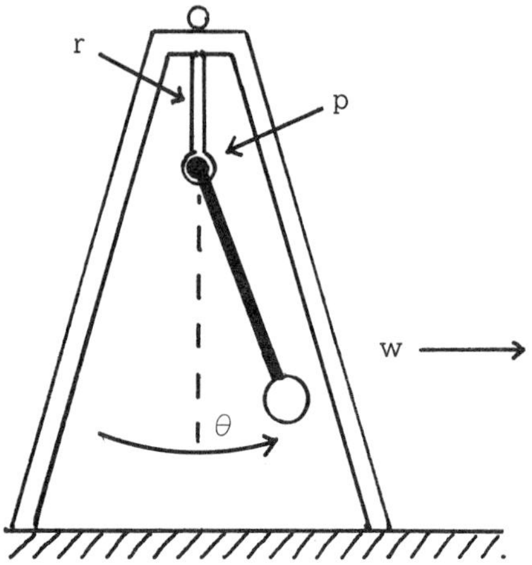

Fig. VI. 2. Pendulum

or, in first order form with

$$x^1 = \theta, \quad x^2 = \dot{\theta}$$

$$\begin{pmatrix} \dot{x}^1 \\ \dot{x}^2 \end{pmatrix} = \begin{pmatrix} 0 & 1 \\ -1 & -1 \end{pmatrix} \begin{pmatrix} x^1 \\ x^2 \end{pmatrix} + \begin{pmatrix} 0 \\ 1 \end{pmatrix} w . \qquad (7.2)$$

Let us think of this pendulum as, say, a device used in surveying to determine the vertical direction. It is used in an outdoor environment and is slightly affected by wind velocity w, even on a nearly calm day. (Alternatively, one could think of the supporting structure being subject to vibration, etc.)

If the mean wind velocity is known, a correction can be made to take that into account. So we assume the disturbance w has zero mean.

Is it reasonable to assume that w is white noise? Probably not. That assumption amounts to supposing that there is no correlation at all between wind velocity at an instant t_1 and wind velocity at another instant, t_2, however close to t_1. That is, contrary to experience and intuition. Wind velocity can only meaningfully be regarded as "colored noise", wherein $\text{cov}(w(t_1, t_2)) = W(t_1, t_2)$. A treatment of this type of disturbance lies outside the scope of our present efforts. It is also safe to say that real-life problems are not often treated in this framework. Nevertheless, an artifice can be employed which has much the same effect, while still permitting us to work within the white noise framework. Let us suppose the wind velocity is $w = w(t)$ and, rather than w itself being white noise, w is though of as being "driven" by white noise v according to the equation

$$\dot{w} + \alpha w = v, \qquad \alpha > 0 . \tag{7.3}$$

For small values of α this amounts, approximately, to supposing that it is the rate of change of wind velocity, rather than the velocity itself, which is to be thought of as white noise. We take $\alpha > 0$ in order to keep $\text{cov}(w(t), w(t))$ bounded for all t.

If v has steady state covariance V (a 1×1 matrix here, i. e., a scalar), the steady state covariance, W, of w, satisfies (cf. (2. 39))

$$-2\alpha W + V = 0 ,$$

giving

$$W = \frac{1}{2\alpha} V.$$

This artifice can be extended to higher levels. One could replace (7. 3) by

$$\ddot{w} + \gamma\dot{w} + \alpha w = v, \qquad \gamma > 0, \ \alpha > 0 \tag{7.4}$$

and obtain an even smoother process. We will not pursue this here.

The complete system, assuming the model (7. 3) for wind velocity, is obtained by combining (7. 2) and (7. 3):

$$\begin{pmatrix} x^1 \\ x^2 \\ w \end{pmatrix} = \begin{pmatrix} 0 & 1 & 0 \\ -1 & -1 & 1 \\ 0 & 0 & -\alpha \end{pmatrix} \begin{pmatrix} x^1 \\ x^2 \\ w \end{pmatrix} + \begin{pmatrix} 0 \\ 0 \\ 1 \end{pmatrix} v . \tag{7.5}$$

If we let $x = (x^1, x^2, w)^*$, then

$$\lim_{t \to \infty} \operatorname{cov}(x(t), x(t)) = X$$

where (cf. (2.39))

$$AX + XA^* + CVC^* = 0 \tag{7.6}$$

$$A = \begin{pmatrix} 0 & 1 & 0 \\ -1 & -1 & 1 \\ 0 & 0 & -\alpha \end{pmatrix}, \quad C = \begin{pmatrix} 0 \\ 0 \\ 1 \end{pmatrix}. \tag{7.7}$$

It should be noted that A has a block triangular structure

$$A = \begin{pmatrix} A_1^1 & A_2^1 \\ 0 & A_2^2 \end{pmatrix},$$

A_1^1 in the present instance being 2×2, A_2^2 1×1. If (7.4) were used, both A_1^1 and A_2^2 would be 2×2. Taking symmetry into account, we write

$$X = \begin{pmatrix} X_1^1 & X_2^1 \\ (X_2^1)^* & X_2^2 \end{pmatrix}$$

and, from the form of C in (7.7),

$$CVC^* = \begin{pmatrix} 0 & 0 \\ 0 & V \end{pmatrix}.$$

We find that

$$A_2^2 X_2^2 + X_2^2 (A_2^2)^* + V = 0, \tag{7.8}$$

$$A_1^1 X_2^1 + X_2^1 (A_2^2)^* + A_2^1 X_2^2 = 0, \tag{7.9}$$

$$A_1^1 X_1^1 + X_1^1 (A_1^1)^* + A_2^1 (X_2^1)^* + X_2^1 (A_2^1)^* = 0. \tag{7.10}$$

Solution of the larger system $AX + XA^* + CVC^* = 0$ is facilitated by carrying it out in these three steps. In our example this gives

$$X_2^2 = W = \frac{1}{2\alpha} V , \tag{7.11}$$

$$X_2^1 = \begin{pmatrix} \dfrac{1}{2\alpha + 2\alpha^2 + 2\alpha^3} V \\ \\ \dfrac{1}{2 + 2\alpha + 2\alpha^2} V \end{pmatrix} , \tag{7.12}$$

$$X_1^1 = \begin{pmatrix} \dfrac{1}{2\alpha + 2\alpha^2 + 2\alpha^3} V + \dfrac{1}{2 + 2\alpha + 2\alpha^2} V & 0 \\ \\ 0 & \dfrac{1}{2 + 2\alpha + 2\alpha^2} V \end{pmatrix} \tag{7.13}$$

It should be noted that $W = X_2^2$ depends only on the ratio V/α but V and α affect X_2^1 and X_1^1 in a more complicated manner. It follows that some care should be used in the selection of both α and V. This is an important question in its own right but will not be discussed further here because its treatment lies outside the scope of our present efforts.

For the moment, let us suppose we have identified (to serve as an example)

$$\alpha = 1, \; V = .15.$$

Then, from (7.11) - (7.13)

$$X_2^2 = .075 \tag{7.14}$$

$$X_2^1 = \begin{pmatrix} .025 \\ .025 \end{pmatrix}, \quad X_1^1 = \begin{pmatrix} .05 & 0 \\ 0 & .025 \end{pmatrix} . \tag{7.15}$$

The question arises now as to how this is to be interpreted. The equation (7.15) tells us that

$$\lim_{t\to\infty} \operatorname{cov}(x^1(t), x^1(t)) = \lim_{t\to\infty} \operatorname{cov}(\theta(t), \theta(t)) = .05$$

$$\lim_{t\to\infty} \operatorname{cov}(x^2(t), x^2(t)) = \lim_{t\to\infty} \operatorname{cov}(\dot{\theta}(t), \dot{\theta}(t)) = .025 .$$

The surveyor using our hypothetical instrument to determine the vertical direction might ask: how often will θ exceed one degree? We can answer his question, if we assume, following our discussion in Sections 1 and 2, that θ and $\dot{\theta}$ are normally distributed. Here, since $\lim_{t\to\infty} \operatorname{cov}(\theta(t), \dot{\theta}(t)) = 0$, $\theta(t)$ and $\dot{\theta}(t)$ are, in the limit, independent. The probability densities are therefore (cf. (1.18))

For $\theta = \nu^2 = .05$, $\nu = .22360$

$$\sigma_{0,.22360}(\theta) = \frac{1}{.22360\sqrt{2\pi}} \exp\left(\frac{-\theta^2}{2 \times .05}\right) ,$$

For $\dot{\theta} = \nu^2 = .025$, $\nu = .15811$

$$\sigma_{0,.15811}(\dot{\theta}) = \frac{1}{.15811\sqrt{2\pi}} \exp\left(\frac{-(\dot{\theta})^2}{2 \times .025}\right) .$$

Assuming that θ is measured in radians, one degree corresponds to $\theta = .01745$. The probability that (at any given moment after the steady state situation has been reached within our accuracy limits) θ exceeds .01745 is

$$2 \int_{.01745}^{\infty} \sigma_{.22360}(\theta) d\theta . \tag{7.16}$$

The tables are ordinarilygiven for $\sigma = 1$. Letting $\theta = .22360\psi$ ($= \nu\psi$), (7.16) becomes

$$2 \int_{.07804}^{\infty} \frac{1}{\sqrt{2\pi}} e^{-\psi^2/2} d\psi .$$

Using the four place tables in [13] we find this probability to be

$$P = \{|\theta| > .01745\} = .9378 ,$$

a figure which will clearly be unacceptable to any serious surveyor.

How can we re-design the instrument to make it more acceptable? Returning to (7.1) we introduce a control variable u:

$$\ddot{\theta} + \dot{\theta} + \theta = w + u . \tag{7.17}$$

Then (7.5) is replaced by (still taking $\alpha = 1$)

$$\begin{pmatrix} \dot{x}^1 \\ x^2 \\ w \end{pmatrix} = \begin{pmatrix} 0 & 1 & 0 \\ -1 & -1 & 1 \\ 0 & 0 & -1 \end{pmatrix} \begin{pmatrix} x^1 \\ x^2 \\ w \end{pmatrix} + \begin{pmatrix} 0 \\ 1 \\ 0 \end{pmatrix} u + \begin{pmatrix} 0 \\ 0 \\ 1 \end{pmatrix} v. \tag{7.18}$$

Now full state feedback would involve setting

$$u = k_1 x^1 + k_2 x^2 + k_3 w . \tag{7.19}$$

While one can conceive of various devices to measure the wind velocity w, it is clear that the incorporation of such a device would involve serious complications. One therefore wishes to restrict consideration to feedback relations of the form

$$u = k_1 x^1 + k_2 x^2 . \tag{7.20}$$

This corresponds to

$$u = Hx,$$

$$K_1 = (k_1, k_2), \quad H = \begin{pmatrix} 1 & 0 & 0 \\ 0 & 1 & 0 \end{pmatrix},$$

a case of limited observation as discussed in Chapter III and Section 5 of the present chapter.

The implementation of (7.20) is very simple indeed. Closing the loop, (7.17) is replaced by

$$\ddot{\theta} + (1 - k_2)\dot{\theta} + (1 - k_1)\theta = w .$$

Assuming $1 - k_1$, $1 - k_2$ turn out to be positive, closed loop control just amounts to an adjustment of the friction at p(perhaps using some sort of viscous damping device) and an adjustment of the mass m.

We may obtain an improved system by choosing $K = \hat{K}$ to solve a system of the form (5.39) - (5.41). However, it is not efficient to use those equations directly in this case. We explain.

The system (7.18) is a special case of a more general "block triangular" system

$$\begin{pmatrix} \dot{\xi} \\ \eta \end{pmatrix} = \begin{pmatrix} A_\xi^\xi & A_\eta^\xi \\ 0 & A_\eta^\eta \end{pmatrix} \begin{pmatrix} \xi \\ \eta \end{pmatrix} + \begin{pmatrix} B \\ 0 \end{pmatrix} u + \begin{pmatrix} O \\ C \end{pmatrix} v \tag{7.21}$$

with A_2^2 a stability matrix. Setting

$$u = K\xi$$

gives the closed loop form

$$\begin{pmatrix} \dot{\xi} \\ \eta \end{pmatrix} = \begin{pmatrix} A_\xi^\xi + BK & A_\eta^\xi \\ 0 & A_\eta^\eta \end{pmatrix} \begin{pmatrix} \xi \\ \eta \end{pmatrix} + \begin{pmatrix} O \\ C \end{pmatrix} v . \tag{7.22}$$

The observation is just

$$\omega = H\begin{pmatrix} \xi \\ \eta \end{pmatrix} = (I, 0)\begin{pmatrix} \xi \\ \eta \end{pmatrix} = \xi . \tag{7.23}$$

Let us agree to use a cost integrand

$$\xi^* W\xi + \xi^* Ru + u^* R^* \xi + u^* Uu = (\xi^*, \eta^*) \begin{pmatrix} W & 0 \\ 0 & 0 \end{pmatrix} \begin{pmatrix} \xi \\ \eta \end{pmatrix}$$

$$+ (\xi^*, \eta^*)\begin{pmatrix} R \\ 0 \end{pmatrix} u + u^* (R^*, 0)\begin{pmatrix} \xi \\ \eta \end{pmatrix} + u^* Uu . \tag{7.24}$$

We partition matrices $P = \mathrm{cov}(\begin{pmatrix} \xi \\ \eta \end{pmatrix}, \begin{pmatrix} \xi \\ \eta \end{pmatrix})$ in a manner compatible with (7.22):

$$P = (P_\xi, P_\eta) = \begin{pmatrix} P^\xi \\ P^\eta \end{pmatrix} = \begin{pmatrix} P^\xi_\xi & P^\xi_\eta \\ P^\eta_\xi & P^\eta_\eta \end{pmatrix}.$$

Then the equations (5. 39), (5. 40) reduce to

$$\hat{K} = -U^{-1}(B^*[Q^\xi_\xi(\hat{K})P^\xi_\xi(\hat{K}) + Q^\xi_\eta(\hat{K})P^\eta_\xi(\hat{K})] + R^*P^\xi_\xi(\hat{K}))P^\xi_\xi(\hat{K})^{-1}$$

$$= -U^{-1}(B^*[Q^\xi_\xi(\hat{K}) + Q^\xi_\eta(\hat{K})P^\eta_\xi(\hat{K})P^\xi_\xi(\hat{K})^{-1}] + R^*) \tag{7.25}$$

$$(A^\xi_\xi + B\hat{K})^* Q^\xi_\xi(\hat{K}) + Q^\xi_\xi(\hat{K})(A^\xi_\xi + B\hat{K}) + W + R\hat{K} + \hat{K}^*R^* + \hat{K}^*U\hat{K} = 0 \tag{7.26}$$

$$(A^\xi_\xi + B\hat{K})^* Q^\xi_\eta(\hat{K}) + Q^\xi_\eta(\hat{K})A^\eta_\eta + Q^\xi_\xi(\hat{K})A^\xi_\eta = 0 \tag{7.27}$$

To obtain the reduced form of (7. 41), we note that the counterpart of (5. 37) in this case is

$$M = \begin{pmatrix} 0 & 0 \\ 0 & CVC^* \end{pmatrix}.$$

This choice gives, in place of (7. 41),

$$(A^\xi_\xi + B\hat{K})P^\xi_\xi(\hat{K}) + P^\xi_\xi(\hat{K})(A^\xi_\xi + B\hat{K})^* + A^\xi_\eta P^\eta_\xi(\hat{K}) + P^\xi_\eta(\hat{K})(A^\xi_\eta)^* = 0 \tag{7.28}$$

$$(A^\xi_\xi + B\hat{K})P^\xi_\eta(\hat{K}) + P^\xi_\eta(\hat{K})A^\eta_\eta + A^\xi_\eta P^\eta_\eta(\hat{K}) = 0 \tag{7.29}$$

$$(P^\eta_\xi(\hat{K}) = P^\xi_\eta(\hat{K})^*) \tag{7.30}$$

$$A^\eta_\eta P^\eta_\eta(\hat{K}) + P^\eta_\eta(\hat{K})A^{\eta *}_\eta + CVC^* = 0\,. \tag{7.31}$$

Note that (7. 31) is independent of $\hat{K}$ in fact, and may be solved once and for all. The end result, as far as iterative solution is concerned, five equations, (7. 25), (7. 26), (7. 27), (7. 28), (7. 29) whose solution involves manipulation of matrices smaller than those appearing in (5. 39) - (5. 41).

In Exercise 11 the reader is invited to obtain $(\hat{k}_1, \hat{k}_2)$ for our pendulum problem in a manner pleasing to our hypothetical "surveyor". A full numerical treatment here would lead us too far astray. For a starting value of $K_0 = (k_{10}, k_{20})$, however, we suggest the optimal control for (7.17) when w is assumed to be white noise. Since θ is what concerns us, we take

$$W = \begin{pmatrix} 1 & 0 \\ 0 & 0 \end{pmatrix}, \qquad R = \begin{pmatrix} 0 \\ 0 \end{pmatrix}.$$

The quadratic matrix equation is (cf. IV - 3)

$$\begin{pmatrix} 0 & -1 \\ 1 & -1 \end{pmatrix}\begin{pmatrix} q_1 & q_2 \\ q_2 & q_3 \end{pmatrix} + \begin{pmatrix} q_1 & q_2 \\ q_2 & q_3 \end{pmatrix}\begin{pmatrix} 0 & 1 \\ -1 & -1 \end{pmatrix} + \begin{pmatrix} 1 & 0 \\ 0 & 0 \end{pmatrix} - U^{-1}\begin{pmatrix} q_2^2 & q_2q_3 \\ q_2q_3 & q_3^2 \end{pmatrix} = 0.$$

The solution is, in part,

$$q_2 = -U + \sqrt{U^2 + U}$$

$$q_3 = -U + \sqrt{U^2 + 2Uq_2} = -U + \sqrt{U^2 + 2U(-U + \sqrt{U^2 + U})} = -U + \sqrt{2U\sqrt{U^2+U} - U^2}$$

giving

$$k_1 = 1 - \sqrt{1 + 1/U}$$
$$k_{20} = 1 - \sqrt{2\sqrt{1 + 1/U} - 1} \tag{7.32}$$

Just as an example, let us take $U = \frac{1}{8}$. Then

$$k_{10} = -2, \quad k_{20} = 1 - \sqrt{5}\ . \tag{7.33}$$

Returning to (7.11), (7.12), (7.13) with A_1^1 now replaced by

$$A_1^1 + BK_0 = \begin{pmatrix} 0 & 1 \\ -3 & -\sqrt{5} \end{pmatrix}$$

and again taking $\alpha = 1$, $V = .15$, we still have (7.11), and hence (7.14), but (7.15) is replaced by

$$\begin{pmatrix} 0 & 1 \\ -3 & -\sqrt{5} \end{pmatrix} X_2^1 + X_2^1(-1) + \begin{pmatrix} 0 \\ 1 \end{pmatrix} .075 = 0$$

from which we have

$$X_2^1 = \begin{pmatrix} \dfrac{.075}{4+\sqrt{5}} \\ \dfrac{.075}{4+\sqrt{5}} \end{pmatrix} = \begin{pmatrix} .01202 \\ .01202 \end{pmatrix} .$$

and

$$\begin{pmatrix} 0 & 1 \\ -3 & -\sqrt{5} \end{pmatrix} X_1^1 + X_1^1 \begin{pmatrix} 0 & -3 \\ 1 & -\sqrt{5} \end{pmatrix} + \begin{pmatrix} 0 & .01202 \\ .01202 & .02404 \end{pmatrix} = 0$$

giving

$$X_1^1 = \begin{pmatrix} .00579 & 0 \\ 0 & .00537 \end{pmatrix} .$$

We now have

$$\nu^2 = .00579, \quad \nu = .07609 .$$

The probability that $|\theta|$ exceeds one degree is now

$$2\int_{.01745}^{\infty} \sigma_{.07609}(\theta)d\theta = 2\int_{.22933}^{\infty} \frac{1}{\sqrt{2\pi}} e^{-\psi^2/2} d\psi = .8185, \tag{7.34}$$

still hardly acceptable, but better than before.

Now suppose we take $U = \frac{1}{99}$. We then have, from (7.32),

$$k_{10} = -9, \quad k_{20} = -3.35889 . \tag{7.35}$$

The matrix $A_1^1\ \delta\ BK_0$ now becomes

$$\begin{pmatrix} 0 & 1 \\ -10 & -4.35889 \end{pmatrix}$$

with the result that

$$X_2^1 = \begin{pmatrix} .00488 \\ .00488 \end{pmatrix},$$

$$X_1^1 = \begin{pmatrix} .00060 & 0 \\ 0 & .00111 \end{pmatrix}.$$

Now

$$\nu^2 = .00060, \quad \nu = .02449 .$$

The probability that $|\theta|$ exceeds one degree is

$$2\int_{.01745}^{\infty} \sigma_{.02449}(\theta)d\theta = 2\int_{.71253}^{\infty} \frac{1}{\sqrt{2\pi}} e^{-\psi^2/2} d\psi = .4762 .$$

The result is clearly becoming more acceptable but needs further improvement. In Exercise 11 the reader is invited to see how each of the results (7.34) and (7.36) can be improved upon by solving the system (7.27) - (7.29), specialized to this problem, using (7.33) and (7.35) as starting values, respectively.

A more thorough examination of the question of evaluation of control policies will be undertaken at the beginning of the next chapter.

Bibliographical Notes, Chapter VI

No actual use is made of the concepts of Lebesgue measure, or measurable functions and sets in general, in our work here but completeness requires some gesture in that direction. Reference [1] contains a classical treatment of Lebesgue's integration theory while [2] presents a more abstract, up to date development of the theory of measurable functions and sets.

[1] E. C. Titchmarsh: "The Theory of Functions", Second Edition, Oxford University Press, 1939.

[2] W. Rudin: "Real and Complex Analysis", McGraw-Hill Book Co., New York, 1966.

Reference [3] could almost be called "the" standard treatment of elementary probability theory. It contains a marvellous collection of examples. Strictly speaking, it treats only discrete probability distributions but limiting cases, involving the familiar normal and Poisson density functions are treated also. Reference [4] is a more recent book, especially to be valued here for (among many other things) its extensive treatment of the normal distribution in E^n and for an excellent discussion of the concept of independence of random variables, which we have neglected somewhat in our brief introductory treatment.

[3] W. Feller: "An Introduction to Probability Theory and its Applications", Third Edition, John Wiley and Sons, Inc., New York, 1968.

[4] B. Harris: "Theory of Probability", Addison Wesley, Reading, Mass., 1966.

Suggested references in functional analysis, including theory of Hilbert spaces:

[5] G. Bachman and L. Narici: "Functional Analysis", Academic Press, New York, London, 1966.

[6] F. Riesz and B. Sz.-Nagy: "Functional Analysis", Frederick Ungar Pub. Co., New York, 1955 (Translated from 2nd French Edition by L. Boron.)

[7] S. Axsäter: "Sub-optimal time-variable feedback control of linear dynamic systems with random inputs", Int. J. Contr., 4 (1966), pp. 549-566.

Discussions of Newton's method for solution of systems of non-linear equations may be found in

[8] L. Collatz: "Functional Analysis and Numerical Methematics", Academic Press, New York, London, 1966.

The references devoted to the statistical theory of linear systems are almost beyond counting. Reference [9] is the original expository treatment, [10] gives a rather general treatment and [11] contains numerous papers devoted to the subject and an extensive bibliography.

[9] R. E. Kalman and R. S. Bucy: "New results in linear prediction and filtering theory", J. Basic Eng., (Trans. ASME, Ser. D) 83 (1961), pp. 95-100.

[10] R. E. Kalman, P. L. Falb and M. A. Arbib: "Topics in Mathematical System Theory", McGraw-Hill Book Co., New York, 1969.

[11] Special Issue on Linear-Quadratic-Gaussian Problem, IEEE Transactions on Automatic Control, AC-16 (1971).

A discussion of the meaning of the term "white noise" and related topics appears in

[12] L. Arnold: "Stochastic Differential Equations: Theory and Applications", John Wiley and Sons, Inc., New York, 1974.

[13] R. S. Burington: "Handbook of Mathematical Tables and Formulas", Handbook Pub., Inc., Sandusky, O., 1957.

The control theory literature abounds with proofs and pseudo-proofs of the separation principle - many of them almost impossible to wade through". One of the most readable is found in reference [14] while [15], [16] contain quite general results. Reference [14] is also to be prized as an excellent exposition of the whole subject of control theory from the engineering standpoint, as is [15] from the mathematical standpoint.

[14] H. Kwakernaak and R. Sivan: "Linear Optimal Control Systems", Wiley-Interscience, New York, 1972.

[15] W. Fleming and R. Rishel: "Deterministic and Stochastic Optimal Control", Springer-Verlag, New York, Heidelberg, Berlin, 1975.

[16] W. M. Wonham: "On the separation theorem of stochastic control", SIAM J. Control, 6 (1968), pp. 312-326.

[17] B. D. O. Anderson and J. B. Moore: "Linear Optimal Control", Prentice-Hall, Englewood Cliffs, N. J., 1971.

Exercises, Chapter VI

1. Use the change of variable $\xi = Ax$ together with the standard result for transformation of multiple integrals (see any good text on advanced calculus, II - [6], e.g.) to establish the formula (2.33).

2. Electromagnetic radiation, including visible light, is termed "white" if all frequencies are present in equal intensity. In the theory of stochastic processes the notion of intensity at given frequency is expressed in terms of the power spectrum. Suppose $v \in L_p^2[t_0, t_1]$ with a given probability distribution - which we need not say anything further about here - and suppose v has zero mean in the sense that $E(v, a)_{L_p^2[t_0,t_1]} = 0$ for all $a \in L_p^2[t_0, t_1]$. For real $\omega \geq 0$ define

$$\sum(\omega, v) = E(\| \int_{t_0}^{t_1} e^{i\omega t} v(t)dt \|) .$$

Equivalently

$$\sum(\omega, v) = [\operatorname{Tr} \operatorname{cov} (\int_{t_0}^{t_1} e^{i\omega t} v(t)dt, \int_{t_0}^{t_1} e^{i\omega t} v(t)dt)]^{1/2} .$$

Show, using Definition 2.3 and preceding material that if v is white noise in $L_p^2[t_0, t_1]$ then $\Sigma(\omega, v)$ is independent of ω, i.e. the power spectrum is a constant function of ω.

3. Modify Proposition 5.2 to obtain a uniqueness result for the time varying optimal feedback law (4.10) under assumptions analogous to (5.28)-(5.30). Then show by example that the condition analogous to (5.30) ($\dot{x} = (A(t) + B(t)\hat{K}(t))x + C(t)v$ is controllable - with respect to v - on $[t_0, t_1]$) or else the condition $X_0 > 0$ are actually needed - i.e., that uniqueness may not obtain if both of these conditions are not met. Hint: Set $K(t) = \hat{K}(t) + \delta K(t)$. Then (4.21) becomes

$$Q_{\hat{K}+\delta K}(\tau) - Q_{\hat{K}}(\tau) = \int_{\tau}^{t_1} \Phi(t, \tau)^* \delta K(t)^* U(t) \delta K(t) \Phi(t, \tau)\, dt .$$

(Note, however, that $\Phi(t, \tau)$ depends on δK.) Then use formula (4.23) with $K = \hat{K} + \delta K$ and try to find δK such that $J(K) - J(\hat{K}) = 0$.

4. Show that Proposition 5.2 may fail if (5.30) is not true and give a numerical example. Essentially the same procedure as suggested in 3 applies here also.

5. Show that if (5.43), (5.44) are replaced by the coupled eqations (difficult to solve in general)

$$(A + BK_{i+1})P_i + P_i(A + BK_{i+1})^* + M = 0,$$

$$K_{i+1} = -U^{-1}(B^*Q_iP_iH^* + R^*P_iH^*)(HP_iH^*)^{-1}$$

then $\operatorname{Tr} Q_i$ decreases monotonically as i increases. Show by example that this property is not generally true for (5.42) - (5.44). Also show by example that even if the above modifications are made, it is not generally true that $Q_{i+1} \leq Q_i$. (See the section on output feedback in [17] for pertinent material and references.)

6. Obtain a result dual to the uniqueness result of Theorem 4.1 - to the effect that $\hat{L}(t)$, as defined by (6.15), is the unique solution of the optimal filtering problem, i.e., minimization of (6.14), if $\tilde{Z} > 0$ or if $\tilde{Y}(t) = J(t)^*Y_0(t)J(t)$, $Y_0(t) > 0$, $\dot{x} = (A(t) - \hat{L}(t)H(t))x$, $\omega = J(t)x$ observable on $[t_0, t_1]$.

7. Modify the proof of Theorem 6.1, in much the same way as the proof of Theorem 4.1 is modified to give Theorem 5.1, to prove Theorem 6.2. Analyze conditions for uniqueness as in Exercise 6 above.

8. If the differential equation

$$-\frac{dQ}{dt} = A(t)^*Q + QA(t) + W(t) - (QB(t) + R(t))U(t)^{-1}(B(t)^*Q + R(t)^*) = 0,$$

with $Q(t_1) = V > 0$, is multiplied on the right and on the left by $Q(t)^{-1}$, and if the structure of $W(t)$, $R(t)$ and the formula for $\frac{dQ^{-1}}{dt}$ are taken into account, and if t is replaced by $s = t_0 + (t_1 - t)$, an equation for Q^{-1} having the same form as (6.15) for P results. Define an optimal filtering problem for which Q^{-1} plays the role of P and construct a duality theory distinct from that developed in Section 6 and exhibited in the table following Theorem 6.2. Construct a comparable table for your new duality relationship.

9. Modify the statement and proof of Theorem 6. 3 to obtain the separation theorem for combined control and filtering of time varying systems on a finite interval $[t_0, t_1]$. Comparison of the proofs of Theorems 4. 1 and 5. 1 will facilitate replacement of the various linear matrix equations of the proof of Theorem 6. 3 by the parallel linear matrix differential equations. The analog of Proposition 3. 3 is the relationship (4. 16), which allows $J(K)$ to be expressed either in terms of the solution $X(t)$ of (4. 3) or the solution $Q_K(t)$ of (4. 12). A comparable relationship will be needed involving the differential equation analogous to (6. 46) and the differential equation analogous to (6. 48).

10. Consider the scalar control system

$$\dot{x} = ax + bu + d\nu$$

the scalar feedback law and cost

$$u = ky, \quad J(k, \ell) = \int_0^\infty [w(x(t))^2 + (u(t))^2]\, dt,$$

and the scalar state estimator

$$\dot{y} = (a + bk)y + \ell h(\omega - hy)$$

where

$$\omega = hx + \tilde{j}\nu + \mu .$$

The disturbances ν, μ are scalar white noise with scalar covariances $\tilde{N}, \tilde{M}$, respectively, $\tilde{M} > 0$. For such a system $2n - 1 = 2 \cdot 1 - 1 = 1$ and (6. 57) yields two equations in the two unknowns k, ℓ. The first gives (cf. (6. 58))

$$\ell = (\hat{k}\hat{\ell})/\ell .$$

Substitute this expression for ℓ in the equation (6. 57) corresponding to $j = 1$ and obtain, after multiplication by k, a cubic polynomial equation in k. Try to find values of a, b, d, ... etc., such that this polynomial has roots k for which $k \neq k$, $\ell \neq \hat{\ell}$ and, if possible, such that $a + bk < 0$, $a - \ell h < 0$. What are your conclusions and/or conjectures relative to uniqueness of the minimum $\hat{k}$, $\hat{\ell}$ of $J(k, \ell)$?

11. Use the method (5. 42) - (5. 43), decomposed in agreement with (7. 25) - (7. 31), to attempt to reduce the expected limiting (as $t \to \infty$) instantaneous value of the cost (7. 24) with the values for W, R, U, α, V, given following (7. 31). Use the values (7. 33) as starting values for the procedure. Note that this is a restricted feedback situation with H as displayed in (7. 23). Repeat, if time and circumstances permit, using $U = 1/99$ and the starting values (7. 35).

12. Essentially the same as 11. but allow feedback

$$u = k_1 y^1 + k_2 y^2 + k_3 y^3$$

where $y^* = (y^1, y^2, y^3)$ is an optimal estimate of (x^1, x^2, w) obeying (7. 18), constructed following Theorem 6. 2 with $\nu = v$, $\tilde{N} = V = 1$, $\tilde{M} = .1$, $\tilde{J} = 0$. Use $U = 1/8$, $1/99$ as in 11. Try to find out how small $\tilde{M}$, U must be taken so that the probability that θ exceeds one degree is reduced to .1.

13. Re-develop the conjugate gradient minimization technique (see III- [1] or V-[3]), using the matrix inner product (3. 1), so that it applies to an objective function F(K), $F : R^{mn} \to R^1$, K an $m \times n$ matrix. Specialize your method to the cost function F(K) defined following (5. 36) and use it in place of the method (5. 42) - (5. 44) to re-do problem 11. Compare results. Obviously this should be considered a very ambitious assignment involving both theoretical and computational work of a significant order of difficulty.

14. Modify Theorems 6. 1 and 6. 2 by constructing optimal state estimators for systems

$$\dot{x} = \tilde{A}(t)x + f(t) + D(t)\nu , \qquad \dot{x} = \tilde{A}x + Fz + D\nu ,$$

where, in the first case, f(t) is a known function on $[t_0, t_1]$ and the estimation problem is posed on $[t_0, t_1]$ as in Theorem 6. 1, while in the second case the estimation problem is posed as in Theorem 6. 2 and z satisfies an equation

$$\dot{z} = Cz + C_0\nu + C_1\mu ,$$

C being a stability matrix of appropriate dimension. All that is required in each case is a minor modification of (6. 11).

CHAPTER VII

ASPECTS OF LINEAR SYSTEM DESIGN

1. Preliminaries. In fabricating a control law for an actual working plant, the control designer must usually take a rather wide range of criteria into account. Asymptotic stability of some sort of equilibrium or steady state operating condition is usually the *sine qua non* of the analysis but it is by no means the whole story. In this chapter we propose to study a variety of properties which in various applications it is desirable that the controlled plant should possess. Attempting to reproduce such properties in the closed loop system leads to a class of what might be termed "non-standard" linear quadratic optimal control problems which do not conform to the standard problem discussed in Chapters IV - VI. They correspond more closely to the limited or "output" feedback problem discussed in VI- 5 and the methods of solution are much the same as outlined there for that problem.

The basis for all of these criteria is the statistical analysis of Chapter VI, which affords a means for generating scalar costs independent of particular initial states. We begin by summarizing the development of these scalar cost criteria in the context of a system

$$\dot{x} = Ax + Bu, \quad x \in E^n, \; u \in E^m, \tag{1.1}$$

with the assumption of stabilizability. We employ a performance criterion of the form

$$J(x_0, u) = \int_0^\infty [\,(Fx(t) + Gu(t))^* \tilde{W}(Fx(t) + Gu(t)) + u(t)^* \tilde{U} u(t)\,]\, dt, \tag{1.2}$$

where $\tilde{W}$ and $\tilde{U}$ are symmetric positive definite $q \times q$ and $m \times m$ matrices, respectively, and

$$y(t) = Fx(t) + Gu(t) \tag{1.3}$$

represents the q-dimensional response, or output, to be regulated, it being always assumed that the pair (F, A) is observable (or at least (see IV-Exercise 14) detectable - we will not emphasize this weaker hypothesis).

Given a linear feedback relation

$$u = Kx\ , \tag{1.4}$$

K being an $m \times n$ matrix such that $A + BK$ is a stability matrix, we have seen that the cost (1.4) may then be represented in the form

$$J(x_0, K) = x_0^* Q(K) x_0\ ,$$

where $Q(K)$ is the unique symmetric positive definite (non-negative in the case (F, A) detectable) solution of the Liapounov equation

$$(A + BK)^* Q(K) + Q(K)(A + BK) + (F + GK)^* \tilde{W}(F + GK) + K^* \tilde{U} K = 0\ , \tag{1.5}$$

or, equivalently,

$$(A + BK)^* Q(K) + Q(K)(A + BK) + W + RK + K^* R^* + K^* \tilde{U} K = 0\ . \tag{1.6}$$

The control law

$$u = \hat{K}x = -\,U^{-1}(B^* Q + R^*)x\ , \tag{1.7}$$

with Q the unique positive definite solution of the quadratic matrix equation (cf. IV - (3.13))

$$A^* Q + QA + W - (QB + R)U^{-1}(B^* Q + R^*) = 0 \tag{1.8}$$

is optimal in the sense that

$$J(x_0, \hat{K}) = x_0^* Q(K)x_0 = x_0^* Qx_0 \leq x_0^* Q(K)x_0 = J(x_0, K)$$

for every initial state $x_0 \in E^n$, equivalently, the matrix

$$Q(K) - Q(\hat{K})$$

is non-negative for any $m \times n$ feedback matrix K such that $A + BK$ is a stability matrix.

Despite the strength and simplicity of this result, which we have discussed in detail in Chapters IV - VI, it is not wholly satisfactory in certain respects. Consider, for example, the following problem. Suppose a stabilizing feedback matrix K is given; perhaps corresponding to a control which is already being used in a system (1. 1). One could contemplate replacing K by $\hat{K}$ - in practice this would amount to changing certain parameters in the analog or digital device being used to regulate the system in question. Before doing this it is natural to ask, "how much better is the control law (1. 7) as compared with the control law $u = Kx$? ". This is not at all answered by the fact that $Q(K) - Q(\hat{K})$ is non-negative and no direct answer can be given even upon inspection of this matrix in detail. The question "how much ? " in some sense requires a single number, i. e. , a scalar, for an answer and $Q(K) - Q(\hat{K})$ is a matrix. Such a matrix can very well be large in some respects and small in other respects; indeed, for $\|x_0\| = 1$, the quadratic form $x_0^*[Q(K) - Q(\hat{K})]x_0$ takes on all values μ in the interval $[\mu_1, \mu_n]$, where μ_1 and μ_n are the smallest and largest (real, nonnegative) eigenvalues of the symmetric nonnegative matrix $Q(K) - Q(\hat{K})$. The need thus arises for a scalar indication of performance related in a natural way to the data available in a given situation and the feedback matrix being employed to regulate the system.

The scalar performance indicator favored in practice is provided by the statistical theory of Chapter VI. In VI - (5. 17) we have the expression

$$\operatorname{Tr} Q(K)(X_0 + \frac{1}{\mu} CVC^*) \equiv \operatorname{Tr} Q(K)\, M \tag{1. 9}$$

which, in our interpretation of VI - Section 5, is the expected value of the cost (1. 2) for the control generated by (1. 4), assuming that the initial state

$$x_0 = x(0) \tag{1.10}$$

is a random vector with covariance

$$\operatorname{cov}(x_0, x_0) = X_0 \tag{1.11}$$

and that (1. 1) is augmented to include the effects of an exponentially discounted white noise disturbance: i. e.,

$$\dot{x} = Ax + Bu + Cv, \tag{1.12}$$

v being steady state white noise with covariance matrix $e^{-\mu t}V$. Letting $\mu \to \infty$ gives in the limit

$$\operatorname{Tr} Q(K)X_0\ , \tag{1.13}$$

the expected value of the cost (1. 2) for an undisturbed system (1. 1) with random initial vector (1. 10), (1. 11). Similarly

$$\operatorname{Tr} Q(K)\, CVC^* \tag{1.14}$$

can be interpreted as the expected value of the cost (1. 2) for $X_0 = 0$, $\mu = 1$, or, following the discussion in Chapter IV, as the limiting expected value of the cost integrand $(Fx(t) + Gu(t))^*\tilde{W}(Fx(t) + Gu(t)) + u(t)) + u(t)^*\tilde{U}u(t)$ when we set $u = Kx$ so that

$$\dot{x} = (A + BK)x + Cv.$$

For $\mu > 0$, (1. 9) can be interpreted as above in connection with (1. 2), (1. 12) or simply as a linear combination of the cost (1. 13) associated with (1. 1) and the cost (1. 14) associated with (1. 12). In any event (1. 9) (or (1. 13), (1. 14)) provides a scalar criterion for performance comparison, provided X_0, C and V have been identified. In cases of complete ignorance, which arise all too often, it is common to take $X_0 = C = V = I$ and, multiplying

(1.9) by $(1 + \frac{1}{\mu})^{-1}$, to use just

$$\text{Tr } Q(K)I = \text{Tr } Q(K)$$

as the scalar performance indicator.

An example involving cost functionals of the form (1.14) is presented at the end of Section 2.

2. Multiple Control Objectives. Because a wide variety of criteria must normally be taken into account it is useful to carry the analysis of Section 1 somewhat further. The cost (1.2) is composed of two parts, namely the response cost

$$J_y(x_0, u) = \int_0^\infty (Fx(t) + Gu(t))^* \tilde{W} (Fx(t) + Gu(t))dt = (\text{cf.}(1.3)) = \int_0^\infty y(t)^* \tilde{W} y(t)dt \tag{2.1}$$

and the control cost

$$Ju(x_0, u) = \int_0^\infty u(t)^* \tilde{U} u(t)dt . \tag{2.2}$$

When a control law $u = Kx$ is used, we have in place of (2.1), (2.2), respectively,

$$J_y(x_0, K) = x_0^* Q_y(K) x_0$$

and

$$J_u(x_0, K) = x_0^* Q_u(K) x_0$$

where $Q_y(K)$, $Q_u(K)$ solve the respective Liapounov equations (cf. VI-(5.12))

$$(A + BK)^* Q_y(K) + Q_y(K)(A + BK) + (F + GK)^* \tilde{W} (F + GK) = 0 , \tag{2.3}$$

$$(A + BK)^* Q_u(K) + Q_u(K)(A + BK) + K^* \tilde{U} K = 0 . \tag{2.4}$$

We have, of course,

$$Q(K) = Q_y(K) + Q_u(K) . \tag{2.5}$$

Associated with these, just as before, are scalar response and control costs (cf. (1.9))

$$\begin{aligned} &\operatorname{Tr} Q_y(K) M , \\ &\operatorname{Tr} Q_u(K) M . \end{aligned} \tag{2.6}$$

Very frequently the response (1.3) consists of a number of subsidiary responses:

$$y(t) = \begin{pmatrix} y_1(t) \\ y_2(t) \\ \vdots \\ y_N(t) \end{pmatrix} = \begin{pmatrix} F_1 x(t) + G_1 u(t) \\ F_2 x(t) + G_2 u(t) \\ \vdots \\ F_N x(t) + G_N u(t) \end{pmatrix}$$

which may be scalar or, in general, vectors of dimensions $q_1, q_2, \ldots, q_N$ with $q_1 + q_2 + \ldots + q_N = q$, and, correspondingly, $\tilde{W}$ may be taken block diagonal:

$$\tilde{W} = \begin{pmatrix} \tilde{W}_1 & 0 & \cdots & 0 \\ 0 & \tilde{W}_2 & \cdots & 0 \\ \vdots & \vdots & \ddots & \\ 0 & 0 & \cdots & \tilde{W}_N \end{pmatrix} \tag{2.7}$$

each $\tilde{W}_k$ being of dimension $q_k \times q_k$. We then have

$$\begin{aligned} J_y(x_0, u) &= \sum_{k=1}^{N} \int_0 y_k(t)^* \tilde{W}_k y_k(t)\, dt \\ &= \sum_{k=1}^{N} \int_0 (F_k x(t) + G_k u(t))^* \tilde{W}_k (F_k x(t) + G_k u(t)) dt \end{aligned} \tag{2.8}$$

and, when a control law $u = Kx$ is used, we have, correspondingly

$$J_y(x_0, K) = \sum_{k=1}^{N} x_0^* Q_k(K) x_0, \quad Q_y(K) = \sum_{k=1}^{N} Q_k(K)$$

where

$$(A+BK)^* Q_k(K) + Q_k(K)(A+BK) + (F_k + G_k K)^* \tilde{W}_k (F_k + G_k K) = 0, \quad k = 1,2,\ldots, N. \tag{2.9}$$

The responses $y_k(t)$ will often represent individual quantities of physical, economic or other intrinsic interest, depending on the context in which the system (1.1) (or (1.12), or (1.15)) is viewed. In evaluating a control law $u = Kx$ the separate costs associated with these responses will often need to be taken into account. As before, this is most conveniently done in terms of the individual scalar costs

$$\operatorname{Tr} Q_k(K) M, \quad k = 1, 2, \ldots, N. \tag{2.10}$$

Solving the linear-quadratic optimal control problem, i.e., finding the K which solves

$$\min_{K \in \mathscr{S}(A, B)} \{\operatorname{Tr} Q(K) M\}$$

(here and subsequently $\mathscr{S}(A, B) = \{K, m \times n \,|\, A + BK \text{ is a stability matrix}\}$ corresponding to particular initially specified weighting matrices $\tilde{W}$, $\tilde{U}$ may not lead to satisfactory values for all of the costs (2.10) - some may not be small enough, others may be smaller than necessary, etc. It is not at all easy to select $\tilde{W}$, $\tilde{U}$, a priori, so as to achieve the required balance among these frequently conflicting objectives. Thus we require a means for changing the weighting matrices $\tilde{W}$, $\tilde{U}$ in a systematic manner so as to effect the desired "trade-offs" among the costs (2.10) and the control cost (2.6).

It goes almost without saying that the control $u \in E^m$ might also be partitioned into individually significant components

$$u = \begin{pmatrix} u_1 \\ u_2 \\ \vdots \\ u_\nu \end{pmatrix}, \quad u_k \in E^{m_k}, \quad m_1 + m_2 + \ldots + m_\nu = m,$$

representing separate actuators on control devices and separate control costs

$$J_\nu(x_0, u) = \int_0^\infty u_k(t)^* U_k u_k(t)dt, \qquad k = 1, 2, \ldots, \nu$$

corresponding to a block diagonal weighting matrix

$$\tilde{U} = \begin{pmatrix} U_1 & 0 & \cdots & 0 \\ 0 & U_2 & \cdots & 0 \\ \vdots & \vdots & \ddots & \vdots \\ 0 & 0 & \cdots & U_\nu \end{pmatrix}$$

may be defined with (cf. (2.8))

$$\sum_{k=1}^{\nu} \hat{J}_k(x_0, u) = J_u(x_0, u) . \tag{2.11}$$

When a feedback law $u = Kx$ is used, it can be decomposed as

$$u = \begin{pmatrix} u_1 \\ u_2 \\ \vdots \\ u_\nu \end{pmatrix} = \begin{pmatrix} K_1 \\ K_2 \\ \vdots \\ K_\nu \end{pmatrix} x$$

and we have

$$\hat{J}_k(x_0, u) = x_0^* \hat{Q}_k(K) x_0, \qquad k = 1, 2, \ldots, \nu,$$

where, again for $k = 1, 2, \ldots, \nu$,

$$(A + BK)^* \hat{Q}_k(K) + \hat{Q}_k(K)(A + BK) + K_k^* U_k K_k = 0 . \tag{2.12}$$

The associated scalar costs are (cf. (2. 6), (2. 10))

$$\operatorname{Tr} \hat{Q}_k(K)M, \qquad k = 1, 2, \ldots, \nu . \tag{2.13}$$

We will not study the question of individual control costs here. The interested reader may wish to develop this by carrying out Exercise 1 in a manner parallel to the material to be presented in the rest of this section relative to individual response costs.

Suppose, instead of (2. 7), we define $\tilde{W}$ by

$$\tilde{W} = \tilde{W}(\sigma) = \begin{pmatrix} \sigma_1 \tilde{W}_1 & 0 & \cdots & 0 \\ 0 & \sigma_2 \tilde{W}_2 & \cdots & 0 \\ \vdots & \vdots & & \vdots \\ 0 & 0 & \cdots & \sigma_N \tilde{W}_N \end{pmatrix} \tag{2.14}$$

for a vector $\sigma = (\sigma_1, \sigma_2, \ldots, \sigma_N)$ of non-negative scalars and let $Q = Q(\sigma)$ solve (1. 8) with this value of $\tilde{W}$ and compute $K = \hat{K}(\sigma)$ from (1. 7) with Q replaced by $Q(\sigma)$, i. e.,

$$K = \hat{K}(\sigma) = - U^{-1}(B^* Q(\sigma) + R(\sigma)) \tag{2.15}$$

($R(\sigma) = F^* \tilde{W}(\sigma) G$). For the individual responses y_k, $k = 1, 2, \ldots, N$, however, we define the matrices

$$Q_k(\sigma) = Q_k(\hat{K}(\sigma))$$

by (2. 9), wherein $\tilde{W}_k$ is not multiplied by σ_k. The individual response cost matrices are thus computed with respect to fixed weighting matrices $\tilde{W}_k$ but the feedback matrix $K = \hat{K}(\sigma)$ is computed using a weighting matrix $\tilde{W} = \tilde{W}(\sigma)$ which reflects a certain differential weighting of the individual responses. The weighting matrix $\tilde{U}$ will be kept fixed. Thus the $Q_k(\sigma)$ satisfy

$$(A + B\hat{K}(\sigma))Q_k(\sigma) + Q_k)(\sigma)(A + B\hat{K}(\sigma)) + (F_k + G_k\hat{K}(\sigma))^*\tilde{W}_k(F_k + G_k\hat{K}(\sigma)) = 0 \tag{2.16}$$

and the earlier relationship (2. 8) is now replaced by the weighted sum identity

$$Q(\sigma) = \sum_{k=1}^{N} \sigma_k Q_k(\sigma).+ Q_u(\sigma) . \tag{2.17}$$

The corresponding scalar costs, computed as before, are defined by

$$c_0(\sigma) \equiv \text{Tr } Q(\sigma)M , \tag{2.18}$$

$$c_k(\sigma) \equiv \text{Tr } Q_k(\sigma)M, \qquad k = 1, 2, \ldots, N, \tag{2.19}$$

$$c_u(\sigma) \equiv \text{Tr } Q_u(\sigma)M , \tag{2.20}$$

and (2. 17) implies

$$c_0(\sigma) = \sum_{k=1}^{N} \sigma_k c_k(\sigma) + c_n(\sigma) . \tag{2.21}$$

We expect that $c_k(\sigma)$, $k = 1, 2, \ldots, N$, to bear some sort of inverse relationship to the σ_k; each $c_k(\sigma)$ tending to decrease as the weight σ_k on the response y_k is increased in (2. 14), (2. 17). Let us see what can be developed along these lines.

We set ourselves the task of determining how the $c_k(\sigma)$, $k = 1,2,\ldots,N$, vary with the components σ_j, $j = 1, 2, \ldots, N$ of σ. From (2. 19) we evidently have

$$\frac{\partial c_k}{\partial \sigma_j} = \text{Tr}\frac{\partial Q_k}{\partial \sigma_j} M, \qquad k, j = 1, 2, \ldots, N.$$

Differentiating (2. 16) with respect to σ_j we have

$$(A + B\hat{K}(\sigma))^*\frac{\partial Q_k}{\partial \sigma_j} + \frac{\partial Q_k}{\partial \sigma_j}(A + B\hat{K}(\sigma)) + \frac{\partial \hat{K}}{\partial \sigma_j}^*[B^*Q_k(\sigma) + G_k^*\tilde{W}_k(F_k + G_k\hat{K}(\sigma))]$$

$$+ [Q_k(\sigma)B + (F_k + G_k\hat{K}(\sigma))^*\tilde{W}_k G_k]\frac{\partial \hat{K}}{\partial \sigma_j} = 0 .$$

It follows from VI - Proposition 3. 3 that

$$\frac{\partial c_k}{\partial \sigma_j} = \mathrm{Tr}\,\frac{\partial Q_k}{\partial \sigma_j} M = \mathrm{Tr}\, M \frac{\partial Q_k}{\partial \sigma_j}$$

$$= \mathrm{Tr}\, P(\sigma)\left(\frac{\partial \hat{K}^*}{\partial \sigma_j} H_k(\sigma)^* + H_k(\sigma)\frac{\partial \hat{K}}{\partial \sigma_j}\right) = 2\,\mathrm{Tr}\, P(\sigma) H_k(\sigma)\frac{\partial \hat{K}}{\partial \sigma_j} \tag{2.22}$$

where $P(\sigma)$ satisfies the "adjoint" Liapounov equation

$$(A + B\hat{K}(\sigma))P(\sigma) + P(\sigma)(A + B\hat{K}(\sigma))^* + M = 0 \tag{2.23}$$

and

$$H_k(\sigma) \equiv Q_k(\sigma)B + (F_k + G_k\hat{K}(\sigma))^* \tilde{W}_k G_k\ . \tag{2.24}$$

It should be noted that $P(\sigma)$ is independent of either k of j and $H_k(\sigma)$ is independent of j.

When $\tilde{W} = \tilde{W}(\sigma)$ is given by (2. 14) the matrices $W(\sigma) \equiv F^*\tilde{W}(\sigma)F$, $R(\sigma) \equiv F^*\tilde{W}(\sigma)G$, and $U(\sigma) \equiv \tilde{U} + G^*\tilde{W}(\sigma)G$ all depend on $\sigma = (\sigma_1, \sigma_2, \ldots, \sigma_N)$. Differentiating, we have

$$\frac{\partial W}{\partial \sigma_j} = \frac{\partial}{\partial \sigma_j}\left(\sum_{k=1}^{N} \sigma_k F_k^* \tilde{W}_k F_k\right) = F_j^* \tilde{W}_j F_j \equiv W_j \tag{2.25}$$

$$\frac{\partial R}{\partial \sigma_j} = \frac{\partial}{\partial \sigma_j}\left(\sum_{k=1}^{N} \sigma_k F_k^* \tilde{W}_k F_k\right) = F_j^* \tilde{W}_j G_j \equiv R_j \tag{2.26}$$

$$\frac{\partial U}{\partial \sigma_j} = \frac{\partial}{\partial \sigma_j}\left(\tilde{U} + \sum_{k=1}^{N} \sigma_k G_k^* \tilde{W}_k G_k\right) = G_j^* \tilde{W}_j G_j \equiv U_j\ . \tag{2.27}$$

Then from (2. 15) we have

$$\frac{\partial \hat{K}}{\partial \sigma_j} = -\,U(\sigma)^{-1}\left(B^*\frac{\partial Q}{\partial \sigma_j} + \frac{\partial R^*}{\partial \sigma_j}\right) + U(\sigma)^{-1}\frac{\partial U}{\partial \sigma_j}U(\sigma)^{-1}\left(B^*Q(\sigma) + R(\sigma)^*\right)$$

$$= -\,U(\sigma)^{-1}\left(B^*\frac{\partial Q}{\partial \sigma_j} + G_j^*\tilde{W}_j F_j\right) - U(\sigma)^{-1}G_j^*\tilde{W}_j G_j \hat{K}(\sigma) \tag{2.28}$$

from which we see that (2. 22) becomes

$$\frac{\partial c_k}{\partial \sigma_j} = -2\,\mathrm{Tr}\,P(\sigma)H_k(\sigma)U(\sigma)^{-1}[G_j^*\tilde{W}_jG_j\hat{K}(\sigma) + B^*\frac{\partial Q}{\partial \sigma_j} + \frac{\partial R^*}{\partial \sigma_j}]$$

$$= -2\,\mathrm{Tr}\,P(\sigma)H_k(\sigma)U(\sigma)^{-1}[G_j^*\tilde{W}_jG_j\hat{K}(\sigma) + B^*\frac{\partial Q}{\partial \sigma_j} + G_j^*\tilde{W}_jF_j] \;. \tag{2.29}$$

Note that in (2. 28) we have used the differentiation formula

$$\frac{\partial (U(\sigma))^{-1}}{\partial \sigma_j} = -U(\sigma)^{-1}\frac{\partial U}{\partial \sigma_j}U(\sigma)^{-1}(= -U(\sigma)^{-1}U_jU(\sigma)^{-1}) \;. \tag{2.30}$$

The next requirement is an equation for $\frac{\partial}{\partial \sigma_j}(Q(\sigma))$. Using (2. 30) again, we differentiate (1. 8) (with $Q = Q(\sigma)$, W_j, R and U as described before (2. 25)), using (2. 25) - (2. 27) to obtain

$$A^*\frac{\partial Q}{\partial \sigma_j} + \frac{\partial Q}{\partial \sigma_j}A + W_j - (\frac{\partial Q}{\partial \sigma_j}B + R_j)U^{-1}(B^*Q + R^*)$$

$$-(QB+R)U^{-1}(B^*\frac{\partial Q}{\partial \sigma_j} + R_j) + (QB+R)U^{-1}U_jU^{-1}(B^*Q+R^*) = 0$$

which, bringing (2. 15) into play again, can be rearranged to read

$$(A+B\hat{K}(\sigma))^*\frac{\partial Q}{\partial \sigma_j} + \frac{\partial Q}{\partial \sigma_j}(A+B\hat{K}(\sigma)) + W_j + R_j\hat{K}(\sigma) + \hat{K}(\sigma)^*R_j + \hat{K}(\sigma)^*U_j\hat{K}(\sigma) = 0 \;,$$

or, from (2. 25) - (2. 27),

$$(A+B\hat{K}(\sigma))^*\frac{\partial Q}{\partial \sigma_j} + \frac{\partial Q}{\partial \sigma_j}(A+B\hat{K}(\sigma)) + (F_j + G_j\hat{K}(\sigma))^*\tilde{W}_j(F_j + G_j\hat{K}(\sigma)) = 0 \;.$$

Referring back to (2. 16) we find, remarkably, that

$$\frac{\partial Q}{\partial \sigma_j} = Q_j(\sigma) \;. \tag{2.31}$$

Using (2. 31) in (2. 29) together with (2. 24) gives

$$\frac{\partial c_k}{\partial \sigma_j} = -2 \operatorname{Tr} P(\sigma)H_k(\sigma)U(\sigma)^{-1}H_k(\sigma)^* . \tag{2.32}$$

This formula shows that the n^2 entries $\frac{\partial c_k}{\partial \sigma_j}$ of the Jacobian matrix of the c_k with respect to the σ_j can be computed, assuming the $Q_k(\sigma)$ have already been obtained in evaluating the individual costs c_k via (2.19), by solution of just one additional Liapounov equation, namely the equation (2.23) for $P(\sigma)$. Thus the complete process of evaluation of the $c_k(\sigma)$ and computation of the partial derivatives $\partial c_k/\partial \sigma_j$ involves the solution of just $N + 1$ Liapounov equations altogether, namely, (2.19) for $k = 1, 2, \ldots, N$, and (2.23).

We can see from (2.32) that $\partial c/\partial \sigma \equiv (\frac{\partial c_k}{\partial \sigma_j})$ is -2 times the Gramian (I- [1]), with respect to the trace inner product, of the vectors $U(\sigma)^{-1/2}H_k(\sigma)^*P(\sigma)^{1/2}$, $k = 1, 2, \ldots, N$ in E^{mn}. Thus the Jacobian $\frac{\partial c}{\partial \sigma}$ is symmetric and non-positive; negative definite if these matrices, considered as vectors in E^{mn}, are linearly independent.

An interesting identity can be obtained by differentiating (2.17) with respect to σ_j:

$$\frac{\partial Q}{\partial \sigma_j} = \sum_{k=1}^{N} \sigma_k \frac{\partial Q}{\partial \sigma_j} + Q_j(\sigma) + \frac{\partial Q_u}{\partial \sigma_j} .$$

Using (2.31) and (2.18) - (2.20), this gives

$$\sum_{k=1}^{N} \sigma_k \frac{\partial Q_k}{\partial \sigma_j} + \frac{\partial Q_u}{\partial \sigma_j} = 0, \qquad j = 1, 2, \ldots, N, \tag{2.33}$$

$$\sum_{k=1}^{N} \sigma_k \frac{\partial c_k}{\partial \sigma_j} + \frac{\partial c_u}{\partial \sigma_j} = 0, \qquad j = 1, 2, \ldots, N. \tag{2.34}$$

In terms of the gradients row vectors $\frac{\partial c_k}{\partial \sigma}$, (2.34) reads

$$\sum_{k=1}^{N} \sigma_k \frac{\partial c_k}{\partial \sigma} + \frac{\partial c_n}{\partial \sigma} = 0 . \tag{2.35}$$

This identity can be interpreted in several ways. It shows that the Kuhn-Tucker conditions (V- [2]) are satisfied at $\sigma_1, \sigma_2, \ldots, \sigma_N$ for the problem

$$\min_{\hat{\sigma}} \quad c_u(\hat{\sigma})$$

subject to the constraints

$$c_k(\hat{\sigma}) \le c_k(\sigma) .$$

Another way of looking at this is to say that no variation in σ will simultaneously reduce all of $c_1, c_2, \ldots, c_N, c_u$.

The identity (2. 35) can also be viewed as a restatement of the almost obvious fact that multiplication of all of the weighting matrices $\sigma_1\tilde{W}_1, \ldots, \sigma_N\tilde{W}_N$, $\tilde{U}$ by a positive constant will leave each of the costs $c_1, \ldots, c_N, c_u$ invariant. Indeed, such a multiplication can easily be seen from (2. 15) to leave $\hat{K}$ invariant and thus $Q_1, \ldots, Q_N, Q_u$ remains constant also.

The identities (2. 33), (2. 34) with (2. 20) show that no separate computations of Q_u, c_u, $\frac{\partial Q_u}{\partial \sigma_j}$, $\frac{\partial c_u}{\partial \sigma_j}$ are ever necessary.

The symmetry of the Jacobian $\partial c/\partial \sigma$ can be deduced in another way. Since $\frac{\partial Q}{\partial \sigma_j} = Q_j(\sigma)$, we have, for the aggregate weighted cost $c_0(\sigma)$ given by (2. 18),

$$\frac{\partial c_0}{\partial \sigma_k} = \frac{\partial}{\partial \sigma_k}(Q(\sigma)M) = Q_k(\sigma)M = c_k(\sigma)$$

and therefore

$$\frac{\partial c_k(\sigma)}{\partial \sigma_j} = \frac{\partial^2 c_0}{\partial \sigma_k \partial \sigma_j} ,$$

that is, $\partial c/\partial \sigma$ is the Hessian matrix of the scalar function $c_0(\sigma)$ - and hence is necessarily symmetric from the identity

$$\frac{\partial^2 c_0}{\partial \sigma_k \partial \sigma_j} = \frac{\partial^2 c_0}{\partial \sigma_j \partial \sigma_k} ,$$

which follows from the fact that $c_0(\sigma)$ is twice continuously differentiable with respect to σ.

The non-positivity of $\partial c/\partial \sigma$, which we have seen to be the Hessian,

$\partial^2 c_0/\partial\sigma^2$, of $c_0(\sigma)$, implies that $c_0(\sigma)$ is a concave function of σ. (See V-[2] for definitions.) if $\partial c/\partial\sigma$ is nonsingular at some point $\hat{\sigma}$, then $\partial^2 c_0/\partial\sigma^2$ is negative definite there and $c_0(\sigma)$ is strictly concave in a neighborhood of $\hat{\sigma}$. From this we can prove the following theorem.

<u>Theorem 2.1.</u> <u>Let</u> $\hat{\sigma}$ <u>and</u> $\tilde{\sigma}$ <u>be vectors in</u> E^N <u>with non-negative components,</u> $\|\tilde{\sigma}\|_{E^N} = 1$. <u>Consider the line segment</u>

$$\{\sigma \mid \sigma = \hat{\sigma} + \epsilon\tilde{\sigma}, \quad \epsilon \geq 0\} .$$

Then the directional derivative

$$\frac{\partial c_0}{\partial\sigma}(\hat{\sigma} + \epsilon\tilde{\sigma}) \equiv \frac{\partial c_0}{\partial\sigma}(\hat{\sigma} + \epsilon\tilde{\sigma})\tilde{\sigma} = \sum_{k=1}^{N} \tilde{\sigma}_k c_k(\hat{\sigma} + \epsilon\tilde{\sigma}) \equiv c(\hat{\sigma} + \epsilon\tilde{\sigma}) ,$$

<u>has the property that</u>

$$\lim_{\epsilon\to\infty} \frac{\partial c_0}{\partial\tilde{\sigma}}(\hat{\sigma} + \epsilon\tilde{\sigma})$$

<u>exists and is non-negative.</u> <u>Thus the cost</u> $\tilde{c}(\hat{\sigma} + \epsilon\tilde{\sigma})$ <u>tends to a non-negative limit as</u> $\epsilon \to \infty$.

<u>Proof.</u> The concavity of $c_0(\sigma)$ shows that $\frac{\partial c_0}{\partial\sigma}(\hat{\sigma} + \epsilon\tilde{\sigma})$ is non-increasing as ϵ increases. Since $\frac{\partial c_0}{\partial\tilde{\sigma}}(\hat{\sigma} + \epsilon\tilde{\sigma}) \geq 0$ for all $\epsilon > 0$, the existence of the limit follows from the familiar monotone convergence theorem. Q. E. D.

<u>Corollary 2.2.</u> <u>If</u>

$$\lim_{\epsilon\to\infty} \tilde{c}(\hat{\sigma} + \epsilon\tilde{\sigma}) = \lim_{\epsilon\to\infty} \frac{\partial c_0}{\partial\tilde{\sigma}}(\hat{\sigma} + \epsilon\tilde{\sigma}) = \gamma > 0$$

<u>then</u>

$$\lim_{\epsilon\to\infty} \frac{c_0(\hat{\sigma} + \epsilon\tilde{\sigma})}{\epsilon} = \gamma .$$

<u>Proof.</u> Immediate, by ℓ'Hospital's rule. Q. E. D.

It does not follow that $\lim_{\epsilon \to \infty} c_0(\hat{\sigma} + \epsilon\tilde{\sigma}) = 0$ when $\gamma = 0$. For the system

$$\dot{x} = x + u$$

$$y = x$$

the relevant nonlinear equation is

$$2q + (\hat{\sigma} + \epsilon\tilde{\sigma}) - q^2 = 0$$

which has the positive solution

$$q(\sigma + \epsilon\tilde{\sigma}) = \frac{2 + \sqrt{4 + 4(\hat{\sigma} + \epsilon\tilde{\sigma})}}{2}$$

and, for $M = 1$, $c_0(\hat{\sigma} + \epsilon\tilde{\sigma})$ grows like $\sqrt{\epsilon}$ even though, as is easily verified, $\gamma = 0$ in this case. It can be shown that this is typical.

Suppose we begin with a vector of weights $\sigma = (\sigma_1, \sigma_2, \ldots, \sigma_N)$, thereby realizing individual response costs $c_1(\sigma)$, $c_2(\sigma), \ldots, c_N(\sigma)$. We may very well regard these as unsuitable for any number of reasons. For changes $\delta\sigma_1, \delta\sigma_2, \ldots, \delta\sigma_N$ in these weights the changes $\delta c_1, \delta c_2, \ldots, \delta c_N$ in the costs are, to first order in σ:

$$\delta c \equiv \begin{pmatrix} \delta c_1 \\ \vdots \\ \delta c_N \end{pmatrix} \begin{pmatrix} \frac{\partial c_1}{\partial \sigma_1} & \cdots & \frac{\partial c_1}{\partial \sigma_N} \\ \vdots & & \vdots \\ \frac{\partial c_N}{\partial \sigma_1} & & \frac{\partial c_N}{\partial \sigma_N} \end{pmatrix} \begin{pmatrix} \delta\sigma_1 \\ \vdots \\ \delta\sigma_N \end{pmatrix} = \frac{\partial c}{\partial \sigma}\,\delta\sigma \quad . \tag{2.36}$$

If $\partial c/\partial\sigma$ is nonsingular, (2.36) may be solved for $\delta\sigma$:

$$\delta\sigma = \left(\frac{\partial c}{\partial \sigma}\right)^{-1} \delta c \; .$$

If values $\tilde{c}_1, \tilde{c}_2, \ldots, \tilde{c}_N$ for the individual response costs are considered desirable, we may move from $c(\sigma) = (c_1(\sigma), c_2(\sigma), \ldots, c_N(\sigma))$ toward $\tilde{c} = (\tilde{c}_1, \tilde{c}_2, \ldots, \tilde{c}_N)$ by setting

$$\delta c = \epsilon(\tilde{c} - c(\sigma)), \qquad \epsilon > 0,$$

and setting

$$\delta\sigma = \epsilon\left(\frac{\partial c}{\partial \sigma}\right)^{-1}(\tilde{c} - c(\sigma)) \, . \tag{2.37}$$

If $\tilde{c} - c(\sigma)$ is small, $\epsilon = 1$ may yield an adequate result in one step. If not, it may be necessary to carry out a series of steps with ϵ small. It will be noted that iterative use of (2. 37) with $\epsilon = 1$ is, of course, just Newton's method for solving

$$c(\sigma) = \tilde{c} \, .$$

In the event that $\frac{\partial c}{\partial \sigma}$ is singular, (2. 36) serves to describe those first order changes in c which can be realized by changes in the weighting pattern.

A variety of modifications of the above might be envisioned. In many cases we would not want the control cost to get too high, leading us to impose an inequality constraint

$$c_u(\sigma) \le \tilde{c}_u$$

for some positive $\tilde{c}_u$. One might then try to minimize $\|c(\sigma) - \tilde{c}\|^2_{E^N}$ subject to this constraint. Another scenario might involve a minimax type of criterion, i. e., determine σ so as to minimize the maximum of the $c_k(\sigma)$, $k = 1, 2, \ldots, N$. We will not discuss any of these in detail but Exercises 2, 3 at the end of the chapter may be of interest.

For an example we return to the mass-spring, or "suspension" system introduced in IV - Sections 1, 3. The system is

$$\begin{pmatrix} \dot{x}^1 \\ \dot{x}^2 \end{pmatrix} = \begin{pmatrix} 0 & 1 \\ -1 & 0 \end{pmatrix} \begin{pmatrix} x^1 \\ x^2 \end{pmatrix} + \begin{pmatrix} 0 \\ 1 \end{pmatrix} u \tag{2.38}$$

and the responses of interest are

$$\text{position} \equiv y^1 = x^1 \qquad = (1, 0)x + 0u$$

$$\text{velocity} \equiv y^2 = x^2 = \quad = (0, 1)x + 0u$$

$$\text{acceleration} \equiv y^3 = -x^1 + u = (-1, 0)x + 1u \ .$$

We will take (cf. (2.7)) $N = 2$,

$$y_1 = \begin{pmatrix} y^1 \\ y^2 \end{pmatrix} = \begin{pmatrix} 1 & 0 \\ 0 & 1 \end{pmatrix} x + 0u = F_1 x + G_1 u, \qquad \tilde{W}_1 = \begin{pmatrix} 1 & 0 \\ 0 & 1 \end{pmatrix}, \tag{2.39}$$

$$y_2 = y^3 = (-1, 0)x + 1u = F_2 x + G_2 u, \qquad \tilde{W}_2 = 1 \ . \tag{2.40}$$

Following the work at the end of IV - Section 3, we identify α, β appearing there as σ_1, σ_2 of this section. We will call then α, β in the work below. We take $\tilde{U}$ (= δ in IV - (3.37)) = 1 throughout.

Let us write

$$\sigma = \begin{pmatrix} \alpha \\ \beta \end{pmatrix}.$$

In IV - Section 3 we have computed the solution $Q(\sigma) = \begin{pmatrix} q_1(\sigma), & q_2(\sigma) \\ q_2(\sigma), & q_3(\sigma) \end{pmatrix}$

for general α, β with the result (for $\delta = 1$)

$$q_1 = \sqrt{(1+\beta)(-2+2\sqrt{1+\alpha+\beta+\alpha\beta}+\alpha)}\ \frac{\sqrt{1+\alpha+\beta+\alpha\beta}}{1+\beta}$$

$$q_2 = -1 + \sqrt{1+\alpha+\beta+\alpha\beta}$$

$$q_3 = \sqrt{(1+\beta)(-2+2\sqrt{1+\alpha+\beta+\alpha\beta}+\alpha)} \ .$$

We begin our analysis with the weighting vector

$$\sigma = \begin{pmatrix} \frac{1}{2} \\ \frac{1}{2} \end{pmatrix}, \ \text{i.e.,}\ \sigma_1 = \alpha = \tfrac{1}{2}, \ \alpha_2 = \beta = \tfrac{1}{2} \ .$$

This gives

$$Q(\tfrac{1}{2}, \tfrac{1}{2}) = \begin{pmatrix} 1.5000 & .5000 \\ .5000 & 1.5000 \end{pmatrix}$$

The individual response cost matrices are $Q_1(\frac{1}{2}, \frac{1}{2})$, $Q_2(\frac{1}{2}, \frac{1}{2})$, $Q_u(\frac{1}{2}, \frac{1}{2})$. From (2.9), (2.12)

$$(A + B\hat{K}(\sigma))^* Q_k(\sigma) + Q_k(\sigma)(A + B\hat{K}(\sigma)) + (F_k + G_k\hat{K}(\sigma))^* \tilde{W}_k (F_k + G_k\hat{K}(\sigma)) = 0 \quad (2.41)$$

$$(A + B\hat{K}(\sigma))^* Q_u(\sigma) + Q_u(\sigma)(A + B\hat{K}(\sigma)) + \hat{K}(\sigma)^* \tilde{U} \hat{K}(\sigma) = 0 . \quad (2.42)$$

In our case

$$A = \begin{pmatrix} 0 & 1 \\ -1 & 0 \end{pmatrix}, \quad B = \begin{pmatrix} 0 \\ 1 \end{pmatrix}, \quad \tilde{W}_1 = \begin{pmatrix} 1 & 0 \\ 0 & 1 \end{pmatrix}, \quad \tilde{W}_2 = 1, \ \tilde{U} = 1,$$

$$F_1 = \begin{pmatrix} 1 & 0 \\ 0 & 1 \end{pmatrix}, \quad G_1 = 0, \quad F_2 = (-1, 0), \quad G_2 = 1, \quad U = 1+\beta = \frac{3}{2}$$

$$\hat{K}(\tfrac{1}{2}, \tfrac{1}{2}) = \text{(see IV - Section 3)}$$

$$= \left(-\frac{\sqrt{1+\alpha+\beta+\alpha\beta}}{1+\beta} + 1\right) x^1 - \sqrt{\frac{-2+2\sqrt{1+\alpha+\beta+\alpha\beta+\alpha\beta} + \alpha}{1+\beta}}\, x^2$$

$$= (0, -1)\begin{pmatrix} x^1 \\ x^2 \end{pmatrix} \quad (2.43)$$

$$(A + B\hat{K}(\tfrac{1}{2}, \tfrac{1}{2})) = \begin{pmatrix} 0 & 1 \\ -1 & -1 \end{pmatrix}$$

$$(F_1 + G_1\hat{K}(\tfrac{1}{2}, \tfrac{1}{2}))^* \tilde{W}_1 (F_1 + G_1\hat{K}(\tfrac{1}{2}, \tfrac{1}{2})) = \begin{pmatrix} 1 & 0 \\ 0 & 1 \end{pmatrix}$$

$$(F_2 + G_2\hat{K}(\tfrac{1}{2}, \tfrac{1}{2}))^* \tilde{W}_2 (F_2 + G_2\hat{K}(\tfrac{1}{2}, \tfrac{1}{2})) = \begin{pmatrix} -1 \\ -1 \end{pmatrix} \cdot 1 \cdot (-1, -1) = \begin{pmatrix} 1 & 1 \\ 1 & 1 \end{pmatrix}$$

$$\hat{K}(\tfrac{1}{2}, \tfrac{1}{2})^* \tilde{U} \hat{K}(\tfrac{1}{2}, \tfrac{1}{2}) = \begin{pmatrix} 0 \\ -1 \end{pmatrix} \cdot 1 \cdot (0, -1) = \begin{pmatrix} 0 & 0 \\ 0 & 1 \end{pmatrix} .$$

With these data we can solve the Liapounov equations (2.41), (2.42) to give

$$Q_1(\tfrac{1}{2}, \tfrac{1}{2}) = \begin{pmatrix} 1.5 & .5 \\ .5 & 1 \end{pmatrix}$$

$$Q_2(\tfrac{1}{2}, \tfrac{1}{2}) = \begin{pmatrix} .5 & .5 \\ .5 & 1 \end{pmatrix}$$

$$Q_u(\tfrac{1}{2}, \tfrac{1}{2}) = \begin{pmatrix} .5 & 0 \\ 0 & .5 \end{pmatrix}$$

and we verify equation (2.17) immediately:

$$\tfrac{1}{2} Q_1(\tfrac{1}{2}, \tfrac{1}{2}) + \tfrac{1}{2} Q_2(\tfrac{1}{2}, \tfrac{1}{2}) + Q_u(\tfrac{1}{2}, \tfrac{1}{2}) = Q(\tfrac{1}{2}, \tfrac{1}{2}) .$$

Next we note that

$$\left.\begin{aligned} Q(1, 0) &= \begin{pmatrix} 1.9123 & .4142 \\ .4142 & 1.3522 \end{pmatrix} \\ Q(0, 1) &= \begin{pmatrix} .9102 & .4142 \\ .4142 & 1.2872 \end{pmatrix} \end{aligned}\right\} \tag{2.44}$$

From (2.31) we should have, approximately

$$\begin{aligned} Q(1, 0) &\approx Q(\tfrac{1}{2}, \tfrac{1}{2}) + \tfrac{1}{2} \frac{\partial Q}{\partial \alpha}(\tfrac{1}{2}, \tfrac{1}{2}) - \tfrac{1}{2} \frac{\partial Q}{\partial \beta}(\tfrac{1}{2}, \tfrac{1}{2}) \\ &= Q(\tfrac{1}{2}, \tfrac{1}{2}) + \tfrac{1}{2} Q_1(\tfrac{1}{2}, \tfrac{1}{2}) - \tfrac{1}{2} Q_2(\tfrac{1}{2}, \tfrac{1}{2}) = \begin{pmatrix} 2 & .5 \\ .5 & 1.5 \end{pmatrix} \end{aligned}$$

$$\begin{aligned} Q(0, 1) &\approx Q(\tfrac{1}{2}, \tfrac{1}{2}) - \tfrac{1}{2} \frac{\partial Q}{\partial \alpha}(\tfrac{1}{2}, \tfrac{1}{2}) + \tfrac{1}{2} \frac{\partial Q}{\partial \beta}(\tfrac{1}{2}, \tfrac{1}{2}) \\ &= Q(\tfrac{1}{2}, \tfrac{1}{2}) - \tfrac{1}{2} Q_1(\tfrac{1}{2}, \tfrac{1}{2}) + \tfrac{1}{2} Q_2(\tfrac{1}{2}, \tfrac{1}{2}) = \begin{pmatrix} 1 & .5 \\ .5 & 1.5 \end{pmatrix} , \end{aligned}$$

approximations, which do bear comparison with (2.44) considering that $\delta\sigma = (\frac{1}{2}, \frac{1}{2})$ and $\delta\sigma = (-\frac{1}{2}, \frac{1}{2})$ are fairly large perturbations.

Returning to (2.38), we now suppose our "suspension" system to be subject to "road shocks" taking the form of white noise v:

$$\begin{pmatrix} \dot{x}^1 \\ \dot{x}^2 \end{pmatrix} = \begin{pmatrix} 0 & 1 \\ -1 & 0 \end{pmatrix} \begin{pmatrix} x^1 \\ x^2 \end{pmatrix} + \begin{pmatrix} 0 \\ 1 \end{pmatrix} u + \begin{pmatrix} 0 \\ 1 \end{pmatrix} v . \tag{2.45}$$

We suppose the white noise disturbance v has 1×1 covariance matrix $V = 1$. Since $C = \binom{0}{1}$ we have

$$CVC^* = \begin{pmatrix} 0 \\ 1 \end{pmatrix} \cdot 1 \cdot (0, 1) = \begin{pmatrix} 0 & 0 \\ 0 & 1 \end{pmatrix}.$$

We will take (cf. (1.14))

$$M = CVC^* = \begin{pmatrix} 0 & 0 \\ 0 & 1 \end{pmatrix}. \tag{2.46}$$

Then we can evaluate various control schemes as they relate to the individual responses (2.39), (2.40). Beginning with $\sigma = (\frac{1}{2}, \frac{1}{2})$ we have

$$c_1(\tfrac{1}{2}, \tfrac{1}{2}) = \mathrm{Tr} \begin{pmatrix} 0 & 0 \\ 0 & 1 \end{pmatrix} Q_1(\tfrac{1}{2}, \tfrac{1}{2}) = \mathrm{Tr} \begin{pmatrix} 0 & 0 \\ 0 & 1 \end{pmatrix} \begin{pmatrix} 1.5 & .5 \\ .5 & 1 \end{pmatrix} = 1$$

$$c_2(\tfrac{1}{2}, \tfrac{1}{2}) = \mathrm{Tr} \begin{pmatrix} 0 & 0 \\ 0 & 1 \end{pmatrix} Q_2(\tfrac{1}{2}, \tfrac{1}{2}) = \mathrm{Tr} \begin{pmatrix} 0 & 0 \\ 0 & 1 \end{pmatrix} \begin{pmatrix} 1.5 & .5 \\ .5 & 1 \end{pmatrix} = 1 .$$

Using the optimal feedback matrices computed in IV - Section 3, i.e.,

$$\hat{K}(1, 0) = (-\sqrt{2} + 1, \ -\sqrt{(2\sqrt{2} - 1)}) ,$$

$$\hat{K}(0, 1) = (-\frac{1}{\sqrt{2}} + 1, \ -\sqrt{(\sqrt{2} - 1)}) ,$$

and solving (2.41) in each case we obtain

$$Q_1(1,0) = \begin{pmatrix} 1.3708 & .3536 \\ .3536 & .6312 \end{pmatrix}$$

$$Q_2(1,0) = \begin{pmatrix} .7395 & .7071 \\ .7071 & 1.1990 \end{pmatrix}$$

$$Q_1(0,1) = \begin{pmatrix} 1.7813 & .7071 \\ .7071 & 1.8756 \end{pmatrix}$$

$$Q_2(0,1) = \begin{pmatrix} .3884 & .3536 \\ .3536 & .8711 \end{pmatrix}$$

Then

$$c_1(1,0) = \mathrm{Tr}\begin{pmatrix} 0 & 0 \\ 0 & 1 \end{pmatrix} Q_1(1,0) = .63123 \tag{2.47}$$

$$c_2(1,0) = \mathrm{Tr}\begin{pmatrix} 0 & 0 \\ 0 & 1 \end{pmatrix} Q_2(1,0) = 1.19902 \tag{2.48}$$

$$c_1(0,1) = \mathrm{Tr}\begin{pmatrix} 0 & 0 \\ 0 & 1 \end{pmatrix} Q_1(0,1) = 1.87557 \tag{2.49}$$

$$c_2(0,1) = \mathrm{Tr}\begin{pmatrix} 0 & 0 \\ 0 & 1 \end{pmatrix} Q_2(0,1) = .87113 \tag{2.50}$$

These computations enable us to answer the question posed following IV - (3.43). Does $\alpha = 1$, $\beta = 0$ give lower "energy" cost than $\alpha = 0$, $\beta = 1$? Yes, $.6312 < 1.8756$. Does $\alpha = 0$, $\beta = 1$ give lower acceleration cost (i.e., better ride quality) than $\alpha = 1$, $\beta = 0$? Yes, $.8711 < 1.1990$.

For $\alpha = \beta = \frac{1}{2}$ we have $c_1(\frac{1}{2}, \frac{1}{2}) = 1$, $c_2(\frac{1}{2}, \frac{1}{2}) = 1$. Let us ask this question: can we reduce c_2 somewhat while keeping c_1 at the value 1? What combination of weights, if any, will realize such an objective? In order to attempt a resolution of this we compute $\partial c_1/\partial\alpha$, $\partial c_2/\partial\alpha$, $\partial c_1/\partial\beta$,

$\partial c_2/\partial\beta$ at the point $\sigma = (\frac{1}{2}, \frac{1}{2})$. First we compute $P(\sigma)$ from (2.23) with M as in (2.46), giving

$$P(\tfrac{1}{2}, \tfrac{1}{2}) = \begin{pmatrix} .5 & 0 \\ 0 & .5 \end{pmatrix} .$$

We have, from (2.24), (2.38), (2.39), (2.40),

$$H_1(\tfrac{1}{2}, \tfrac{1}{2}) = Q_1(\tfrac{1}{2}, \tfrac{1}{2})\begin{pmatrix} 0 \\ 1 \end{pmatrix} = \begin{pmatrix} .5 \\ 1 \end{pmatrix}$$

$$H_2(\tfrac{1}{2}, \tfrac{1}{2}) = Q_2(\tfrac{1}{2}, \tfrac{1}{2})\begin{pmatrix} 0 \\ 1 \end{pmatrix} + ((-1, 0) + 1(0, -1))^* \cdot 1 \cdot 1 = \begin{pmatrix} .5 \\ 1 \end{pmatrix} + \begin{pmatrix} -1 \\ -1 \end{pmatrix} = \begin{pmatrix} -.5 \\ 0 \end{pmatrix}$$

and we have

$$U(\tfrac{1}{2}, \tfrac{1}{2})^{-1} H_1(\tfrac{1}{2}, \tfrac{1}{2})^* P(\tfrac{1}{2}, \tfrac{1}{2}) = \frac{2}{3}(.5 \quad 1)\begin{pmatrix} .5 & 0 \\ 0 & .5 \end{pmatrix} = (.16667,\ .33333) \equiv Y_1(\tfrac{1}{2}, \tfrac{1}{2}) \tag{2.51}$$

$$U(\tfrac{1}{2}, \tfrac{1}{2})^{-1} H(\tfrac{1}{2}, \tfrac{1}{2})^* P(\tfrac{1}{2}, \tfrac{1}{2}) = \frac{2}{3}(-.5,\ 0)\begin{pmatrix} .5 & 0 \\ 0 & .5 \end{pmatrix} = (-.16667,\ 0) \equiv Y_2(\tfrac{1}{2}, \tfrac{1}{2}) . \tag{2.52}$$

Then, from (2.32), (2.51), (2.52)

$$\frac{\partial c_1}{\partial \alpha} = -2\,\mathrm{Tr}\,(Q_1(\tfrac{1}{2}, \tfrac{1}{2})B + F_1^* \tilde{W}_1 G_1) Y_1(\tfrac{1}{2}, \tfrac{1}{2})$$

$$= -2\,\mathrm{Tr}\begin{pmatrix} 1.5 & .5 \\ .5 & 1 \end{pmatrix}\begin{pmatrix} 0 \\ 1 \end{pmatrix}(.16667,\ .33333) = -.83333$$

$$\frac{\partial c_1}{\partial \beta} = -2\,\mathrm{Tr}(Q_2(\tfrac{1}{2}, \tfrac{1}{2})B + F_2^* \tilde{W}_2 G_2 + G_2^* \tilde{W}_2 G_2 \hat{K}(\tfrac{1}{2}, \tfrac{1}{2})) Y_1(\tfrac{1}{2}, \tfrac{1}{2})$$

$$= -2\,\mathrm{Tr}\left(\begin{pmatrix} .5 & .5 \\ .5 & 1 \end{pmatrix}\begin{pmatrix} 0 \\ 1 \end{pmatrix} + \begin{pmatrix} -1 \\ -1 \end{pmatrix}\right)(.16667,\ .33333) = .16667$$

$$\frac{\partial c_2}{\partial \alpha} = -2\,\mathrm{Tr}\,(Q_1(\tfrac{1}{2},\tfrac{1}{2})B + F_1^* \tilde{W}_1 G_1)\,Y_2(\tfrac{1}{2},\tfrac{1}{2})$$

$$= -2\,\mathrm{Tr}\begin{pmatrix} 1.5 & .5 \\ .5 & 1 \end{pmatrix}\begin{pmatrix} 0 \\ 1 \end{pmatrix}(-.16667,\ 0) = .16667$$

$$\frac{\partial c_2}{\partial \beta} = -2\,\mathrm{Tr}\,(Q_2(\tfrac{1}{2},\tfrac{1}{2})B + F_2^* \tilde{W}_2 G_2 + G_2^* \tilde{W}_2 G_2 \hat{K}(\tfrac{1}{2},\tfrac{1}{2}))\,Y_2(\tfrac{1}{2},\tfrac{1}{2})$$

$$= -2\,\mathrm{Tr}\left(\begin{pmatrix} .5 & .5 \\ .5 & 1 \end{pmatrix}\begin{pmatrix} 0 \\ 1 \end{pmatrix} + \begin{pmatrix} -1 \\ -1 \end{pmatrix}\right)(-.16667,\ 0) = -.16667 \ .$$

Thus

$$\frac{\partial(c_1, c_2)}{\partial(\alpha, \beta)}(\tfrac{1}{2},\tfrac{1}{2}) = \begin{pmatrix} -.83333 & .16667 \\ .16667 & -.16667 \end{pmatrix}\left(\text{actually}\begin{pmatrix} -\frac{5}{6} & \frac{1}{6} \\ \frac{1}{6} & -\frac{1}{6} \end{pmatrix}\right) \tag{2.53}$$

The inverse is readily seen to be

$$\frac{\partial(c_1, c_2)}{\partial(\alpha, \beta)}^{-1} = \begin{pmatrix} -\frac{3}{2} & -\frac{3}{2} \\ -\frac{3}{2} & -\frac{15}{2} \end{pmatrix} . \tag{2.54}$$

We want to reduce c_2 while keeping c_1 constant. Thus we want

$$\delta c = \begin{pmatrix} \delta c_1 \\ \delta c_2 \end{pmatrix} = \begin{pmatrix} 0 \\ -1 \end{pmatrix}$$

and this gives

$$\delta_\sigma = \begin{pmatrix} \delta\alpha \\ \delta\beta \end{pmatrix} = \left(\frac{\partial(c_1, c_2)}{\partial(\alpha, \beta)}\right)^{-1} = \begin{pmatrix} -\frac{3}{2} & -\frac{3}{2} \\ -\frac{3}{2} & -\frac{15}{2} \end{pmatrix}\begin{pmatrix} 0 \\ -1 \end{pmatrix} = \begin{pmatrix} \frac{3}{2} \\ \frac{15}{2} \end{pmatrix} .$$

We take (cf. (2.37)) $\epsilon = .1$ giving new values of α and β

$$\alpha = \tfrac{1}{2} + .15 = .65, \qquad \beta = \tfrac{1}{2} + .75 = 1.25 \ .$$

The new feedback matrix, from (2.43), is

$$K = (.14365, \quad -1.05484) \ .$$

We then solve (2.41) for $k = 1, 2$ to give

$$Q_1(.65, \ 1.25) = \begin{pmatrix} 1.49581 & .58387 \\ .58387 & 1.02752 \end{pmatrix},$$

so that (cf. (1.14))

$$c_1(.65, \ 1.25) = 1.02752$$

and

$$Q_2(.65, \ 1.25) = \begin{pmatrix} .34760 & .42817 \\ .42817 & .93333 \end{pmatrix},$$

and thus

$$c_2(.65, \ 1.25) = .93333 \ .$$

We see that c_2 has been reduced by .06667 while c_1 has been increased by only .02752. For smaller ϵ the results are, relatively, closer to that predicted by the gradients. For $\epsilon = .01$,

$$c_1(.515, \ .575) = 1.00050, \qquad (\delta c_1 = \ .00050)$$

$$c_2(.515, \ .575) = \ .99051, \qquad (\delta c_2 = -.00949) \ .$$

3. Model Following. Let us suppose that a linear control system

$$\dot{x} = Ax + Bu + Cv, \quad x \in E^n, \ u \in E^m, \ v \in E^p, \tag{3.1}$$

and an output or response

$$y = Fx + Gu, \qquad y \in E^q \tag{3.2}$$

are given, with the usual assumption that (A, B) is stabilizable and (F, A) is observable (or, perhaps, detectable). We suppose v is a white noise disturbance with $p \times p$ covariance matrix $e^{-\mu t}V$, $\mu > 0$. Rather than posing a control objective solely in terms of (3.1) and (3.2), we now complicate the matter by introducing a second linear control system, subject to the same white noise disturbance v,

$$\dot{\xi} = A_0\xi + C_0 v, \qquad \xi \in E^{n_0}, \tag{3.3}$$

with A_0 a stability matrix, and a second response

$$\eta = F_0\xi, \qquad \eta \in E^q. \tag{3.4}$$

We will suppose that (3.3), (3.4) is an "ideal" system in the sense that solutions of (3.3) lead to outputs (3.4) which behave in a desirable manner relative to some criterion of performance. Further, we suppose that there is some natural way to associate states ξ for (3.3) with states x for (3.1), given in the form

$$\xi \sim Hx + Ju, \quad H \;\; n_0 \times n, \quad J \;\; n_0 \times m. \tag{3.5}$$

(In most applications J will be equal to 0.) We suppose also that there is a natural way in which the outputs (3.4) and (3.2) should be compared, i.e., y should be compared with $\hat{y} = \hat{H}\eta$. But, since this leads to $\hat{y} = \hat{H}F_0\xi$, in effect replacing F_0 by $\hat{H}F_0$, we may as well assume that y and η have the same dimension, as we have in (3.2) and (3.4), and take $\hat{H} = I$ so that y and η may be compared directly. We then pose the

<u>Optimal Model Following Problem.</u> Let v be white noise with $p \times p$ covariance matrix $e^{-\mu t}V$, as already stated, let $x(0) = x_0$ be a random vector with

$$\operatorname{cov}(x_0, x_0) = X_0,$$

and let

$$\xi(0) = \xi_0 = Hx_0 + Ju(0).$$

Determine a linear feedback control law

$$u = \hat{K}x, \quad \hat{K}\ m \times n,$$

such that, relative to the cost (E ~ expected value)

$$J(x_0, K) = E(\int_0^\infty [(y(t) - \eta(t))^* \tilde{W}(y(t) - \eta(t)) + u(t)^* \tilde{U}u(t)]dt) \quad (3.6)$$

we have

$$J(x_0, \hat{K}) = \min_{K \in \mathscr{S}(A, B)} \{J(x_0, K)\}.$$

The idea here is to choose $\hat{K}$ so that the system (3.1), (3.2) follows the model (3.3), (3.4) in the sense that $y(t) - \eta(t)$ is kept "small" in an optimal manner, as defined by (3.6), which also includes the term $u(t)^* \tilde{U}u(t)$ to preclude excessive control effort.

In order to illustrate the type of application which we have in mind, we offer the following example of a small engine regulator. In Figure VII-1. we show, schematically an engine, E, feed by a throttle, T, from a carburetor, C, the engine and throttle being connected by a regulating mechanism, R, with an external "nominal" throttle setting being provided by the engine operator, O. The engine speed is given by S and the throttle angle by θ.

If no regulator R is present, the amount of fuel-air mixture passing through the throttle is determined by the area of the throttle opening and the vacuum pressure induced by the speed of the motor; for simplicity let us suppose that the amount of fuel-air mixture per unit time is given by the product $a(\theta)p(S)$, where $a(\theta)$ and $p(S)$ are increasing functions of their arguments. The ignition of the fuel-air mixture tends to accelerate the engine, while the external load on the engine, normally an increasing function, $\ell(S)$, of the speed, S, tends to slow it. For our purposes we will suppose

$$\dot{S} = c\,a(\theta)p(S) - d\ell(S), \quad (3.7)$$

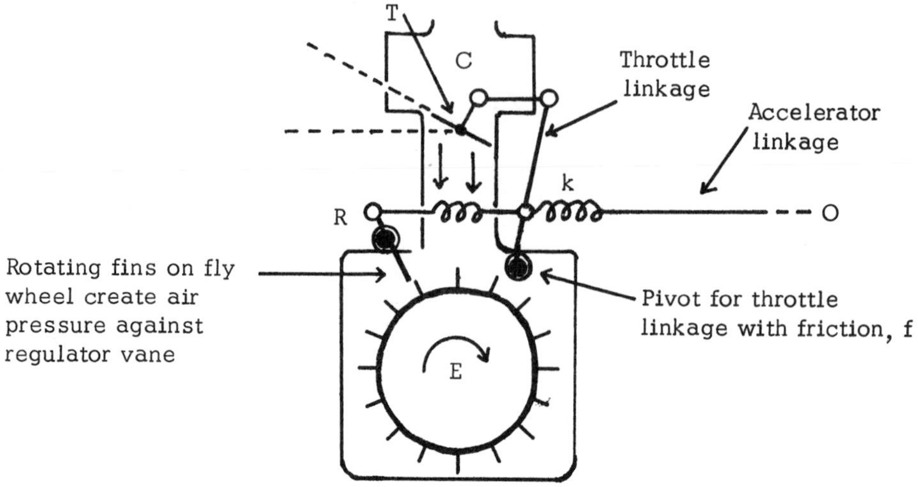

Fig. VII-1. Engine Regulator-Schematic

where c and d are positive constants. For a given throttle setting, θ_0, the equilibrium speed S_0 is obtained by solving

$$c\,a(\theta_0)p(S_0) - d\ell(S_0) = 0 . \tag{3.8}$$

We assume the function p and ℓ are such that a solution S_0 exists. Putting

$$S = S_0 + s$$
$$\theta = \theta_0 + \psi$$

we find that for small speed and throttle variations s and ψ

$$\dot{s} = c\,a(\theta_0 + \psi)p(S_0 + s) - d\ell(S_0 + s)$$
$$\approx c\,a(\theta_0)p'(S_0)s + c\,p(S_0)a'(\theta_0)\psi - d\ell'(S_0)s \equiv \alpha s + \beta\psi,$$

where

$$\alpha = c\,a(\theta_0)p'(S_0) - d\ell'(S_0),$$
$$\beta = c\,p(S_0)a'(\theta_0).$$

If the throttle were rigidly set, so that $\psi \equiv 0$, we would have, as linearized dynamics

$$\dot{s} = \alpha s .$$

For some loading conditions and throttle settings it can happen that $\alpha > 0$. The equilibrium speed S_0 is then unstable; the motor tends to either race or die, to the accompanying great frustration of the operator. Thus the throttle is not ordinarily set rigidly. The throttle linkage is connected to a spring of constant k which in turn is connected to the operator's "accelerator linkage". We then have, taking friction into account, the equation of motion for the throttle variation,

$$m\ddot{\psi} + f\dot{\psi} + ek\psi = u ,$$

for some $e, f > 0$ depending on the parameters of the throttle linkage. We assume the mass, m, involved in the throttle mechanism is negligible, leaving us with

$$\dot{\psi} = -\frac{ek}{f}\psi + \frac{1}{f}u ,$$

and, by rescaling u, we have

$$\dot{\psi} = -\gamma\psi + u ,$$

$$\gamma = \frac{ek}{f} > 0 .$$

Here u is the regulating force, which we have not yet discussed. For $u = 0$ we have

$$\begin{pmatrix} \dot{s} \\ \dot{\psi} \end{pmatrix} = \begin{pmatrix} \alpha & \beta \\ 0 & -\gamma \end{pmatrix} \begin{pmatrix} s \\ \psi \end{pmatrix} , \tag{3.9}$$

which is still unstable for $\alpha > 0$. The regulator R adds a force u, determined by the engine speed variation s in a variety of possible ways; typically we would have, after linearization,

$$u = k_1 s, \qquad k_1 < 0 . \tag{3.10}$$

However, the original spring constant, k, or the friction, f, could also be varied, so we may augment (3. 10) to

$$u = k_1 s + k_2 \psi, \tag{3.11}$$

leading to a closed loop system

$$\begin{pmatrix} \dot{s} \\ \psi \end{pmatrix} = \begin{pmatrix} \alpha & \beta \\ k_1 & -\gamma + k_2 \end{pmatrix} \begin{pmatrix} s \\ \psi \end{pmatrix} \tag{3.12}$$

obtained when the feedback relation (3. 11) is substituted in the system

$$\begin{pmatrix} \dot{s} \\ \psi \end{pmatrix} = \begin{pmatrix} \alpha & \beta \\ 0 & -\gamma \end{pmatrix} \begin{pmatrix} s \\ \psi \end{pmatrix} + \begin{pmatrix} 0 \\ 1 \end{pmatrix} u , \tag{3.13}$$

which is of the form (3. 1)(with $v = 0$ so far). The response with which we are concerned here is the speed variation

$$y = s = (1, 0) \begin{pmatrix} s \\ \psi \end{pmatrix} . \tag{3.14}$$

Ideally we would like the speed variation, s, to be exponentially damped at some rate depending on the application. This leads to comparison with the system

$$\dot{\sigma} = - \alpha_0 \sigma \tag{3.15}$$

for some $\alpha_0 > 0$. The systems (3. 13) and (3. 15) correspond to (3. 1) and (3. 3), respectively, and (3. 14) corresponds to (3. 2). Corresponding to (3. 4) we set

$$\eta = \sigma . \tag{3.16}$$

The relationship (3. 5) corresponds in our case to

$$\sigma \sim (1, 0) \begin{pmatrix} s \\ \psi \end{pmatrix} ,$$

so that the initial state relationship in the Optimal Model Following Problem becomes

$$\sigma(0) = \sigma_0 = s_0 = s(0).$$

External disturbances will often arise due to load variations, leading us to augment (3.13) to

$$\begin{pmatrix} \dot{s} \\ \dot{\psi} \end{pmatrix} = \begin{pmatrix} \alpha & \beta \\ 0 & -\gamma \end{pmatrix} \begin{pmatrix} s \\ \psi \end{pmatrix} + \begin{pmatrix} 0 \\ 1 \end{pmatrix} u + \begin{pmatrix} 1 \\ 0 \end{pmatrix} v \tag{3.17}$$

where v is scalar white noise with covariance V. The final objective is to select $(k_1, k_2) = K$ so as to minimize the expected value of

$$\int_0^\infty [\tilde{W}(s(t) - \sigma(t))^2 + \tilde{U}u(t)^2]\,dt, \quad \tilde{W} > 0, \quad \tilde{U} > 0 \text{ (scalar)}. \tag{3.18}$$

Having introduced our example as motivation, let us return once more to the theoretical problem. We restrict attention to linear feedback control laws

$$u = Kx, \quad K \ m \times n,$$

yielding a closed loop system

$$\dot{x} = (A + BK)x + Cv \tag{3.19}$$

and a response

$$y = (F + GK)x. \tag{3.20}$$

For consideration of how well $y(t)$ follows the response (3.4) of the system (3.3) we are led to consider the response

$$y(t) - \eta(t) = (F + GK, -F_0)\begin{pmatrix} x \\ \xi \end{pmatrix}$$

for the composite system

$$\begin{pmatrix} \dot{x} \\ \dot{\xi} \end{pmatrix} = \begin{pmatrix} A+BK & 0 \\ 0 & A_0 \end{pmatrix} \begin{pmatrix} x \\ \xi \end{pmatrix} + \begin{pmatrix} C \\ C_0 \end{pmatrix} v. \tag{3.21}$$

To begin our analysis, let us consider the somewhat special case wherein (cf. (3.2), (3.3)) $n_0 = q$ and F_0 is nonsingular. We introduce a new variable, the response difference

$$z = \eta - y = F_0\xi - (F + GK)x \,. \tag{3.22}$$

Since we are identifying ξ with $Hx + Ju$ = (in view of $u = Kx$) = $(H + JK)x$, the initial state for z is

$$z(0) = z_0 = F_0(H + JK)x_0 - (F + GK)x_0 \,. \tag{3.23}$$

The differential equation satisfied by z is

$$\begin{aligned}\dot{z} &= F_0\dot{\xi} - (F + GK)\dot{x} = F_0(A_0\xi + C_0 v) \\ &\quad - (F + GK)((A + BK)x + Cv) \,. \end{aligned} \tag{3.24}$$

Since F_0 is assumed nonsingular, (3.22) gives

$$\xi = F_0^{-1}z + F_0^{-1}(F + GK)x$$

and (3.24) becomes

$$\begin{aligned}\dot{z} &= F_0[A_0(F_0^{-1}z + F_0^{-1}(F+GK)x) + C_0 v] - (F + GK)((A+BK)x + Cv) \\ &= F_0A_0F_0^{-1}z + [F_0A_0F_0^{-1}(F+GK) - (F+GK)(A+BK)]x + [F_0C_0 - (F+GK)C]v\end{aligned}$$

or, with $C_1 \equiv F_0C$,

$$\dot{z} = A_1 z + [A_1(F + GK) - (F+GK)(A+BK)]x + [C_1 - (F+GK)C]v \,. \tag{3.25}$$

The response difference is seen to depend upon the feedback matrix K in three ways: via the initial state (3.23), via the disturbance $[C_1 - (F + GK)C]v$ in (3.25), and via the coupling term

$$w = [A_1(F+GK) - (F+GK)(A+BK)]x \tag{3.26}$$

whereby the system (3.1) influences z.

Let us postpone the question of optimal model following briefly and consider that of perfect model following. We ask: under what circumstances will it be true that

$$z(t) \equiv 0, \quad t \geq 0 \ ? \tag{3.27}$$

Since the initial state z_0 is given by (3.23), if (3.27) is to be true for arbitrary x_0 we require

$$H + JK = F_0^{-1}(F + GK), \tag{3.28}$$

which we can assume if we take

$$H = F_0^{-1}F, \quad J = F_0^{-1}F \tag{3.29}$$

(i.e., our state identification process agrees with the outputs which we wish to compare). Then we have $z(0) = 0$ and (3.25) gives

$$z(t) = \int_0^t e^{A_1(t-s)}([A_1(F+GK)-(F+GK)(A+BK)]x(s)+[C_1-(F+GK)C]\,v(s))ds \tag{3.30}$$

In order that $z(t) \equiv 0$ for any x_0 we clearly require, in addition to (3.28),

$$A_1(F+GK) - (F+GK)(A+BK) = 0, \tag{3.31}$$

$$C_1 - (F+GK)C = 0. \tag{3.32}$$

Perfect model following is thus possible if and only if K can be selected so that (3.28), (3.31), (3.32) are all true. It is a rare and fortuitous situation indeed when this is the case. For our small engine example

$$A_1(F+GK)-(F+GK)(A+BK) = -\alpha_0(1,0) - (1,0)\begin{pmatrix} \alpha & \beta \\ k_1 & -\gamma+k_2 \end{pmatrix} = (-\alpha_0-\alpha, \ -\beta) \tag{3.33}$$

and the feedback coefficients k_1 and k_2 have no effect whatever on this quantity. Assuming $\beta \neq 0$, perfect model following is impossible here.

If we suppose (3.28) to be true and neglect the term $[C_1-(F+GK)C]\,v$ in (3.25), the formula (3.30) does suggest an "ad hoc" approach for determination of a "good" (generally not optimal) feedback matrix $\hat{K}_1$. We reason that $z(t)$ will be small if the response

$$w(t) = [A_1(F+GK)-(F+GK)(A+BK)]\,x(t) = [(A_1F-FA)+(A_1G-FB)K-GKA-GKBK]\,x(t)$$

$$\equiv (F_1 + G_1(K))x(t) \tag{3.34}$$

is small for solutions $x(t)$ of

$$\dot{x} = (A+BK)x + Cv, \quad x(0) = x_0, \quad \mathrm{cov}(x_0, x_0) = X_0, \quad \mathrm{cov}(v(t), v(t)) = e^{-\mu t}V.$$

As in Chapter VI, this leads to the problem

$$\min_{K \in \mathcal{S}(A, B)} \mathrm{Tr}\, MX(K), \quad M = X_0 + \frac{1}{\mu} CVC^*, \tag{3.35}$$

where

$$(A+BK)^*X(K)+X(K)(A+BK)+(F_1+G_1(K))^*\tilde{W}(F_1+G_1(K)) + K^*\tilde{U}K = 0, \tag{3.36}$$

$\tilde{W}$ and $\tilde{U}$ being suitably chosen weighting matrices. If $G = 0$,

$$F_1 + G_1(K) = A_1F - FA - FBK \equiv F_1 + G_1K$$

and the solution of (3. 35) is

$$\hat{K}_1 = -U^{-1}B^*[Q_1 + R_1^*],$$

where Q_1 is the unique symmetric positive definite solution of

$$A^*Q_1 + Q_1A + W_1 - (Q_1B+R_1)U_1^{-1}(B^*Q_1 + R_1^*) = 0,$$

$$W_1 = F_1^*\tilde{W}F_1, \quad U_1 = \tilde{U} + G_1^*\tilde{W}G_1, \quad R_1 = F_1^*\tilde{W}G_1.$$

If $G \neq 0$, setting the gradient (cf. VI- Section II) of $\mathrm{Tr}\, MX(K)$ with respect to $K = 0$ at $\hat{K}_1$ leads to a cubic matrix equation for $\hat{K}_1$ which will ordinarily not lead to a closed form expression for $\hat{K}_1$ in terms of $X(\hat{K}_1)$. The interested reader may wish to explore this further in Exercise 5 at the end of this chapter.

In our small motor example, let us take (cf. (3. 17), (3. 15))

$$\alpha = \beta = \gamma = \alpha_0 = \tilde{W} = 1, \quad \tilde{U} = \frac{1}{4}, \quad X_0 = \begin{pmatrix} 1 & 0 \\ 0 & 0 \end{pmatrix}, \quad V = 0. \tag{3.37}$$

and (cf. (3.14))

$$F_0 = 1, \quad F = H = (1,0), \quad G = J = 0, \quad A_1 = -1. \tag{3.38}$$

Then (3.13), (3.14) give

$$A = \begin{pmatrix} 1 & 1 \\ 0 & -1 \end{pmatrix}, \quad B = \begin{pmatrix} 0 \\ 1 \end{pmatrix},$$

$$F_1 = A_1 F - FA = \text{(cf. (3.33))} = (-2,-1),$$

$$W_1 = F_1^* \tilde{W} F_1 = \begin{pmatrix} -2 \\ -1 \end{pmatrix} \cdot 1(-2, -1) = \begin{pmatrix} 4 & 2 \\ 2 & 1 \end{pmatrix},$$

$$G_1 = -FB = (-1, 0)\begin{pmatrix} 0 \\ 1 \end{pmatrix} = 0,$$

$$U = \tilde{U} = \frac{1}{4},$$

and our quadratic matrix equation is

$$\begin{pmatrix} 1 & 0 \\ 1 & -1 \end{pmatrix} \tilde{Q} + \tilde{Q} \begin{pmatrix} 1 & 1 \\ 0 & -1 \end{pmatrix} + \begin{pmatrix} 4 & 2 \\ 2 & 1 \end{pmatrix} - \tilde{Q} \begin{pmatrix} 0 & 0 \\ 0 & 4 \end{pmatrix} \tilde{Q} = 0$$

which may be solved to give

$$\tilde{Q} = \begin{pmatrix} 3.2361 & 1.6180 \\ 1.6180 & .8090 \end{pmatrix}, \quad \tilde{K} = -4(0,1)Q_1 = (-6.4722, -3.2361).$$

We postpone the rigorous assessment of this particular feedback control until we have completed the rest of the theoretical development. We do make the following observations, however. The eigenvalues of the closed loop system

$$\begin{pmatrix} \dot{s} \\ \psi \end{pmatrix} = \begin{pmatrix} 1 & 1 \\ \hat{k}_1 & -1+\hat{k}_2 \end{pmatrix} \begin{pmatrix} s \\ \psi \end{pmatrix} = \begin{pmatrix} 1 & 1 \\ -6.4722 & -4.2361 \end{pmatrix}$$

may be computed (to four decimal places) to be

$$\lambda_1 = -1, \quad \lambda_2 = -2.2361 . \tag{3.39}$$

It is noteworthy, and we will comment on this, that one of these agrees with the single eigenvalue, -1, of the model system (3.15) while the other is considerably to the left of that.

Inspection of the term (3.34) whereby the system (3.1) is coupled into the equations (3.25) governing the model following error enables one to make an heuristic argument for the way in which the eigenvalues of $A + BK$ should behave in order for $w(t)$, and hence $z(t)$, to be small. One possibility, of course, since $x(t)$ satisfies $\dot{x} = (A + BK)x$, is for the eigenvalues of $A+BK$ to move to the far left of the complex plane. But one can show without difficulty that not all components of $x(t)$ tend to zero in $L_n^2[0,\infty)$ when this happens. There is another way in which the feedback matrix K can cause $w(t)$, or certain components of $w(t)$, to become small, and that is for the matrix

$$F_1 + G_1(K) = A_1(F+GK) - (F+GK)(A+BK) \tag{3.40}$$

to be small in some sense. If we define the operator S_K on $q \times n$ matrices F by

$$S_K(F) = A_1F - F(A + BK)$$

then (3.40) is just $S_K(F + GK)$.

From III - Section 1 we know that the eigenvalues of S_K have the form $\lambda_i - \mu_j$, where λ_i is an eigenvalue of A_1 (equivalently $A_0 = F_0^{-1}A_1F_0$) and $\mu_j = \mu_j(K)$ is an eigenvalue of $A + BK$. The operator S_K has a non-trivial null space, $\mathcal{N}(T_K)$, in the space E^{qn} of $q \times n$ matrices just in case $\lambda_i = \mu_j$ for some i, j. If it were true that $F + GK \in \mathcal{N}(T_K)$, the matrix (3.40) would be zero. Ordinarily this is too much to expect but it is found [4] that if one lets $\tilde{U} \to 0$ in (3.36) (weighting w(t) more heavily) that some of the eigenvalues of $A + BK$ move toward some of the eigenvalues of A_1 so that T_K becomes small on a certain subspace of R^{qn}. Correspondingly, K acts to bring all or part of $F + GK$ into this subspace. The remaining eigenvalues of $A + BK$ move leftward in the complex plane.

For a more complete treatment we must tackle the real problem "by the horns". Assuming we use feedback control of the form $u = Kx$, and, for the moment, restricting the state identification to one of the form $\xi = Hx$ (i.e., $J = 0$), we have the system (3.21), i.e.,

$$\begin{pmatrix} \dot{x} \\ \dot{\xi} \end{pmatrix} = \begin{pmatrix} A+BK & 0 \\ 0 & A_0 \end{pmatrix} \begin{pmatrix} x \\ \xi \end{pmatrix} + \begin{pmatrix} C \\ C_0 \end{pmatrix} v ,$$

and, using the theory of VI-Section 5, the expected value of the cost, (3.6), becomes

$$\mathrm{Tr} \begin{pmatrix} M & N \\ N^* & M_0 \end{pmatrix} \begin{pmatrix} X & Y \\ Y^* & Z \end{pmatrix} = \mathrm{Tr}(MX + NY^*) + \mathrm{Tr}(N^*Y + M_0 Z) \tag{3.41}$$

where, from the statement of the Optimal Model Following Problem,

$$\begin{pmatrix} M & N \\ N^* & M_0 \end{pmatrix} = \begin{pmatrix} X_0 + \frac{1}{\mu} CVC^* & X_0 H^* + \frac{1}{\mu} CVC_0^* \\ HX_0 + \frac{1}{\mu} C_0 VC^* & HX_0 H^* + \frac{1}{\mu} C_0 VC_0^* \end{pmatrix} \tag{3.42}$$

and $\begin{pmatrix} X & Y \\ Y^* & Z \end{pmatrix}$ is the unique $(n + n_0) \times (n + n_0)$ solution of

$$\begin{pmatrix} A+BK & 0 \\ 0 & A_0 \end{pmatrix}^* \begin{pmatrix} X & Y \\ Y^* & Z \end{pmatrix} + \begin{pmatrix} X & Y \\ Y^* & Z \end{pmatrix} \begin{pmatrix} A+BK & 0 \\ 0 & A_0 \end{pmatrix}$$

$$+ \begin{pmatrix} (F+GK)^* \tilde{W} (F+GK) & -(F+GK)^* \tilde{W} F_0 \\ -F_0^* \tilde{W} (F+GK) & F_0^* \tilde{W} F_0 \end{pmatrix}$$

$$+ \begin{pmatrix} K^* \tilde{U} K & 0 \\ 0 & 0 \end{pmatrix} = 0 .$$

For $X = X(K)$, $Y = Y(K)$ and Z we have separate equations

$$(A+BK)^* X(K) + X(K)(A+BK) + (F+GK)^* \tilde{W}(F+GK) + K^* \tilde{U} K = 0, \tag{3.43}$$

$$(A+BK)^* Y(K) + Y(K) A_0 - (F+GK)^* \tilde{W} F_0 = 0, \tag{3.44}$$

$$A_0^* Z + Z A_0 + F_0^* \tilde{W} F_0 = 0. \tag{3.45}$$

It is clear that Z is independent of K and plays no role in the determination of an optimal $\hat{K}$ for which the cost (3. 41) is minimized. It must be included, of course, when that cost is computed for a given K. Since $\operatorname{Tr} N^* Y = \operatorname{Tr} Y^* N = \operatorname{Tr} NY^*$ our optimal model following problem becomes

$$\min_{K \in \mathscr{S}(A, B)} \{\operatorname{Tr}(MX(K) + 2N^* Y(K))\}. \tag{3.46}$$

Letting $Q(K)$ and $P(K)$ satisfy the adjoint matrix equations

$$(A+BK)Q(K) + Q(K)(A+BK)^* + M = 0, \tag{3.47}$$

$$A_0 P(K) + P(K)(A+BK)^* + N^* = 0, \tag{3.48}$$

and putting $K = \hat{K} + \epsilon \delta K$, the necessary condition for $\hat{K}$ to solve (3. 46) is that, for all $m \times n$ matrices δK,

$$\frac{\partial}{\partial \epsilon} \operatorname{Tr}(MX(\hat{K} + \epsilon\delta K) + 2N^* Y(\hat{K} + \epsilon \delta K)) = 0$$

at $\epsilon = 0$. Here $X(\hat{K} + \epsilon \delta K)$ and $Y(\hat{K} + \epsilon \delta K)$ are the solutions of (3. 43) and (3. 44), respectively, for $K = \hat{K} + \epsilon \delta K$. Differentiating (3. 43) and (3. 44) with respect to ϵ, the necessary condition is

$$\operatorname{Tr}(MX_\epsilon(\delta K) + 2N^* Y_\epsilon(\delta K)) = 0, \tag{3.49}$$

where $X_\epsilon(\delta K) = \frac{\partial}{\partial \epsilon} X(\hat{K} + \epsilon \delta K)\Big|_{\epsilon = 0}$, $Y_\epsilon(\delta K) = \frac{\partial}{\partial \epsilon} Y(\hat{K} + \epsilon \delta K)\Big|_{\epsilon = 0}$ satisfy

$$(A\ B\hat{K})^* X_\epsilon(\delta K) + X_\epsilon(\delta K)(A+B\hat{K}) + \delta K^* B^* X(\hat{K}) + X(\hat{K}) B \delta K + \delta K^* G^* \tilde{W} F + F^* \tilde{W} G \delta K$$

$$+ \delta K^* (\tilde{U} + G^* \tilde{W} G)\hat{K} + \hat{K}^* (\tilde{U} + G^* \tilde{W} G) \delta K = 0 \tag{3.50}$$

and

$$(A + B\hat{K})^* Y_\epsilon(\delta K) + Y_\epsilon(\delta K) A_0 - \delta K^* B^* Y(\hat{K}) - \delta K^* G^* \tilde{W} F_0 = 0. \tag{3.51}$$

Using the properties of the trace stated in VI - Propositions 3.2, 3.3, the condition that (3.49) should be true for all δK, subject to the constraints (3.50) and (3.51), is seen to be equivalent to

$$\operatorname{Tr} \delta K^* [(B^* X(\hat{K}) + G^* \tilde{W} F + (\tilde{U} + G^* \tilde{W} G)K)Q(\hat{K}) + (B^* Y(\hat{K}) - G^* \tilde{W} F_0)P(\hat{K})] = 0, \tag{3.52}$$

for all $m \times n$ matrices δK. This implies that the coefficient of δK^* in (3.52) is zero and, assuming $Q(K)$ to be positive definite (which is true if $(M, A + B\hat{K})$ is an observable pair), multiplication on the right by $Q(\hat{K})^{-1}$ yields the final form of the necessary condition

$$\hat{K} = -U^{-1}[B^* X(\hat{K}) + G^* \tilde{W} F + (B^* Y(\hat{K}) - G^* \tilde{W} F_0)P(\hat{K})Q(\hat{K})^{-1}], \tag{3.53}$$

$$U = \tilde{U} + G^* \tilde{W} G. \tag{3.54}$$

The same result follows from Axsäter's result described in VI - 5 but the above partitioned form is more direct and easier to use in this specialized instance.

If the state identification takes the general form (3.5), i.e., $J \neq 0$, then the matrix in (3.42) does not have the form shown there but, rather, depends on K; viz.:

$$\begin{pmatrix} M & N \\ N^* & M_0 \end{pmatrix} = \begin{pmatrix} M & N(K) \\ N(K)^* & M_0(K) \end{pmatrix} = \begin{pmatrix} X_0 + \frac{1}{\mu} CVC^* & X_0(H+JK)^* + \frac{1}{\mu} CVC_0^* \\ (H+JK)X_0 + \frac{1}{\mu} C_0VC^* & (H+JK)X_0(H+JK)^* + \frac{1}{\mu} C_0VC_0^* \end{pmatrix}. \tag{3.55}$$

The optimization problem (3.46) is replaced by

$$\min_{K \in \mathscr{S}(A, B)} \{\operatorname{Tr}(MX(K) + 2N(K)^* Y(K) + M_0(K)Z)\}$$

and the condition for $\hat{K}$ to be optimal is

$$\operatorname{Tr}(MX_\epsilon(\delta K) + 2N_\epsilon(\delta K)^* Y(\hat{K}) + 2N(\hat{K})^* Y_\epsilon(\delta K) + M_{0,\epsilon}(\delta K)Z) = 0$$

where, from (3.55),

$$N_{\epsilon}(\delta K) = \frac{\partial}{\partial \epsilon} N(\hat{K} + \epsilon \delta K)\Big|_{\epsilon = 0} = X_0 \delta K^* J^*$$

$$M_{0,\epsilon}(\delta K) = \frac{\partial}{\partial \epsilon} M_0(\hat{K} + \epsilon \delta K)\Big|_{\epsilon = 0} = J\delta K X_0 (H + J\hat{K})^* + (H + J\hat{K})X_0 \delta K^* J^* .$$

The necessary condition (3. 52) for optimality is now replaced by

$$\operatorname{Tr} \delta K^* [(B^* X(\hat{K}) + G^* \tilde{W} F + (\tilde{U} + G^* \tilde{W} G)\hat{K})Q(\hat{K}) + (B^* Y(\hat{K}) - G^* \tilde{W} F_0)P(\hat{K})$$
$$+ J^* Y(\hat{K})^* X_0 + J^* Z(H + J\hat{K})X_0] = 0 . \qquad (3.56)$$

Because $\hat{K}$ appears twice in this equation it will be seen that it is not easy to obtain an analytic expression for $\hat{K}$ in terms of $X(\hat{K})$, $Y(\hat{K})$, $P(\hat{K})$, $Q(\hat{K})$ and the other quantities appearing in (3. 56). However, when the coefficient of δK^* is set equal to zero the resulting equation, which has the form

$$(\tilde{U} + G^* \tilde{W} G)\hat{K} Q(\hat{K}) + J^* Z J \hat{K} X_0 = D(\hat{K})$$

for some $m \times n$ matrix $D(\hat{K})$ whose formula is readily derived from (3. 56), is, in fact, a necessary condition which must be satisfied by an optimal feedback matrix $\hat{K}$. When $X_0 = 0$ this simplifies to

$$\hat{K} = (\tilde{U} + G^* \tilde{W} G)^{-1} D(\hat{K})\, Q(\hat{K})^{-1}$$

and a variant of Axsäter's method (see VI- 5) can be attempted for approximation of $\hat{K}$:

$$K_{i+1} = (\tilde{U} + G^* \tilde{W} G)^{-1} D(K_i) Q(K_i)^{-1} .$$

This method is suggested for use in Exercise 6 below. One may also use gradient type methods which apply whether or not $X_0 = 0$.

Now let us apply the theory which led to (3. 53) to our small motor regulator. We have noted that the problem of minimizing the cost (3. 18) with the constraints (3. 15), (3. 17) is a problem of limited state feedback, as briefly discussed in VI - 5. We will use a variant of Axsäter's method which we discussed there in connection with that problem, altered slightly to take into account the fact that we have made use of the special block diagonal structure of (3. 21) or (3. 15), (3. 17)) here) to obtain the equations (3. 43) - (3. 45). In this context it can be verified, using the fact that the matrix multiplying δK^* in (3. 52) is the (matrix) gradient of the cost with respect to K, that Axsäter's method consists in generating matrices K_i,

X_i, Y_i, Q_i, P_i according to the following scheme, where K_0 is assumed to be an initially given stabilizing feedback matrix.

$$(A+BK_i)^* X_i + X_i(A+BK_i) + (F+GK_i)^* \tilde{W}(F+GK_i) + K_i^* \tilde{U} K_i = 0, \tag{3.57}$$

$$(A+BK_i)^* Y_i + Y_i A_0 - (F+GK_i)^* \tilde{W} F_0 = 0, \tag{3.58}$$

$$(A+BK_i)Q_i + Q_i(A+BK_i)^* + M = 0, \tag{3.59}$$

$$A_0 P_i + P_i(A+BK_i)^* + N^* = 0, \tag{3.60}$$

$$K_{i+1} = -U^{-1}[B^* X_i + G^* \tilde{W} F + (B^* Y_i - G^* \tilde{W} F_0) P_i Q_i^{-1}], \tag{3.61}$$

with U given by (3. 54). As noted in VI-5, one cannot <u>guarantee</u> convergence even if K_0 is close to a solution $\hat{K}$ of (3. 53) but, with that reservation in mind, the method nevertheless often works very well.

We begin, for our small motor regulator, with the feedback matrix K_0 obtained by our earlier "ad hoc" approach, i. e.,

$$K_0 = (-6.4722, \quad -3.2361).$$

The relevant data matrices for this example are listed in (3. 37), (3. 38), ff., note that here A_1 (cf. (3. 25)) $= A_0$. The matrices M and N^* (cf. (3. 37), (3. 42)) are

$$M = X_0 \equiv \begin{pmatrix} 1 & 0 \\ 0 & 0 \end{pmatrix}$$

$$N^* = HX_0 = (1, 0)\begin{pmatrix} 1 & 0 \\ 0 & 0 \end{pmatrix} = (1, 0)$$

and we have

$$M_0 = HX_0H^* = 1.$$

The matrix Z (here 1×1) is constant throughout, since (3. 45) does not involve the feedback matrix. We have

$$(-1)Z + (-1)Z + 1 = 0, \quad \text{i. e.,} \quad Z = \frac{1}{2}. \tag{3.62}$$

Application of the method (3. 57) - (3. 61) yields the following table.

i	K_i	X_i		Y_i	Q_i		P_i^*	Total Cost (3.41)	Model Following Cost
0	(-6.4722, -3.2361)	3.7361	1.4635	-.8090	1.3944	-1.8944	.8090	2.6181	.2764
		1.4635	.65451	-.1545	-1.8944	2.8944	-1		
1	(-5.2491, -2.4271)	3.7318	1.4623	-.8434	1.5340	-2.0340	.8434	2.5450	.3471
		1.4623	.6416	-.1905	-2.0340	3.1153	-1		
2	(-5.1447, -2.3509)	3.7359	1.4664	-.8457	1.5440	-2.0440	.8457	2.5445	.3526
		1.4664	.6438	-.1944	-2.0440	3.1382	-1		
⋮	⋮	⋮		⋮	⋮		⋮	⋮	⋮

Table VII-1

Axsäter's Method Applied to Small Motor Regulator

The reader is invited to carry out the rest of the computations necessary to carry the table to four decimal place convergence in Exercise 7 at the end of this chapter.

It will be noted that, while the total cost, given by (3.41), is decreasing, the model following cost is actually increasing. The earlier "ad hoc" method achieved a smaller model following cost but at the expense of greater control cost. We could reduce the model following cost if we replaced $\tilde{U} = \frac{1}{4}$ by a smaller number, say $\tilde{U} = \frac{1}{8}$, but the control cost inevitably would rise. (The control cost is defined by $\operatorname{Tr} MX_u$, where for any feedback matrix K,

$$(A+BK)^* X_u + X_u (A+BK) + K^* \tilde{U}_0 K = 0 , \tag{3.63}$$

$\tilde{U}_0$ being a standard weighting matrix - in our example we would take $\tilde{U}_0 = 1$ to provide a uniform comparison of control costs for optimal feedback matrices computed with weighted matrices $\tilde{U} = \frac{1}{4}\tilde{U}_0 = \frac{1}{4}$, $\tilde{U} = \frac{1}{8}\tilde{U}_0 = \frac{1}{8}$, etc.). See Exercise 8.

The eigenvalues of

$$A + BK_2 = \begin{pmatrix} 1 & 1 \\ -5.1447 & -3.3509 \end{pmatrix}$$

are

$$\lambda_1 = -1.1755 + .6420i$$

$$\lambda_2 = -1.1755 - .6420i$$

which helps to explain (by comparison with (3.39) and taking the discussion there following into account) the inferior model following performance of K_2 as compared with K_0. The method (3.46) gives greater weight to the control cost then does (3.35) even though we have taken $\tilde{U} = \frac{1}{4}$ in both cases. It is difficult to conjecture as to whether this would be true generally.

A somewhat more elaborate model following scheme can be envisioned which would involve feedback on the "model state" ξ as well as the plant state x. We form the composite system (3.21) again:

$$\begin{pmatrix} \dot{x} \\ \xi \end{pmatrix} = \begin{pmatrix} A & 0 \\ 0 & A_0 \end{pmatrix} \begin{pmatrix} x \\ \xi \end{pmatrix} + \begin{pmatrix} B \\ 0 \end{pmatrix} u + \begin{pmatrix} C \\ C_0 \end{pmatrix} v \; .$$

Assuming the pair (A, B) to be stabilizable, our a priori assumption that A_0 is a stability matrix shows (3.21) to be stabilizable. Assuming v to be white noise with covariance $e^{-\mu t}V$ and assuming a random initial vector

$$x(0) = x_0, \quad \text{cov}\,(x_0, x_0) = X_0$$

and setting

$$\xi(0) = Hx(0)$$

(we assume from the start that $J = 0$ here (cf. 3.5))) we can seek for the control law

$$u = Kx + K_0 \xi, \quad K \;\; m \times n, \quad K_0 \;\; m \times n_0$$

for which the cost (3.6) is minimized. Since the feedback relation is unrestricted in this case, the theory of Chapters IV and VI applies to characterize the optimal feedback matrices $\hat{K}$ and $\hat{K}_0$. The quadratic matrix equation IV-(3.13) here takes the partitioned form

$$\begin{pmatrix} A^* & 0 \\ 0 & A_0^* \end{pmatrix} \begin{pmatrix} Q & P \\ P^* & Q_0 \end{pmatrix} + \begin{pmatrix} Q & P \\ P^* & Q_0 \end{pmatrix} \begin{pmatrix} A & 0 \\ 0 & A_0 \end{pmatrix} + \begin{pmatrix} F^*\tilde{W}F & -F^*\tilde{W}F_0 \\ -F_0^*\tilde{W}F & F_0^*\tilde{W}F_0 \end{pmatrix}$$

$$- \left[\begin{pmatrix} Q & P \\ P^* & Q_0 \end{pmatrix} \begin{pmatrix} B \\ 0 \end{pmatrix} + \begin{pmatrix} F^*\tilde{W}G \\ -F_0^*\tilde{W}G \end{pmatrix} \right] U^{-1} \left[(B^*, 0) \begin{pmatrix} Q & P \\ P & Q_0 \end{pmatrix} \right.$$

$$\left. + \, (G^*\tilde{W}F, \; -G^*\tilde{W}F_0) \right] = 0 \, ,$$

where, as usual, $U = \tilde{U} + G^*\tilde{W}G$. From this we can obtain the separate equations

$$A^*Q + QA + F^*\tilde{W}F - (QB + F^*\tilde{W}G)U^{-1}(B^*Q + G^*\tilde{W}F) = 0 \tag{3.64}$$

$$A^*P + PA_0 - F^*\tilde{W}F_0 - (QB + F^*\tilde{W}G)U^{-1}(B^*P - G^*\tilde{W}F_0) = 0 \tag{3.65}$$

$$A_0^* Q_0 + Q_0 A_0 + F_0^* \tilde{W} F_0 - (P^* B - F_0^* \tilde{W} G) U^{-1} (B^* P - G^* \tilde{W} F_0) = 0 . \quad (3.66)$$

These equations are solved in the order shown. The solvability of (3. 65) is immediate when it is realized that (3. 64) has the same form as VI- (3. 13), so that $A - BU^{-1}(B^* Q + G^* \tilde{W} F)$ is a stability matrix, and (3. 65) is rewritten as

$$(A - BU^{-1}(B^* Q + G^* \tilde{W} F))^* P + PA_0 - F^* \tilde{W} F_0 + (QB + F^* \tilde{W} G) U^{-1} G^* \tilde{W} F_0 = 0 .$$

The resulting optimal feedback matrices are (whatever X_0 and V may be, from Chapter VI)

$$\hat{K} = - U^{-1}(B^* Q + G^* \tilde{W} F), \quad \hat{K}_0 = -U^{-1}(B^* P - G^* \tilde{W} F_0), \quad (3.67)$$

and the final closed loop system is then

$$\begin{pmatrix} \dot{x} \\ \dot{\xi} \end{pmatrix} = \begin{pmatrix} A + B\hat{K} & B\hat{K}_0 \\ 0 & A_0 \end{pmatrix} \begin{pmatrix} x \\ \xi \end{pmatrix} + \begin{pmatrix} C \\ C_0 \end{pmatrix} v . \quad (3.68)$$

At this point an objection is immediately apparent. How is ξ to be measured, since the model system (3. 3) has no "real" existence? In fact, even if (3. 3) were to be constructed electronically, the solutions could not be developed without knowing $v(t)$, which is ordinarily out of the question.

Various schemes may be envisioned for getting around this difficulty. If the resulting system

$$\dot{x} = (A + B\hat{K})x + B\hat{K}_0 \xi + Cv \quad (3.69)$$

does, indeed, follow

$$\xi = A_0 \xi + C_0 v \quad (3.70)$$

quite closely, one could attempt the artifice of treating $(F + G\hat{K})x$ as if it were $F_0 \xi$ or an observation on $F_0 \xi$ corrupted by noise. In the latter case (3. 70) would be replaced by a state estimator for ξ (see III- 4)

$$\dot{\hat{\xi}} = A_0 \xi + L((F + G\hat{K})x - F_0 \hat{\xi}) \quad (3.71)$$

with L chosen, if possible, so that $A_0 - LF_0$ is a stability matrix with eigenvalues suitably to the left of those of A_0 or $A + B\hat{K}$, and (3.68) would be replaced by

$$\begin{pmatrix} \dot{x} \\ \dot{\hat{\xi}} \end{pmatrix} = \begin{pmatrix} A + B\hat{K} & B\hat{K}_0 \\ L(F + G\hat{K}) & A_0 - LF_0 \end{pmatrix} \begin{pmatrix} x \\ \hat{\xi} \end{pmatrix} \tag{3.72}$$

The system (3.71) would be realized electronically. The usefulness of this artifice must be regarded as problematical, as, indeed, is the asymptotic stability of (3.72) in general. If one can solve the quadratic matrix equation

$$(A - LF_0)T - T(A + B\hat{K}) - TBK_0T + L(F + GK) = 0$$

for the $n_0 \times n$ matrix T, then with

$$\eta = \hat{\xi} - Tx$$

one has

$$\begin{pmatrix} \dot{x} \\ \dot{\eta} \end{pmatrix} = \begin{pmatrix} A + B(K_0 + \hat{K}_0 T) & B\hat{K}_0 \\ 0 & A_0 - LF_0 - TB\hat{K}_0 \end{pmatrix} \begin{pmatrix} x \\ \eta \end{pmatrix}$$

and asymptotic stability obtains if the two matrices $A + B(K_0 + \hat{K}_0 T)$, $A_0 - LF_0 - TB\hat{K}_0$ both have only eigenvalues with negative real parts.

Another artifice of this type consists in posing the problem

$$\min E\left(\int_0^\infty [(\dot{y}(t) - \dot{\eta}(t))^* \tilde{W}(\dot{y}(t) - \dot{\eta}(t)) + u(t)^* \tilde{U} u(t)]\, dt\right)$$

initially, then assuming

$$u = Kx$$

so that

$$\dot{y} = F\dot{x} + G\dot{u} = (F + GK)\dot{x} = (F + GK)(A + BK)x .$$

Then

$$\dot{\eta} = F_0 \dot{\xi} = F_0 A_0 \xi$$

is replaced by the corresponding formula which results if we have $\xi = (H + JK)x$, i. e.

$$\dot{\eta} = F_0 A_0 (H + JK)x.$$

The problem is then

$$\min_K \operatorname{Tr} MQ(K), \quad M = X_0 + \frac{1}{\mu} CVC^*,$$

subject to the constraint

$$(A + BK)^* Q(K) + Q(K)(A + BK) + ((F + GK)(A + BK) - F_0 A_0 (H+JK))^* \tilde{W}((F+GK)(A+BK)$$
$$- F_0 A_0 (H + JK)) + K^* \tilde{U} K = 0 .$$

As usual, this problem is most easily treated if $G = 0$, in which case it has the standard form discussed in Chapters IV and VI. The reader is invited to apply this method to our small engine problem in Exercise 9.

4. Tracking a Command Input

Many control systems involve, in addition to state, control and disturbance variables, a "command" input variable, which we call w here. This input is determined by the plant operator and is distinguished from the "regulating" control u in the sense that while u is normally determined automatically by feedback, w is used "open loop" to move the system from one operating state to another. The general constant coefficient linear control system in this context may be written as

$$\dot{x} = Ax + Bu + Cv + Dw, \quad x \in E^n, \quad u \in E^m, \quad v \in E^p, \quad w \in E^\ell. \tag{4.1}$$

In many cases $B = D$ and u and w act additively in the combined input $B(u + w)$, but this need not necessarily be the case.

For the moment let us take $u = 0$, $v = 0$ and suppose $w \in E^\ell$ to be a constant vector. Thus

$$\dot{x} = Ax + Dw. \tag{4.2}$$

Supposing A to be nonsingular and setting

$$x_w = -A^{-1}Dw$$

we have, since w and x_w are constant

$$(x - x_w)^{\bullet} = \dot{x} = Ax + Dw = A(x + A^{-1}Dw) = A(x - x_w) , \qquad (4.3)$$

and we see that x_w is the unique equilibrium point for (4.1), asymptotically stable just in case A is a stability matrix.

But now suppose we take $w = w(t)$ to be a function of time, piecewise differentiable (at least). Then with

$$x_{w(t)} = -A^{-1}Dw(t)$$

we find that

$$(x - x_{w(t)})^{\bullet} = \dot{x} + A^{-1}D\dot{w}(t)$$

$$= Ax + Dw(t) + A^{-1}D\dot{w}(t) = A(x - x_{w(t)}) + A^{-1}D\dot{w}(t) . \qquad (4.4)$$

Thus, while $x - x_w = 0$ is a constant solution of our system when w is constant, it is not in general true that $x - x_{w(t)} = 0$ is a solution of (4.3) for variable $w(t)$.

In applications $x_w = -A^{-1}Dw$ represents an operating condition for the system which the plant operator selects by choosing w. This operating condition often has to be varied with time in a continuous manner, leading to consideration of the following question: if $w = w(t)$ is given as a piecewise differentiable function of time, how well does the actual operating state $x(t)$ "track" the nominal operating state $x_{w(t)} = -A^{-1}Dw(t)$? If A is not a stability matrix the answer will usually be that $x(t)$ and $x_{w(t)}$ diverge from each other as time advances, i. e., $x(t)$ does not track $x_{w(t)}$ at all. If A is a stability matrix there are some easy conclusions. If after some time, t_1, we have

$$w(t) \equiv w_1, \quad t \geq t_1,$$

then for $t \geq t_1$, arguing as we did in (4.3), we have

$$(x - x_{w_1})^{\bullet} = A(x - x_{w_1}),$$

and, since A is a stability matrix

$$x(t) - x_{w_1} = e^{A(t-t_1)}(x(t_1) - x_{w_1}) \to 0, \quad t \to \infty ,$$

so that the "new" operating condition $x = x_{w_1}$ is approached asymptotically. This by itself, however, is ordinarily not enough for good performance.

Since the command input $w(t)$, and hence $x_{w(t)}$, is given, one's first attempt to improve tracking performance might be to set

$$u = Kx + \hat{K}w, \quad K \;\; m \times n, \quad \hat{K} \;\; m \times \ell .$$

Since the most elementary requirement is that x_w should be a solution when w is constant, we need

$$Kx_w + \hat{K}w = (-KA^{-1}D + \hat{K})w = 0 .$$

If this is to be true for all w we must have $\hat{K} = KA^{-1}D$ and we may as well assume from the start that

$$u = K(x - x_w) = Kx + KA^{-1}D_w . \qquad (4.5)$$

With this choice we have, for constant w,

$$(\dot{x - x_w}) = \dot{x} = Ax + Bu + Dw = A(x - x_w) + BK(x - x_w) = (A + BK)(x - x_w)$$

so that $x = x_w$ is still an asymptotically stable equilibrium point if $A + BK$ is a stability matrix.

With these preliminaries out of the way, we consider the disturbed closed loop system with time varying command input $w(t)$:

$$\dot{x} = Ax + BK(x - x_{w(t)}) + Dw(t) + Cv, \qquad (4.6)$$

wherein we will assume v to be white noise with covariance $e^{-\mu t}V$ as in earlier work. Defining the new variable

$$z(t) = x(t) - x_w(t)$$

we have

$$\dot{z} = (A + BK)z + A^{-1}D\dot{w}(t) + Cv . \qquad (4.7)$$

We will see shortly that the feedback relation (4.5) is too naive in many applications. But we are not ready to motivate this yet so we proceed to pose the

Optimal Tracking Problem. Determine an $m \times n$ feedback matrix $\hat{K}$ such that $A + B\hat{K}$ is a stability matrix and such that $\hat{K}$ solves the problem (E = expected value)

$$\min_{K \in \mathscr{S}(A, B)} E(J(v, w, K)),$$

$J(v, w, K)$ to be specified below. Herein we assume the system and command vectors have initial states

$$x(0) = x_0,$$

$$w(0) = w_0,$$

so that

$$z(0) = x_0 - x_{w_0} \equiv z_0 . \tag{4.8}$$

The relevant output or response is defined to be

$$y = F(x - x_w) + Gu = (F + GK)z$$

and the cost $J(v, w, K)$ is defined by

$$J(v, w, K) = \int_0^\infty [(F(x - x_w) + Gu)^* \tilde{W} (F(x - x_w) + Gu) + u(t)^* U u(t)] \, dt$$

$$= \int_0^\infty z(t)^* [(F + GK)^* \tilde{W} (F + GK) + K^* UK] \, z(t) \, dt , \tag{4.9}$$

it being understood that $z(t)$ satisfies (4.7), (4.8).

It will be evident that the problem is not yet well posed because the statistics of the data x_0, v, w determining the system's evolution have not yet been prescribed. For x_0, v we give the usual

$$\operatorname{cov}(x_0, x_0) = X_0, \tag{4.10}$$

and stipulate v to be white noise with covariance $e^{-\mu t} V$ (modulo all the reservations and explanations of VI-5).

It is difficult to give statistics for "arbitrary" command inputs w. Certainly it is unrealistic to suppose that w is white noise. What is commonly done is to restrict w to a class of functions determined by a finite number of parameters and then give the statistics of those parameters. For example, w might be constructed as a linear combination of "basic" inputs with coefficients having certain statistical properties. This will be our basic program. It turns out that we have to divide our analysis into two parts at the outset, according to whether or not

$$\int_0^\infty \|\dot{w}(t)\|^2 \, dt < \infty \quad . \tag{4.11}$$

If (4.11) is true, it can be shown, applying the Laplace transform to (4.7), that the expected value of

$$\int_0^\infty \|z(t)\|^2 \, dt$$

is finite, so that the expected value of the cost (4.9) is finite. If (4.11) is not true, special precautions have to be taken to assure finite cost. We will first study cases for which (4.11) is true.

In all our work here we make use of the following scheme. We introduce an auxiliary ℓ_1-dimensional vector, which we will call w_1, which includes among its components those of w, or else, more generally,

$$w = H_1 w_1, \qquad H_1 \quad \ell \times \ell_1, \tag{4.12}$$

and we will suppose that

$$\dot{w}_1 = A_1 w_1,$$

for some $\ell_1 \times \ell_1$ matrix A_1, on certain subintervals $I_1, I_2, \ldots, I_N$ of $[0, \infty)$ such that

$$[0, \infty) = I_1 \cup I_2 \cup \ldots \cup I_N \, .$$

On each interval I_k (4.7) can be replaced by

$$\dot{z} = (A+BK)z + D_1\dot{w}_1 + Cv = (A+BK)z + D_1A_1w_1 + Cv \tag{4.14}$$

$$\dot{w}_1 = A_1w_1 , \tag{4.15}$$

where (cf. (4.12))

$$D_1 = DH_1 .$$

For example, if one wishes as command inputs on $I_k = [t_{k-1}, t_k]$ the class of all ℓ-vector polynomials of the form

$$w(t) = p(t) = p_0(t-t_{k-1})^{\sigma} + p_1(t-t_{k-1})^{\sigma-1} + \ldots + p_{\sigma-1}(t-t_{k-1}) + p_{\sigma}$$

we would take $w_1 \in E^{(\sigma+1)\ell}$

$$w_1 = \begin{pmatrix} w \\ \dot{w} \\ \vdots \\ w^{(\sigma-1)} \\ w^{\sigma} \end{pmatrix} \qquad \text{(in this case } H_1 = (I, 0, \ldots, 0))$$

$$A_1 = \begin{pmatrix} 0 & I & \cdots & 0 & 0 \\ 0 & 0 & \cdots & 0 & 0 \\ \vdots & \vdots & & \vdots & \vdots \\ 0 & 0 & \cdots & 0 & I \\ 0 & 0 & \cdots & 0 & 0 \end{pmatrix}, \quad w_1(t_{k-1}) = \begin{pmatrix} p_{\sigma} \\ p_{\sigma-1} \\ \vdots \\ (\sigma-1)!\, p_1 \\ \sigma!\, p_0 \end{pmatrix} .$$

We will study one of these cases in detail - that of a so-called "ramp" input. For vectors w_0, w_T we define

$$w(t) = \begin{cases} (1-\frac{t}{T})w_0 + \frac{t}{T}w_T, & 0 \leq t < T \\ w_T, & t \geq T \end{cases} .$$

Thus the command input is moved in a linear manner from its initial setting, w_0, at time $t = 0$, to its terminal setting, w_T, at time $t = T$, remaining

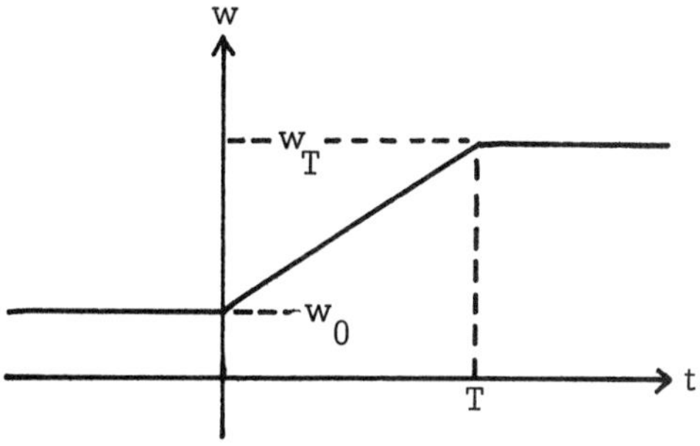

Fig. VII-2. Ramp Input

at w_T thereafter. Note that this is a case where (4.11) is true. Note also, since (4.7) depends only on $\dot{w}$, which is independent of w_0 here, that we may replace w_0 by 0 without loss of generality.

In this case we let

$$w_1 = \begin{pmatrix} w \\ \dot{w} \end{pmatrix}, \quad \dot{w}_1 = \begin{pmatrix} 0 & I \\ 0 & 0 \end{pmatrix} w_1 \equiv A_1 w_1 \tag{4.16}$$

$$H_1 = (I, 0), \quad D_1 = DH_1 = (D, 0) . \tag{4.17}$$

The vector function $w_1(t)$ satisfies (4.16) on both of

$$I_1 = [0, T], \quad I_2 = [T, \infty)$$

with initial conditions

$$w_1(0) = \begin{pmatrix} 0 & \equiv & 0 \\ \frac{1}{T} w_T & & \frac{1}{T} w_T \end{pmatrix}, \tag{4.18}$$

To avoid excessive complexity here we will let

$$v(t) \equiv 0 .$$

Thus the system of interest is (cf. (4.14), (4.15))

$$\dot{z} = (A + BK)z + D_1 A_1 w_1 \tag{4.19}$$

$$\dot{w}_1 = A_1 w_1 \tag{4.20}$$

with D_1, A_1 as identified in (4.16), (4.17).

We will define

$$S(K) = A + BK \tag{4.21}$$

$$\Sigma(K) = \begin{pmatrix} A+BK & D_1 A_1 \\ 0 & A_1 \end{pmatrix} . \tag{4.22}$$

Assuming

$$x(0) = A^{-1} D w_0 = 0$$

we have $z(0) = 0$ and the initial state for (4.16) has covariance matrix (cf. (4.18))

$$\begin{pmatrix} 0 & 0 \\ 0 & T^{-2} W_1 \end{pmatrix} , \quad W_1 = \text{cov}(w_T, w_T) .$$

For such an initial state the portion of the cost (cf. (4.9)) attributable to the interval $[0, T]$ is

$$\text{Tr} \begin{pmatrix} 0 & 0 \\ 0 & T^{-2} W_1 \end{pmatrix} \int_0^T e^{\Sigma(K)^* t} \Phi(K) e^{\Sigma(K) t} dt ,$$

$$\Phi(K) = \begin{pmatrix} (F+GK)^* \tilde{W} (F+GK) + K^* \tilde{U} K & 0 \\ 0 & 0 \end{pmatrix} .$$

We define

$$X(K, T) = \int_0^T e^{\Sigma(K)^* t} \Phi(K) e^{\Sigma(K) t} dt . \tag{4.23}$$

Multiplying on the left by $\Sigma(K)^*$, on the right by $\Sigma(K)$ and integrating we find that

$$\Sigma(K)^* X(K, T) + X(K, T) \Sigma(K) + \Phi(K) - e^{\Sigma(K)^* T} \Phi(K) e^{\Sigma(K) T} = 0 . \tag{4.24}$$

On the interval $[T, \infty)$ $\dot{w}(t) \equiv 0$ and thus

$$\dot{z} = (A + BK)z \ .$$

The cost attributable to $[T, \infty)$ is then

$$\mathrm{cov}(z(T),\ z(T))Z(K)$$

where (cf. (4. 21))

$$S(K)^* Z(K) + Z(K)S(K) + H(K) = 0 \ , \tag{4.25}$$

$$H(K) = (F + GK)^* \tilde{W}(H + GK) + K^* \tilde{U}K \ . \tag{4.26}$$

The matrix cov (z(T), z(T)) may be identified from the fact that

$$\begin{pmatrix} \mathrm{cov}(z(T),\ z(T)) & \mathrm{cov}(z(T),\ w_1(T)) \\ \mathrm{cov}(w_1(T), z(T)) & \mathrm{cov}(w_1(T), w_1(T)) \end{pmatrix} = e^{\Sigma(K)T} \begin{pmatrix} 0 & 0 \\ 0 & T^{-2}W_1 \end{pmatrix} e^{\Sigma(K)^* T} \ .$$

The optimization problem may thus be stated

$$\min_{K \in \mathcal{S}(A, B)} \mathrm{Tr} \begin{pmatrix} 0 & 0 \\ 0 & T^{-2}W_1 \end{pmatrix} \left(X(K, T) + e^{\Sigma(K)^* T} \begin{pmatrix} Z(K) & 0 \\ 0 & 0 \end{pmatrix} e^{\Sigma(K)T} \right) , \tag{4.27}$$

subject to X(K, T) satisfying (4. 24) and Z(K) satisfying (4. 25).

As it stands, the solution of (4. 27) is quite difficult because differentiation of the matrix

$$e^{\Sigma(K)T}$$

with respect to K is required. This is not at all impossible, of course, in view of the formula

$$\frac{d}{d\epsilon}\left(e^{\Sigma(K + \epsilon \delta K)T}\right)\Big|_{\epsilon = 0} = \int_0^T e^{\Sigma(K)(T - s)} \begin{pmatrix} B\delta K & 0 \\ 0 & 0 \end{pmatrix} e^{\Sigma(K)s}\, ds \ , \tag{4.28}$$

but it is rather difficult since representation of the integral (4. 28) in convenient and usable form is usually not possible.

We will not pursue the solution of (4. 27) for fixed T. Rather, we will offer the following argument. It is rarely the case that a command input is introduced into a plant during a time interval $[0, T]$ which never varies in length. Usually intervals of varying lengths will arise, with certain relative frequencies, i.e., probabilities. While we make no claim to "naturalness" - only to mathematical convenience - we believe the reader will agree that in many applications it is not unreasonable to assume that T has associated with it a probability density function of the form

$$p_\mu(T) = \frac{\mu^3}{2} T^2 e^{-\mu T} . \tag{4.29}$$

Graphs of $p(T)$ for $\mu = 1$ and $\mu = 4$ are shown in Figure VII-3.

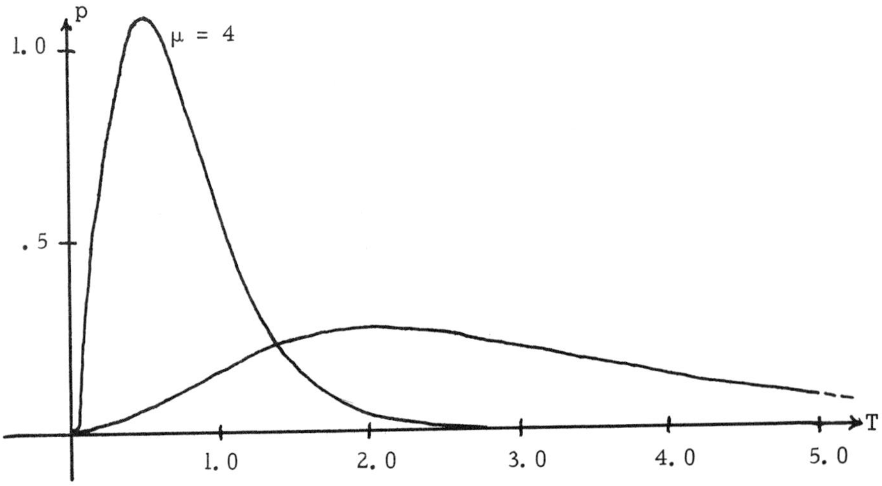

Fig. VII-3. The distribution $p_\mu(T)$, $\mu = 1, 4$

It will be noted that

$$\int_0^\infty p_\mu(T)dT = 1,$$

$$E(T) = \int_0^\infty T p(T)dT = \frac{\mu^3}{2} \int_0^\infty T^3 e^{-\mu T} dT = \frac{\mu^3}{2} \frac{6}{\mu^4} = \frac{3}{\mu}$$

so that small values of μ correspond to large average values of T and large values of μ correspond to small average values of T. The standard deviation is

$$\frac{\mu^3}{2}\int_0^\infty (T-\frac{3}{\mu})^2\, p(T)dT = \frac{\mu^3}{2}\int_0^\infty T^4 e^{-\mu T}dT - 3\mu^2\int_0^\infty T^3 e^{-\mu T}dT$$

$$+\frac{9}{2}\mu\int_0^\infty T^2 e^{-\mu T}dT = \frac{3}{\mu^2}$$

If we compute the expected value of the cost in (4. 27) for T a random variable in $[0,\infty)$ with the probability density $p_\mu(T)$ given by (4. 29), that expected value becomes

$$\mathrm{Tr}\begin{pmatrix} 0 & 0 \\ 0 & W_1 \end{pmatrix} \frac{\mu^3}{2}[\int_0^\infty e^{-\mu T} X(K, T)dT$$

$$+\int_0^\infty e^{-\mu T} e^{\Sigma(K)^* T}\begin{pmatrix} Z(K) & 0 \\ 0 & 0 \end{pmatrix} e^{\Sigma(K)T} dt] \quad . \tag{4.30}$$

We will treat the terms separately. From (4. 23) we see that if we set

$$\hat{X}(K,\mu) = \frac{\mu^3}{2}\int_0^\infty e^{-\mu T} X(K, T)\, dT$$

we have

$$\begin{aligned}
\hat{X}(K,\mu) &= \frac{\mu^3}{2}\int_0^\infty\int_0^T e^{\Sigma(K)^* t}\,\Phi(K) e^{\Sigma(K)t} dt\, e^{-\mu T} dT \\
&= \frac{\mu^3}{2}\int_0^\infty e^{\Sigma(K)^* t}\,\Phi(K) e^{\Sigma(K)t}\int_t^\infty e^{-\mu T} dTdt \\
&= \frac{\mu^2}{2}\int_0^\infty e^{\Sigma(K)^* t}\,\Phi(K) e^{\Sigma(K)t} e^{-\mu t} dt \\
&= \frac{\mu^2}{2}\int_0^\infty e^{(\Sigma(K)-\frac{\mu}{2}I)^* t}\,\Phi(K)\, e^{(\Sigma(K)-\frac{\mu}{2}I)t} dt = \frac{\mu^2}{2} X(K,\mu)
\end{aligned}$$

where, since $\Sigma(K) - \frac{\mu}{2}I$ is a stability matrix,

$$(\Sigma(K) - \frac{\mu}{2}I)^* X(K,\mu) + X(K,\mu)(\Sigma(K) - \frac{\mu}{2}I) + \Phi(K) = 0 \quad . \tag{4.31}$$

Thus in (4. 30) the first term may be replaced by

$$\frac{\mu^2}{2}\,\mathrm{Tr}\begin{pmatrix} 0 & 0 \\ 0 & W_1 \end{pmatrix} X(K,\mu)$$

where $X(K, \mu)$ is the unique solution of (4. 31).

Next we focus attention on the term

$$\frac{\mu^3}{2}\int_0^\infty e^{-\mu T} e^{\Sigma(K)^* T}\begin{pmatrix} Z(K) & 0 \\ 0 & 0 \end{pmatrix} e^{\Sigma(K) T}\, dt \equiv \frac{\mu^2}{2} Y(K, \mu)$$

and we readily see that

$$(\Sigma(K) - \frac{\mu}{2}I)^* \hat{Y}(K,\mu) + \hat{Y}(K,\mu)(\Sigma(K) - \frac{\mu}{2}I) + \begin{pmatrix} \mu Z(K) & 0 \\ 0 & 0 \end{pmatrix} = 0 . \tag{4.32}$$

Let $Y(K,\mu)$ be partitioned in the form

$$Y(K,\mu) = \begin{pmatrix} \hat{Y}_1(K,\mu) & \hat{Y}_2(K,\mu) \\ \hat{Y}_2(K,\mu)^* & \hat{Y}_3(K,\mu) \end{pmatrix} .$$

We find, using (4. 22), (4. 32), that

$$(S(K) - \frac{\mu}{2}I)^* \hat{Y}_1(K,\mu) + \hat{Y}_1(K,\mu)(S(K) - \frac{\mu}{2}I) + \mu Z(K) = 0 \tag{4.33}$$

$$(S(K) - \frac{\mu}{2}I)^* \hat{Y}_2(K,\mu) + \hat{Y}_2(K,\mu)(A_1 - \frac{\mu}{2}I) + \hat{Y}_1(K,\mu)D_1A_1 = 0 \tag{4.34}$$

$$(A_1 - \frac{\mu}{2}I)^* \hat{Y}_3(K,\mu) + \hat{Y}_3(K,\mu)(A_1 - \frac{\mu}{2}I) + A_1^* D_1^* Y_2(K,\mu) + \hat{Y}_2(K,\mu)^* D_1A_1 = 0 \tag{4.35}$$

Partitioning $X(K,\mu)$ similarly and using (4. 22), (4. 31):

$$(S(K) - \frac{\mu}{2}I)^* X_1(K,\mu) + X_1(K,\mu)(S(K) - \frac{\mu}{2}I) + H(K) = 0 \tag{4.36}$$

$$(S(K) - \frac{\mu}{2}I)^* X_2(K,\mu) + X_2(K,\mu)(A_1 - \frac{\mu}{2}I) + X_1(K,\mu)D_1A_1 = 0 \tag{4.37}$$

$$(A_1 - \frac{\mu}{2}I)^* X_3(K,\mu) + X_2(K,\mu)(A_1 - \frac{\mu}{2}I) + A_1^* D_1^* X_2(K,\mu) + X_2(K,\mu)^* D_1A_1 = 0 . \tag{4.38}$$

Adding (4. 33) to (4. 36) and then subtracting (4. 25) we find that

$$(S(K) - \frac{\mu}{2}I)^* (X_1(K,\mu) + \hat{Y}_1(K,\mu) - Z(K)) + (X_1(K,\mu) + \hat{Y}_1(K,\mu) - Z(K))(S(K) - \frac{\mu}{2}I) = 0 ,$$

from which we conclude that

$$X_1(K,\mu) + \hat{Y}_1(K,\mu) = Z(K) . \tag{4.39}$$

Then renaming

$$X_2(K,\mu) + \hat{Y}_2(K,\mu) \equiv Y(K,\mu) \tag{4.40}$$

$$X_3(K,\mu) + Y_3(K,\mu) \equiv W(K,\mu) \tag{4.41}$$

the expected cost (4. 30) becomes

$$\frac{\mu^2}{2} \operatorname{Tr} W_1 W(K,\mu)$$

where, from (4. 21), (4. 26), (4. 25), (4. 40), (4. 41), (4.34), (4. 35), (4. 37), (4. 38),

$$(A + BK)^* Z(K) + Z(K)(A + BK) + (F + GK)^* \tilde{W}(F + GK) + K^* \tilde{U} K = 0 \tag{4.42}$$

$$(A + BK - \frac{\mu}{2} I)^* Y(K,\mu) + Y(K,\mu)(A_1 - \frac{\mu}{2} I) + Z(K) D_1 A_1 = 0 \tag{4.43}$$

$$(A_1 - \frac{\mu}{2} I)^* W(K,\mu) + W(K,\mu)(A_1 - \frac{\mu}{2} I) + A_1^* D_1^* Y(K,\mu) + Y(K,\mu)^* D_1 A_1 = 0 . \tag{4.44}$$

From this point of view the optimal tracking problem becomes

$$\min_{K \in \mathscr{S}(A,B)} \{\operatorname{Tr} W_1 W(K,\mu)\} \tag{4.45}$$

subject to the constraints (4. 42), (4. 43), (4. 44).

The necessary conditions for a feedback matrix $\hat{K}$ to solve (4. 45) may be developed as follows. We define

$$W_\epsilon(\hat{K},\mu) = \frac{d}{d\epsilon} W(\hat{K} + \epsilon \delta K) \Big|_{\epsilon = 0}$$

and $Y_\epsilon(\hat{K},\mu)$, $Z_\epsilon(\hat{K})$ similarly. Clearly we reqire

$$\operatorname{Tr} W_1 W_\epsilon(\hat{K},\mu) = 0 \tag{4.46}$$

for all choices of δK. Differentiating (4. 44) with $K = \hat{K} + \epsilon \delta K$ and using the by now familiar adjoint relations, (4.46) becomes

$$\mathrm{Tr}\ P(\mu)[A_1^* D_1^* Y_\epsilon(\hat{K},\mu) + Y_\epsilon(\hat{K},\mu)^* D_1 A_1] = 0, \tag{4.47}$$

where

$$(A_1 - \frac{\mu}{2} I)P(\mu) + P(\mu)(A_1 - \frac{\mu}{2} I)^* + W_1 = 0 . \tag{4.48}$$

Now (4.47) can be rewritten as

$$2\ \mathrm{Tr}\ P(\mu) A_1^* D_1^* Y_\epsilon(\hat{K},\mu) = 0 \tag{4.49}$$

and, differentiating (4.43) and using the adjoint relation again, this becomes (dividing (4.49) by 2)

$$\mathrm{Tr}\ R(\hat{K},\mu)[\delta K^* B^* Y(\hat{K},\mu) + Z_\epsilon(\hat{K}) D_1 A_1] = 0 \tag{4.50}$$

where

$$(A_1 - \frac{\mu}{2} I)R(\hat{K},\mu) + R(\hat{K},\mu)(A + B\hat{K} - \frac{\mu}{2} I)^* + P(\mu) A_1^* D_1^* = 0 .$$

The term $\mathrm{Tr}\ R(\hat{K},\mu) Z_\epsilon(K) D_1 A_1$ in (4.50) can be further treated, differentiating (4.42) and again using the adjoint relations, to show that

$$\begin{aligned} \mathrm{Tr}\ R(\hat{K},\mu) Z_\epsilon(\hat{K}) D_1 A_1 &= \mathrm{Tr}\ D_1 A_1 R(\hat{K},\mu) Z_\epsilon(\hat{K}) \\ &= 2\ \mathrm{Tr}\ Q(\hat{K},\mu)[\delta K^* B^* Z(\hat{K}) + \delta K^* G^* \tilde{W} F + \delta K^* U \hat{K}] , \end{aligned} \tag{4.51}$$

where

$$(A + BK)Q(\hat{K},\mu) + Q(\hat{K},\mu)(A + BK)^* + D_1 A_1 R(\hat{K},\mu) = 0 , \tag{4.52}$$

and, as usual

$$U = \tilde{U} + G^* \tilde{W} G .$$

Substituting (4.51) into (4.50) we have

$$\mathrm{Tr}\ \delta K^* [B^* Y(\hat{K},\mu) R(\hat{K},\mu) + 2B^* Z(\hat{K}) Q(\hat{K},\mu) + 2G^* \tilde{W} F Q(\hat{K},\mu) + 2U\hat{K} Q(\hat{K},\mu)] = 0 \tag{4.53}$$

and the necessary condition for optimality becomes

$$M(K,\mu) \equiv B^*Y(\hat{K},\mu)R(\hat{K},\mu) + 2B^*Z(\hat{K})Q(\hat{K},\mu) + 2G^*\tilde{W}FQ(\hat{K},\mu) + 2UKQ(\hat{K},\mu) = 0 . \tag{4.54}$$

If $Q(\hat{K},\mu)$ is nonsingular, which will often be true but is hard to guarantee in advance because of the nonsymmetric form of the inhomogeneous term $D_1A_1R(\hat{K},\mu)$ in (4. 52), (4. 54) can be solved for $\hat{K}$:

$$\hat{K} = -U^{-1}[B^*Z(\hat{K}) + G^*\tilde{W}F + B^*Y(\hat{K},\mu)R(\hat{K},\mu)Q(\hat{K},\mu)^{-1}]$$

To approximate $\hat{K}$ one can attempt a variant of Axsäter's method, employed as in Section 3, or one can use a steepest descent method whereby

$$K_{i+1} = K_i + \epsilon\delta K_i$$

with (cf. (4. 54))

$$\delta K_i = -M(K_i,\mu) .$$

The latter can be used even if $Q(K_i,\mu)$ should turn out to be singular at some stage. A problem of this type is suggested to the reader as Exercise 4 at the end of this chapter.

The distribution $p_\mu(T)$ is special to problems for which (cf. following (4. 22)), the covariance matrix for the initial state has the form T^{-2} times a fixed matrix. If this matrix involves different powers of T the situation becomes much more complicated and other distributions for T must be used if T is to be eliminated by a comparable averaging process. Some variations on the analysis carried out above can be developed.

A somewhat modified version of the tracking problem has much in common with model following. We consider again the plant

$$\dot{x} = Ax + Bu + Cv + Dw \tag{4.55}$$

and a model plant (with A_0 a stability matrix)

$$\dot{\xi} = A_0\xi + C_0v + D_0w . \tag{4.56}$$

It will be supposed again that

$$w = H_1 w_1$$

where, as in (4.13)

$$\dot{w}_1 = A_1 w_1$$

on subintervals as described there. It will be supposed that (4.56) responds to w in an "ideal" fashion insofar as $F_0(\xi(t) + A_0^{-1}D_0 w(t))$ is concerned. Since A_0 is already a stability matrix, the composite system (4.55), (4.56) is stabilizable via feedback

$$u = Kx, \tag{4.57}$$

or, if electronic (or other) realization of (4.56) is contemplated, one might use

$$u = Kx + K_0 \xi . \tag{4.58}$$

The tracking criterion might now be expressed in terms of the difference

$$(F + GK)(x(t) + A^{-1}Dw) - F_0(\xi(t) + A_0^{-1}D_0 w(t)) , \tag{4.59}$$

(or, if (4.58) is used, perhaps

$$(F+GK)(x(t) + A^{-1}Dw) + (GK_0 - F_0)(\xi(t) + A_0^{-1} D_0 w(t)).$$

This is simply an enlarged version of the problem we have already studied if we replace the system (4.1) by the composite system (4.55), (4.56), the equilibrium (for constant inputs) $-A^{-1}Dw$ by $-\begin{pmatrix} A^{-1}Dw \\ A_0^{-1} D_0 w \end{pmatrix}$, the vector $Z(t) = x(t) + A^{-1}Dw$ by $\zeta(t) = \begin{pmatrix} x(t) + A^{-1}Dw \\ \xi(t) + A_0^{-1} Dw \end{pmatrix}$, the feedback $u(t) = K(x(t) + A^{-1}Dw(t))$ by $K(x(t) + A^{-1}Dw(t)) + K_0(\xi(t) + A_0^{-1}D_0 w(t))$ and the cost (4.9) by

$$\int_0^\infty \zeta(t)^* \left[\begin{pmatrix} (F+GK)^* \\ (GK_0 - F_0)^* \end{pmatrix} \tilde{W}(F+GK, GK_0 - F_0)\zeta(t) + \begin{pmatrix} K^* \\ K_0^* \end{pmatrix} \tilde{U}(K, K_0) \right] \zeta(t)dt$$

($K_0 = 0$ if (4.57) is used instead of (4.58)).

The analysis proceeds just as before in this enlarged setting. This type of approach can also be used without any reference to $A^{-1}Dw$ or $A_0^{-1}D_0w$, replacing, e.g., (4. 59) by

$$(F + GK)x(t) - F_0\xi(t)$$

and making comparable alterations elsewhere.

Ultimately a tracking problem is much like a model following problem, the system

$$\dot{w}_1 = A_1w_1$$

serving in place of (3. 3) and the quantity $-A^{-1}Dw(t) = -A^{-1}DH_1w_1(t)$ in place of (3. 4). But we do not in general assume that A_1 is a stability matrix, we may (as in the ramp input case) have $\dot{w}_1 = A_1w_1$ only on certain intervals with discontinuities in $\dot{w}_1$ where these intervals meet, and, finally, the command input $w(t) = H_1w_1(t)$ is, by its very nature, a real and known quantity whereas the model state, $\xi(t)$, in Section 3 is not.

The tracking problem which we have just finished considering is in the class of problems for which (4. 11) holds, i. e.

$$\int_0^\infty \|\dot{w}(t)\|^2\,dt < \infty$$

and in that case the cost

$$\int_0^\infty z(t)^*[(F + GK)^*\tilde{W}(F + GK)]\,z(t)dt < \infty \tag{4.60}$$

as long as $A + BK$ is a stability matrix. For such problems it is reasonable to seek for an optimal feedback control in the class of feedback controls of the form

$$u(t) = Kz(t) = K(x(t) - x_{w(t)}) \tag{4.61}$$

since we know that any stabilizing control gives finite cost.

But now let us consider a rather different sort of problem. We will start with an example. Let us suppose that an antenna is set up on an equatorial mount to receive signals from a satellite in equatorial orbit. The satellite appears above the horizon at $t = 0$ and for $0 \leq t \leq T$ makes an angle

$$\psi(t) = \psi_0 + \psi_1 t \tag{4.62}$$

with the local meridian. Here T is the time when it disappears below the horizon again. Assuming a high orbit, T will be very large by comparison with other constants in the system, so we replace it by ∞. The antenna angle, $\theta(t)$, will be assumed to satisfy

$$I\ddot{\theta} + \gamma\dot{\theta} = \tau(t)$$

where I is the moment of inertia about the mount, γ is the coefficient of friction for the particular mode of mounting, and $\tau(t)$ is the torque supplied by an electric motor to turn the antenna. We may suppose this torque proportional to an applied voltage and, rescaling if necessary, we do not have to distinguish between the two. Letting $x^1 = \theta$, $x^2 = \dot{\theta}$ we have, with $\alpha = \frac{\gamma}{I}$ and $\hat{u}(t) = \tau(t)/I$,

$$\dot{\begin{pmatrix} x^1 \\ x^2 \end{pmatrix}} = \begin{pmatrix} 0 & 1 \\ 0 & -\alpha \end{pmatrix} \begin{pmatrix} x^1 \\ x^2 \end{pmatrix} + \begin{pmatrix} 0 \\ 1 \end{pmatrix} \hat{u}(t) . \tag{4.63}$$

Our analysis proceeds in several stages. We let

$$\hat{u}(t) = u_0(t) + u(t) + w(t) .$$

We set

$$u_0(t) = -\beta x^1, \quad \beta > 0 ,$$

to achieve a system with nonsingular "A" matrix:

$$\dot{\begin{pmatrix} x^1 \\ x^2 \end{pmatrix}} = \begin{pmatrix} 0 & 1 \\ -\beta & -\alpha \end{pmatrix} \begin{pmatrix} x^1 \\ x^2 \end{pmatrix} + \begin{pmatrix} 0 \\ 1 \end{pmatrix} u + \begin{pmatrix} 0 \\ 1 \end{pmatrix} w .$$

When $u = 0$ and $w = w_0$ is constant, the equilibrium point x_{w_0} is easily seen to be the point $x^1_{w_0} = w_0/\beta$, $x^2_{w_0} = 0$.

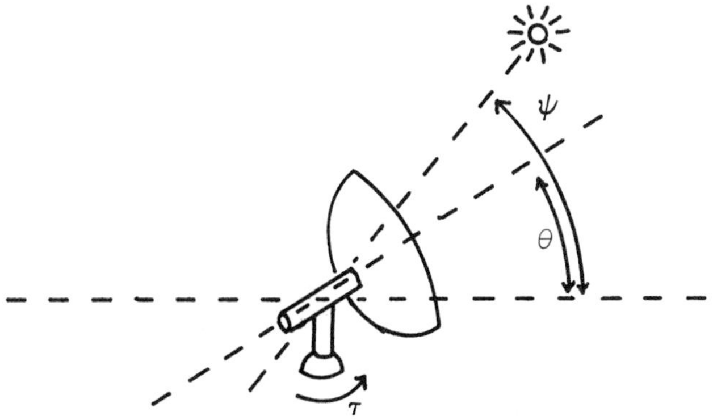

Fig. VII-4. Antenna

If, for a hypothetically fixed satellite angle ψ, we want this equilibrium point to be $x^1_{w_0} = \psi$, $x^2_{w_0} = 0$, we must set

$$w_0 = \beta\psi .$$

If we extend this directly to a varying satellite angle $\psi(t)$ we obtain a time varying command input

$$w(t) = \beta\psi(t)$$

and our system becomes

$$\begin{pmatrix} \dot{x}^1 \\ x^2 \end{pmatrix} = \begin{pmatrix} 0 & 1 \\ -\beta & -\alpha \end{pmatrix}\begin{pmatrix} x^1 \\ x^2 \end{pmatrix} + \begin{pmatrix} 0 \\ 1 \end{pmatrix} u + \begin{pmatrix} 0 \\ \beta \end{pmatrix} \psi . \qquad (4.64)$$

Thus, referring to (4.1), we have

$$A = \begin{pmatrix} 0 & 1 \\ -\beta & -\alpha \end{pmatrix}, \quad B = \begin{pmatrix} 0 \\ 1 \end{pmatrix}, \quad D = \begin{pmatrix} 0 \\ \beta \end{pmatrix} .$$

Let us set (cf. (4.5))

$$u(t) = k_1(x^1(t) - \psi(t)) + k_2 x^2(t) \qquad (4.65)$$

with k_1, k_2 arbitrary, except that

$$A + BK = \begin{pmatrix} 0 & 1 \\ -\beta + k_1 & -\alpha + k_2 \end{pmatrix} \qquad (4.66)$$

is required to be a stability matrix. Then setting (cf. (4.9))

$$F = (1,0), \qquad G = 0$$

and

$$\begin{pmatrix} z^1 \\ z^2 \end{pmatrix} = \begin{pmatrix} x^1 \\ x^2 \end{pmatrix} - \begin{pmatrix} \psi \\ 0 \end{pmatrix},$$

the quantity which we are interested in is

$$Fz = (1,0)\left(\begin{pmatrix} x^1 \\ x^2 \end{pmatrix} - \begin{pmatrix} \psi \\ 0 \end{pmatrix}\right) = x^1 - \psi = \theta - \psi,$$

the difference between the satellite angle and the antenna angle. With u given by (4.65) we can obtain from (4.64) the system of equations satisfied by z (cf. (4.7))

$$\begin{pmatrix} \dot{z}^1 \\ z^2 \end{pmatrix} = \begin{pmatrix} 0 & 1 \\ -\beta+k_1 & -\alpha+k_2 \end{pmatrix}\begin{pmatrix} z^1 \\ z^2 \end{pmatrix} - \begin{pmatrix} \dot{\psi} \\ 0 \end{pmatrix} = \text{(cf. (4.62))}$$

$$= \begin{pmatrix} 0 & 1 \\ -\beta+k_1 & -\alpha+k_2 \end{pmatrix}\begin{pmatrix} z^1 \\ z^2 \end{pmatrix} - \begin{pmatrix} \psi_1 \\ 0 \end{pmatrix}. \tag{4.67}$$

The presence of a constant vector as the homogeneous term on the right hand side of (4.67) suggests that we look for a particular solution of (4.67) in the form of a constant vector. That solution, of course, is

$$\begin{pmatrix} 0 & 1 \\ -\beta+k_1 & -\alpha+k_2 \end{pmatrix}^{-1}\begin{pmatrix} \psi_1 \\ 0 \end{pmatrix} = \frac{1}{\beta-k_1}\begin{pmatrix} -\alpha+k_2 & -1 \\ \beta-k_1 & 0 \end{pmatrix}\begin{pmatrix} \psi_1 \\ 0 \end{pmatrix} = \begin{pmatrix} \frac{-\alpha+k_2}{\beta-k_1}\psi_1 \\ \psi_1 \end{pmatrix}.$$

Since we are requiring that $\begin{pmatrix} 0 & 1 \\ -\beta+k_1 & -\alpha+k_2 \end{pmatrix}$ should be a stability matrix, all solutions of (4.67) tend to this constant solution as $t \to \infty$. We conclude therefore that

$$\lim_{t\to\infty} (\theta(t)-\psi(t)) = \lim_{t\to\infty} F\left(\begin{pmatrix} z^1(t) \\ z^2(t) \end{pmatrix} - \begin{pmatrix} \psi(t) \\ 0 \end{pmatrix}\right) = \frac{-\alpha+k_2}{\beta-k_1}\psi_1$$

and this cannot be zero, since $-\alpha+k_2$ must be negative if the system matrix in (4.67) is to be a stability matrix. It can also be seen that $\beta-k_1$ must be positive. So we conclude that

$$\lim_{t\to\infty} (\theta(t)-\psi(t)) = \frac{-\alpha+k_2}{\beta-k_1}\psi_1 < 0$$

for <u>any</u> k_1, k_2 satisfying the stability requirements. Thus the antenna angle will <u>always</u> lag the satellite angle for large t by an amount which does not tend to zero as $t\to\infty$. Consequently

$$\int_0^\infty (\theta(t)-\psi(t))^2 dt = \infty$$

in all cases - no finite response cost can be obtained with a control of the form (4.65). We see, then, that in cases where (4.11) is not true it is not generally adequate to use a control of the form (4.5) (for (4.65) has this form in our example) in order to achieve finite cost.

In order to find out what is necessary, we return to the theoretical development and consider the plant

$$\dot{x} = Ax + Bu + Dw \qquad (4.68)$$

and suppose that the response of interest takes the form (we suppose G (cf. (4.9)) $= 0$ henceforth)

$$y = F(x + A^{-1}Dw) \qquad (4.69)$$

where the command input

$$w = H_1 w_1 \qquad (4.70)$$

is itself a response associated with the system

$$\dot{w}_1 = A_1 w_1 . \qquad (4.71)$$

The composite plant is, with $D_1 = DH_1$,

$$\begin{pmatrix} \dot{x} \\ \dot{w}_1 \end{pmatrix} = \begin{pmatrix} A & D_1 \\ 0 & A_1 \end{pmatrix} \begin{pmatrix} x \\ w_1 \end{pmatrix} + \begin{pmatrix} B \\ 0 \end{pmatrix} u \tag{4.72}$$

and (4.69) is just

$$y = F(x + A^{-1}D_1w_1), \qquad D_1 = DH_1 .$$

An arbitrary feedback control law

$$u = K_1x + K_2w_1 \tag{4.73}$$

can clearly be put in the form

$$u = K_1(x + A^{-1}D_1w_1) + K_2w_1 . \tag{4.74}$$

Introducing

$$z = x + A^{-1}D_1w_1 \tag{4.75}$$

and assuming u takes the form (4.74) we find that

$$\dot{z} = (A + BK_1)z + (BK_2 + A^{-1}D_1)\dot{w}_1 = (A + BK_1)z + (BK_2 + A^{-1}D_1)A_1w_1$$

and the closed loop, composite system for z and w_1 is seen to be

$$\begin{pmatrix} \dot{z} \\ \dot{w}_1 \end{pmatrix} = \begin{pmatrix} A + BK_1 & A_2 + BK_2 \\ 0 & A_1 \end{pmatrix} \begin{pmatrix} z \\ w_1 \end{pmatrix} \tag{4.76}$$

with

$$A_2 = A^{-1}D_1A_1 . \tag{4.77}$$

The response of interest (cf. (4.69)) is now

$$y = Fz . \tag{4.78}$$

Now let us assume the following: that the eigenvalues of A_1 all lie to the right of the eigenvalues of $A + BK_1$ in the complex plane. (In our satellite tracking problem

$$A_1 = \begin{pmatrix} 0 & 1 \\ 0 & 0 \end{pmatrix}$$

so this assumption is automatically satisfied if $A + BK_1$ is a stability matrix.) Following a procedure similar to that used in III-4, 5 where the reduced order observer and the transfer matrix were discussed, we let

$$z = \zeta + Tw_1 \tag{4.79}$$

and compute that

$$\dot{\zeta} = \dot{z} - T\dot{w}_1 = (A+BK_1)z + (A_2 + BK_2)w_1 - TA_1w_1$$

$$= (A+BK_1)\zeta + [(A+BK_1)T - TA_1 + A_2 + BK_2]\, w_1 \,. \tag{4.80}$$

From our assumption on the eigenvalues of A_1 we can solve

$$(A + BK_1)T - TA_1 + A_2 + BK_2 = 0 \tag{4.81}$$

for the transfer matrix T. Then, with this choice of T,

$$\dot{\zeta} = (A + BK_1)\zeta .$$

Since the eigenvalues of $A + BK_1$ lie to the left of those of A_1, the term Tw_1 will tend to be dominant in (4. 79) as $t \to \infty$. Now the response (4. 78) becomes

$$y = F\zeta + FTw_1 \quad (= Fz) \tag{4.82}$$

and it is reasonable to say that Fx tracks $-FA^{-1}D_1w_1$ if the tracking error, Fz, tends to zero more rapidly than the norm of any solution of (4. 71). Since we have (4. 82), this requires that

$$FT = 0 \,. \tag{4.83}$$

When A_1 is not a stability matrix, (4. 81), (4. 83) is necessary if we are to have $\int_0^\infty z(t)^* F^* \tilde{W} Fz(t)dt < \infty$. This prompts us to formally state the

<u>Consistency Condition for Tracking.</u> We have

$$\| F(x(t) + A^{-1}Dw_1 \| = \mathcal{O}(\| \hat{w}_1(t) \|), \qquad t \to \infty, \tag{4.84}$$

<u>for every non-zero solution</u> $\hat{w}_1$ <u>of</u> (4. 71) <u>if</u> K_1 <u>and</u> K_2 <u>are such that the solution</u> T <u>of</u> (4. 81) <u>obeys</u> (4. 83).

In view of our discussion of transfer matrices in relationship to ordinary transfer functions in III-5 this condition is analogous to the consistency condition more frequently used in the literature ([4], VI-[17]) involving the standard transfer function associated with the Laplace transform.

Let us apply this consistency condition to our satellite tracking problem. We pick the problem up at the point (4. 64), i. e.,

$$\begin{pmatrix} \dot{x}^1 \\ x^2 \end{pmatrix} = \begin{pmatrix} 0 & 1 \\ -\beta & -\alpha \end{pmatrix} \begin{pmatrix} x^1 \\ z^2 \end{pmatrix} + \begin{pmatrix} 0 \\ 1 \end{pmatrix} u + \begin{pmatrix} 0 \\ \beta \end{pmatrix} \psi .$$

Letting $\psi = w^1$, $\dot{\psi} = w^2$, $w_1 = \begin{pmatrix} w^1 \\ w^2 \end{pmatrix}$ satisfies

$$\begin{pmatrix} \dot{w}^1 \\ w^2 \end{pmatrix} = \begin{pmatrix} 0 & 1 \\ 0 & 0 \end{pmatrix} \begin{pmatrix} w^1 \\ w^2 \end{pmatrix}, \text{ i. e., } A_1 = \begin{pmatrix} 0 & 1 \\ 0 & 0 \end{pmatrix} .$$

We let (cf. (4. 74))

$$u(t) = (k_1, k_2) \begin{pmatrix} x^1 - w^1 \\ x^2 \end{pmatrix} + (k_3, k_4) \begin{pmatrix} w^1 \\ w^2 \end{pmatrix} . \tag{4.85}$$

Then, letting

$$\begin{pmatrix} z^1 \\ z^2 \end{pmatrix} = \begin{pmatrix} x^1 - w^1 \\ x^2 \end{pmatrix}$$

we obtain (cf. (4. 75))

$$\begin{pmatrix} \dot{z}^1 \\ z^2 \end{pmatrix} = \begin{pmatrix} 0 & 1 \\ -\beta + k_1 & -\alpha + k_2 \end{pmatrix} \begin{pmatrix} z^1 \\ z^2 \end{pmatrix} + \begin{pmatrix} 0 & -1 \\ k_3 & k_4 \end{pmatrix} \begin{pmatrix} w^1 \\ w^2 \end{pmatrix} . \tag{4.86}$$

In this case, as may be verified directly from the definition, (4. 77),

$$A_2 = \begin{pmatrix} 0 & -1 \\ 0 & 0 \end{pmatrix}, \quad BK_2 = \begin{pmatrix} 0 & 0 \\ k_3 & k_4 \end{pmatrix} .$$

With $A_1 = \begin{pmatrix} 0 & 1 \\ 0 & 0 \end{pmatrix}$, $T = \begin{pmatrix} T_1^1 & T_2^1 \\ T_1^2 & T_2^2 \end{pmatrix}$, (4.81) is

$$\begin{pmatrix} 0 & 1 \\ -\beta+k_1 & -\alpha+k_2 \end{pmatrix}\begin{pmatrix} T_1^1 & T_2^1 \\ T_1^2 & T_2^2 \end{pmatrix} - \begin{pmatrix} T_1^1 & T_2^1 \\ T_1^2 & T_2^2 \end{pmatrix}\begin{pmatrix} 0 & 1 \\ 0 & 0 \end{pmatrix} + \begin{pmatrix} 0 & -1 \\ k_3 & k_4 \end{pmatrix} = 0$$

which may be solved to give

$$T = \begin{pmatrix} \dfrac{k_3}{\beta - k_1} & \dfrac{(-\alpha+k_2)\left(\dfrac{k_3}{\beta-k_1} + 1\right) + k_4}{\beta - k_1} \\ 0 & \dfrac{k_3}{\beta-k_1} + 1 \end{pmatrix}$$

so that, with $F = (1, 0)$, the requirement (4.83) becomes

$$FT = \begin{pmatrix} \dfrac{k_3}{\beta-k_1}, & \dfrac{(-\alpha+k_2)\left(\dfrac{k_3}{\beta-k_1} + 1\right) + k_4}{\beta - k_1} \end{pmatrix} = 0,$$

giving

$$k_3 = 0, \quad -\alpha + k_2 + k_4 = 0 \quad (k_4 = \alpha - k_2). \tag{4.87}$$

When these conditions are satisfied the antenna will, indeed, track the satellite provided k_1 and k_2, which may still be independently chosen, are such that the system matrix in (4.86) is a stability matrix. Clearly k_1 and k_2 may now be adjusted to vary the rate at which the antenna "homes in" on the satellite. A clearer indication of the manner in which the conditions (4.87) work is obtained by returning to (4.64), (4.85), the latter becoming

$$u(t) = (k_1, k_2)\begin{pmatrix} x^1 - w^1 \\ x^2 \end{pmatrix} + (0, \alpha - k_2)\begin{pmatrix} w^1 \\ w^2 \end{pmatrix}. \tag{4.88}$$

Letting

$$\begin{pmatrix} \hat{z}^1 \\ \hat{z}^2 \end{pmatrix} = \begin{pmatrix} x^1 - w^1 \\ x^2 - w^2 \end{pmatrix},$$

(4. 64) gives (using $\dot{w}^1 = w^2$, $\dot{w}^2 = 0$)

$$\begin{pmatrix} \hat{z}^1 \\ \hat{z}^2 \end{pmatrix}^{\bullet} = \begin{pmatrix} 0 & 1 \\ -\beta + k_1 & -\alpha + k_2 \end{pmatrix} \begin{pmatrix} \hat{z}^1 \\ \hat{z}^2 \end{pmatrix} \tag{4.89}$$

and we conclude that if the matrix in (4. 86), (4. 89) is a stability matrix, $\theta - \psi = x^1 - w^1 = \hat{z}^1$ tends to zero exponentially as $t \to \infty$ and

$$\int_0^\infty (\theta(t) - \psi(t))^2 \, dt > \infty.$$

In a simple application such as this one we might have guessed at the start that the variables $\hat{z}^1, \hat{z}^2$ were more pertinent than z^1, z^2 used in (4.86), but in more complicated applications comparable divination is often out of the question. Even then it should be noted that in (4. 88) $u(t)$ does not have the form $u(t) = \hat{k}_1 \hat{z}^1 + \hat{k}_2 \hat{z}^2$. In the final analysis the consistency condition is a non-trivial requirement which cannot be ignored and frequently cannot be guessed at in some intuitive fashion.

We will discuss now a general scheme for solving tracking problems of this sort in such a way as to meet an optimality criterion. In doing so, we keep in mind the following.

(i) The conditions (4. 81), (4. 83) must be satisfied if our system is to track in the sense (4. 84). Indeed, if A_1 is not a stability matrix, (4. 81), (4. 83) must be true, in general, if we are to have

$$\int_0^\infty z(t)^* F^* \tilde{W} F z(t) dt < \infty . \tag{4.90}$$

(ii) Even when (i) is satisfied, the control cost

$$\int_0^\infty u(t)^* \tilde{U} u(t) dt$$

will ordinarily not be finite if A_1 is not a stability matrix.

That (ii) is, indeed, the case can be seen from (4. 73), which, with z given by (4. 75) and ζ given by (4. 79) may be written

$$u = K_1 z + K_2 w_1 = K_1 \zeta + (K_1 T + K_2) w_1 . \tag{4.91}$$

Since $\|w_1(t)\|$ does not in general tend to zero as $t \to \infty$ when A_1 is not a stability matrix, we cannot expect $\|u(t)\|$ to tend to zero. Indeed, in the antenna problem a non-zero torque is constantly required to overcome the frictional forces.

A further point which we should make is this. In principal (4.81), (4.83) can be used analytically to eliminate certain feedback coefficients, as in our very simple example, but in practice this is out of the question since T will very frequently be a complicated rational function of the entries of the matrices K_1 and K_2.

Having thus issued our "caveat lector" we proceed to propose a method for optimal (in a sense to be made precise) solution of this tracking problem.

With reference to the system (4.76) and the variables z and w_1, let us assume

$$\mathrm{cov}(z(0), z(0)) = Z, \qquad \mathrm{cov}(w_1(0), w_1(0)) = W_1 .$$

We thus establish statistics for the initial tracking error and for the initial state of the "tracked" system (4.71). Going over to the variables ζ, w_1, ζ defined by (4.79), and assuming

$$\mathrm{cov}\ (z(0), w_1(0)) = 0$$

we have

$$\mathrm{cov}(\zeta(0), \zeta(0)) = Z + TW_1T^* .$$

Assuming that T is the solution of (4.81), (4.80) becomes

$$\dot{\zeta} = (A + BK_1)\zeta \tag{4.92}$$

and our system consists of (4.92) and (4.71). The tracking cost integrand in (4.90) is now

$$z^*F^*\tilde{W}Fz = \zeta^*F^*\tilde{W}F\zeta + 2\zeta^*F^*\tilde{W}FTw_1 + w_1^*T^*F^*\tilde{W}FTw_1 .$$

We are looking for T such that (4.81) and (4.83) are both true. Consequently, in measuring performance it will be enough to use the cost integrand

$$\zeta^* F^* \tilde{W} F \zeta \; .$$

Thus that part of the total cost associated with the tracking error (4. 90) and the partial control term $K_1 \zeta$ in the formula (4. 91) for u may conveniently be formulated as

$$\mathrm{Tr}[\,(Z + TW_1 T^*)X\,] \tag{4.93}$$

where

$$(A + BK_1)^* X + X(A + BK_1) + F^* \tilde{W} F + K_1^* \tilde{U} K_1 = 0 \; . \tag{4.94}$$

(We could also adjoin a disturbance term Cv to (4. 92) and use $Z + TW_1 T^* + \frac{1}{\mu} CVC^*$ in (4. 93), as in earlier applications. Additionally one could modify the inhomogeneous term in (4. 94) to $(F + GK_1)^* \tilde{W}(F + GK_1) + K_1^* \tilde{U} K_1$ but it is unlikely that this would be pertinent since $K_1 \zeta \neq u$ in general.)

The remaining part of the control is (cf. (4. 91))

$$(K_1 T + K_2)\, w_1$$

and this does not in general lie in $L_m^2[\,0, \infty)$ since we do not require A_1 to be a stability matrix. This part of the control may be compared to the constant effort required in the satellite tracking problem to overcome friction. This type of situation is commonly treated by "discounting" (see, e. g.[1]). For $\nu > 0$ we form the cost integral

$$\int_0^\infty e^{-2\nu t} w_1(t)^* (K_1 T + K_2)^* \tilde{U} (K_1 T + K_2) w_1(t) dt \tag{4.95}$$

with w_1 satisfying (4. 71). Observing that $e^{-\nu t} w_1$ satisfies

$$\frac{d}{dt}(e^{-\nu t} w_1) = (A_1 - \nu I)(e^{-\nu t} w_1) \; ,$$

the expected value of the cost (4. 95) is

$$\mathrm{Tr}\; W_1 Y \; ,$$

where, assuming ν chosen large enough so that $A_1 - \nu I$ is a stability matrix,

$$(A_1 - \nu I)^* Y + Y(A_1 - \nu I) + (K_1 T + K_2)^* \tilde{U}(K_1 T + K_2) = 0 \; . \tag{4.96}$$

We pose our optimal tracking problem as follows:

$$\min_{\substack{K_1, K_2 \\ K_1 \in \mathscr{S}(A, B)}} \{\mathrm{Tr}\,(Z + TW_1T^*)X + \mathrm{Tr}\, W_1Y\} \tag{4.97}$$

subject to the constraints (4. 81), (4. 83), (4. 94), (4. 96).

We remark that the emphasis placed on the two terms in (4. 97) may be varied by different choices of ν.

There are various ways in which one may attempt to solve this problem. If initial matrices $K_{1,0}$, $K_{2,0}$ can be found so that (4. 81), (4. 83) are true and $A + BK_{1,0}$ is a stability matrix, it is possible to move along the hypersurface defined by the constraints (4. 81), (4. 83), (4. 94), (4. 96) in a direction decreasing the cost (4. 97). A logical choice here is the gradient projection technique, described, e. g. in [2], [3]. We will indicate, briefly, how this is done in the example to follow.

The first requirement is to find K_1, K_2 such that $K_1 \in \mathscr{S}(A, B)$, i. e., $A + BK_1$ is a stability matrix, and (4. 81), (4. 83) hold. We will confine attention here to the case wherein the dimension of the response y (cf. (4. 69)) is less than or equal to the dimension of the control u. The number of entries in the matrix FT is then less than or equal to the number of entries in the matrix K_2. Selecting K_1 so that $A + BK_1$ is a stability matrix and keeping K_1 fixed, the equations (4. 81), (4. 83) constitute a system of linear equations in the entries of K_2, the number of equations being less than or equal to the number of unknowns. In our satellite tracking example these equations are obtained from the entries in the first row of T, shown following (4. 86), and reduce to the form (4. 87).

Consider changes in the two feedback matrices:

$$K_1 \to K_1 + \delta K_1, \quad K_2 \to K_2 + \delta K_2. \tag{4.98}$$

This gives rise to a change $T \to T + \delta T$ and, from (4. 81), we have

$$(A + BK_1)\,\delta T - \delta T A_1 + B\delta K_1 T + B\delta K_2 = 0 \;. \tag{4.99}$$

The requirement that the motion take place in a direction tangent, in the

K_1, K_2 matrix space, to the surface described by (4.81), is that

$$F\delta T = 0 \ .$$

This is equivalent to (ℓ = dim w, q = dim y = Fz)

$$\mathrm{Tr}\, E_j^i F\delta T = 0, \quad i = 1, 2, \ldots, \ell_1, \quad j = 1, 2, \ldots, q, \tag{4.100}$$

But, using VI - Proposition 3.3, (4.100) is the same as

$$\mathrm{Tr}\, E_j^i\, F\delta T = \mathrm{Tr}\, P_j^i(B\delta K_1 T + B\delta K_2) = \mathrm{Tr}\, T P_j^i\, B\delta K_1 + \mathrm{Tr}\, P_j^i\, B\delta K_2 = 0 \tag{4.101}$$

where the $\ell \times n$ matrices P_j^i satisfy

$$P_j^i(A + BK_1) - A_1 P_j^i + E_j^i F = 0, \quad i = 1, 2, \ldots, \ell_1, \quad j = 1, 2, \ldots, q. \tag{4.102}$$

Thus, in order to move tangentially with respect to the surface described by (4.81), (4.83) one computes the matrices P_j^i via (4.102), and one requires the "directions" $\delta K_1, \delta K_2$ to satisfy (4.101). In so doing, of course, it is desired to decrease the cost (4.97). Our next task is to study the effect of the changes (4.98) on that cost. Denoting the changes in T, X and Y induced through (4.98) by δT, δX, δY, the change in the cost (4.97) is

$$\mathrm{Tr}(Z + TW_1T^*)\delta X + \mathrm{Tr}\, W_1(2T^*X\delta T + \delta Y). \tag{4.103}$$

Again using VI-Proposition 3.3 together with (4.94) and (4.96) we have

$$(A + BK_1)^*\delta X + \delta X(A + BK_1) + \delta K_1^*(B^*X + \tilde{U}K_1) + (XB + K_1^*\tilde{U})\ K_1 = 0 \tag{4.104}$$

$$(A_1 - \nu I)^*\delta Y + \delta Y(A_1 - \nu I) + (T^*\delta K_1^* + \delta T^* K_1^* + \delta K_2^*)\tilde{U}(K_1T + K_2)$$
$$+ (K_1T + K_2)^*\tilde{U}(\delta K_1 T + K_1\delta T + \delta K_2) = 0\ , \tag{4.105}$$

and we see that the change (4.103) can be expressed in the form

$$2\,\mathrm{Tr}\, P(XB + K_1^*\tilde{U})\delta K_1 + 2\,\mathrm{Tr}\, Q(K_1T + K_2)^*\tilde{U}(\delta K_1 T + K_1\delta T + \delta K_2)$$
$$+ 2\,\mathrm{Tr}\, W_1 T^* X\delta T$$

where

$$(A + BK_1)P + P(A + BK_1)^* + Z + TW_1T^* = 0 , \tag{4.106}$$

$$(A_1 - \nu I)Q + Q(A_1 - \nu I)^* + W_1 = 0 . \tag{4.107}$$

Dividing by 2 and rearranging, this cost becomes

$$\mathrm{Tr}(P(XB + K_1^*\tilde{U}) + TQ(K_1T + K_2)^*\tilde{U})\delta K_1 + \mathrm{Tr}\ Q(K_1T + K_2)^*\tilde{U}\delta K_2$$
$$+ \mathrm{Tr}[\ Q(K_1T + K_2)^*\tilde{U}K_1 + W_1T^*X]\ \delta T . \tag{4.108}$$

The last term in (4.108), using (4.99) and VI- Proposition 3.3 again, is

$$\mathrm{Tr}\ R(B\delta K_1\ T + B\delta K_2) \tag{4.109}$$

where

$$R(A + BK_1) - A_1R + Q(K_1T + K_2)^*UK_1 + W_1T^*X = 0 . \tag{4.110}$$

Substituting (4.109) for the last term in (4.108), the cost change finally becomes

$$\mathrm{Tr}\ (P(XB + K_1^*\tilde{U}) + TQ(K_1T + K_2)^*\tilde{U} + TRB)\delta K_1$$
$$+ \mathrm{Tr}(Q(K_1T + K_2)^*\tilde{U} + RB)\delta K_2 . \tag{4.111}$$

The gradient projection method ([2], [3]) finds the steepest descent direction (for the cost (4.97)) tangent to the surface described by (4.81). It can be shown (V- [2], Chap. IX) that this can be obtained by setting

$$(\delta K_1, \delta K_2) = \epsilon[\alpha_0 (B^*X+\tilde{U}K_1+\tilde{U}(K_1T+K_2)QT^*+B^*R^*T^*,\ B^*R^*+\tilde{U}(K_1T+K_2)Q)$$
$$+ \sum_{i,j} \alpha_j^i(B^*(P_j^i)^*T^*,\quad B^*(P_j^i)^*)] \tag{4.112}$$

The unknown α_0, α_{ij} are determined by substituting the coefficient of ϵ in (4.112) into (4.111), and requiring that quantity to equal -1, and into (4.101). The result takes the form

$$G \begin{pmatrix} \alpha_1^0 \\ \alpha_1^1 \\ \vdots \\ \alpha_q^\ell \end{pmatrix} = \begin{pmatrix} -1 \\ 0 \\ \vdots \\ 0 \end{pmatrix} \tag{4.113}$$

where G can be seen to be the Gram matrix of the (matrix) vectors multiplying the α_0, α_j^i in (4.112).

After new values $\tilde{K}_1, \tilde{K}_2$ are obtained by this process, the equation (4.83) is only satisfied to first order in ϵ. The values of $\tilde{K}_1, \tilde{K}_2$ must then be adjusted so that (4.81), (4.83) hold exactly. In most cases this can be done by an $\mathcal{O}(\epsilon^2)$ adjustment of $\tilde{K}_2$. If ϵ is small, (4.97) will still be reduced, as compared with its values for K_1, K_2, and the new $\tilde{K}_1, \tilde{K}_2$ satisfy the consistency conditions (4.81), (4.83). The process is repeated until the minimum (or a local minimum) is found, it being signalled by G becoming singular (the gradient of the cost then being a linear combination of the constraint gradients, so that the Lagrange multiplier condition holds [3]).

We will carry out this process for one step in the case of our satellite tracking system in order to clarify what has, quite possibly, been a less than illuminating discussion up to this point.

We begin by setting (cf. (4.63) ff.) $\alpha = \beta = 1$ so that

$$A = \begin{pmatrix} 0 & 1 \\ -1 & -1 \end{pmatrix}, \quad B = \begin{pmatrix} 0 \\ 1 \end{pmatrix}.$$

We have also (cf. (4.86)

$$A_1 = \begin{pmatrix} 0 & 1 \\ 0 & 0 \end{pmatrix}, \quad A_2 = \begin{pmatrix} 0 & -1 \\ 0 & 0 \end{pmatrix}.$$

Here A is already stable so, for simplicity we will take $k_1 = k_2 = 0$. Then from (4.87)

$$k_3 = 0, \quad k_4 = \alpha - k_2 = 1 - 0 = 1.$$

We have

$$F = (1, 0).$$

Computing T from (4.81), in this case

$$\begin{pmatrix} 0 & 1 \\ -1 & -1 \end{pmatrix} T - T \begin{pmatrix} 0 & 1 \\ 0 & 0 \end{pmatrix} + \begin{pmatrix} 0 & -1 \\ 0 & 1 \end{pmatrix}$$

gives

$$T=\begin{pmatrix} 0 & 0 \\ 0 & 1 \end{pmatrix}$$

and clearly $FT = 0$.

We will set (see material before (4.92))

$$Z = \begin{pmatrix} 1 & 0 \\ 0 & 0 \end{pmatrix}, \quad W_1 = \begin{pmatrix} 0 & 0 \\ 0 & 1 \end{pmatrix} .$$

We take (cf. (4.96)) $\nu = 1$ and compute X and Y from (4.94), (4.96) with $\tilde{W} = 1$, $\tilde{U} = 1$:

$$\begin{pmatrix} 0 & 1 \\ -1 & -1 \end{pmatrix}^* X + X\begin{pmatrix} 0 & 1 \\ -1 & -1 \end{pmatrix} + \begin{pmatrix} 1 & 0 \\ 0 & 0 \end{pmatrix} = 0 ,$$

$$\begin{pmatrix} -1 & 1 \\ 0 & -1 \end{pmatrix}^* Y + Y\begin{pmatrix} -1 & 1 \\ 0 & -1 \end{pmatrix} + \begin{pmatrix} 0 & 0 \\ 0 & 1 \end{pmatrix} = 0 ,$$

giving

$$X = \begin{pmatrix} 1 & .5 \\ .5 & .5 \end{pmatrix}, \quad Y = \begin{pmatrix} 0 & 0 \\ 0 & .5 \end{pmatrix} .$$

The cost is computed from (4.97) to be, in this case

$$\mathrm{Tr}\begin{pmatrix} 1 & 0 \\ 0 & 1 \end{pmatrix} X + \mathrm{Tr}\begin{pmatrix} 0 & 0 \\ 0 & 1 \end{pmatrix} Y = 2 .$$

To attempt to reduce this cost, the first step is to compute the P_j^i defined by (4.102). In this case $\ell_1 = 2$, $q = 1$ and

$$E_1^1 F = \begin{pmatrix} 1 \\ 0 \end{pmatrix}(1, 0) = \begin{pmatrix} 1 & 0 \\ 0 & 0 \end{pmatrix}, \quad E_1^2 F = \begin{pmatrix} 0 \\ 1 \end{pmatrix}(1, 0) = \begin{pmatrix} 0 & 0 \\ 1 & 0 \end{pmatrix}$$

and we compute, from (4.102)

$$P_1^1\begin{pmatrix} 0 & 1 \\ -1 & 1 \end{pmatrix} - \begin{pmatrix} 0 & 1 \\ 0 & 0 \end{pmatrix} P_1^1 + \begin{pmatrix} 1 & 0 \\ 0 & 0 \end{pmatrix} = 0 ,$$

$$P_1^2\begin{pmatrix}0 & 1\\ -1 & 1\end{pmatrix} - \begin{pmatrix}0 & 1\\ 0 & 0\end{pmatrix}P_1^2 + \begin{pmatrix}0 & 0\\ 1 & 0\end{pmatrix} = 0$$

giving

$$P_1^1 = \begin{pmatrix}-1 & 1\\ 0 & 0\end{pmatrix}, \quad P_1^2 = \begin{pmatrix}0 & 1\\ -1 & 1\end{pmatrix}.$$

Then we compute the matrices P, Q and R from (4.106), (4.107), (4.110), respectively. In our case these equations become

$$\begin{pmatrix}0 & 1\\ -1 & -1\end{pmatrix}P + P\begin{pmatrix}0 & 1\\ -1 & -1\end{pmatrix}^* + \begin{pmatrix}1 & 0\\ 0 & 1\end{pmatrix} = 0$$

$$\begin{pmatrix}-1 & 1\\ 0 & -1\end{pmatrix}Q + Q\begin{pmatrix}-1 & 1\\ 0 & -1\end{pmatrix}^* + \begin{pmatrix}0 & 0\\ 0 & 1\end{pmatrix} = 0$$

giving

$$P = \begin{pmatrix}1.5 & -.5\\ -.5 & 1\end{pmatrix}, \quad Q = \begin{pmatrix}.25 & .25\\ .25 & .5\end{pmatrix}.$$

Then R is computed from (4.110). In this case $K_1 = 0$ and the equation is, using our value for X,

$$R\begin{pmatrix}0 & 1\\ -1 & -1\end{pmatrix} - \begin{pmatrix}0 & 1\\ 0 & 0\end{pmatrix}R + \begin{pmatrix}0 & 0\\ .5 & .5\end{pmatrix} = 0,$$

giving

$$R = \begin{pmatrix}.5 & 0\\ 0 & .5\end{pmatrix}.$$

We are now able to calculate the coefficient matrices in (4.112). They are as follows:

of α_0 : (.5 1.5 .25 1)

of α_1^1 : (0 0 1 0)

of α_1^2 : (0 1 1 1)

and this gives the Gram matrix

$$G = \begin{pmatrix} 3.5625 & .25 & 2.75 \\ .25 & 1 & 1 \\ 2.75 & 1 & 3 \end{pmatrix} .$$

Solving the equation (4.113) we have

$$\alpha_0 = -2.6667 \qquad \alpha_1^1 = -2.5667 \qquad \alpha_1^2 = 3.3333 ,$$

giving

$$(\delta K_1, \delta K_2) = \epsilon(-2.6667(.5 \quad 1.5 \quad .25 \quad 1) - 2.6667(0 \quad 0 \quad 1 \quad 0)$$

$$+ 3.3333(0 \quad 1 \quad 1 \quad 1)) = \epsilon(-1.3333, \quad -.6667, \quad 0, \quad .6667) .$$

The size and form of the entries here suggest $\epsilon = .15$ as a good choice. The new values of k_1, k_2, k_3, k_4 are

$$\tilde{k}_1 = 0 + .15 \times 1.3333 = -.2$$

$$\tilde{k}_2 = 0 + .15 \times (-.6667) = -.1$$

$$\tilde{k}_3 = 0 + .15 \times .0 = .0$$

$$\tilde{k}_4 = 1 + .15 \times .6667 = 1.1 .$$

That these values satisfy the consistency conditions (4.87) is verified readily. Those conditions are satisfied exactly in this particularly simple instance because they are, in fact, linear (even though the entries of T are rational functions of k_1, k_2, k_3, k_4). In a more complicated problem this fortuitous situation would not obtain and the condition (4.81), (4.83) would be satisfied only to first order in ϵ. The value of $\tilde{K}_2$ would need to be adjusted slightly or, in some cases, $\tilde{K}_1, \tilde{K}_2$ might have to be adjusted simultaneously to give (4.83) exactly. A method for carrying out this readjustment procedure is described in Chapter IX of V-[2].

To study the effect on the cost (4.97), we solve (4.81), (4.94), (4.96) with the new feedback values. The results are

$$\tilde{T} = \begin{pmatrix} 0 & 0 \\ 0 & 1 \end{pmatrix} \quad \text{(again)},$$

$$\tilde{X} = \begin{pmatrix} .93484 & .43333 \\ .43333 & .39848 \end{pmatrix},$$

$$\tilde{Y} = \begin{pmatrix} 0 & 0 \\ 0 & .5 \end{pmatrix} \quad \text{(again)}.$$

The new cost is

$$\mathrm{Tr}\begin{pmatrix} 1 & 0 \\ 0 & 1 \end{pmatrix}\tilde{X} + \mathrm{Tr}\begin{pmatrix} 0 & 0 \\ 0 & 1 \end{pmatrix}\tilde{Y} = 1.83332$$

which is, indeed, reduced. It should be noted in this particularly simple example that when (4.87) is true, T will always be equal to $\begin{pmatrix} 0 & 0 \\ 0 & 1 \end{pmatrix}$ and $K_1T + K_2 = (0, k_2) + (0, k_4) = (0, k_2 + k_4) = (0, \alpha) = (0, 1)$ and thus, from (4.96), Y will always be $\begin{pmatrix} 0 & 0 \\ 0 & .5 \end{pmatrix}$. The method acts to reduce $\mathrm{Tr}\,(Z + T^* W_1 T)X = \mathrm{Tr}\,X$ in this case, at the same time maintaining (4.87). In higher dimensional examples with less special structure the action would be harder to follow.

5. Cost Sensitivity Reduction

The general problem of sensitivity reduction and the related question which we will call "robustness" (and which we treat in Section 6) is distinguished by pervasive importance throughout control engineering and pervasive lack of agreement as to what it really means. Its importance derives from the fact that a control engineer almost never has a fixed, perfectly known plant to work with. In practice the system varies due to temperature changes, altitude changes, different loading conditions, linearization of a nonlinear plant about different equilibria, imperfect data used in determining plant parameters, and a host of other possible variations, known or unknown, which affect the dynamics and/or specification of the system. Apposite to these are the criteria of performance which must be taken into account, such as stability, performance with respect to a cost criterion, etc. The basic question is how each of the former affects each of the latter and how different control strategies can reduce performance variations in the face of plant deviations or uncertainties. The confusion usually arises because the term sensitivity is used without specifying which performance criteria and which plant variations are to be studied in relation to each other. A second source of confusion arises when it is unclear whether performance variations are the main question or whether the maintenance of some minimal level of performance throughout the range of plant variations is the object of prime concern. In our treatment we (more or less arbitrarily) designate the first of these as a sensitivity question and the second as a question of robustness. Sensitivity, to be discussed in this section, and robustness, in the section to follow, can be discussed in terms of the general system.

$$\dot{x} = A(\alpha)x + B(\alpha)u + C(\alpha)v + D(\alpha)w, \tag{5.1}$$

x, u, v, w representing n, m, q and ℓ dimensional state, (automatic) control, disturbance and command variables, respectively. The matrices $A(\alpha)$, $B(\alpha)$, $C(\alpha)$, $D(\alpha)$ are assumed to depend smoothly upon a parameter vector p whose range, $\mathcal{R}(\alpha) \subseteq E^r$, serves to describe the plant variations under consideration. In some cases $C(\alpha)$ and/or $D(\alpha)$ may be suppressed in the analysis. Along with (5.1) we have outputs or responses

$$y = F(\alpha)x + G(\alpha)u + H(\alpha)w \tag{5.2}$$

deemed to be of interest in the analysis. If $D(\alpha)$ is suppressed, $L(\alpha)$ will also be suppressed.

Typically the control analysis is based on a "nominal" plant corresponding to some particular value of α, without loss of generality $\alpha = 0$. Setting

$$A(0) = A_0,\ B(0) = B_0,\ C(0) = C_0,\ D(0) = D_0,\ F(0) = F_0,\ G(0) = G_0,$$
$$H(0) = H_0,$$

the nominal plant and nominal response are

$$\dot{x} = A_0x + B_0u + C_0v + D_0w, \tag{5.3}$$

$$y = F_0x + G_0u + H_0w. \tag{5.4}$$

We will suppose that a quadratic cost is given

$$J(x_0, u, v, w, \alpha) = \int_0^\infty [y(t)^* \tilde{W} y(t) + u(t)^* \tilde{U} u(t)]\, dt \ . \tag{5.5}$$

The basic question is how J varies with α and how measures may be taken to reduce this variation.

For our work here we will assume (5.1), (5.2) specialized to the form

$$\mathcal{P}_\alpha : \begin{cases} \dot{x} = A(\alpha)x + B(\alpha)u + Cv \\ y = F(\alpha)x + G(\alpha)u \end{cases} \tag{5.6}$$

with random initial state x_0, $\operatorname{cov}(x_0, x_0) = X_0$, and v a white noise disturbance with $e^{-\mu t}V$. We will suppose the feedback control

$$u = K_0 x \tag{5.7}$$

is determined, as in Chapters IV and VI so as to minimize the expected cost

$$J(X_0, K_0, V, 0) = \mathcal{E} J(x_0, u, v, 0) \tag{5.8}$$

associated with the nominal plant and response

$$\mathcal{P}_0 : \begin{cases} \dot{x} = A_0 x + B_0 u + Cv , \\ y = F_0 x + G_0 u \end{cases} \tag{5.9}$$

It is ordinarily not possible to analyze the infinitely many plants $\mathcal{P}_\alpha$ as designated by (5.6) for $\alpha \in \mathcal{R}(\alpha)$; a finite set of representatives must be selected. Often $\mathcal{R}(\alpha)$ is a polyhedron in E^r and, under such circumstances $\alpha = 0$ might correspond to some interior point while the vertices of the polyhedron might be selected as the representative parameter variations. However this is done, we consider N representative perturbed plants

$$\mathcal{P}_\nu : \begin{cases} \dot{\xi}_\nu = A_\nu \xi_\nu + B_\nu u + Cv \\ \eta_\nu = F_\nu \xi_\nu + G_\nu u \end{cases} , \quad \nu = 1,2,\ldots,N, \tag{5.10}$$

$$A_\nu = A(\alpha_\nu), \quad B_\nu = B(\alpha_\nu), \quad F_\nu = F(\alpha_\nu), \quad G_\nu = G(\alpha_\nu),$$

corresponding to the N representative parameter values $\alpha = \alpha_1, \alpha_2, \dots, \alpha_N$.

Our goal is to find a single fixed form feedback relation

$$u = Kx \tag{5.11}$$

which simultaneously keeps the cost $J(X_0, K, V, 0)$ small, hopefully not too much larger than $J(X_0, K_0, V, 0)$ discussed earlier, and keeps the N cost differences

$$J(X_0, K, V, \alpha_\nu) - J(X_0, K, V, 0) \tag{5.12}$$

small - corresponding to small performance deviations in the face of the parameter variations $\alpha_0 (= 0) \to \alpha_\nu$, $\nu = 1, 2, \dots, N$. It turns out, however, that the differences (5.12) are not convenient to work with directly because they lead, in the analysis, to cost terms which are not, in general, non-negative. Squaring (5.12) leads to expressions which are too complicated.

Let $x(t)$, $y(t)$ denote the trajectory and response associated with the nominal plant (5.9) and let $\xi_\nu(t)$, $\eta_\nu(t)$ denote the corresponding entities for (5.10). In each case we suppose

$$x(0) = \xi_\nu(0) = x_0, \quad \nu = 1, 2, \dots, N,$$

and the control is determined by $u = Kx$, $u_\nu = K\xi_\nu$, etc. Then, assuming all costs finite, $|2ab| \le \theta^2 a^2 + \frac{1}{\theta^2} b^2$ gives

$$\begin{aligned}
&|J(x_0, u_\nu, v, \alpha_\nu) - J(x_0, u, v, 0)| \\
&= \left| \int_0^\infty [\eta_\nu(t)^* \tilde{W} \eta_\nu(t) + u_\nu(t)^* \tilde{U} u_\nu(t)]\, dt - \int_0 [y(t)^* \tilde{W} y(t) + u(t)^* \tilde{U} u(t)]\, dt \right| \\
&= \left| 2 \int_0^\infty [y(t)^* \tilde{W} (\eta_\nu(t) - y(t)) + u(t)^* \tilde{U} (u_\nu(t) - u(t))]\, dt \right. \\
&\quad \left. + \int_0^\infty [(\eta_\nu(t) - y(t))^* \tilde{W} (\eta_\nu(t) - y(t)) + (u_\nu(t) - u(t))^* \tilde{U} (u_\nu(t) - u(t))]\, dt \right| \\
&\le \theta^2 J(x_0, u, v, 0) + \frac{\theta^2 + 1}{\theta^2} \delta J_\nu(x_0, u, u_\nu, v), \quad \theta > 0, \text{ arbitrary},
\end{aligned} \tag{5.13}$$

where

$$\delta J_\nu(x_0,u,u_\nu,v) = \int_0^\infty [(\eta_\nu(t)-y(t))^*\tilde{W}(\eta_\nu(t)-y(t)) + (u_\nu(t)-u(t))^*\tilde{U}(u_\nu(t)-u(t))]\,dt\ . \tag{5.14}$$

It follows that (5.12) can be kept small if the nominal plant cost $J(x_0,u,v,0)$ and the "response, control difference cost" $\delta J_\nu(x_0, u, u_\nu, v)$ are both kept small. With the expected nominal cost defined by (5.8) and the expected response, control difference cost defined by

$$\delta J_\nu(X_0, K, V) = \mathcal{E}(\delta J_\nu(x_0, u, u_\nu, v)) \tag{5.15}$$

we are in a position to state the

Optimal Sensitivity Reduction Problem for Finite Parameter Variations. Determine an $m \times n$ feedback matrix $\hat{K}$ solving the problem

$$\min_K (\sigma_0 J(X_0, K, V) + \sum_{\nu=1}^N \sigma_\nu \delta J_\nu(X_0, K, V)) \tag{5.16}$$

with the constraint that $A_0 + B_0K$, $A_\nu + B_\nu K$, $\nu = 1, 2, \ldots, N$, shall be stability matrices for all K under consideration.

The weights $\sigma_0, \sigma_1, \ldots, \sigma_N$ may, in view of (5.13), take the form $\sigma_0 = 1 + \epsilon_0 N\theta^2$, the "1" reflecting an initial weight placed on the nominal plant cost, $\sigma_\nu = \epsilon_0(\theta^2 + 1)/\theta^2$, $\nu = 1, 2, \ldots, N$. As ϵ_0 is varied, more or less weight is placed on cost variation as opposed to nominal plant cost. Alternatively, any other weighting scheme might be used, e.g., the plant variations $\alpha_0(= 0) \rightarrow \alpha_\nu$ might be viewed as independent random fluctuations and the σ_ν might be chosen corresponding to the probability that such a parameter variation will occur. It is also clear that by scaling, we can always reduce to the case $\sigma_0 = 1$.

In some applications it may be desirable to deal only with the response cost

$$J_0(x_0, u, v, 0) = \int_0^\infty y(t)^*\tilde{W}y(t)dt$$

and the corresponding response costs for the perturbed plants. Under such circumstances (5.16) would be replaced by

$$\min_{K}(\sigma_0 J(X_0, K, V) + \hat{\sigma}_0 J_0(X_0, K, V) + \sum_{\nu=1}^{N} \sigma_\nu J_{0,\nu}(X_0, K, V)) \tag{5.17}$$

where $J_{0,\nu}$ differs from (5.14) by elimination of the term $(u_\nu(t) - u(t))^* \tilde{U}(u_\nu(t) - u(t))$.

The above sensitivity reduction problem is conveniently studied via the composite system

$$\dot{\xi} = (\mathcal{A} + \mathcal{B}_K)\xi$$

and the composite response

$$\eta = (\mathcal{F} + \mathcal{G}_K)\xi ,$$

where, in block decomposition

$$\mathcal{A} = \text{diag}\,(A_0, A_1, A_2, \ldots, A_N)$$

$$\mathcal{B}_K = \text{diag}\,(B_0K, B_1K, B_2K, \ldots, B_NK)$$

$$\mathcal{F} + \mathcal{G}_K = \begin{pmatrix} F_0 + G_0K & 0 & 0 & \cdots & 0 \\ -F_0 - G_0K & F_1 + G_1K & 0 & \cdots & 0 \\ -F_0 - G_0K & 0 & F_2 + G_2K & \cdots & 0 \\ \vdots & \vdots & \vdots & & \vdots \\ -F_0 - G_0K & 0 & 0 & \cdots & F_N + G_NK \end{pmatrix} .$$

The problem becomes

$$\min_{K} \ \text{Tr}\ \mathcal{M}\mathcal{X} , \tag{5.18}$$

subject to the restriction that $\mathcal{A} + \mathcal{B}_K$ is a stability matrix, and

$$(\mathcal{A} + \mathcal{B}_K)^* \mathcal{X} + \mathcal{X}(\mathcal{A} + \mathcal{B}_K) + (\mathcal{F} + \mathcal{G}_K)^* \tilde{\mathcal{W}}(\sigma)(\mathcal{F} + \mathcal{G}_K) + \mathcal{H}_K^* \tilde{\mathcal{U}}(\sigma)\mathcal{H}_K = 0 \tag{5.19}$$

$$\left.\begin{aligned} \tilde{\mathcal{W}}(\sigma) &= \text{diag}(\sigma_0\tilde{W}, \sigma_1\tilde{W}, \sigma_2\tilde{W}, \ldots, \sigma_N\tilde{W}) \\ \tilde{\mathcal{U}}(\sigma) &= \text{diag}(\sigma_0\tilde{U}, \sigma_1\tilde{U}, \sigma_2\tilde{U}, \ldots, \sigma_N\tilde{U}) \end{aligned}\right\} \tag{5.20}$$

and

$$\mathcal{H}_K = \begin{pmatrix} K & 0 & 0 & \cdots & 0 \\ -K & K & 0 & \cdots & 0 \\ -K & 0 & K & \cdots & 0 \\ \vdots & \vdots & \vdots & & \vdots \\ -K & 0 & 0 & \cdots & K \end{pmatrix}$$

Finally, $\mathcal{M}$ is an $n(N+1) \times n(N+1)$ matrix, all of whose $n \times n$ blocks are $X_0 + \frac{1}{\mu} CVC^* \equiv M$.

The block diagonal form of $\mathcal{A} + \mathcal{B}_K$ and the special forms of $\mathcal{F} + \mathcal{G}_K$, $\mathcal{H}_K$ enable us to see that $\mathcal{X}$ can be partitioned in the form

$$\mathcal{X} = \begin{pmatrix} \hat{\sigma} Q_0 & -\sigma_1 R_1 & -\sigma_2 R_2 & \cdots & -\sigma_N R_N \\ -\sigma_1 R_1^* & \sigma_1 Q_1 & 0 & \cdots & 0 \\ -\sigma_2 R_2^* & 0 & \sigma_2 Q_2 & \cdots & 0 \\ \vdots & \vdots & \vdots & & \vdots \\ -\sigma_N R_N^* & 0 & 0 & \cdots & \sigma_N Q_N \end{pmatrix}$$

permitting us to partition (5.19) in the more manageable form of $2N+1$ equations for the $n \times n$ matrices $Q_0, Q_1, \ldots, Q_N, R_1, R_2, \ldots, R_N$ (here

$$\hat{\sigma} = \sigma_0 + \sum_{\nu=1}^{N} \sigma_\nu); \tag{5.21}$$

$$(A_0 + B_0K)^* Q_0(K) + Q_0(K)(A_0 + B_0K) + (F_0 + G_0K)^* \tilde{W}(F_C + G_0K) + K^* \tilde{U} K = 0 \tag{5.22}$$

$$(A_\nu + B_\nu K)^* Q_\nu(K) + Q_\nu(K)(A_\nu + B_\nu K) + (F_\nu + G_\nu K)^* \tilde{W}(F_\nu + G_\nu K) + K^* \tilde{U} K = 0,$$
$$\nu = 1, 2, \ldots, N, \tag{5.23}$$

$$(A_0 + B_0K)^* R_\nu(K) + R_\nu(K)(A_\nu + B_\nu K) + (F_0 + G_0 K)^* \tilde{W}(F_\nu + G_\nu K) + K^* \tilde{U} K = 0,$$
$$\nu = 1, 2, \ldots, N. \tag{5.24}$$

The cost (5.18), for a given K, using the symmetry of M, is

$$\mathrm{Tr}\, M\Big(\hat{\sigma} Q_0(K) + \sum_{\nu=1}^{N} \sigma_\nu [Q_\nu(K) - 2R_\nu(K)]\Big) \tag{5.25}$$

The usual sort of analysis, replacing K by $\hat{K} + \epsilon\delta K$, differentiating the resulting equations with respect to ϵ at $\epsilon = 0$, and using the adjoint equations, establishes the following. If

$$(A_0 + B_0K)Y_0(K) + Y_0(K)(A_0 + B_0K)^* + M = 0, \tag{5.26}$$

$$(A_\nu + B_\nu K)Y_\nu(K) + Y_\nu(K)(A_\nu + B_\nu K)^* + M = 0, \quad \nu = 1,2,\dots, N, \tag{5.27}$$

$$(A_\nu + B_\nu K)Z_\nu(K) + Z_\nu(K)(A_0 + B_0K)^* + M = 0, \quad \nu = 1,2,\dots, N, \tag{5.28}$$

then the necessary condition for optimality of $\hat{K}$ is that for every δK

$$\begin{aligned}
\mathrm{Tr}\,\{&\hat{\sigma} Y_0(\delta K^* B_0^* Q_0 + Q_0 B_0 \delta K + [\delta K^* G_0^* \tilde{W} F_0 + F_0^* \tilde{W} G_0 \delta K + \delta K^* U_0 \hat{K} + \hat{K}^* U_0 \delta K])\\
&+ \sum_{\nu=1}^{N} \sigma_\nu Y_\nu (\delta K^* B_\nu^* Q_\nu + Q_\nu B_\nu \delta K + [\delta K^* G_\nu^* \tilde{W} F_\nu + F_\nu^* \tilde{W} G_\nu \delta K + \delta K^* U_\nu \hat{K} + \hat{K}^* U_\nu \delta K])\\
&- 2\sum_{\nu=1}^{N} \sigma_\nu Z_\nu (\delta K^* B_0^* R_\nu + R_\nu B_\nu \delta K + [\delta K^* G_0^* \tilde{W} F_\nu + F_0^* \tilde{W} G_\nu \delta K + \delta K^* U_\nu \hat{K} + \hat{K}^* U_\nu^* \delta K])\} = 0.
\end{aligned} \tag{5.29}$$

Here $Y_0 = Y_0(\hat{K})$, $Q_0 = Q_0(\hat{K})$, etc., and

$$U_0 = \tilde{U} + G_0^* \tilde{W} G_0,$$

$$U_\nu = \tilde{U} + G_\nu^* \tilde{W} G_\nu,$$

$$\hat{U}_\nu = \tilde{U} + G_0^* \tilde{W} G_\nu.$$

Using the properties of the trace and the symmetry of Y_0, Q_0, Y_ν, Q_ν, $\nu = 1, 2, \dots, N$, (note however that Z_ν, R_ν, $\nu = 1, 2, \dots, N$, are not, in general, symmetric) (5.29) becomes

$$\mathrm{Tr}\, \delta K^* (E_0(\hat{K}) + E_1(\hat{K})) = 0 \tag{5.30}$$

where

$$E_0(K) = 2[\sigma B_0^* Q_0(K) Y_0(K) + \hat{\sigma} G_0^* \tilde{W} F_0 Y_0(K) + \sum_{\nu=1}^{N} \sigma_\nu (B_\nu^* Q_\nu(K) Y_\nu(K) + \sigma_\nu G_\nu^* \tilde{W} F_\nu Y_\nu(K))$$

$$- \sum_{\nu=1}^{N} \sigma_\nu (B_0^* R_\nu(K) Z_\nu(K) + B_\nu^* R_\nu(K)^* Z_\nu(K)^* + (G_0^* \tilde{W} F_\nu + G_\nu^* \tilde{W} F_0))] \qquad (5.31)$$

and

$$E_1(K) = 2[\hat{\sigma} U_0 K Y_0(K) + \sum_{\nu=1}^{N} \sigma_\nu U_\nu K Y_\nu(K) - \sum_{\nu=1}^{N} \sigma_\nu (\hat{U}_\nu K Z_\nu(K) + \hat{U}_\nu K Z_\nu(K)^*)]. \qquad (5.32)$$

As (5.30) must be true for all δK, the necessary condition, which must be satisfied by an optimal $\hat{K}$, becomes

$$E_0(\hat{K}) + E_1(\hat{K}) = 0. \qquad (5.33)$$

A variant of Axsäter's method (VI-5) is obtained, as usual, by solving

$$E_1(K_{i+1}) = - E_0(K_i) \qquad (5.34)$$

for K_{i+1}. In the present circumstance a closed form solution is not possible and (5.34) has to be treated as mn linear equations in the corresponding number of unknown entries of the $m \times n$ matrix K_{i+1}. Even then convergence is hard to guarantee without special assumptions on the size of $\sigma_1, \sigma_2, \ldots, \sigma_N$ relative to σ_0.

This is as far as we will pursue the problem for the case of finite parameter variations. That the method (5.34), or a gradient type method

$$K_i \to K_i + \delta K_i \qquad (5.35)$$

$$\delta K_i = - \epsilon (E_0(K_i) + E_1(K_i)), \quad \epsilon > 0, \text{ small}, \qquad (5.36)$$

can be employed to obtain a (possibly non-unique) result is clear. In general the procedure is quite possibly more complicated than many circumstances call for.

A slightly more tractible special case arises when the control term in the response is invariant: $G(\alpha) \equiv G_0$, i.e., $G_0 = G_1 = G_2 = \ldots = G_N$. Then (5.33) gives

$$\hat{K} = - U_0^{-1} E_0(\hat{K}) (2\hat{\sigma} Y_0(\hat{K}) + \sum_{\nu=1}^{N} \sigma_\nu (2Y_\nu(\hat{K}) - Z_\nu(\hat{K}) - Z_\nu(\hat{K})^*))^{-1} \qquad (5.37)$$

and (5. 34) becomes

$$\hat{K}_{i+1} = -U_0^{-1}E_0(K_i)(2\hat{\sigma}Y_0(K_i) + \sum_{\nu=1}^{N} \sigma_\nu (2Y_\nu(K_i) - Z_\nu(K_i) - Z_\nu(K_i)^*))^{-1} . \tag{5.38}$$

A different, but related, problem is obtained when we study the case of small variations

$$\alpha_0 (= 0) \to \epsilon\alpha_\nu , \qquad \epsilon \text{ real, small.}$$

The basic idea here is very simple. We consider the infinitesimally perturbed closed loop plants and responses

$$\mathcal{P}_{\nu,\epsilon} \begin{cases} \dot{\xi}_\nu = (A(\epsilon\alpha_\nu) + B(\epsilon\alpha_\nu)K)\xi_\nu \equiv (A_\nu(\epsilon) + B_\nu(\epsilon)K)\xi_\nu , \\ \eta_\nu = (F(\epsilon\alpha_\nu) + G(\epsilon\alpha_\nu)K)\xi_\nu = (F_\nu(\epsilon) + G_\nu(\epsilon)K)\xi_\nu , \end{cases} \qquad \nu = 1,2,\ldots, N \tag{5.39}$$

and define $\delta J_\nu(x_0, u, u_\nu, v, \epsilon)$, $\delta J_\nu(X_0, K, V, \epsilon)$ as in (5. 14), (5. 15), replacing α_ν by ϵ_ν in each case, i. e., replacing (5. 10) by (5. 39). Now it can be seen (cf. (5. 25)) that

$$\delta J_\nu(X_0, K, V, \epsilon) = \mathrm{Tr}\, M (Q_0 + Q_\nu(\epsilon) - 2R_\nu(\epsilon)) , \tag{5.40}$$

where $Q_\nu(\epsilon)$, $R_\nu(\epsilon)$ satisfy (5. 23), (5. 24), but with A_ν replaced by $A_\nu(\epsilon)$, B_ν replaced by $B_\nu(\epsilon)$, as defined by (5. 39). Let us expand $Q_\nu(\epsilon)$, $R_\nu(\epsilon)$ with respect to ϵ:

$$Q_\nu(\epsilon) = Q_{\nu,0} + \epsilon Q_{\nu,1} + \epsilon^2 Q_{\nu,2} + \mathcal{O}(\epsilon^3), \qquad \nu = 1, 2, \ldots , N, \tag{5.41}$$

$$R_\nu(\epsilon) = R_{\nu,0} + \epsilon R_{\nu,1} + \epsilon^2 R_{\nu,2} + \mathcal{O}(\epsilon^3), \qquad \nu = 1, 2, \ldots , N. \tag{5.42}$$

It is immediately apparent from (5. 22) - (5. 24) that for any K

$$Q_{\nu,0}(K) = R_{\nu,0}(K) = Q_0(K) . \tag{5.43}$$

Letting

$$A_\nu(\epsilon) = A_0 + \epsilon A_{\nu,1} + \epsilon^2 A_{\nu,2} + \mathcal{O}(\epsilon^3) , \tag{5.44}$$

$$B_\nu(\epsilon) = B_0 + \epsilon B_{\nu,1} + \epsilon^2 B_{\nu,2} + \mathcal{O}(\epsilon^3) , \tag{5.45}$$

$$F_\nu(\epsilon) = F_0 + \epsilon F_{\nu,1} + \epsilon^2 F_{\nu,2} + \mathcal{O}(\epsilon^3), \tag{5.46}$$

$$G_\nu(\epsilon) = G_0 + \epsilon G_{\nu,1} + \epsilon^2 G_{\nu,2} + \mathcal{O}(\epsilon^3), \tag{5.47}$$

(here $A_{\nu,1} = \frac{\partial A(\alpha)}{\partial \alpha}\Big|_{\alpha=0} \alpha_\nu$, $A_{\nu,2} = \frac{1}{2}\alpha_\nu^* \frac{\partial^2 A(\alpha)}{\partial \alpha^2}\Big|_{\alpha=0} \alpha_\nu$, etc.) replacing the indicated quantities in (5.23), (5.24) by (5.41), (5.42), (5.44) - (5.47) we find by collecting like terms that the equations (5.22) - (5.24) yield, first of all for the coefficients of ϵ (we suppress the argument, K, of $Q_\nu(K)$, $R_\nu(K)$)

$$\begin{aligned}&(A_0 + B_0K)^* Q_{\nu,1} + Q_{\nu,1}(A_0 + B_0K) + A_{\nu,1} + B_{\nu,1}K)^* Q_0 + Q_0(A_{\nu,1} + B_{\nu,1}K)\\ &\quad + (F_0 + G_0K)^*\tilde{W}(F_{\nu,1} + G_{\nu,1}K) + (F_{\nu,1} + G_{\nu,1}K)^*\tilde{W}(F_0 + G_0K) = 0,\end{aligned} \tag{5.48}$$

$$\begin{aligned}&(A_0 + B_0K)^* R_{\nu,1} + R_{\nu,1}(A_0 + B_0K) + Q_0(A_{\nu,1} + B_{\nu,1}K)\\ &\quad + (F_0 + G_0K)^*\tilde{W}(F_{\nu,1} + G_{\nu,1}K) = 0.\end{aligned} \tag{5.49}$$

Adding (5.48) to minus the sum of (5.49) and its transpose we conclude that for any K

$$Q_{\nu,1}(K) - R_{\nu,1}(K) - R_{\nu,1}(K)^* = 0. \tag{5.50}$$

Going to the coefficients of ϵ^2 we find that

$$\begin{aligned}&(A_0 + B_0K)^* Q_{\nu,2} + Q_{\nu,2}(A_0 + B_0K) + (A_{\nu,1} + B_{\nu,1}K)^* Q_{\nu,1} + Q_{\nu,1}(A_{\nu,1} + B_{\nu,1}K)\\ &\quad + (A_{\nu,2} + B_{\nu,2}K)^* Q_0 + Q_0(A_{\nu,2} + B_{\nu,2}K) + (F_0 + G_0K)^*\tilde{W}(F_{\nu,2} + G_{\nu,2}K)\\ &\quad + (F_{\nu,2} + G_{\nu,2}K)^*\tilde{W}(F_0 + G_0K) + (F_{\nu,1} + G_{\nu,1}K)^*\tilde{W}(F_{\nu,1} + G_{\nu,1}K)\\ &\quad = 0, \qquad \nu = 1, 2, \ldots, N,\end{aligned} \tag{5.51}$$

$$\begin{aligned}&(A_0 + B_0K)^* R_{\nu,2} + R_{\nu,2}(A_0 + B_0K) + R_{\nu,1}(A_{\nu,1} + B_{\nu,1}K) + Q_0(A_{\nu,2} + B_{\nu,2}K)\\ &\quad + (F_0 + G_0K)^*\tilde{W}(F_{\nu,2} + G_{\nu,2}K) = 0, \qquad \nu = 1, 2, \ldots, N.\end{aligned} \tag{5.52}$$

We define the "ν-th second order response variation matrix":

$$V_\nu(K) = Q_{\nu,2}(K) - R_{\nu,2}(K) - R_{\nu,2}(K)^* \tag{5.53}$$

and find from (5. 50) - (5. 52) that it satisfies

$$(A_0 + B_0K)^* V_\nu(K) + V_\nu(K)(A_0 + B_0K) + (A_{\nu,1} + B_{\nu,1}K)^* R_{\nu,1}(K)^* + R_{\nu,1}(K)(A_{\nu,1} + B_{\nu,1}K)$$
$$+ (F_{\nu,1} + G_{\nu,1}K)^* \tilde{W}(F_{\nu,1} + G_{\nu,1}K) = 0, \qquad \nu = 1, 2, \ldots, N. \tag{5.54}$$

Using (5. 40), (5. 41), (5. 42), (5. 50), (5. 53), (5. 54) we see that to second order in ϵ

$$\delta J_\nu(X_0, K, V, \epsilon) = \mathrm{Tr}\, M(\epsilon^2 V_\nu(K) + \mathcal{O}(\epsilon^3)) \tag{5.55}$$

and the cost (5. 25) reduces to

$$\mathrm{Tr}\, M(\sigma_0 Q_0(K) + \epsilon^2 \sum_{\nu=1}^{N} \sigma_\nu V_\nu(K) + \mathcal{O}(\epsilon^3)) .$$

It should be noted from (5. 54) that the matrices $A_{\nu,2}$, $B_{\nu,2}$, $F_{\nu,2}$, $G_{\nu,2}$ play no role in forming the V_ν, $\nu = 1, 2, \ldots, N$.

We absorb the coefficient ϵ^2 in the weighting factors σ_ν, ignore the term $\mathcal{O}(\epsilon^3)$, and propose the

Response Variation Method for Sensitivity Reduction. Determine an $m \times n$ feedback matrix K solving the problem

$$\min_K \{\mathrm{Tr}\, M(\sigma_0 Q_0(K) + \sum_{\nu=1}^{N} \sigma_\nu V_\nu(K))\} \equiv \mathrm{Tr}\, M\, P_\sigma(K) \tag{5.56}$$

with the constraint that $A_0 + B_0K$ shall be a stability matrix for all K under consideration, i. e. , that $K \in \mathcal{S}(A_0, B_0)$.

The principal computational advantage to be expected in use of this method derives from the fact that the coefficient matrix $A_0 + B_0K$ occurs in (5. 22) and also in (5. 49), (5. 54) for $\nu = 1, 2, \ldots, N$. Combining (5. 22), (5. 54) we see that $P_\sigma(K)$, defined by (5. 56), satisfies

$$(A_0+B_0K)^* P_\sigma(K) + P_\sigma(K)(A_0+B_0K) + \sigma_0[(F_0+G_0K)^*\tilde{W}(F_0+G_0K) + K^*\tilde{U}K]$$
$$+ \sum_{\nu=1}^{N} \sigma_\nu [(A_{\nu,1}+B_{\nu,1}K)^* R_{\nu,1}(K)^* + R_{\nu,1}(K)(A_{\nu,1}+B_{\nu,1}K) + (F_{\nu,1}+G_{\nu,1}K)^*\tilde{W}(F_{\nu,1}+G_{\nu,1}K)] = 0, \tag{5.57}$$

The complete system of constraints involves (5. 22), (5. 49) and (5. 57), since the $R_{\nu,1}$ appear in (5. 57) and, via (5. 49), depend in turn on Q_0, which satisfies (5. 22).

The usual process of replacing K by $\hat{K} + \epsilon\delta K$, taking the derivative at $\epsilon = 0$ and using the adjoint equations provides necessary conditions which must be satisfied by an optimal feedback matrix $\hat{K}$. The condition

$$\frac{d}{d\epsilon} \operatorname{Tr} M P_\sigma(\hat{K} + \epsilon\delta K)\Big|_{\epsilon = 0} = 0$$

becomes, with $Y_0(K)$ defined by (5. 26)

$$\begin{aligned} 2 \operatorname{Tr} Y_0(\hat{K})[&\delta K^* B_0^* P_\sigma(\hat{K}) + \sigma_0(\delta K^* G_0^* \tilde{W} F_0 + \delta K^* G_0^* \tilde{W} G_0 \hat{K} + \delta K^* \tilde{U} \hat{K}) \\ &+ \sum_{\nu=1}^{N} \sigma_\nu (\delta K^* G_{\nu,1}^* \tilde{W} F_{\nu,1} + \delta K^* G_{\nu,1}^* \tilde{W} G_{\nu,1} \hat{K} \\ &+ \delta K^* B_{\nu,1}^* R_{\nu,1}(\hat{K})^* + \frac{d}{d\epsilon} R_{\nu,1}(\hat{K} + \epsilon\delta K)\Big|_{\epsilon = 0} (A_{\nu,1} \quad B_{\nu,1}\hat{K})] \, . \end{aligned} \tag{5. 58}$$

The term

$$\operatorname{Tr} Y_0(\hat{K}) \frac{d}{d\epsilon} R_{\nu,1}(\hat{K} + \epsilon\delta K)\Big|_{\epsilon = 0} (A_{\nu,1} + B_{\nu,1}\hat{K}) \tag{5. 59}$$

is the one which causes the most complication here. Differentiation of (5. 49) with $K = \hat{K} + \epsilon\delta K$ shows that

$$\begin{aligned} (A_0 + B_0\hat{K})^* &\frac{d}{d\epsilon} R_{\nu,1}(\hat{K} + \epsilon\delta K)\Big|_{\epsilon = 0} + \frac{d}{d\epsilon} R_{\nu,1}(\hat{K} + \epsilon\delta K)\Big|_{\epsilon = 0} (A_0 + B_0\hat{K}) \\ &+ \delta K^* B_0^* R_{\nu,1}(\hat{K}) + R_{\nu,1}(\hat{K}) B_0 \delta K + \frac{d}{d\epsilon} Q_0(\hat{K} + \epsilon\delta K)(A_{\nu,1} + B_{\nu,1}\hat{K}) \\ &+ Q_0 B_{\nu,1} \delta K + \delta K^* G_0^* \tilde{W} F_{\nu,1} + F_0^* \tilde{W} G_{\nu,1} \delta K \\ &+ \delta K^* G_0^* \tilde{W} G_{\nu,1} \hat{K} + \hat{K}^* G_0^* \tilde{W} G_{\nu,1} \delta K = 0 \, . \end{aligned} \tag{5. 60}$$

The fact that (5. 59) involves δK implicitly via (5. 60) makes it impossible to give any equation expressing $\hat{K}$ explicitly in terms of $Y_0(\hat{K})$, $P_\sigma(\hat{K})$, etc. If we let

$$(A + B_0\hat{K}) Y_{1,\nu}(\hat{K}) + Y_{1,\nu}(\hat{K})(A + B_0\hat{K})^* + (A_{\nu,1} + B_{\nu,1}\hat{K}) Y_0(\hat{K}) = 0 \, , \quad \nu = 1, 2, \ldots, N \, ,$$

then (5.58) becomes, dividing by 2,

$$\begin{aligned}\operatorname{Tr} \delta K^* \{[B_0^* P_\sigma(\hat{K}) &+ \sigma_0 (G_0^* \tilde{W} F_0 + G_0^* \tilde{W} G_0 \hat{K} + \tilde{U}\hat{K}) \\ &+ \sum_{\nu=1}^{N} \sigma_\nu (G_{\nu,1}^* \tilde{W} F_{\nu,1} + G_{\nu,1}^* \tilde{W} G_{\nu,1} \hat{K} + B_{\nu,1}^* R_{\nu,1}(\hat{K})^*] Y_0(\hat{K}) \\ &+ [\sum_{\nu=1}^{N} \sigma_\nu (B_0^* R_{\nu,1}(\hat{K}) + G_0^* \tilde{W} F_{\nu,1} + G_0^* \tilde{W} G_{\nu,1} \hat{K}) Y_{1,\nu}(\hat{K})] \\ &+ [\sum_{\nu=1}^{N} \sigma_\nu (B_0^* R_{\nu,1}(\hat{K})^* + G_{\nu,1}^* \tilde{W} F_0 + G_{\nu,1}^* \tilde{W} G_0 \hat{K} + B_{\nu,1}^* Q_0(\hat{K})) Y_{1,\nu}(\hat{K})^*] \\ &+ \sum_{\nu=1}^{N} \sigma_\nu (A_{\nu,1} + B_{\nu,1} \hat{K}) Y_{1,\nu}(\hat{K}) \frac{d}{d\epsilon} Q_0(\hat{K} + \epsilon \delta K) \} = 0 .\end{aligned}$$

Then further setting

$$(A_0 + B_0 \hat{K}) Y_{2,\nu}(\hat{K}) + Y_{2,\nu}(\hat{K})(A_0 + B_0 \hat{K})^* + (A_{\nu,1} + B_{\nu,1} \hat{K}) Y_{1,\nu}(\hat{K}) = 0, \quad \nu = 1,2,\dots, N$$

and differentiating (5.22), with Q_0 replaced by $Q_0(\hat{K} + \epsilon\delta K)$, at $\epsilon = 0$, we finally have (5.58) in the form

$$\operatorname{Tr} \delta K^* E(\hat{K}) = 0$$

and the necessary condition for optimality of K is

$$E(\hat{K}) = 0$$

where

$$\begin{aligned}E(K) = B_0^* P_\sigma(K) &+ \sigma_0 (G_0^* \tilde{W} F_0 + G_0^* \tilde{W} G_0 K + \tilde{U} K) \\ &+ \sum_{\nu=1}^{N} \sigma_\nu [(G_{\nu,1}^* \tilde{W} F_{\nu,1} + G_{\nu,1}^* \tilde{W} G_{\nu,1} K + B_{\nu,1}^* R_{\nu,1}(K)^*) Y_0(K) \\ &+ (B_0^* R_{\nu,1}(K) + G_0^* \tilde{W} F_{\nu,1} + G_0^* \tilde{W} G_{\nu,1} K) Y_{1,\nu}(K) \\ &+ (B_0^* R_{\nu,1}(K)^* + G_{\nu,1}^* \tilde{W} F_0 + G_{\nu,1}^* \tilde{W} G_0 K + B_{\nu,1}^* Q_0(K)) Y_{1,\nu}(K)^* \\ &+ (B_0^* Q_0(K) + G_0^* \tilde{W} F_0 + G_0^* \tilde{W} G_0 K + \tilde{U} K) Y_{2,\nu}(K) \\ &+ (B_0^* Q_0(K) + G_0^* \tilde{W} F_0 + G_0^* \tilde{W} G_0 K + \tilde{U} K) Y_{2,\nu}(K)^* .\end{aligned} \qquad (5.61)$$

The gradient method (see, e. g. [3]) is

$$K_i \to K_i + \epsilon\delta K_i \equiv K_{i+1}, \quad \epsilon \text{ is small},$$

$$\delta K_i = -E(K_i).$$

A variant of Axsäter's algorithm (or equally well the Newton, Kleinman method), consists in approximating $\hat{K}$ by solving

$$E_1(K_i, K_{i+1}) + E_0(K_i) = 0, \tag{5.62}$$

where

$$E_1(K_i, K_{i+1}) = \sigma_0(G_0^*\tilde{W}G_0K_{i+1} + \tilde{U}K_{i+1})$$
$$+\sum_{\nu=1}^{N}\sigma_\nu[(G_{\nu,1}^*\tilde{W}G_{\nu,1}K_{i+1})Y_0(K_i) + (G_0^*\tilde{W}G_{\nu,1}K_{i+1})Y_{1,\nu}(K_i)$$
$$+ (G_{\nu,1}^*\tilde{W}G_0K_{i+1})Y_{1,\nu}(K_i)^* + (G_0^*\tilde{W}G_0K_{i+1} + \tilde{U}K_{i+1})Y_{2,\nu}(K_i)$$
$$+ (G_0^*\tilde{W}G_0K_{i+1} + \tilde{U}K_{i+1})Y_{2,\nu}(K_i)^*]$$

and the definition of $E_0(K_i)$ should be clear. It will be noted that if $G(\alpha) \equiv G(0) = G_0$, so that the $G_{\nu,1}$ are all zero, we have, from (5.62),

$$K_{i+1} = -U^{-1}E_0(K_i)(\sigma_0 I + \sum_{\nu=1}^{N}\sigma_\nu(Y_{2,\nu}(K_i) + Y_{2,\nu}(K_i)^*))^{-1}$$

(with U defined by $U = \tilde{U} + G_0^*\tilde{W}G_0$), provided the inverse appearing on the right does, in fact, exist. This will certainly be true if the σ_ν are small relative to σ_0.

An approximation, which makes sense if the σ_ν are small relative to σ_0 and which, in fact, often works fairly well even if this is not the case, consists in replacing $Q_0(K)$ in the formula (5.49) for the $R_{\nu,1}(K)$, by $Q_0(K_0)$, where K_0, as described earlier, is the optimal feedback for the nominal plant (5.9) with respect to the cost (5.5). If we do this, $R_{\nu,1} = R_{\nu,1}(K_0)$ does not vary with K and we obtain the

<u>Simplified Response Variation Weighting Method</u>. Determine $\tilde{K} \in \mathcal{J}(A_0, B_0)$ so as to

$$\min_{K \in \mathscr{S}(A_0, B_0)} \operatorname{Tr} M \tilde{P}_\sigma (K) \tag{5.63}$$

where

$$(A_0 + B_0K)^* \tilde{P}_\sigma (K) + \tilde{P}_\sigma (K)(A_0 + B_0K) + \sigma_0[(F_0 + G_0K)^* \tilde{W}(F_0 + G_0K) + K^*\tilde{U}K]$$
$$+ \sum_{\nu=1}^{N} \sigma_\nu[(A_{\nu,1} + B_{\nu,1}K)^* R_{\nu,1}(K_0)^* + R_{\nu,1}(K_0)(A_{\nu,1} + B_{\nu,1}K)$$
$$+ (F_{\nu,1} + G_{\nu,1}K)^* \tilde{W}(F_{\nu,1} + G_{\nu,1}K)] = 0 \tag{5.64}$$

where $R_{\nu,1}(K_0)$ satisfies (5.49) with $Q_0 \equiv Q_0(K_0)$.

For this simplified problem the condition for optimality of $\tilde{K}$, i.e.,

$$\frac{d}{d\epsilon} \operatorname{Tr} M \tilde{P}_\sigma (\tilde{K} + \epsilon \delta K)\Big|_{\epsilon = 0} = 0$$

for all δK, becomes

$$2 \operatorname{Tr} \delta K^* [B_0^* \tilde{P}_\sigma (\tilde{K}) + \sigma_0 (G_0^* \tilde{W} F_0 + G_0^* \tilde{W} G_0 \tilde{K} + \tilde{U} \tilde{K})$$
$$+ \sum_{\nu=1}^{N} \sigma_\nu (B_{\nu,1}^* R_{\nu,1}(K_0)^* + G_{\nu,1}^* \tilde{W} F_{\nu,1} + G_{\nu,1}^* \tilde{W} G_{\nu,1} \tilde{K})] M = 0$$

and we have, as the necessary condition

$$\tilde{K} = -(\sigma_0 U + \sum_{\nu=1}^{N} \sigma_\nu G_{\nu,1}^* \tilde{W} G_{\nu,1})^{-1} [B_0^* \tilde{P}_\sigma (\tilde{K}) + \sigma_0 G_0^* \tilde{W} F_0$$
$$+ \sum_{\nu=1}^{N} \sigma_\nu (B_{\nu,1}^* R_{\nu,1}(K_0)^* + G_{\nu,1}^* \tilde{W} F_{\nu,1})] . \tag{5.65}$$

The iteration method

$$K_{i+1} = - (\sigma_0 U + \sum_{\nu=1}^{N} \sigma_\nu G_{\nu,1}^* \tilde{W} G_{\nu,1})^{-1} [B_0^* \tilde{P}_\sigma (\tilde{K}) + \sigma_0 G_0^* \tilde{W} F_0$$
$$+ \sum_{\nu=1}^{N} \sigma_\nu (B_{\nu,1}^* R_{\nu,1}(K_0)^* + G_{\nu,1}^* \tilde{W} F_{\nu,1})] \tag{5.66}$$

works very effectively in this case.

It is likely that in most applications the simplified response variation weighting method will be adequate and the greatly increased computational requirements associated with the first two methods will not be justifiable.

With regard to comparison between the first offered "Optimal Sensitivity Reduction Problem for Finite Parameter Variations" and the second offered "Response Variation Weighting Method" we can say that the second is computationally simpler only if advantage is taken of the fact that in the latter the homogeneous parts of all Liapounov equations are the same - involving only $A_0 + B_0K$ rather than involving $A_\nu + B_\nu K$, $\nu = 1, 2, \ldots, N$, also. If a "transformational method" such as is described in V-[4] is used to solve the Liapounov equations, this can result in significant reduction of computational effort.

Throughout this section we have shown the system matrices as functions of the parameter vector α, viz.: $A(\alpha)$, $B(\alpha)$, $F(\alpha)$, $G(\alpha)$. Quite commonly however, no such representations will be available; all one may have is the nominal system (5.9), i.e., $\dot{x} = A_0x + B_0u$, $y = F_0x + G_0u$ and certain perturbed systems (cf. (5.10))

$$\begin{aligned} \dot{\xi}_\nu &= (A_0 + \delta A_\nu)\xi_\nu + (B_0 + \delta B_\nu)u \\ \eta_\nu &= (F_0 + \delta F_\nu)\xi_\nu + (G_0 + \delta G_\nu)u \end{aligned} \qquad \nu = 1, 2, \ldots, N. \quad (5.67)$$

The method for finite parameter variations may, of course, be used but one of the other methods, particularly the simplified response variation method, is likely to be more attractive. Defining

$$\alpha = (\alpha^1, \alpha^2, \ldots, \alpha^N)^*$$

we can set

$$A(\alpha) = A_0 + \sum_{\nu=1}^{N} \alpha^\nu \delta A_\nu, \text{ etc.},$$

and we will have (cf. (5.44) - (5.47))

$$A_{\nu,1} = \delta A_\nu, \; B_{\nu,1} = \delta B_\nu, \text{ etc.}$$

Suppose this approach were applied to the crane of Chapter I in the level hold operating condition. Taking $L = \sqrt{2}$, we study the nominal case

$$\psi = 45^\circ, \quad \ell = 1$$

and a single perturbed case

$$\psi = 60^{\circ}, \quad \ell_1 = \frac{1}{\sqrt{2}} . \tag{5.68}$$

For the nominal system we have (cf. III, Ex. 1)

$$A_0 = \begin{pmatrix} 0 & 1 & 0 & 0 \\ -1 & 0 & 0 & 0 \\ 0 & 0 & 0 & 1 \\ 0 & 0 & 0 & 0 \end{pmatrix}, \quad B_0 = \begin{pmatrix} 0 \\ -1 \\ 0 \\ 1 \end{pmatrix}$$

and the perturbations which yield the system corresponding to the linearization about (5.68) are

$$\delta A = \begin{pmatrix} 0 & 0 & 0 & 0 \\ -.4142 & 0 & 0 & 0 \\ 0 & 0 & 0 & 0 \\ 0 & 0 & 0 & 0 \end{pmatrix}, \quad \delta B = 0 .$$

The responses of interest are the horizontal displacement and velocity of the load. At the nominal configuration the linearized responses correspond to

$$F_0 = \begin{pmatrix} 1 & 0 & 1 & 0 \\ 0 & 1 & 0 & 1 \end{pmatrix}, \quad G_0 = 0$$

and the perturbations corresponding to (5.68) are

$$\delta F = \begin{pmatrix} -.2929 & 0 & -.2929 & 0 \\ 0 & -.2929 & 0 & -.2929 \end{pmatrix}, \quad \delta G = 0 .$$

The first step is to compute the optimal control for the nominal system relative to the cost (we take $\tilde{W} = I$, $\tilde{U} = I = 1$)

$$\int_0^{\infty} x^* F_0^* F_0 x + (u)^2 \, dt .$$

This corresponds, of course, to $u = K_0 x$ with

$$K_0 = -B_0^* Q_0 ,$$

$$A_0^* Q_0 + Q_0 A_0 + F_0^* F_0 - Q_0 B_0 B_0^* Q_0 = 0 .$$

The result is computed* as

$$Q_0 = \begin{pmatrix} 2.7158 & -.3110 & .6148 & -1.5846 \\ -.3110 & 3.4988 & 2.2736 & 4.1136 \\ .6148 & 2.2736 & 2.7472 & 3.2736 \\ -1.5846 & 4.1136 & 3.2736 & 6.8608 \end{pmatrix}$$

and, from IV -(3.17), we have the optimal feedback matrix for the nominal system

$$K_0 = (1.2736, \ -.6148, \ -1.0000, \ -2.7472).$$

The matrix $R_{\nu,1}(K_0)$ which here is just a single matrix which we will call $R_{1,0}$ is obtained in the case of the simplified response variation method (cf. (5.63), (5.64)ff) by solving

$$(A_0 + B_0 K)^* R_{1,0} + R_{1,0}(A_0 + B_0 K_0) + Q_0 \delta A + F_0^* \delta F = 0 \tag{5.69}$$

and then the response variation matrix $V_{\nu,1}(K_0) \equiv V_{1,0}$ is computed from (cf. (5.54))

$$(A_0 + B_0 K)^* V_{1,0} + V_{1,0}(A_0 + B_0 K) + \delta A^* R_{1,0}^* + R_{1,0} \delta A + \delta F^* \delta F = 0 \tag{5.70}$$

with the result

$$V_{1,0} = \begin{pmatrix} .9537 & .0406 & .1621 & -.3178 \\ .0406 & 1.9675 & .8040 & 2.1032 \\ .1621 & .8040 & .4417 & .8469 \\ -.3178 & 2.1032 & .8469 & 2.4271 \end{pmatrix} . \tag{5.71}$$

To attempt to improve sensitivity, i. e., to reduce the response variation, we determine $\tilde{K}$ so as to solve the problem (5.63). For the present

* These and other computations listed below were performed by S. Neftci under the auspices of Honeywell, Inc.

case the equation for P_σ has the form, for general K and with $\sigma_0 = 1$,

$$(A_0 + B_0K)^* P_\sigma(K) + P_\sigma(K)(A_0 + B_0K) + F_0^* F_0 + K^* K + \sigma_1[\delta A^* R_{1,0}^* + R_{1,0}\delta A + \delta F^* \delta F] = 0.$$

Taking

$$\sigma = 2$$

the optimal $\tilde{K}$ is, in this case,

$$\tilde{K} = -B^* Q_\sigma \quad (\equiv -B^* P_\sigma(\tilde{K}))$$

where Q_σ solves the Riccati equation

$$A_0^* Q_\sigma + Q_\sigma A_0 + F_0^* F_0 + 2[\delta A^* R_{1,0}^* + R_{1,0}\delta A + \delta F^* \delta F] - Q_\sigma B_0 B_0^* Q_\sigma = 0,$$

giving

$$P_\sigma(\tilde{K}) = Q_\sigma = \begin{pmatrix} 4.3653 & -.1972 & .9542 & -2.0363 \\ -.1972 & 7.2277 & 3.7727 & 8.1092 \\ .9542 & 3.7727 & 3.5705 & 4.8551 \\ -2.0363 & 8.1092 & 4.8551 & 11.4080 \end{pmatrix},$$

$$\tilde{K} = (1.8391, \quad -.8815, \quad -1.0824, \quad -3.2987).$$

To test this result, $R_{\nu,1}(\tilde{K}) \equiv R_1$ is computed, replacing K_0 in (5.69) by $\tilde{K}$, and Q_0 by the solution Q_1 of

$$(A_0 + B\tilde{K})^* Q_1 + Q_1(A_0 + B\tilde{K}) + F_0^* F_0 + K_1^* K_1 = 0, \tag{5.72}$$

and then $V_{\nu,1}(\tilde{K}) \equiv V_1$ is computed, replacing $R_{1,0}$ by R_1 in (5.70). The result is

$$V_1 = \begin{pmatrix} .6250 & .0772 & .1824 & -.1026 \\ .0772 & 1.6911 & .6664 & 1.8274 \\ .1824 & .6664 & .3765 & .7061 \\ -.1026 & 1.8274 & .7061 & 2.0544 \end{pmatrix} \tag{5.73}$$

which can indeed, be seen to be reduced from $V_{1,0}$ as given by (5.71). Of course the nominal performance suffers somewhat. This is given by

$$Q_1 = \begin{pmatrix} 2.8953 & -.3397 & .6041 & -1.7158 \\ -.3397 & 3.6322 & 2.3437 & 4.2548 \\ .6041 & 2.3437 & 2.7861 & 3.3469 \\ -1.7158 & 4.2548 & 3.3469 & 7.0704 \end{pmatrix} \tag{5.74}$$

but while the relative or "percentage" improvement of V_1 over $V_{1,0}$ is substantial, the relative deterioration of Q_1 as compared with Q_0 is fairly minor. Greater reduction of the response variation can be obtained by taking $\sigma > 2$ with, of course, some further deterioration in nominal performance.

Perhaps one of the strong points of the analysis presented in this section is the idea of the response variation and the associated response variation matrix. The latter provides a convenient, quantitative measure of sensitivity with respect to parameter variations which we may expect to prove useful in many applications. The multiple objective methods of Section 1 may, if desired, be used to attempt to decrease V with constraints on either response or control cost. See the exercises for other applications.

6. Robustness and Maneuverability

We again consider a linear control system

$$\dot{x} = A(\alpha)x + B(\alpha)u \tag{6.1}$$

defined for $\alpha \in \mathcal{A}$, which serves to describe the set of admitted parameter variations. We will say, qualitatively, that such a system is robust with respect to a given measure of performance if that measure remains at satisfactory levels throughout the range, $\mathcal{A}$, of admitted parameter variations.

Obviously a wide range of possible performance criteria might be admitted for discussion. We will confine our discussion to some material bearing on robustness of the stability property. We assume that the nominal plant corresponds to $\alpha = 0$:

$$\dot{x} = A_0x + B_0u = A(0)x + B(0)u, \tag{6.2}$$

we assume that this system is stabilizable and that, by some means, a feedback matrix K has been selected so that

$$S_0 \equiv A_0 + B_0 K \tag{6.3}$$

is a stability matrix. The nominal closed loop system is then

$$\dot{x} = S_0 x = (A_0 + B_0 K)x. \tag{6.4}$$

For a given parameter variation $0 \to \alpha$ we define

$$\left.\begin{aligned} \delta A &= A(\alpha) - A_0 , \\ \delta B &= B(\alpha) - B_0 . \end{aligned}\right\} \tag{6.5}$$

If the feedback relation $u = Kx$ continues to be used in the system (6. 1), i. e., in

$$\dot{x} = (A_0 + \delta A)x + (B_0 + \delta B)u \tag{6.6}$$

we realize the perturbed closed loop system

$$\dot{x} = (A_0 + \delta A + (B_0 + \delta B)K)x = (S_0 + \delta S)x, \quad \delta S = \delta A + \delta BK . \tag{6.7}$$

We then have the following result.

Theorem 6. 1. *Let P be a symmetric positive definite matrix and let Z be the solution of the Liapounov equation*

$$S_0^* Z + ZS_0 + P = 0 . \tag{6.8}$$

Then $S_0 + \delta S$ remains a stability matrix if

$$P - \delta S^* Z - Z\delta S > 0 \tag{6.9}$$

which in turn is assured if

$$\delta S P^{-1} \delta S^* < \frac{1}{4} Z^{-1} P Z^{-1} \tag{6.10}$$

and these are the best possible estimates of their kind.

Proof. The equation (6. 8) is equivalent to

$$(S_0 + \delta S)^* Z + Z(S_0 + \delta S) + P - \delta S^* Z - Z\delta S = 0 . \tag{6.11}$$

Since S_0 is a stability matrix and P is symmetric positive definite, Z is symmetric and positive definite. Then, if $P - \delta S^* Z - Z\delta S$ is positive definite, it follows from III, Theorem 1.5 that $S_0 + \delta S$ is a stability matrix if (6.9) holds.

A condition equivalent to (6.9) is, clearly

$$Z^{-1}PZ^{-1} - Z^{-1}\delta S^* - \delta S Z^{-1} > 0 \tag{6.12}$$

With $P^{\frac{1}{2}}$ the positive definite square root of P, this inequality may be rewritten in the form

$$Z^{-1}PZ^{-1} - Z^{-1}P^{\frac{1}{2}}P^{-\frac{1}{2}}\delta S^* - \delta S P^{-\frac{1}{2}}P^{\frac{1}{2}}Z^{-1} > 0 .$$

Then we note that

$$0 \leq (\tfrac{1}{\sqrt{2}}Z^{-1}P^{\frac{1}{2}} \pm \sqrt{2}\,\delta S P^{-\frac{1}{2}})(\tfrac{1}{\sqrt{2}}Z^{-1}P^{\frac{1}{2}} \pm \sqrt{2}\,\delta S P^{-\frac{1}{2}})^*$$

$$= \tfrac{1}{2}Z^{-1}PZ^{-1} \pm Z^{-1}\delta S^* \pm \delta S Z^{-1} + 2\delta S P^{-1}\delta S^*$$

so that

$$\pm (Z^{-1}\delta S + \delta S Z^{-1}) \leq \tfrac{1}{2}Z^{-1}PZ^{-1} + 2\delta S P^{-1}\delta S^* .$$

Therefore

$$Z^{-1}PZ^{-1} - Z^{-1}\delta S^* - \delta S Z^{-1} \geq \tfrac{1}{2}Z^{-1}PZ^{-1} - 2\delta S P^{-1}\delta S^* \tag{6.13}$$

and it follows that (6.12) obtains if the right hand side of (6.13) is positive, which is the same as (6.10).

That this is the best possible estimate of its kind follows from the fact that if we take

$$\delta S = \mu Z^{-1}P \tag{6.14}$$

then (6.9) and (6.12) are both violated just in case $\mu > \frac{1}{2}$. Substituting (6.14) into (6.11) gives

$$(S_0 + \mu Z^{-1}P)^* Z + Z(S_0 + \mu Z^{-1}P) + (1 - 2\mu)P = 0 ,$$

from which it follows that $S_0 + \mu Z^{-1}P$ has purely imaginary eigenvalues (see Exercise 14 below) for $\mu = \frac{1}{2}$ and is completely unstable for $\mu > \frac{1}{2}$.

Q. E. D.

The condition (6. 9) is sharper than (6. 10): (6. 9) may hold is some cases when (6. 10) does not. It is also easier to compute with in many cases.

The foregoing result can be used to obtain estimates on how far the eigenvalues of a stability matrix S_0 are separated from the imaginary axis.

Corollary 6. 2. *If Z satisfies (6. 8) then the eigenvalues λ_i of S_0 satisfy*

$$\mathrm{Re}\,(\lambda_i) \le -\mu_1/2, \tag{6.15}$$

where μ_1 is the largest positive number such that

$$P - \mu_1 Z > 0. \tag{6.16}$$

Equivalently, μ_1 is the smallest eigenvalue of $L^* Z^{-1} L$ whenever $LL^* = P$.

Proof. Setting $\delta S = \lambda I$, (6. 9) obtains just in case

$$P - 2\,\mathrm{Re}\,\lambda\, Z > 0$$

which is true just in case $\mathrm{Re}\,\lambda < \mu_1/2$, μ_1 as described in (6. 16). So $S_0 + \lambda I$ remains a stability matrix if $\mathrm{Re}\,\lambda < \mu_1/2$, and this implies (6. 15). The last result follows from rewriting (6. 16) in the form

$$LL^* - \mu_1 Z > 0 \Rightarrow I - \mu_1 L^{-1} Z (L^*)^{-1} > 0$$

and then multiplying on the right and on the left by the positive symmetric square root of $L^* Z^{-1} L$.

Q. E. D.

Theorem 6. 1 can be adapted to provide an algorithm for determining computationally the largest positive number $\hat{\epsilon}$ such that the matrix $S_0 + \epsilon\delta S$ remains a stability matrix if $0 \le \epsilon < \hat{\epsilon}$. Solving (6. 8) for $Z \equiv Z_0$, we determine the largest value of ϵ_0 such that

$$P - \epsilon_0(\delta S^* Z_0 + Z_0 + S) \ge 0. \tag{6.17}$$

This can be accomplished by simultaneous diagonalization of the symmetric matrices P and $\delta S^* Z_0 + Z_0 \delta S$, the first of which is positive definite. We review the procedure. Since P is symmetric positive definite, it may be expressed in the form $P = LL^*$, where L is nonsingular and lower triangular (see, e.g., V-[1]). Multiplying (6.17) on the left by L^{-1}, on the right by $(L^*)^{-1}$, it becomes

$$I - \epsilon_0 L^{-1}(\delta S^* Z_0 + Z_0 \delta S)(L^*)^{-1} > 0 . \tag{6.18}$$

Since $L^{-1}(\delta S^* Z_0 + Z_0 \delta S)(L^*)^{-1}$ is symmetric, it can be diagonalized by an orthogonal matrix Q_0 $(Q_0^* = Q_0^{-1})$. Multiplying (6.18) on the left by Q_0^* and on the right by Q_0, it becomes

$$Q_0^* Q_0 - \epsilon_0 Q_0^* L^{-1}(\delta S^* Z_0 + Z_0 \delta S)(L^*)^{-1} Q_0 > 0$$

or

$$I - \epsilon_0 \operatorname{diag}(\mu_{0,1}, \mu_{0,2}, \ldots, \mu_{0,n}) > 0,$$

whence

$$\epsilon_0 = 1/\max \{\mu_{0,1}, \mu_{0,2}, \ldots, \mu_{0,n}\}.$$

The next step is to replace S_0 by

$$S_1 = S_0 + \epsilon_0 \delta S ,$$

and it may be seen from (6.17) that the eigenvalues of S_1 have non-positive real parts. We attempt to solve

$$S_1^* Z_1 + Z_1 S_1 + P = 0 .$$

If S_1 has an eigenvalue with zero real part, this will fail and $\hat{\epsilon} = \epsilon_0$. Otherwise $Z_1 > 0$ is obtained and we may proceed to find the largest ϵ_1 such that

$$P - \epsilon_1 (\delta S^* Z_1 + Z_1 \delta S) \geq 0$$

just as above. Continuing this process we obtain sequences $\{S_k\}$, $\{Z_k\}$, of matrices and a sequence $\{\epsilon_k\}$ of scalars. For each k, the eigenvalues of $S_k \equiv S_0 + (\epsilon_0 + \epsilon_1 + \ldots + \epsilon_{k-1})\delta S$ have non-positive real parts. If for any k there is an eigenvalue of S_k with zero real part, we will not be able to compute Z_k and we shall have $\hat{\epsilon} = \epsilon_0 + \epsilon_1 + \ldots + \epsilon_k$. Otherwise it may be seen that

$$\hat{\epsilon} = \sum_{k=0}^{\infty} \epsilon_k \quad \text{(which may } = +\infty\text{)}.$$

This result was obtained in [5] by M. Eslami. If $\hat{\epsilon} < \infty$, the equations

$$S_k^* Z_k + Z_k S_k + P = 0, \quad S_k = S_0 + (\sum_{\ell=0}^{k-1} \epsilon_k)\, \delta S$$

become progressively more ill-conditioned as $k \to \infty$ and the process will have to stop for some finite value of k due to numerical difficulties.

A number of variations on this procedure are possible to accommodate differing computational capabilities. In some cases no routine for simultaneous diagonalization will be available. Rather than finding the largest value of ϵ_k such that

$$P - \epsilon_k(\delta S^* Z_k + Z_k \delta S) \geq 0, \tag{6.19}$$

which is the essential feature of the k-th step of the method just described, one may find the smallest eigenvalue, λ_{min}, of the matrix P and the largest eigenvalue, μ_{max}, of the matrix $\delta S^* Z_k + Z_k \delta S$. Since

$$P \geq \lambda_{min} I, \quad \mu_{max} I \geq \delta S^* Z_k + Z_k \delta S,$$

we have

$$P - \eta_k(\delta S^* Z_k + Z_k \delta S) \geq 0 \tag{6.20}$$

for

$$\eta_k = \frac{\lambda_{min}}{\mu_{max}} .$$

It can be shown (see Exercise 15 below) that $\eta_0 + \eta_1 + \eta_2 + \ldots$ will still converge to $\hat{\epsilon}$, but the convergence will ordinarily be slower than for the ϵ_k determined by (6.19). The procedure (6.20) can be used whenever a routine for computing the eigenvalues of a symmetric matrix is available.

A further modification involves the use of (6.10), or, equivalently,

$$Z \delta S P^{-1} \delta S^* Z < \tfrac{1}{4} P .$$

One determines the smallest eigenvalue λ_{min}, of P and the largest eigenvalue, ν_{max}, of $Z\delta SP^{-1}\delta S^* Z$. Then

$$\tfrac{1}{4}P > \tfrac{1}{4}\lambda_{min}, \quad Z_k \delta SP^{-1}\delta S^* Z_k < \nu_{max},$$

and with

$$\theta_k = \tfrac{1}{2}\sqrt{\frac{\lambda_{min}}{\nu_{max}}} \tag{6.21}$$

we have

$$\theta_k^2 Z_k \delta SP^{-1}\delta S^* Z_k \leq \tfrac{1}{4}P \tag{6.22}$$

and $S_k + \theta\delta S$ is a stability matrix for $\theta < \theta_k$. This method, with $P = I$, is particularly useful if no general eigenvalue routines are available, because the largest eigenvalue of $Z_k \delta S\delta S^* Z_k \equiv A_k$ can be obtained iteratively by choosing a (more or less) arbitrary vector y_0 with $\|y_0\| = 1$, computing

$$z_0 = A_k y_0$$

and setting

$$y_1 = \frac{z_0}{\|z_0\|} \quad (\text{thus } \|y_1\| = 1).$$

In the general step $\|y_j\| = 1$ and

$$z_j = A_k y_j$$

$$y_{j+1} = \frac{z_j}{\|z_j\|}.$$

It can be shown (V - [1]) that

$$\nu_{max} = \lim_{j \to \infty} \frac{\|z_j\|}{\|y_j\|}$$

if y_0 is not orthogonal to the eigenspace of A_k corresponding to the eigenvalue ν_{max}. Here $\lambda_{min} = 1$ since $P = I$ and we have

$$\theta_k = \frac{1}{2\sqrt{\nu_{max}}} \tag{6.23}$$

in place of (6.21). It will still be true that $\theta_0 + \theta_1 + \theta_2 + \ldots$ converges to $\hat{\epsilon}$.

Preparatory to an example illustrating one of the uses of this robustness analysis, we digress to a short discussion of distortion in feedback signals. It will be convenient to consider complex systems and solutions in this discussion.

We have represented the nominal closed loop system in the form

$$\dot{x} = S_0 x = (A_0 + B_0 K)x, \tag{6.24}$$

which corresponds, of course, to setting the control $u = Kx$. In practice the measurement of the state x in order to form Kx and, once that expression is formed mathematically or electronically, the process of implementing the control $u = Kx$, are both subject to unavoidable distortion due to noise, use of dynamic devices in the measurement or signal propagation devices, electrical resistance, etc. Hence the control actually implemented is not precisely Kx.

We provide an example (admittedly rather contrived) to show what we mean. Consider a controlled linear oscillator with small damping force and moderate restoring force, say

$$\ddot{x} + \delta\dot{x} + x = u, \qquad \delta > 0, \text{ small},$$

or, with $x^1 = x$, $x^2 = \dot{x}$,

$$\begin{pmatrix} \dot{x}^1 \\ \dot{x}^2 \end{pmatrix} = \begin{pmatrix} 0 & 1 \\ -1 & -\delta \end{pmatrix} \begin{pmatrix} x^1 \\ x^2 \end{pmatrix} + \begin{pmatrix} 0 \\ 1 \end{pmatrix} u . \tag{6.25}$$

A crude way to improve the regulation of such a system is the following. Since the acceleration, $\dot{x}^2 = \ddot{x}$, can be sensed by physical means, one could set

$$u = (1 - \delta)\dot{x}^2 \quad (\text{equiv. } u = (1 - \tfrac{1}{\delta})x^1 + (\delta - 1)x^2)$$

to give a closed loop system

$$\begin{pmatrix} \dot{x}^1 \\ \dot{x}^2 \end{pmatrix} = \begin{pmatrix} 0 & 1 \\ -\frac{1}{\delta} & -1 \end{pmatrix} \begin{pmatrix} x^1 \\ x^2 \end{pmatrix}, \tag{6.26}$$

in which the eigenvalues

$$\lambda = \frac{-1 \pm i\sqrt{\frac{4}{\delta} - 1}}{2}$$

have (for $\delta \leq 4$) real part $-\frac{1}{2}$ instead of real part $-\delta/2$ as in (6.25)(for $\delta \leq 2$). This would seem to provide damping on the order of $e^{-t/2}$ for arbitrarily small positive values of δ.

But how is acceleration sensed? Ordinarily by means of an accelerometer. Physically an accelerometer is just another damped linear oscillator mounted on the plant which we wish to regulate. Schematically the configuration has the form indicated in Fig. VII-5. The accelerometer is just another damped mass-spring system mounted on the one which we wish to control. The damping and restoring forces in the accelerometer depend only

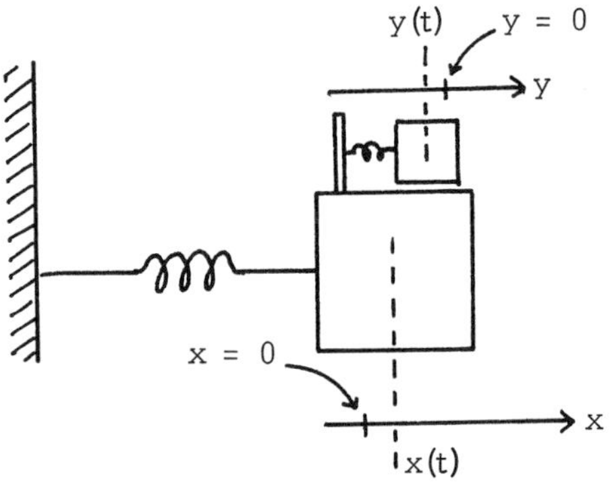

Fig. VII-5. Accelerometer attached to linear oscillator

on the velocity and displacement of the accelerometer mass, m, with respect to its equilibrium position, y = 0, but the total acceleration must be computed with respect to an inertial frame of reference, leading to the equation of motion

$$m(\ddot{y} + \ddot{x}) + \beta\dot{y} + \alpha y = 0 \tag{6.27}$$

or, with $y^1 = y$, $y^2 = \dot{y}$,

$$\begin{pmatrix} \dot{y}^1 \\ \dot{y}^2 \end{pmatrix} = \begin{pmatrix} 0 & 1 \\ -\frac{\alpha}{m} & -\frac{\beta}{m} \end{pmatrix} \begin{pmatrix} y^1 \\ y^2 \end{pmatrix} + \begin{pmatrix} 0 \\ -\dot{x}^2 \end{pmatrix} . \tag{6.28}$$

Assuming $\beta > 0$, $\alpha > 0$, (6.27) is asymptotically stable and a constant acceleration $\dot{x}^2$ would cause $y = y^1$ to approach the equilibrium value

$$y^1 = -\frac{m}{\alpha}\dot{x}^2 .$$

One can show that, if $\dot{x}^2$ does not vary too rapidly, y^1 will not vary too much from $-\frac{m}{\alpha}\dot{x}^2$ when β/m is large compared with $\sqrt{(\beta/m)^2 - 4(\alpha/m)}$.

But let us be more specific. If, e.g., $\dot{x}^2$ has the (for convenience, complex) form

$$\dot{x}^2 = e^{(\rho + i\sigma)t}, \tag{6.29}$$

then all solutions of (6.27) will tend to the particular solution

$$y(t) = \frac{-m}{m(\rho + i\sigma)^2 + \beta(\rho + i\sigma) + \alpha} e^{(\rho + i\sigma)t},$$

which we may rewrite as

$$y(t) = \left(\frac{1}{\frac{m}{\alpha}(\rho + i\sigma)^2 + \frac{\beta}{\alpha}(\rho + i\sigma) + 1}\right)\left(-\frac{m}{\alpha}\dot{x}^2\right) \equiv re^{i\psi}\left(-\frac{m}{\alpha}\dot{x}^2\right). \tag{6.30}$$

Suppose the accelerometer reading, $y(t)$, is accepted as $-\frac{m}{\alpha}\dot{x}^2$. Then we would set

$$u = -\frac{\alpha}{m}(1 - \delta)y. \tag{6.31}$$

But if we use (6.31) when (6.30) obtains, it corresponds in the limit, as $t \to \infty$, to

$$u(t) = -\frac{\alpha}{m}(1 - \delta)re^{i\psi}\left(-\frac{m}{\alpha}\dot{x}^2(t)\right) = re^{i\psi}(1 - \delta)\dot{x}^2(t) \tag{6.32}$$

rather than $u(t) = (1 - \delta)\dot{x}^2(t)$. That is, the dynamics of the accelerometer lead to the asymptotic distorting factor $re^{i\psi}$ in the feedback relation if $\dot{x}^2$ is assumed to have the form (6.29). This distortion may be expected, under certain circumstances, to lead to de-stabilization rather than improved stabilization. In fact, this is a real concern because for small values of δ the closed loop system (6.26) has solutions of the form

$$e^{\left(-\frac{1}{2} \pm i\sqrt{\frac{4}{\delta} - 1}\right)t}$$

which correspond to large values of σ in (6. 29), i. e., (6. 26) has high frequency oscillatory solutions. As $\sigma \to \infty$ we see from (6. 30) that

$$r \to \frac{\alpha}{m\sigma^2} = \frac{\alpha\delta}{m(4 - \delta)}$$

$$e^{i\psi} \to -1 \quad (\psi \to \pi).$$

Thus for small values of δ the feedback signal more closely approximates

$$-\frac{\alpha\delta}{m(4-\delta)}(1 - \delta)\dot{x}^2$$

than it does $(1 - \delta)\dot{x}^2$ - the feedback signal is reduced in amplitude and changed in sign. This casts doubt on the whole procedure, leading us to expect that stability may not be improved in the manner suggested by (6. 26) and, in fact, the system might be unstable.

Let us return to the abstract system

$$\dot{x} = A_0 x + B_0 u,$$

and suppose that one intends a feedback relation

$$u = Kx,$$

leading to the asymptotically stable closed loop system

$$\dot{x} = (A_0 + B_0 K)x \equiv S_0 x \tag{6. 33}$$

but, due to distortions similar to those just discussed in our example, one has something like

$$u = re^{i\psi} Kx$$

with the resulting closed loop system

$$\dot{x} = (A_0 + re^{i\psi} B_0 K)x = (S_0 + \delta S)x \tag{6. 34}$$

where

$$\delta S = (re^{i\psi} - 1)B_0 K.$$

We may then ask the following question: for what values of r and ψ does (6. 33) remain asymptotically stable? If we arbitrarily set $\psi = 0$ so that

$\delta S = rB_0K$, we would expect to obtain some answer such as " (6. 33) remains asymptotically stable if

$$r_0 < r < r_1"$$

where $r_0 < 1$, $r_1 > 1$. We may refer to the interval (r_0, r_1) as the "gain tolerance interval", since r is referred to in the engineering literature as a gain. The number r_1 is often called the gain margin (see VI-[17]). If, on the other hand, we keep $r = 1$ but allow ψ to vary from 0 we would expect an answer such as "(6. 33) remains asymptotically stable if

$$|\psi| < \psi_0 " .$$

The quantity ψ_0 is called the phase margin (see VI-[17] again).

Applying the criterion of Theorem 6. 1, the system (6. 33) remains asymptotically stable if (cf. (6. 9))

$$P - \delta S^* Z - Z\delta S > 0 ,$$

and with $\delta S = (re^{i\psi} - 1)B_0K$ this becomes

$$P - (re^{-i\psi} - 1)K^*B^*Z - (re^{i\psi} - 1)ZBK = P + (1 - r\cos\psi)(K^*B^*Z + ZBK)$$

$$- ir\sin\psi(K^*B^*Z - ZBK) \equiv P_1 + iP_2 > 0. \qquad (6. 35)$$

Whether this is or is not the case for given r, ψ may be determined by simultaneous (unitary) diagonalization of P and $(re^{-i\psi} - 1)K^*B^*Z + (re^{i\psi} - 1)ZBK$. But it is difficult to determine, in this general case, the range of values of r and ψ for which (6. 34) remains asymptotically stable because we cannot, in general, simultaneously diagonalize the three hermitian matrices $P, K^*B^*Z + ZBK$, and $i(K^*B^*Z - ZBK)$.

A particularly nice example arises when K is determined so as to minimize a cost functional

$$\int_0^\infty [x(t)^* W x(t) + u(t)^* U u(t)]\, dt ,$$

W $n \times n$ symmetric,nonnegative with (W, A) observable, U $m \times m$

symmetric positive definite. In this case

$$K = - U^{-1}B_0^*Q \tag{6.36}$$

where (cf. IV-(3.13))

$$A_0^*Q + QA_0 + W - QB_0U^{-1}B_0^*Q = 0. \tag{6.37}$$

We may rearrange (6.36) to

$$(A_0 - B_0U^{-1}B_0^*Q)^*Q + Q(A_0 - B_0U^{-1}B_0^*Q) + W + QB_0U_0^{-1}B^*Q = 0, \tag{6.38}$$

which agrees with (6.8) if we let

$$Z = Q, \quad S_0 = A_0 + B_0K \quad (K \text{ given by } (6.36)),$$

$$P = W + QB_0U^{-1}B_0^*Q .$$

In this case the inequality (6.35) becomes

$$W + QB_0U^{-1}B_0^*Q + (re^{-i\psi} - 1)QB_0U^{-1}B_0^*Q + (re^{i\psi} - 1)QB_0U^{-1}B_0^*Q$$

$$= W + (2r\cos\psi - 1)QB_0U^{-1}B_0^*Q > 0 . \tag{6.39}$$

Let μ be the largest non-negative number such that $W \geq \mu\, QB_0U^{-1}B_0^*Q$. Then (6.39) is true if

$$\mu + 2r\cos\psi - 1 > 0$$

i.e.,

$$r\cos\psi > \frac{1-\mu}{2} . \tag{6.40}$$

When $\psi = 0$ this is true if

$$\frac{1-\mu}{2} < r < \infty,$$

so that the gain tolerance interval is $(\frac{1-\mu}{2}, \infty)$. When $r = 1$ the condition (6.40) gives

$$|\psi| < \cos^{-1}(\frac{1-\mu}{2}),$$

giving a phase margin of at least $\cos^{-1}(\frac{1}{2}) = \frac{\pi}{3} = 60^\circ$. These results are in agreement with those obtained in VI-[17] by use of the Nyquist diagram, and a slightly more complicated argument than the one which we have presented here. An argument more along our lines, but in a more general setting, is given in [6].

Gain and phase may be studied together by regarding r and ψ as polar coordinates of the (Cartesian) point (ξ, η), i. e.

$$\xi = r \cos \psi, \quad \eta = r \sin \psi.$$

Then (6.40) simply becomes

$$\xi > \frac{1-\mu}{2},$$

the half-plane shown in Fig. VII-6. From this we see easily that

(i) as $r \to +\infty$, the phase margin approaches $\frac{\pi}{2}$;

(ii) as $\psi \to \pm \frac{\pi}{2}$, asymptotic stability may be maintained by increasing r commensurately $\left(r > \frac{1-\mu}{2 \cos \psi}\right)$.

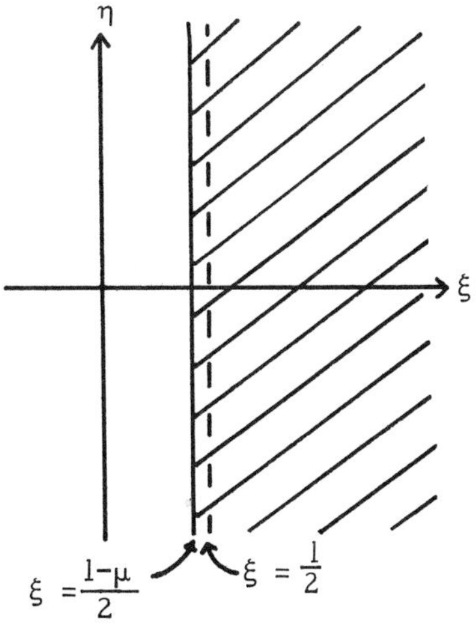

Fig. VII-6.
The half-plane $\xi > \frac{1-\mu}{2}$

In the general one studies the situation where $re^{i\psi}$ in (6.34) is replaced by a diagonal matrix

$$\operatorname{diag}\,(r_1 e^{i\psi_1}, r_2 e^{i\psi_2}, \ldots, r_m e^{i\psi_m})$$

but we will not pursue this here. See [7] for material related to the case $\psi_1 = \psi_2 = \ldots = \psi_m = 0$.

The iterative technique based on repeated use of the inequalities (6.19), (6.20) or (6.22) can be used to analyze the robustness of systems in the presence of specific perturbations - as we have already noted. We will use the linear oscillator - accelerometer system described earlier as an example. Taking $m = 1$, $\alpha = 1$, $\beta = 4$, the system (6.25) with the feedback relation (6.31) based on the accelerometer system (6.28) yields the composite system

$$\dot{x}^1 = x^2,$$
$$\dot{x}^2 = -x^1 - \delta x^2 + u = -x^1 - \delta x^2 - (1-\delta)y^1,$$
$$\dot{y}^1 = y^2,$$
$$\dot{y}^2 = -y^1 - 4y^2 - \dot{x}^2 = x^1 + \delta x^2 - \delta y^1 - 4y^2,$$

or, in matrix form

$$\dot{\begin{pmatrix} x^1 \\ x^2 \\ y^1 \\ y^2 \end{pmatrix}} = \begin{pmatrix} 0 & 1 & 0 & 0 \\ -1 & -\delta & -(1-\delta) & 0 \\ 0 & 0 & 0 & 1 \\ 1 & \delta & -\delta & -4 \end{pmatrix} \begin{pmatrix} x^1 \\ x^2 \\ x^3 \\ x^4 \end{pmatrix} . \tag{6.41}$$

As $\delta \to 0$ the characteristic polynomial tends to $\lambda^4 + 4\lambda^3 + \lambda^2 + 4\lambda - 1$, which has at least one positive real zero, so we conclude the system is unstable for small $\delta > 0$. But when $\delta = 1$ the matrix is lower block triangular; viz.:

$$\begin{pmatrix} 0 & 1 & 0 & 0 \\ -1 & -1 & 0 & 0 \\ 0 & 0 & 0 & 1 \\ 1 & 1 & -1 & -4 \end{pmatrix}$$

and has eigenvalues which are the zeros of $\lambda^2 + \lambda + 1$ and $\lambda^2 + 4\lambda + 1$, all of which have negative real parts. The pertinent question becomes this - for what values of δ is (6.41) asymptotically stable?

The traditional approach to this question is based on recognition of the fact that the instability as $\delta \to 0$ arises from the phase and amplitude distortion of the feedback signal

$$u = (1 - \delta)\dot{x}^2 = (1 - \frac{1}{\delta})x^1 + (\delta - 1)x^2 \tag{6.42}$$

produced when the output, y^1, of the accelerometer is accepted in place of $-m/\alpha\dot{x}^2$, i.e., $-\dot{x}^2$ in the present case. Thus one could attempt a piecemeal analysis by identifying the system (6.26) with (6.33) and the system with feedback distortion, namely (6.25) with

$$u = re^{i\psi}((1 - \frac{1}{\delta})x^1 + (\delta - 1)x^2), \tag{6.43}$$

with the system (6. 34). The values for r and ψ would be obtained from (6. 30) with $m = \alpha = 1$, $\beta = 4$. The traditional Nyquist diagram method, or our method, based on (6. 35), could be used to determine if the closed loop system (6. 25), (6. 43) remains asymptotically stable for the given values of r and ψ.

There are two problems (at least) with this approach. The most obvious is that one either has to do a complete analysis with δ as a parameter, which clearly would be very complex, even for this elementary example, or one would have to repeat the analysis for various values of δ until the point where asymptotic stability ceases to obtain could be approximately identified. This could be quite time consuming. The second, and most serious, problem is that the values of r and ψ are determined from δ through (6. 29), and ρ and σ are computed from the undistorted closed loop system (6. 26). In reality, representation of $\dot{x}^2$ in the form (6. 29) cannot be justified rigorously because $\dot{x}^2$ also depends on the distorted feedback relationship in the composite system.

A far more rigorous and direct approach - and, we believe, ultimately an easier approach - is to treat the four dimensional composite system as a unit. Letting $\epsilon = 1 - \delta$ we have an asymptotically stable system

$$\dot{x} = S_0 x, \qquad S_0 = \begin{pmatrix} 0 & 1 & 0 & 0 \\ -1 & -1 & 0 & 0 \\ 0 & 0 & 0 & 1 \\ 1 & 1 & -1 & -4 \end{pmatrix} \tag{6.43}$$

for $\epsilon = 0$ ($\delta = 1$), and the perturbed system is

$$\dot{x} = (S_0 + \epsilon \delta S)x \tag{6.44}$$

with (cf. (6. 41) with $\delta = 1 - \epsilon$)

$$\delta S = \begin{pmatrix} 0 & 0 & 0 & 0 \\ 0 & 1 & -1 & 0 \\ 0 & 0 & 0 & 0 \\ 0 & -1 & 1 & 0 \end{pmatrix} .$$

In (6.8) we take $P = I \ (4 \times 4)$ and solve that equation with the result

$$Z = Z_0 = \begin{pmatrix} 1.55 & .55 & .1 & .05 \\ .55 & 1.2 & .45 & .15 \\ .1 & .45 & 2.25 & .5 \\ .05 & .15 & .5 & .25 \end{pmatrix}$$

and then

$$Z\delta S\delta S^* Z = \begin{pmatrix} .5 & 1.05 & .4 & -.1 \\ 1.05 & 2.205 & .84 & -.21 \\ .4 & .84 & .32 & -.08 \\ -.1 & -.21 & -.08 & .02 \end{pmatrix} .$$

This matrix has rank 1 and its trace, 3.045, is equal to its largest eigenvalue. Using the method described by (6.21), (6.22) we see that we should take

$$\theta_0 = \frac{1}{2}\sqrt{\frac{1}{3.045}} = .28653$$

and we conclude (6.44) remains asymptotically stable for

$$\epsilon < \theta_0 = .28653.$$

If the method described by (6.20) is used, we form

$$\delta S^* Z + Z\delta S = \begin{pmatrix} 0 & .5 & -.5 & 0 \\ .5 & 2.1 & -.65 & -.1 \\ -.5 & -.65 & -.8 & .1 \\ 0 & -.1 & .1 & 0 \end{pmatrix}$$

It's largest eigenvalue is 2.39499, leading to

$$\eta_0 = \frac{1}{2.39499} = .41753$$

and we conclude (6.44) remains asymptotically stable for

$$\epsilon < \eta_0 = .41753.$$

Continuing, either one of the methods described earlier, applied iteratively, leads to the value for $\hat{\epsilon}$

$$\hat{\epsilon} = .8.$$

If the damping coefficient, $-\beta$, in the accelerometer system (6.28), is set at -16 rather than -4, the value obtained for $\hat{\epsilon}$ is

$$\hat{\epsilon} = .94.$$

These correspond, of course, to $\delta = .2$ and $\delta = .06$, respectively. The accelerometer with the higher damping coefficient gives a composite system which is asymptotically stable for a larger range of values of δ, as one would, of course, suspect.

Theorem 6.1 indicates that robustness of the system (6.5) with respect to parameter variations corresponds (for a fixed P) to small values of Z, which tends to correspond to large negative real parts for the eigenvalues of the matrix $S_0 = A_0 + B_0K$. Viewed in this light it is instructive to relate robustness to the question of maneuverability, by which we mean the degree of, or ease of, controllability. We consider a system (cf. (5.1))

$$\dot{x} = A_0x + B_0u + C_0w, \quad x \in E^n, \ u \in E^m, \ w \in E^{\ell}, \qquad (6.45)$$

where u is an (automatic) control and w a command input to be supplied by the operator of the plant. Setting $u = Kx$, we have (cf. (6.5))

$$\dot{x} = S_0x + C_0w. \qquad (6.46)$$

Now we pose a "maneuver" to be performed by the plant operator, i.e., a control problem: given

$$x(0) = 0$$

and $T > 0$, determine $w(t)$, $0 \leq t \leq T$, so that

$$x(T) = x_1 \in E^n .$$

In discussing maneuverability it makes sense to consider the most efficient way in which a maneuver can be performed. Measuring efficiency by $\|w\|^2_{L^2_\ell[0,T]}$, the optimal command is given by (see II-Theorem 3.12)

$$w_T(t) = C_0^* e^{S_0^*(T-t)} W_T^{-1} x_1 \tag{6.47}$$

where

$$W_T = \int_0^T e^{S_0(T-t)} C_0 C_0^* e^{S_0^*(T-t)} dt \tag{6.48}$$

satisfies, as we have seen, e.g., in Section 3 in connection with the tracking problem

$$S_0 W_T + W_T S_0^* + C_0 C_0^* - e^{S_0 T} CC^* e^{S_0^* T} = 0 . \tag{6.49}$$

The command control cost is defined to be

$$\|w_T\|^2_{L^2_\ell[0,T]} = \int_0^T w_T(t)^* w_T(t) dt = \text{(using (6.47))}$$

$$x_1^* W_T^{-1} x_1 = \operatorname{Tr} X_1 W_T^{-1} ,$$

where we choose to regard the "target" state x_1 as a random vector with

$$\operatorname{cov}(x_1, x_1) = X_1 .$$

If we assume that S_0 is a stability matrix we can allow the time permitted for control to approach infinity. The limiting optimal control cost is

$$\operatorname{Tr} X_1 W^{-1} \tag{6.50}$$

where

$$S_0 W + W S_0^* + C_0 C_0^* = 0 . \tag{6.51}$$

Maneuverability, which corresponds to a small control cost (6.50), is thus seen to be associated with a large matrix W - large in the sense that its smallest eigenvalue is large.

It may be desirable to emphasize short control times T as compared with larger ones. This can be done by forming, for $\mu > 0$,

$$W_\mu = \int_0^\infty \mu e^{-\mu T} W_T \, dT.$$

Carrying out the indicated integration in (6.49),

$$S_0 W_\mu + W_\mu S_0 + C_0 C_0^* - \mu \int_0^\infty e^{(S_0 - (\mu/2)I)T} C_0 C_0^* e^{(S_0^* - (\mu/2)I)T} \, dT = 0.$$

Then setting

$$\hat{W}_\mu = \int_0^\infty e^{(S_0 - (\mu/2)I)T} C_0 C_0^* e^{(S_0^* - (\mu/2)I)T} \, dT$$

we have

$$S_0 W_\mu + W_\mu S_0^* + C_0 C_0^* - \mu \hat{W}_\mu = 0, \tag{6.52}$$

$$(S_0 - (\mu/2)I)\hat{W}_\mu + \hat{W}_\mu (S_0 - (\mu/2)I)^* + C_0 C_0^* = 0. \tag{6.53}$$

Subtracting the second equation from the first we have

$$S_0 (W_\mu - \hat{W}_\mu) + (W_\mu - \hat{W}_\mu) S_0^* = 0$$

and we conclude $W_\mu = \hat{W}_\mu$ satisfies

$$(S_0 - \mu I) W_\mu + W_\mu (S_0 - \mu I)^* + C_0 C_0^* = 0. \tag{6.54}$$

For $\mu_2 > \mu_1 > 0$ we have

$$W_{\mu_2} < W_{\mu_1} < W$$

indicating, as we would expect, that emphasis of shorter intervals leads to a more pessimistic assessment of maneuverability as measured by W_μ. Another interpretation of W_μ is given in Exercise 16 below.

Consider then the problem of the control analyst who wishes to design a system (6.45) with a fixed form linear feedback law $u = Kx$ in such a way that the resulting system (6.46) has the property of being robust as far as asymptotic stability is concerned while, at the same time, it remains maneuverable, i.e., responsive to command inputs w. We have seen, for the former, that for symmetric $P > 0$ the solution Z of

$$S_0^* Z + ZS_0 + P = 0 \tag{6.55}$$

should be small, or equivalently Z^{-1} should be large. For the latter, using $\mu = 0$ here, the solution W of (6.51) should be large, i.e. W^{-1} should be small. These may be regarded as generally conflicting requirements; a highly stable S_0 will tend to make both Z and W small while Z and W will both become large if all of the eigenvalues of S_0 approach the imaginary axis from the left. In fact, if we write (6.55), (6.51) as $T^*(Z) + P = 0$, $T(W) + C_0C_0^* = 0$, VI - Proposition 3.3 gives

$$\operatorname{Tr} WP = \operatorname{Tr}(T^{-1}(C_0C_0^*), P)_{E^{n^2}} = \operatorname{Tr}(C_0C_0^*, (T^*)^{-1}(P))_{E^{n^2}} = \operatorname{Tr} C_0C_0^* Z$$

or, alternatively,

$$\operatorname{Tr} P^{\frac{1}{2}} W P^{\frac{1}{2}} = \operatorname{Tr} C_0^* Z C_0 ,$$

which shows a fairly rigid relationship between W and Z. Nevertheless, within these constraints, the design of systems at once robust and maneuverable may still be contemplated.

Referring back to (6.45), let us consider the special case where $B_0 = C_0$:

$$\dot{x} = A_0 x + C_0(u + w),$$

i.e., a portion of the control effort is determined automatically, the remainder by the plant operator. Suppose u is selected so as to minimize the cost functional

$$\int_0^\infty [\rho x(t)^* P x(t) + u(t)^* u(t)]\, dt, \qquad \rho > 0 \tag{6.56}$$

when $w(t) \equiv 0$. Then, from Chapter IV,

$$u(t) \equiv - C_0^* Q x(t) \equiv \hat{K} x(t) \tag{6.57}$$

where

$$A_0^* Q + QA_0 + \rho P - QC_0C_0^* Q = 0. \tag{6.58}$$

With $\hat{S}_0 = A_0 + C_0\hat{K}$ this is

$$(A_0 - \frac{1}{\rho} C_0 C_0^* Q)^* Q + Q(A_0 - \frac{1}{\rho} C_0 C_0^* Q) + \rho P + Q C_0 C_0^* Q = 0$$

or

$$\hat{S}_0^* Q + Q S_0 + \rho P + Q C_0 C_0^* Q = 0 .$$

Multiplying on the left and on the right by Q^{-1} we have

$$\hat{S}_0 Q^{-1} + Q^{-1}\hat{S}_0^* + \rho Q^{-1} P Q^{-1} + C_0 C_0^* = 0 .$$

We let $\hat{W}$ solve (cf. (6. 51))

$$\hat{S}_0 \hat{W} + \hat{W}\hat{S}_0^* + C_0 C_0^* = 0$$

and then have

$$\hat{S}_0 (Q^{-1} - \hat{W}) + (Q^{-1} - \hat{W}) S_0^* + \rho Q^{-1} P Q^{-1} = 0 . \tag{6.59}$$

Since $Q^{-1} P Q^{-1}$ is positive definite and S_0 is a stability matrix, $Q^{-1} - \hat{W}$ is positive definite and (6. 59) gives

$$\hat{S}_0^* (Q^{-1} - \hat{W})^{-1} + (Q^{-1} - \hat{W})^{-1} \hat{S}_0 + \rho (Q^{-1} - \hat{W})^{-1} Q^{-1} P Q^{-1} (Q^{-1} - \hat{W})^{-1} = 0 ,$$

so that, with

$$\hat{Z} = \rho (Q^{-1} - \hat{W})^{-1} , \tag{6.60}$$

$$\hat{S}_0 \hat{Z} + \hat{Z}\hat{S}_0 + (I - Q\hat{W})^{-1} P (I - Q\hat{W})^{-1} = 0$$

or, with $\hat{P} = (I - Q\hat{W})^{-1} P (I - Q\hat{W})^{-1}$,

$$\hat{S}_0 \hat{Z} + \hat{Z}\hat{S}_0 + \hat{P} = 0 . \tag{6.61}$$

From (6. 60) we have

$$Q = (\hat{W} + \rho Z^{-1})^{-1} . \tag{6.62}$$

Since Q represents a minimal cost matrix, the equation (6. 62) suggests that in keeping Q small the matrices $\hat{W}$ and $\hat{Z}^{-1}$ are in some sense both kept large, that is, the control law (6. 57) already effects a compromise of sorts between robustness and maneuverability. As we let $\rho \to 0$, placing more emphasis on control cost in (6. 56), the minimality of Q is more closely related to that of W^{-1}, so that maneuverability is emphasized as the expense of robustness, while as $\rho \to \infty$ we have the opposite effect.

When we treat a system (6. 45) wherein $B_0 \neq C_0$ it is no longer clear that the approach via the cost functional (6. 56) is still to the point. For (6. 58) is then replaced by

$$A_0^* Q + QA_0 + P - QB_0 B_0^* Q = 0$$

and we can relate this equation only to that of

$$S_0^* \tilde{W} + \tilde{W} S_0 + B_0 B_0^* = 0$$

rather than the equation (6. 51) connected with maneuverability via the control w and its associated input matrix C_0. We are forced to try something different to achieve a balance between robustness of stability and maneuverability in such circumstances.

Having decided to try another approach, we may as well try to make it specific to the parameter variations actually occurring in our system. At the same time we must take care to develop a method which is not too cumbersome to use. This will require a certain amount of manipulation and admittedly ad hoc reasoning. Let us suppose that the nominal plant is

$$\dot{x} = A_0 x + B_0 u$$

when $w \equiv 0$. With $u = Kx$ we have

$$\dot{x} = S_0(K)x, \quad S_0(K) = A_0 + B_0 K.$$

We consider variations in our plant depending on a single parameter α:

$$A_0 \to A_0 + \alpha\delta A, \quad B_0 \to B_0 + \alpha\delta B,$$

giving the perturbed plant

$$\dot{x} = (A_0 + \alpha\delta A)x + (B_0 + \alpha\delta B)u$$

which, with $u = Kx$, becomes

$$\dot{x} = (S_0(K) + \alpha\delta S_0(K))x, \quad \delta S_0(K) = \alpha\delta A + \alpha\delta BK .$$

If we assume that $Z(K)$ is the unique positive definite solution of the Liapounov equation

$$S_0(K)^*Z(K) + Z(K)S_0(K) + P = 0 \tag{6.63}$$

for some positive definite symmetric P, Theorem 6.1 guarantees that $S_0(K)$ remains a stability matrix if (cf. (6.10))

$$\alpha^2 \delta S(K)P^{-1}\delta S(K)^* < \tfrac{1}{4} Z(K)^{-1} P Z(K)^{-1} .$$

This inequality is equivalent to

$$4\alpha^2 P^{-\frac{1}{2}} Z(K)S(K)P^{-1}\delta S(K)^* Z(K)P^{-\frac{1}{2}} < I .$$

This is not guaranteed, but is certainly "encouraged", if we keep the trace of the matrix on the left small. Then taking note of the maneuverability cost (6.50), we state our combined robustness - maneuverability problem in the following form:

$$\min_K \operatorname{Tr}(\beta X_1 W(K)^{-1} + \gamma PZ(K)\delta S(K)P^{-1}\delta S(K)^* Z(K)) \tag{6.64}$$

where β, γ are positive weighting factors, subject to the constraints (6.63) and (cf. (6.51))

$$S_0(K)W(K) + W(K)S_0(K)^* + C_0C_0^* = 0. \tag{6.65}$$

Multiplying (6.65) on the right and on the left by $W(K)^{-1}$, it may be replaced by

$$S_0(K)^*W(K)^{-1} + W(K)^{-1}S_0(K) + W(K)^{-1}C_0C_0^*W(K)^{-1} . \tag{6.66}$$

Replacing K by $K + \epsilon\delta K$ and differentiating (6.64) with respect to ϵ at $\epsilon = 0$ we have

$$\frac{d}{d\epsilon}\operatorname{Tr}(\beta X_1 W(K+\epsilon\delta K)^{-1} + \gamma PZ(K+\epsilon\delta K)\delta S(K+\epsilon\delta K)P^{-1}\delta S(K+\epsilon\delta K)^* Z(K+\epsilon\delta K))\Big|_{\epsilon=0}$$

$$= \operatorname{Tr}(\beta X_1 \frac{dW(K+\epsilon\delta K)^{-1}}{d\epsilon}\Big|_{\epsilon=0} + 2\gamma PZ(K)\delta S(K)P^{-1}\delta S(K)^* \frac{dZ(K+\epsilon\delta K)}{d\epsilon}\Big|_{\epsilon=0}$$

$$+ 2\gamma P^{-1}\delta S(K)^* Z(K)PZ(K)\frac{d\delta S(K+\epsilon\delta K)}{d\epsilon}\Big|_{\epsilon=0}) . \tag{6.67}$$

Differentiating (6.63) and (6.66) (with K replaced by $K + \epsilon\delta K$) with respect to ϵ at $\epsilon = 0$ we have

$$S_0(K)^* \frac{dZ(K+\epsilon\delta K)}{d\epsilon}\Big|_{\epsilon=0} + \frac{dZ(K+\epsilon\delta K)}{d\epsilon}\Big|_{\epsilon=0} S(K) + \delta K^* B^* Z(K) + Z(K)B\delta K = 0 \quad (6.68)$$

and

$$(S_0(K) + C_0C_0^* W(K)^{-1})^* \frac{dW(K+\epsilon\delta K)^{-1}}{d\epsilon}\Big|_{\epsilon=0}$$
$$+ \frac{dW(K+\epsilon\delta K)^{-1}}{d\epsilon}\Big|_{\epsilon=0} (S_0(K) + C_0C_0^* W(K)^{-1})$$
$$+ \delta K^* B^* W(K)^{-1} + W(K)^{-1} B\delta K = 0 \quad (6.69)$$

Let $Q_Z(K)$, $Q_W(K)$ satisfy (noting that $\delta S(K) = \delta A + \delta BK$)

$$S_0(K)Q_Z(K) + Q_Z(K)S_0(K)^* + PZ(K)(\delta A + \delta BK)P^{-1}(\delta A + \delta BK)^* = 0 \quad (6.70)$$

$$(S_0(K) + C_0C_0^* W(K)^{-1})Q_W(K) + Q_W(K)(S_0(K) + C_0C_0^* W(K)^{-1})^* + X_1 = 0. \quad (6.71)$$

Then, using VI-Proposition 3.3, the derivative (6.67) becomes

$$\mathrm{Tr}\,(\beta Q_W(K)[\delta K^* B^* W(K)^{-1} + W(K)^{-1} B\delta K] + 2\gamma Q_Z(K)[\delta K^* B^* Z(K) + Z(K)B\delta K]$$
$$+ 2\gamma P^{-1}(\delta A + \delta BK)^* Z(K)PZ(K)\delta B\delta K) = 0\ .$$

One may use a gradient type algorithm to decrease the cost in (6.64) by taking steps $\epsilon\delta K$, $\epsilon > 0$, with

$$\delta K = -\beta B^* W(K)^{-1} Q_W(K) - 2\gamma B^* Z(K)Q_Z(K) - \gamma\delta B^* Z(K)PZ(K)(\delta A + \delta BK)P^{-1}. \quad (6.72)$$

If δB has rank m (= dim u) the necessary condition for a feedback matrix $K = \hat{K}$ to be optimal is obtained in the form

$$\delta B^* Z(\hat{K})PZ(\hat{K})\delta B\hat{K} = -\delta B^* Z(\hat{K})PZ(\hat{K})\delta A$$
$$- \frac{\beta}{\gamma} B^* W(\hat{K})^{-1} Q_W(\hat{K})P - 2B^* Z(\hat{K})Q_Z(\hat{K})P$$

or

$$\hat{K} = -(\delta B^* Z(\hat{K})PZ(\hat{K})\delta B)^{-1}[\delta B^* Z(\hat{K})PZ(\hat{K})\delta A - \frac{\beta}{\gamma} B^* W(\hat{K})^{-1} Q_W(\hat{K})P - 2B^* Z(\hat{K})Q_Z(\hat{K})P]. \quad (6.73)$$

Equation (6.73) can be substituted in (6.65), (6.66), (6.70) and (6.71) to yield a closed system in which K is eliminated, but its usefulness is doubtful; the gradient method $K \to K + \epsilon \delta K$, with δK given by (6.72) and $\epsilon > 0$, small, can probably be used more effectively.

Bibliographical Notes, Chapter VII

[1] M. J. Beckmann: "Dynamic Programming of Economic Decisions", Economics and Operations Research, Vol. IX, Springer-Verlag, New York, 1968.

It has been seen in Chapter IV that the gradient projection method for minimization in the presence of constraints has application to certain tracking problems. The original reference for this method is [2] below. Expository treatment may be found in [3] and in V-[2], IV-[10]. All references discuss the method as it applies to inequality constraints with the exception of V-[2] which contains a description of the method as it applies to the special case of equality constraints.

[2] J. B. Rosen: "The gradient projection method for nonlinear programming", Part I (Linear constraints), J. SIAM on Appl. Math., 8 (1960), pp. 181-217; Part II (Nonlinear constraints), Ibid., 9 (1961), pp. 514-532.

[3] B. S. Gottfried and J. Weisman: "Introduction to Optimization Theory", Prentice-Hall, Inc., Englewood Cliffs, N. J., 1973.

Tracking problems are discussed extensively in VI-[14] and VI-[17]. An elementary presentation of the consistency conditions for tracking in terms of the transfer function (cf. III-5) appears in

[4] M. Aoki: "Optimal Control and System Theory in Dynamic Economic Analysis", North-Holland Pub. Co., New York, Oxford, Amsterdam, 1976.

The iterative methods for determining the limits of asymptotic stability under perturbation have been developed by M. Eslami in his thesis. He also presents a number of numerical examples.

[5] M. Eslami: "Sensitivity Analysis and Synthesis in Automatic Control Systems", Thesis, Dept. of Electrical Eng., Univ. of Wisconsin-Madison, June, 1978.

[6] M. G. Safonov and M. Athans: "Gain and phase margin for multiloop LQG regulators", IEEE Trans. Aut. Control, Vol. AC-22, (1977), pp. 173-179.

[7] P. K. Wong and M. Athans: "Closed loop structural stability for linear quadratic optimal systems", Ibid., pp. 94-99.

Many of the properties of the linear quadratic optimal regulator, discussed in Chapters IV, V, VI, with reference to design criteria such as sensitivity, etc., are discussed in VI-[17] and in many other places. Important original references are

[8] R. E. Kalman: "When is a linear control system optimal?", J. Basic Eng., Trans. ASME, Ser. D., 86 (1964), pp. 51-60.

[9] J. B. Cruz and W. R. Perkins: "A new approach to the sensitivity problem in multivariable feedback system design", IEEE Trans. Autom. Control, 9 (1964), pp. 216-222.

A variety of methods for design of insensitive controllers are discussed and compared computationally with respect to a particular application in reference [10]. The method discussed in Section 5 is compared on the same basis in reference [10].

[10] C. A. Harvey and R. E. Pope: "Study of Synthesis Techniques for Insensitive Control Systems", NASA Report CR-2803, April 1977.

[11] D. L. Russell: "Response variation weighting methods for cost sensitivity reduction", Honeywell Report MR 12497, May 1978.

Exercises, Chapter VII

We present here a fairly lengthy list of exercises, mostly computational projects which will require, at the very least, the assistance of a small programmable calculator. For the most part these have not been pretested and it may, at times, be necessary for the reader to change the statement of the problem slightly to meet some difficulty which the author has not foreseen. Thus the exercises here should be taken as suggestions for projects to be worked on and not detailed prescriptions for exactly what is to be done.

1. Let (cf. (2.13))

$$\hat{c}_k(\sigma) = \operatorname{Tr} Q_k(K)M, \quad k = 1, 2, \dots, \nu.$$

Develop formulae comparable to (2.32) for the quantities $\frac{\partial \hat{c}_k}{\partial \hat{\sigma}_j}$, the $\hat{\sigma}_j$ playing the same role as described in Section 2, but related to weights on the individual control costs.

2. Let $c_k(\sigma)$, $k = 1, 2, \dots, N$, denote the individual response costs described in Section 2. Develop a method for successively modifying σ_j to $\sigma_{j+1} = \sigma_j + \Delta\sigma_j$ to approximetly solve the minimax problem

$$\min_{\sigma} \{ \max_{k=1,2,\dots,N} \{ c_k(\sigma) \} \}.$$

Your method should take into account the possibility that the maximum might be achieved for several values of k. What are the first order necessary conditions which must be satisfied at a point $\tilde{\sigma}$ which is a solution of this problem?

3. Develop a method for successively modifying σ through steps $\sigma_{j+1} = \sigma_j + \Delta\sigma_j$ so as to solve the problem

$$\min_{\sigma} \| c(\sigma) - \tilde{c} \|^2_{E^N}$$

subject to the single inequality constraint

$$c_u(\sigma) \le \tilde{c}_u .$$

Note that the objective is simply to decrease $\| c(\sigma) - \tilde{c} \|^2_{E^N}$ as long as the constraint is inactive, i. e., as long as $c_u(\sigma) < \tilde{c}_u$. When the constraint becomes "active" ($c_u(\sigma_j) = \tilde{c}_u$ within a given tolerance) the objective is to decrease $\|c(\sigma) - \tilde{c}\|^2_{E^N}$ without increasing $c_u(\sigma)$. Apply your method to the problem of the suspension system discussed at the end of section 2. Take $\tilde{c} = \binom{1}{.5}$, $\tilde{c}_u = c_u(.5, .5)$.

4. In the system (3. 13), i. e.

$$\dot{x} \equiv \begin{pmatrix} \dot{s} \\ \psi \end{pmatrix} = \begin{pmatrix} \alpha & \beta \\ 0 & -\gamma \end{pmatrix}\begin{pmatrix} s \\ \psi \end{pmatrix} + \begin{pmatrix} 0 \\ 1 \end{pmatrix} u \equiv Ax + Bu ,$$

with the data prescribed in (3. 37), (3. 38) ff. Let

$$u = u_0 + u_1 ,$$

where u_0 represents an input by the motor operator and u_1 is now the regulating force. Use the methods of Section 4 to design a feedback control

$$u_1 = k_1 s + k_2 \psi \equiv Kx$$

so that s optimally follows "ramp inputs" u_0 in the sense defined by (4. 45) with $W_1 = 1$. Experiment with the values $\mu = 1$, $\hat{\mu} = 4$ (and other values if you wish). Then take ramp inputs with

$$u_0(t) = 0,\ t \le 0,\quad u_0(t) = 1,\ t \ge 1,\quad \hat{u}_0(t) = 0,\ t \le 0,\quad \hat{u}_0(t) = 1,\ t \ge 4,$$

and graph $s(t)$ along with $s_{u_0}(t) = -F(A + BK)^{-1} D u_0(t)$, i. e.,

$$s_{u_0}(t) = -(1, 0)\begin{pmatrix} 1 & 1 \\ k_1 & -1 + k_2 \end{pmatrix}^{-1}\begin{pmatrix} 0 \\ 1 \end{pmatrix} u_0(t)$$

on the interval $0 \le t \le 2$, $s(t)$ along with $s_{\hat{u}_0}(t)$ (similarly defined) on $0 \le t \le 6$.

5. Consider the suspension system (2. 45), i. e.,

$$\begin{pmatrix} \dot{x}^1 \\ \dot{x}^2 \end{pmatrix} = \begin{pmatrix} 0 & 1 \\ -1 & 0 \end{pmatrix}\begin{pmatrix} x^1 \\ x^2 \end{pmatrix} + \begin{pmatrix} 0 \\ 1 \end{pmatrix} u + \begin{pmatrix} 0 \\ 1 \end{pmatrix} v$$

and let the response of interest be the acceleration

$$y^3 = -x^1 + u \quad \text{(i. e.} \quad F = (-1, 0), \; G = 1).$$

Introduce a model system

$$\dot{\xi} = -\xi + v$$

and associate ξ with y^3. Thus (cf. (3.5) $H = -1$, $J = 1$. Use the method (3.35), (3.36) to determine a feedback control $u = k_1 x^1 + k_2 x^2$ such that y^3 follows ξ optimally in the sense defined by (3.35). Take $\mu = V = 1$, $X_0 = 0$. Initially take $\tilde{W} = \tilde{U} = 1$, but vary these later to obtain improved performance. Note that this is a case where $G \neq 0$ and we do not have a standard linear quadratic regulator problem as in the small motor example of Section 3. Develop a gradient method to solve (3.35), (3.36) in this case, starting with the initial values $k_1 = 0$, $k_2 = -1$.

6. Consider the same model following scenario as in Exercise 5, but this time formulate the problem as in (3.55) ff., and use the variant of Axsäter's method described following (3.56). Note that, as required, $X_0 = 0$ in this case.

7. Extend the table shown following (3.62) until convergence to four decimal places is obtained. Based on your computations, does Axsäter's method converge linearly or quadratically in this case?

8. Generate tables comparable to Table VII-1, but with $\tilde{U} = \frac{1}{4}$ replaced by $\tilde{U} = \frac{1}{8}$, $\tilde{U} = \frac{1}{16}, \ldots$ (as far as you wish). Compare the resulting model following and control costs. See (3.63) ff. for treatment of control cost.

9. Treat the small engine model following problem by the method suggested at the end of Section 3. Use the same data as was used to generate Table VII-1.

10. Consider the five dimensional crane system of Chapter I with (cf. I-(2.8)) $L_1 = \sqrt{2}$, $\psi_0 = 45^\circ$, $\ell_{2,0} = g = m = 1$, $v = 0$,

$$\dot{x} = \dot{\begin{pmatrix} x^1 \\ x^2 \\ x^3 \\ x^4 \\ x^5 \end{pmatrix}} = \begin{pmatrix} 0 & 1 & 0 & 0 & 0 \\ -1 & 0 & 0 & 0 & 0 \\ 0 & 0 & 0 & 1 & 0 \\ 0 & 0 & 0 & 0 & 0 \\ 0 & 0 & 0 & 0 & 0 \end{pmatrix} \begin{pmatrix} x^1 \\ x^2 \\ x^3 \\ x^4 \\ x^5 \end{pmatrix} + \begin{pmatrix} 0 & 1 \\ 0 & -1 \\ 0 & 0 \\ 0 & 1 \\ 1 & 0 \end{pmatrix} \begin{pmatrix} u^1 + w^1 \\ u^2 + w^2 \end{pmatrix}, \equiv Ax + B(u + w)$$

the controls u^1, u^2 here replaced by the sums $u^1 + w^1$, $u^2 + w^2$, where u^1, u^2 are to be determined by linear feedback and w^1, w^2 represent external inputs supplied by the crane operator. Note that, if the system is left as it is, there is no direct algebraci relationship between input pairs $u^1 + w^1$, $u^2 + w^2$ and the horizontal and vertical load coordinates

$$\begin{pmatrix} y^1 \\ y^3 \end{pmatrix} = \begin{pmatrix} 1 & 0 & 1 & 0 & 0 \\ 0 & 0 & -1 & 0 & -1 \end{pmatrix} \begin{pmatrix} x^1 \\ x^2 \\ x^3 \\ x^4 \\ x^5 \end{pmatrix} + \begin{pmatrix} 0 & 0 \\ -1 & 0 \end{pmatrix} \begin{pmatrix} u^1 + w^1 \\ u^2 + w^2 \end{pmatrix} \equiv Fx + G(u + w)$$

because the x^i are obtained by integration of the control inputs and differential equations involving those inputs. But, if the u^1, u^2 are determined by linear feedback $\begin{pmatrix} u^1 \\ u^2 \end{pmatrix} = u = Kx$ with $A + BK$ asymptotically stable, the system becomes

$$\dot{x} = (A + BK)x + Bw$$

and for each external input w there are well defined equilibrium values of y^1, y^3

$$\begin{pmatrix} y^1_w \\ y^3_w \end{pmatrix} \equiv \hat{y}_w = -(F + GK)(A + BK)^{-1} Bw + Gw.$$

Thus there is a simple, one-to-one algebraic relationship between equilibrium values of y^1 and y^3 and the inputs w^1, w^2 just in case $A + BK$ is a stability matrix and $-(F + GK)(A + BK)^{-1}B + G \equiv Y(K)$ is nonsingular.

Determine a 2×5 feedback matrix K_0 such that all eigenvalues of $A + BK$ are equal to -1 and such that $Y(K_0)$ is a nonsingular 2×2 matrix.

Can K_0 be determined so that, additionally, $Y(K_0)$ is diagonal? What would be the implications of such an arrangement?

11. We study the plant developed in Exercise 10 for the case of linear inputs w^1, w^2. With K_0 as defined there, let

$$u = K_0 x + \tilde{u}$$

so that our system becomes

$$\dot{x} = (A + BK_0)x + B\tilde{u} + Bw \equiv A_0 x + B\tilde{u} + Bw .$$

To study linear inputs w we let

$$\dot{w}_1 \equiv \begin{pmatrix} w^1 \\ w^2 \\ w^3 \\ w^4 \end{pmatrix}^{\bullet} = \begin{pmatrix} 0 & 0 & 1 & 0 \\ 0 & 0 & 0 & 1 \\ 0 & 0 & 0 & 0 \\ 0 & 0 & 0 & 0 \end{pmatrix} \begin{pmatrix} w^1 \\ w^2 \\ w^3 \\ w^4 \end{pmatrix} \equiv A_1 w_1 .$$

The objective now is to use the control $\tilde{u}$ so that $(F + GK_0)x + Gw$ tracks $-(F + GK_0)(A + BK_0)^{-1}Bw + Gw$ optimally. We set

$$z = x + (A + BK_0)^{-1} Bw = x - A_0^{-1} D_1 w_1$$

($D_1 = BH_1$, $H_1 = \begin{pmatrix} 1 & 0 & 0 & 0 \\ 0 & 1 & 0 & 0 \end{pmatrix}$) and verify that with (cf. (4.74))

$$\tilde{u} = K_1(x + (A + BK_0)^{-1}Bw) + K_2 w_1$$

we have the composite system

$$\begin{pmatrix} z \\ w_1 \end{pmatrix}^{\bullet} = \begin{pmatrix} A_0 + BK_1 & A_0^{-1} D_1 A_1 + BK_2 \\ 0 & A_1 \end{pmatrix} \begin{pmatrix} z \\ w_1 \end{pmatrix} .$$

Solve the problem (4.97) with A replaced by $A_0 = A + BK_0$, F replaced by $F + GK_0$, $W_1 = I\,(4 \times 4)$, $\tilde{U} = \frac{1}{8} I\ (2 \times 2)$, $\tilde{W} = I\ (2 \times 2)$,

$\nu = 1$. Use the gradient projection method described in Section 4, starting with $K_1 = 0$ as in the example discussed there.

12. Repeat the sensitivity reduction process for the example (5. 68) ff. with $\sigma = 4$ and compare the resulting response variation matrix V_1 and nominal plant cost matrix Q_1, comparing these with the corresponding matrices (5. 73), (5. 74) computed for $\sigma = 2$.

13. Repeat the sensitivity reduction process for the example (5. 68) ff. but use the approach of minimizing the cost (5. 25) by the method (5. 38). ($G(\alpha) \equiv 0$ in thise case.) Use two values of ν, corresponding to $\psi_0 = 60^\circ, 30^\circ$, $\ell_{2,0} = \frac{1}{\sqrt{2}}, \sqrt{\frac{3}{2}}$, respectively, $L = \sqrt{2}$ in each case. Take $\hat{\sigma} = 1$, $\sigma_1 = \sigma_2 = 2$. How would you make comparison with the result of example (5. 68)ff. ?

14. Let S be an $n \times n$ matrix and Z a positive definite symmetric $n \times n$ matrix. Show that if

$$S^* Z + ZS = 0$$

then all eigenvalues of S have zero real part; in fact, S is similar to a matrix which is antihermitian.

15. Prove that for the ϵ_k defined by (6. 19), η_k defined by (6. 20) ff., θ_k defined by (6. 21), we have

$$\lim_{k \to \infty} (\epsilon_0 + \epsilon_1 + \ldots + \epsilon_k) = \lim_{k \to \infty} (\eta_0 + \eta_1 + \ldots + \eta_k) = \lim_{k \to \infty} (\theta_0 + \theta_1 + \ldots + \theta_k) = \hat{\epsilon}$$

where $\hat{\epsilon}$ is the largest real number such that $S + \epsilon \delta S$ is a stability matrix for $0 \leq \epsilon < \hat{\epsilon}$, it being understood that $\hat{\epsilon} = +\infty$ if $S + \epsilon \delta S$ is a stability matrix for all $\epsilon \geq 0$. Hint: show that if $0 \leq \epsilon \leq \tilde{\epsilon}$ is a closed bounded interval such that $S + \epsilon \delta S$ is a stability matrix for all ϵ in this interval, then the solution $Z(\epsilon)$ of

$$(S + \epsilon \delta S)^* Z(\epsilon) + Z(\epsilon)(S + \epsilon \delta S) + P = 0$$

is uniformly bounded on this interval.

16. Consider the problem of steering the system

$$\dot{x} = S_0 x + Cw, \quad S_0 \text{ a stability matrix,}$$

from the zero state at time $t = -\infty$ to the state $x_1 \in E^n$ at time $t = 0$. Show that if the pair (S, C) is controllable, then the control w which minimizes the weighted cost $(\mu > 0)$

$$\int_{-\infty}^{0} e^{-\mu t} \|w(t)\|^2 \, dt$$

and accomplishes the above maneuver is given by

$$\hat{w}(t) = C_0^* e^{-(S_0 - \mu I)^* t} W_\mu^{-1}, \quad t \leq 0,$$

where

$$W_\mu = \int_0^\infty e^{(S_0 - \frac{\mu}{\mu} I)t} C_0 C_0^* e^{(S_0 - \frac{\mu}{\mu} I)^* t} \, dt$$

$$= \int_{-\infty}^{0} e^{-(S_0 - \frac{\mu}{2} I)t} C_0 C_0^* e^{-(S_0 - \frac{\mu}{2} I)^* t} \, dt .$$

Show also that

$$\int_{-\infty}^{0} e^{-\mu t} \|\hat{w}(t)\|^2 dt = x_1^* W_\mu^{-1} x_1 .$$

17. How can the iterative method described in (6.17) ff. be most efficiently adapted to study the following question? We are given a polynomial of n-th degree with coefficients which depend on a (scalar for now) parameter α:

$$p(\lambda, \alpha) = \lambda^n + c_1(\alpha)\lambda^{n-1} + c_2(\alpha)\lambda^{n-2} + \dots + c_n(\alpha)$$

and it is given that the zeros of $p(\lambda, 0)$ all lie in the left half plane. Determine the largest interval (α_0, α_1), $\alpha_0 < 0$, $\alpha_1 > 0$, such that all zeros of $p(\lambda, \alpha)$ lie in the left half plane for $\alpha \in (\alpha_0, \alpha_1)$. Apply the method which you develop to the polynomial

$$p(\lambda, \alpha) = (\lambda + 1)^4 + \alpha\lambda^3 + \alpha\lambda .$$

18. Consider the rod hinged to a movable platform as shown in Fig. III-3. If we assume that the mass density of the rod per unit length is m, the system (3.12) is replaced by

$$\begin{pmatrix} \dot{x}^1 \\ x^2 \\ x^3 \end{pmatrix} = \begin{pmatrix} 0 & \frac{-3m}{m+4} & 0 \\ 0 & 0 & 1 \\ 0 & \frac{6(1+m)}{m+4} & 0 \end{pmatrix} \begin{pmatrix} x^1 \\ x^2 \\ x^3 \end{pmatrix} + \begin{pmatrix} \frac{4}{m+4} \\ 0 \\ -\frac{6}{m+4} \end{pmatrix} u$$

Let us suppose that we continue to use the feedback matrix (3.15) so that

$$u = \frac{15}{9}x^1 + \frac{48}{9}x^2 + \frac{65}{18}x^3 .$$

This stabilizes the system, as we have seen, for $m = 1$. Develop and use a modification of the method described by (6.21), (6.22) and accompanying material to determine the largest range of positive values m for which the system remains asymptotically stable.

19. With $S_0 = A_0 + B_0K$, consider the equations (6.51) and (6.55), i.e.,

$$(A_0 + B_0K)W + W(A_0 + B_0K)^* + C_0C_0^* = 0,$$

$$(A_0 + B_0K)^*Z + Z(A_0 + B_0K) + P = 0 .$$

Let K_0 be a feedback matrix such that $A_0 + B_0K_0$ is a stability matrix. Devise a method for modifying K through steps $K_{i+1} = K_i + \delta K_i$ in such a way that if

$$(A_0 + B_0K_i)W_i + W_i(A_0 + B_0K_i)^* + C_0C_0^* = 0,$$

$$(A_0 + B_0K_i)^*Z_i + Z_i(A_0 + B_0K_i) + P = 0 ,$$

the smallest eigenvalue of the W_i does not decrease (so that, in some sense, maneuverability is not impaired) while the largest eigenvalue of the Z_i is decreased, corresponding to improvement in robustness of stability (in some sense). Consider separately cases where these eigenvalues are simple or multiple. Apply your method to the four dimensional linearized level hold crane system developed in III - Exercise 1 with K_0 the stabilizing feedback matrix of part (c) of that exercise and

$$B_0 = C_0 = \begin{pmatrix} 0 \\ -1 \\ 0 \\ 1 \end{pmatrix}.$$

20. Consider a nominal plant

$$\dot{x} = (A_0 + B_0 K)x + C_0 w$$

and nominal response

$$y = (F_0 + G_0 K)x,$$

and a perturbed plant

$$\dot{\xi} = (A_0 + \delta A + (B_0 + \delta B)K)\xi + (C_0 + \delta C)w$$

and response

$$\eta = (F_0 + \delta F + (G_0 + \delta G)K)x.$$

For $x(0) = \xi(0) = x_0$, $\operatorname{cov}(x_0, x_0) = X_0$, define the "adaptive distance" between the two plants by

$$d_A(K) = \min_{w \in L^2[0,\infty)} E\left(\int_0^\infty [(y(t) - \eta(t))^* W (y(t) - \eta(t)) + w(t)^* U w(t)]\, dt\right)$$

for appropriate positive definite symmetric matrices W and U. Study the dependence of $d_A(K)$ on K and devise a method to solve the problem

$$\min_{K \in \mathscr{S}(A_0, B_0)} d_A(K).$$

INDEX